LIFE TRACES OF THE
GEORGIA COAST

LIFE OF THE PAST James O. Farlow, editor

LIFE TRACES OF THE
GEORGIA COAST

REVEALING THE UNSEEN LIVES OF PLANTS AND ANIMALS

Anthony J. Martin

INDIANA UNIVERSITY PRESS   *Bloomington & Indianapolis*

This book is a publication of

Indiana University Press
601 North Morton Street
Bloomington, Indiana 47404-3797 USA

iupress.indiana.edu

Telephone orders   800-842-6796
Fax orders   812-855-7931

© 2013 by Anthony J. Martin

All rights reserved

No part of this book may be reproduced or utilized in any form or by any means, electronic or mechanical, including photocopying and recording, or by any information storage and retrieval system, without permission in writing from the publisher. The Association of American University Presses' Resolution on Permissions constitutes the only exception to this prohibition.

⊖ The paper used in this publication meets the minimum requirements of the American National Standard for Information Sciences—Permanence of Paper for Printed Library Materials, ANSI Z39.48-1992.

*Manufactured in the
United States of America*

Cataloging information is available from the Library of Congress

1 2 3 4 5 18 17 16 15 14 13

Frontispiece: Alan Campbell

TO MY WIFE
RUTH SCHOWALTER,
*who wholeheartedly embraced my writing this book,
a trace of both you and me.*

Contents

PREFACE AND ACKNOWLEDGMENTS ix

1 Introduction to Ichnology of the Georgia Coast 1
2 History of the Georgia Coast and Its Ichnology 27
3 Tracemaker Habitats and Substrates 83
4 Marginal-Marine and Terrestrial Plants 127
5 Terrestrial Invertebrates 169
6 Marginal-Marine Invertebrates 245
7 Terrestrial Vertebrates, Part I: Fish, Amphibians, Reptiles 331
8 Terrestrial Vertebrates, Part II: Birds and Mammals 371
9 Marginal-Marine and Marine Vertebrates 435
10 Trace Fossils and the Georgia Coast 503
11 Future Studies, Future Traces 569

APPENDIX 599
BIBLIOGRAPHY 605
INDEX 647

Preface & Acknowledgments

WHAT IF WE ALL WOKE UP TOMORROW AND DISCOVERED THAT each fossil shell, bone, impression, or other bodily remains of former life from the geologic past had inexplicably vanished overnight from every museum and every layer of rock? How would we know what life-forms existed before us? How did these plants and animals behave, and how did their behaviors compare to those of modern plants and animals? How did they evolve as their environments and relationships to other organisms changed through time? Did their behaviors also affect other organisms or change their environments, like some modern species?

Fortunately, mass disappearances of body fossils are not likely to happen tomorrow or any other time in the future, so we can rest easy about this imagined scenario. Nonetheless, it is satisfying to know that our knowledge about the history of life is based not just on shells and bones, but augmented handily by trace fossils. Trace fossils include tracks, burrows, nests, scrapings, tooth marks, feces, and other products of plant and animal behavior of the geologic past; the study of modern and fossil traces is called ichnology. Of course, no one actually witnessed any of the behaviors that formed trace fossils, and the vast majority of the plants and animals that made them are extinct. How, then, do we interpret trace fossils? Answering that question is the main goal of this book, giving readers a foundation of knowledge to better diagnose trace fossils, or at least appreciate how ichnologists practice their craft.

What may seem paradoxical, though, is that this book deals with a rather small part of the world—the Georgia barrier islands, in the southeastern United States—and focuses on a thin slice of time—the present. How does this seemingly narrow window of space and time relate to trace fossils studied worldwide, and preserved in rocks representing the past 550 million years? Well, for one, the Georgia coast has been serving as a natural laboratory for the study of modern traces since the 1960s and is recognized worldwide as a special place through this legacy. Moreover, the people who studied these modern traces compared them directly to trace fossils that look very much like the ones from Georgia. Adding to this historical context are the barrier islands themselves, which, unlike many barrier-island systems elsewhere, have been only partially developed or not at all. For example, nearly 40 percent of the acreage of salt marshes in the eastern United States is represented on the Georgia coast, and 11 of the islands are either privately owned or under governmental protection. Consequently, most islands have a wide range of ecosystems—maritime forests, freshwater ponds, salt marshes, coastal dunes, and beaches—containing a huge variety of plants and animals native to the temperate-subtropical climate of coastal Georgia. Even better, some of the Georgia barrier islands and nearby mainland also have trace fossils from the relatively recent Pleistocene Epoch, most of which are less than 50,000 years old. These trace fossils shorten the gap between the past and present, thus easing direct comparisons between the still-living tracemakers and their defunct counterparts. In short, going into the field on the Georgia coast and observing traces of modern plants and animals there can provide the right tools for interpreting the behaviors and environments of the ancient past from nearly anywhere and anytime.

A broader intention of this book is to introduce readers to the principle of uniformitarianism, a scientific worldview that is both affirming and enlightening. Geology has long operated as a science through uniformitarianism, also known by its less polysyllabic synonym, actualism. This principle states that the chemical, physical, and biological processes of today provide insights on what happened during the prehuman history of the earth. This is often neatly summarized by the phrase "The present is the key to the past." Occasionally in ichnology, though, we can reverse this to instead say "The past is the key to the present," as trace fossils sometimes pose puzzles about what organisms might have produced them. These mysteries then prompt ichnologists to go into the field to find modern analogues. In other words, traces made by organisms today show us ways to better understand similar traces preserved in the fossil record, whether we use them to look forward or backward in time. Hence this book encourages all who read it, whether they are geologists, paleontologists, ecologists, biologists, or natural history enthusiasts, to remind themselves that what we see now affects how we perceive the past.

Another theme running throughout the book, and one I will ask the gentle reader to keep in mind, is the importance of careful observations of traces in the field. Such methods have been and will continue to be an integral part of our unraveling of the history of life. A partner to this motif is how good storytelling is an essential part of science, and ichnology is one of the best sciences for producing evocative stories. (Yes, I am biased about this, but still.) One does not even have to travel to the Georgia coast to appreciate the ichnology of that place, as the life traces of plants and animals beckon our attention everywhere, even in urban environments. So although this book presents a systematic approach in its topics—introduction to ichnology, geologic history and ecology of the Georgia barrier islands, tracemakers and their traces, and applications of this knowledge to trace fossils—every chapter begins with a story set in the field, and how observations made in this setting contributed to discoveries, whether personal or more universal.

This book was also written at a time when the methods and roles of science were either being misunderstood or misrepresented, especially with regard to the effects of global climate change. Hence examples of observation-based science, using traces of wild plants and uncaged animals in the context of their environments, can serve as excellent reality checks and test our preconceived perceptions about how plants or animals behave. Also, at the time of this writing, concerns were being raised about how an increasing number of people—especially those of younger generations—are spending more time inside, whether bound by technology or an irrational fear of the outdoors. The message of this book says, go outside, notice the clues left by all of the life surrounding us, and fall in love with what is there for us to discover and ponder, all while using only our highly evolved senses and cognition. My parents, Veronica and Richard, encouraged me as a child to go outside to learn, but also to visit the public library in my hometown of Terre Haute, Indiana, as much as possible. It is possible to do both types of learning—outward and inward—and I appreciate my parents for teaching me this at an early age.

With indebtedness in mind, I must acknowledge those who helped this book to begin, emerge, take form, and come to fruition. James Farlow, editor of the Life of the Past series at Indiana University Press, good-naturedly badgered me for three consecutive years about writing a book like this. Once I succumbed to his encouragement and submitted a book proposal, I was pleasantly surprised to have it accepted. (Although I think he got more than he bargained for with this final version.) He, along with editor Robert Sloan, gave thorough and fair reviews of the manuscript, and generously advised me during its journey to completion. Sarah Wyatt Swanson, editorial assistant at the press, was also remarkably attentive in responding to my pleas for help, or initiated procedures needed for ensuring that this book looks good (and I think it does). Copy editing was expertly handled by Karen Hellekson and Nancy Lightfoot, who polished my prose to a more readable state. Thank you, Jim, Bob, Sarah, Karen, and Nancy.

I didn't know it at the time, but the seeds of *Life Traces of the Georgia Coast* were planted with my first visit to the Georgia coast as a PhD student at the University of Georgia in Athens (UGA) in 1986. Robert Frey, one of the world's experts on trace fossils, was my mentor at UGA, and he introduced me to the voluminous literature on the modern traces of the Georgia barrier islands, much of it written by him and his colleagues. Of the many skills he taught me, though, clear writing was the most valuable and enduring. Nearly every day, as I strive to fulfill this practice, I think of his influence and appreciate him for it. Sadly, his many potential future contributions to ichnology, including a summary of the ichnology of the Georgia coast intended for a popular audience, were stopped with his death in 1992. So I feel that the book you are reading now is the one he should have written. Thank you, Bob.

Two additional former graduate students of Frey are Steve Henderson (Oxford College of Emory University) and Andy Rindsberg (University of West Alabama), who both later became my valued colleagues and treasured friends. Steve and I, through our coteaching of students in the extraordinary natural classrooms of the Georgia barrier islands and elsewhere in the world, have learned much from one another through such interactions and became much better educators through it all. Andy and I have also expanded our minds together, whether from being in the field together on the Georgia barrier islands, looking at trace fossils in many parts of the world, or debating the finer (or blunter) points of ichnology. Andy also reviewed several chapters in this book, delivering incisive yet helpful insights on how to improve these bits of writing. Thank you, Steve and Andy.

My wife, Ruth Schowalter, accompanied me for much of the fieldwork I have done on the Georgia coast during the past decade, and she is to be commended for her extraordinarily steadfast support through those sometimes challenging (but always enlightening) experiences. At home, her fueling me with tea, coffee, food, love, and reassurances during the writing of this book also made this book happen. Thank you, Ruth: I could not have done it without you.

The Georgia coast is blessed with extraordinary scientists and naturalists, and I have been privileged to meet and know many of them through the best medium of all, going into the field together. These people include John "Crawfish" Crawford and Clark Alexander (Skidaway Institute of Oceanography, Skidaway Island); Jim Bitler (Ossabaw Island); Royce Hayes, Jen Hilburn, Gale Bishop, and David Hurst Thomas (St. Catherines Island); Jon Garbisch (UGA Marine Institute, Sapelo Island); Scott Coleman (Little St. Simons and St. Simons Islands); Christa Frangiamore and Steve Newell (Jekyll Island); and Carol Ruckdeschel and V. J. "Jim" Henry (Cumberland Island). Of these people, two died while this book was still in process, V. J. Henry and Jim Bitler. V. J., among the most accomplished and respected of coastal geologists in the southeastern United States, was always a kind and generous mentor to me and many others who are still trying to understand the complex geology of the Georgia coast. As for Jim Bitler, we only spent two days together in the field, but my brief exposure to his enthusiasm for the natural history of Ossabaw and his gift for storytelling inspired me far more than if we had never met. In short, by writing this book, I am following in some rather large footsteps that will likely outlast all of us. My appreciation is extended to those who have gathered knowledge about the Georgia barrier islands through direct experience. Thanks to you all.

Along those lines, this book includes much information that was gathered through skills taught to me by people far outside of the oft-myopic world of academia. Consequently, I am grateful to the instructors of the Tracker School (New Jersey), Wilderness Awareness School (Washington), and A Naturalist's World (Montana) for opening my eyes to the ancient science of tracking as a way of observing and ably augmenting my education in ichnology. Thanks to these and all trackers and tracking instructors for keeping these practices alive.

In the realm of ichnology as a formal science, I am indebted to its most prominent experts who have taught me much about traces, trace fossils, and paleontology of the world outside of Georgia, whether through spirited discussions in the field or afterward. These include, in alphabetical order, Richard Bromley (University of Copenhagen), Luis Buatois (University of Saskatchewan), Tony Ekdale (University of Utah), Kathleen Campbell (University of Auckland), Al Curran (Smith College), Jorge Genise (Museo Paleontológico Egidio Feruglio), Jordi Gilbert (University of Barcelona), Murray Gingras (University of Alberta), Roland Goldring (University of Reading) Murray Gregory (University of Auckland), Stephen Hasiotis (University of Kansas), Martin Lockley (University of Colorado-Denver), Gabriela Mángano (University of Saskatchewan), Renata Guimarães Netto (UNISINOS), George Pemberton (University of Alberta), John Pollard (University of Manchester), Thomas Rich (Museum Victoria), Dolf Seilacher (University of Tübingen), Alfred Uchman (Jagiellonian University), Patricia Vickers-Rich (Monash University), Sally Walker (University of Georgia), and Andreas Wetzel (University of Basel), to name a few. Many thanks to all of you, and to anyone I did not mention in this short list of ichnologists and paleontologists whom I call my friends.

Of course, I would be remiss in not mentioning the support of my institution, Emory University (Atlanta, Georgia), where I have taught many undergraduate students over the course of more than 20 years. Through my teaching these students about the ichnology, geology, and ecology of the Georgia barrier islands, they have helped me to learn more than I ever could have gained on my own. Those people who say, "Those who can't, teach," have obviously never taught and are, quite frankly, clueless, as they are missing out on the spectrum of perspectives provided by students. It is thus no coincidence that some of the stories I tell in this book stem from experiences I shared with students on the Georgia coast. I also wrote this book with the hope that all people reading it will also become my students and will pass on their new knowledge and skills to others through teaching of their own.

Most of the chapters in this book were first heard by two of my colleagues at Emory, Christine Ristaino and Isabel Wilkerson, while they were writing their own books, both of which were very different from mine. Our weekly meetings for more than a year, in which we read aloud our rough drafts of chapters to one another, immensely improved the clarity and timbre of my words, ensuring that the book would be understood and enjoyed by a broad audience. Thank you, Christine and Isabel.

Last, in every ichnological study of mine, I try to acknowledge the tracemakers, and in this instance I want to especially thank the tracemakers of the Georgia coast: the plants, invertebrates, and vertebrates that, through their interacting with the beautiful ecosystems of the Georgia coast, leave their life traces for us to notice, study, and appreciate. Their tracemaking and traces will likely outlast our humanity, so let's be grateful for our living in the same places for now.

LIFE TRACES OF THE
GEORGIA COAST

1

INTRODUCTION TO ICHNOLOGY OF THE GEORGIA COAST

THE MYSTERY OF THE BROKEN BIVALVE

The large clam fell, however improbably, from a great height above the sandy tidal flat. Three fragments of its thick shell lay in front of me: an entire valve and a small part of the other were still connected by a hinge, and the other chunk was nearby. The main part of its fleshy body was now mostly gone, as were several more pieces in this once-living puzzle. Considering that bivalves are not often prone to aerial activities or spontaneously exploding, nor are they likely to will their soft parts to disappear, I wondered what else might have caused such a fatal mishap that fine spring day on Sapelo Island of coastal Georgia (Figure 1.1).

    A glance at the sandy surface revealed a few clues: the impression where the shell bounced on the hard-packed sand surface; the dispersal of shell fragments from that impact site; the remaining soft parts of the bivalve, partially eaten; and the tracks around the semiconsumed remains. Whose tracks? It was two legged, and its tracks started with the feet nearly together, left in front of right, and none before these. After this, it alternated its feet as it walked, left–right and right–left in a tight, linear path to the valve that served its innards on the half shell. The trackmaker's pace aver-

Figure 1.1. The lone, broken, and empty shell of a freshly killed bivalve, the giant Atlantic cockle (*Dinocardium robustum*) on a sand flat at low tide, Sapelo Island, Georgia. How did this happen, who did it, and how are these questions answered by ichnology?

Figure 1.2. *Facing.* A forensic analysis of how the cockle met its end, based on traces left by it and its predator, a laughing gull (*Larus altricilla*). (a) The gull spotted the bivalve from the air, landed, walked up to and extracted it, then flew off with it. (b) The gull landed again, after dropping the bivalve a third time, then trampled the impact spot. The hinged part of the bivalve's shell, which held most of the animal's soft parts, was moved to the left. Not drawn to scale, and some parts abbreviated for the sake of space.

aged about 10 cm (4 in) between each step, and decreased as it approached the bivalve. In these tracks, its digits pointed slightly toward the midline of the body, giving the trackway a pigeon-toed appearance. Its feet came together again and just to one side, followed by much shuffling in the same small area. The tracks were about 5 cm (2 in) long and almost as wide. These indicated feet with three narrow digits, widely spaced but all pointing forward, and a thin, curved line denoted webbing toward the ends of these forward-facing toes. A nub of a toe, its mark rarely seen, was at the rear of the tracks, and faint claw marks accented each digit.

I expanded the search for more clues. Spiraling outward from the shell remains, I looked for other indentations made by the clam bouncing on the sand flat, and found two more. One of these marks was near a shallow excavation, a vertically oriented hole in the sand that matched the external profile of the entire bivalve. This is where the clam had started its airborne exploits. Within a meter (3.3 ft) of the bivalve resting trace were the paired tracks of the same bipedal animal seen at the scene of the bivalve's demise. The same trackway pattern was there, starting with two tracks, changing to slow, pigeon-toed walking, and ending with side-by-side tracks directly in front of where the bivalve had its last moment of peace. Prod marks on one side of the bivalve resting trace showed where the trackmaker had, with much persistence, pried out the recalcitrant mollusk with a hard, sharp-edged digging tool (Figure 1.2).

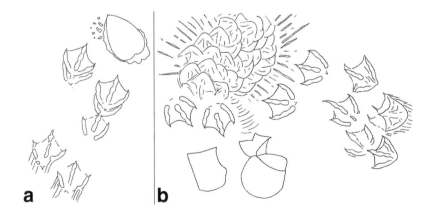

The trackmaker was the laughing gull (*Larus altricilla*), the bivalve was the giant Atlantic cockle (*Dinocardium robustum*), and the traces there on the sand flat clearly told the story of how the cockle met its end through the actions of the gull. Well before my arrival on the scene, millions of years of evolutionary history contributed to a present-day behavior that had preserved the cockle's lineage. That past, though, was now meeting the outcome of natural selection manifested in the gull. The tide had dropped in the preceding few hours, leaving the cockle stranded on a tidal flat drained by nearby runnels and water trickling down between grains of sand. As the tide began to ebb, the cockle used its muscular foot to extend, anchor, and pull the shell into the still-saturated sand—a simple form of burrowing that exposed a bare minimum of its heart-shaped outline. As pore waters continued to drain from the sand flat, sand grains pulled together around the cockle, further securing its position as it waited for the next tidal cycle. Unfortunately for the prey, but fortunately for the predator, the partly buried clam was spotted from the air by the gull, which translated this search image to "food." The gull landed nearby, walked up to the cockle, and wedged it out of its formerly safe burrow with its beak. Having no hands, it used its beak again to grasp the cockle on the edge of its shell and took off with its payload. Perhaps as much as 10 m (33 ft) above the tidal flat, it let go of the bivalve. It did not break. A second time the gull landed, picked up, flew, and dropped its intended meal, but still with no satisfactory results. The third time was the proverbial charm, and the once-protective shell of the hapless clam fractured on the sand flat's surface. The gull's strategy was keen. Through a combination of interstitial moisture and gravity, the low tide had caused the fine-grained sand to pull together in such a way that the intertidal surface became nearly as solid as concrete. Ironically, the same sedimentary circumstances that initially helped to protect the cockle from predators later served as the instrument of its death.

But was this an isolated instance, a singular circumstance of a gull genius? (How difficult it is to say those last two words sequentially!) The shortest possible answer to that question—no—connected to other stories that confirmed this inventive, rapacious behavior in birds. For example, just two years before encountering this scene on the Georgia coast, I saw newly broken shells of freshwater gastropods on a paved road in Everglades National Park. At first I thought cars had crushed them, but I looked closer and noted that the fracture patterns were more from point impacts, and the remaining soft parts held distinctive beak marks. I modified my hypothesis: perhaps these snails were nicked by passing cars and scavenged by birds that saw or smelled the freshly killed food. The final modification of the hypothesis happened as soon as direct observation of a fish crow (*Corvus ossifragus*) provided enlightenment. The crow flew down into the freshwater marsh, emerged with something in its beak, flew up, and let go of the object about 10 m (33 ft) above the road. Crack! Gastropod sashimi was being served within minutes, and the idea that birds would use hard surfaces as tools for gaining food entered my consciousness as an idea to hold, remember, apply, and test elsewhere. Indeed, soon after seeing the results of this avian predation in both Florida and Georgia, I investigated the peer-reviewed literature and found that other scientists had observed and documented the same behavior in various species of gulls, crows, and ravens (Zach 1978; Beck 1980; Kent 1981; Cadée 1989; Gamble and Crisol 2002). In other words, I was not alone in my witnessing of such shell-breaking behaviors. In fact, some of these scientists even proposed how these techniques represented a sort of avian culture, wherein birds of the same species watched a few individuals successfully procure molluscan flesh by dropping shells, then imitated it enough that it spread from there, perhaps over generations (Zach 1978; Gamble and Cristol 2002).

Now let's put on our stylish paleontologist hats and project these modern observations into the future, perhaps long after our species, or that of the cockle and gull, have gone extinct. If just the pieces of the bivalve had been found in the fossil record, would we have known it was the victim of predation? Even if we noticed that its shell had unusual breaks imparted when the animal was still alive, would we have even guessed that its shell was broken by its hitting sediment—not rock—from high above a sand flat? How would we have inferred the identity of the predator, and its methods for acquiring and killing its prey? If only the bivalve's burrow were found, would we have been able to tell that a predator forcefully extracted its occupant? What about the bounce marks made by the bivalve—could we have recognized them for their true identity as traces of predation?

Welcome to the science of ichnology, the study of life traces, and one of the best places in the world to study it, the Georgia coast in the southeastern United States. The Georgia barrier islands collectively reflect a setting for many overlapping cycles of life and death, all leaving clues in sand, mud, plants, shells, and bones for us to read, understand, and project into the past and future—that is, if our senses are trained well enough to discern the many stories told by these traces, and if our imaginations can envision the intersecting lives and behaviors of plants and animals that make these vestiges.

WHAT IS ICHNOLOGY, AND WHAT IS A TRACE?

Ichnology is simply the study of traces (from *ichnos*, "trace," and *logos*, "study"), where a trace is any indirect evidence of an organism exclusive of body parts that also reflect the organism's behavior (Frey 1975; Ekdale et al. 1984; Bromley 1996; Seilacher 2007). Ichnology can be divided into neoichnology, the study of modern traces, and paleoichnology, the study of ancient traces; ancient traces are known more commonly as trace fossils or ichnofossils. The majority of this book is devoted to neoichnology, although toward its end, some of that focus is also applied to the study of trace fossils (Chapter 10). What are some examples of traces? Tracks, trails, burrows, nests, feces, borings, and tooth marks comprise a few types of traces, but virtually any mark left on a medium of some sort (sand, mud, shells, bone, flesh) caused by the behavior of a living organism can constitute a trace (Figure 1.3). In contrast, dead organisms, appropriately enough, do not behave; hence these do not make traces.

Although this might be the gentle reader's first encounter with the word, ichnology is probably one of the oldest sciences known to humankind. This science developed out of our evolutionary past through the recognition and classification of animal tracks and other signs, as well as interpreting animal behavior from these traces, otherwise known simply as tracking (Liebenberg 2001). In a recursive way, humans left their own traces, reflecting their awareness of animal traces, where petroglyphs and other markings on rock perfectly mimic the forms of animal tracks, trackway patterns, and nests. For example, rock art from Australia and South Africa from tens of thousands of years ago includes track iconography identifiable to local species, some of which are now extinct (Flood 1997; Eastwood and Eastwood 2006). The main purposes of such art were likely similar to those of this book: to increase consciousness of tracks and other traces in surrounding environments, as well as to teach future generations about the tracemakers and behaviors that led to the given traces. In other words, ichnology is a part of our human heritage, and learning about it ought to be as natural as following a series of tracks on a sandy beach.

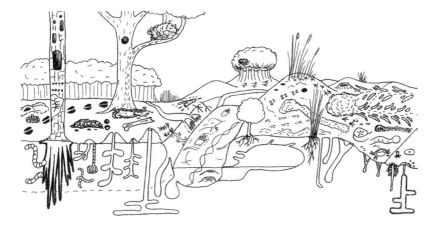

Figure 1.3. A small sampling of the variety of modern traces that one can encounter on the Georgia barrier islands: tracks, trails, burrows, nests, feces, borings, and tooth marks. To find out what these are, please read the rest of the book.

Ichnologically speaking, though, traces are treated as separate from a few other types of indirect signs often associated with traditional tracking. For example, when encountering feathers, fur, skulls, and other body parts in the field, ichnologists do not consider these as traces per se, but as the actual remains of animals. Similarly, a leaf impression seen in an urban sidewalk, formed by a leaf falling onto wet cement, is not a trace, because it only represents a plant body part, rather than actual behavior of the plant. In short, anything constituting a body part and not reflecting behavior from a living organism is not a trace. Nonetheless, body parts can certainly contain traces of the behavior of another tracemaker. For example, the investigative story I used at the start of this chapter illustrates this distinction: the bivalve shell itself is a body part, but the broken shell reflects the predatory behavior of the gull. So it is both a body part (of the bivalve) and a trace (of the gull). Similarly, a scattered assemblage of songbird feathers encountered in the middle of a maritime forest is both a collection of body parts from that songbird and a trace of a raptor feeding on the songbird (Chapter 8). Raccoon feces packed with fiddler crab parts are a trace only of a raccoon, unless one wants to count fiddler crab gullibility and helplessness as traces (Chapters 6, 8). The same leaf impression in a sidewalk mentioned previously may also contain evidence of incisions along its margins and within the leaf, traces of bees, beetles, or other insects that fed on the leaf (Chapter 5). The plant also reacts to these attacks by toughening the tissue around the wounds; this wound response, then, is a trace of the plant's behavior (Chapter 4). In other words, traces are commonly preserved in or otherwise associated with the bodily remains of other organisms, especially traces that involve the consumption of another organism.

A lesson I often use in teaching my students the distinction between traces and other natural phenomena (what is a trace and what is not?) is told from a forensic perspective: When a person drags the dead body of a murder victim across a sandy area, what traces are left, and who is the tracemaker? After some brief debate and argument, the answers emerge. The dead body may have made the drag marks, but the body itself would not have formed the traces (zombie-movie scenarios aside) without the behavior of the person moving it. Hence the body was only a tool (leaving a tool mark), and the resultant drag marks coupled with footprints are traces of the person doing the moving, not of the victim. Furthermore, the tracemaker may not be the murderer; he or she is simply moving the body after someone or something else caused the death (Figure 1.4a). Another "trace or not a trace?" riddle is posed by a stalk of grass rooted in a sand dune that may have been pushed in different directions by the wind, then whipped back to its former position, leaving intricate impressions around the stalk (Figure 1.4b). This movement was not a result of plant behavior but was caused by the wind, so these marks are not traces either. Traces thus must involve behavior by a living tracemaker, and to emphasize this point, I will occasionally refer to them, somewhat redundantly, as life traces. With enough practice and application of healthy skepticism—that is, trying to prove yourself wrong, the trait of a real scientist—the distinction between traces and nontraces will become even more obvious.

Of course, what really makes a trace (or not) is a tracemaker, and a threefold line of integrated inquiry must be used to understand traces: (1) What is (or was) the medium (substrate) hosting the trace?; (2) What aspects of the tracemaker's anatomy were used to make the trace?; and (3) What behaviors by the tracemaker caused its anatomy to interact with materials in a way to leave a trace? In homage to my Catholic upbringing, I often call this mode of study the Holy Trinity of ichnology, but in a more ecumenical spirit, as well as to honor T. E. Lawrence, a fellow appreciator of sandy

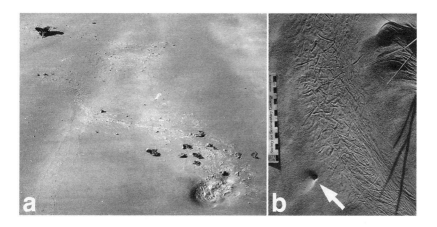

Figure 1.4. Markings that are traces and not traces, depending on behavior: (a) Drag mark caused by black vultures (*Coragyps atratus*) on a beach of Sapelo Island, in which the dead body of a horseshoe crab (*Limulus polyphemus*) was scavenged by the vultures. In this instance, the vultures were the tracemakers, not the dead animal dragged by the vultures. The mark left by the dead limulid's body is not a trace either, for the simple reason that a dead animal does not behave. (b) Stalk of sea oats (*Uniola paniculata*) on a coastal dune, having been moved by the wind, imparted some marks on the sand, but through no behavior of its own, so these are not traces. In contrast, an insect burrow (arrow) is a trace: Jekyll Island. Scale in centimeters.

substrates, I also call it the Three Pillars of Ichnologic Wisdom. Regardless of how one remembers these tenets, this triad can be used a basic guide to understanding traces, from which novices and experts alike will benefit.

THE THREE PILLARS OF ICHNOLOGIC WISDOM: SUBSTRATE, ANATOMY, AND BEHAVIOR

As just mentioned, the gestalt of any trace is defined by the interaction of three factors: (1) the substrate preserving the trace; (2) the anatomy of the tracemaker; and (3) the behavior of the tracemaker. If any of these three are missing, a trace will not be made, nor will it be preserved. Accordingly, we will also not detect it with any of our senses, no matter how well honed our observation skills, catalog of search images, cognitive abilities, or other means of detection.

In sedimentary geology, a substrate is typically some sort of sediment, such as clay, silt, or sand. Likewise, these substrates are the ones we most commonly associate with ichnology. Other substrates include rock, shells, wood, or bones; these types of traces require scratching, breaking, drilling, or other such alterations of these consolidated media. Substrates, however, can be even more ephemeral. For example, I have seen wakes left by alligators in ponds that persisted for several hours after an alligator swam through algae floating on the pond surface. In such instances, the algae comprised the substrate (Figure 1.5). For many animals, pheromones and scent markings are traces used to communicate with other species, creating invisible olfactory landscapes (Chapters 5, 8). For example, owners of domestic dogs and cats are often conscious of how their pets always seem eager to detect scents. These animals also leave their own distinctive calling cards for others of their species through body rubbings, urine, or feces, announcing, "I was here!" Perhaps the most fleeting of traces, though, are sounds, in which compressional waves, imparted on air, received by ears, and translated by brains, vanish instantly after the vibrations alter their substrates, whether these are air, water, wood, or animal bodies. (Or is the brain a substrate changed by the sound, thus preserving it as a more lasting trace? Ponder this thought as it continues to alter your mind.) In ichnology, then, a substrate is any medium that holds the effects of an organism's behavior, however long that effect might last. The Georgia coast and some other coastal areas of the world thus can become ichnological playgrounds, filled with a surfeit of substrates that are constantly shifting with deposition, erosion, and other transformations inherent to coastal systems (Chapters 2, 3). Yet some are also preserved in just the right medium to become part of the geologic record, thereby becoming trace fossils (Chapter 10).

Anatomy refers to any aspect of a tracemaker's body that makes a trace. Normally we might first think of feet as the most important anatomical traits that leave distinctive and identifiable traces. Indeed, trackers of the past made sure they knew the detailed anatomies of a wide variety of animal feet, and modern trackers can easily list these traits: toe counts (how many toes on the front feet, and how many on the rear?); heel pads (two lobes in the rear, or three?); presence or absence of claws (fully retractable, semiretractable, nonretractable?); and other aspects of foot form (Rezendes 1999; Halfpenny and Bruchac 2002; Elbroch 2003). Nonetheless, the anatomies of plants, legless animals, and even fungi can construct magnificent traces. Imagine how the roots of live oaks (*Quercus virginiana*)

Figure 1.5. Ephemeral traces left by alligators (*Alligator mississippiensis*), caused by their swimming along the surface of a pond and disturbing the surface layer of algae: Sapelo Island. These sorts of traces have low potential for preservation in the fossil record, but nonetheless demonstrate how some substrates can unexpectedly hold traces.

push aside sand grains in a forest ecosystem (Chapters 3, 4). Or how the peristaltic body movements of earthworms and their digestive tracts define the qualities of soils in that same ecosystem (Chapter 5). Or how hyphae of fungi probe into wood or soil, seeking nutrients and decomposing materials so that insects can use these substances as food (Chapter 4). Moreover, small and legged animals in great numbers can form complex traces that cumulatively impact entire landscapes or seascapes. For example, ant or termite colonies routinely change subterranean environments on a massive scale through their combined efforts (Chapter 5), just as ghost shrimp (*Callichirus major, Biffarius biformis*) in nearshore environments process cubic meters of sediment through their burrowing and feeding (Chapter 6). The small and rarely seen eastern mole (*Scalopus aquaticus*) digs tunnels that extend for tens of meters (Chapter 8). Armadillos (*Dasypus novemcinctus*) can make dens as deep as 4 meters (13 ft) (Chapter 8). Gopher tortoises (*Gopherus polyphemus*) dig burrows so large and hospitable that hundreds of other species of animals can also reside in them (Chapter 7).

In the case of the latter three tracemakers—moles, armadillos, and gopher tortoises—their anatomies clearly reflect adaptations for digging, as all have relatively large attachments in their shoulder girdles to support the necessary musculature. In some instances, then, a tracemaker and its potential traces can be hypothesized on the basis of its body parts. Moreover, body size, an aspect of anatomy, is a factor when evaluating a trace for its potential tracemakers. An alligator den could never be confused with a ghost crab burrow on the basis of size alone, although some ghost crab burrows overlap with mole burrows in width (Chapters 6, 8). Of course, tracemaker age also affects the size and anatomy of tracemakers. For example, a ghost crab trackway can coincide in size and some aspects of anatomy with a trackway left by a juvenile horseshoe crab; each typically has eight walking leg impressions. Nevertheless, other anatomical distinctions, discernible in their trackways, help to separate the two (Figure 1.6). Last, anatomies reflect evolutionary lineages, and because evolution happens largely as a result of natural selection, anatomies provide information about adaptations, and adaptations in turn affect behavior, no matter how startlingly pliable the latter might be.

Implicit to any trace, however, is that behavior must be transmitting stresses that cause part of an organism's anatomy to leave its mark. In other words, if you had to choose any one of these three facets of ichnology as the most important, behavior would be it. Of course, motivation precedes any behavior, so a list of typical behaviors becomes an easy exercise in empathizing with a tracemaker: feeding, movement, mating, reproducing, resting, fighting, and so on. Nevertheless, behaviors overlap in ways that the primary motivation might be difficult to discriminate. Were the extensive root traces in a sandy soil left by a live oak (*Quercus virginiana*) formed for stability, or to gain more nutrients (Chapter 4)? Was the galloping pattern in the trackway of a whitetail deer (*Odocoileus virginianus*), crossing a clearing in a maritime forest, made because human hunters startled the deer, or it was trying to catch up with others in its family unit (Chapter 8)? Were sandpiper tracks showing it landing on a beach representing its need for rest after a long flight, or did it intend to feed on the invertebrates living

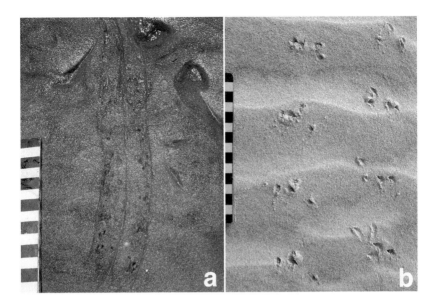

Figure 1.6. Similarity in one aspect of a tracemaker anatomy does not necessarily translate into a similarity of traces, even when the tracemakers have the same number of legs. Also, differences in sediment moisture can affect the appearance of traces, even when the sediment is the same composition and texture. (a) Trackway of a small horseshoe crab (*Limulus polyphemus*) on wet beach sand, in which it used eight walking legs: Sapelo Island. (b) Trackway of a ghost crab (*Ocypode quadrata*), but on dry beach sand, also having used eight walking legs: Sapelo Island. Scale in centimeters in both photos.

in the sand (Chapter 9)? Was a male fiddler crab burrow made to escape predators, or to attract a mate (Chapter 6)? Was a pileated woodpecker cavity, carved from a rotting tree, caused by its feeding on wood-boring insects, or because it was making a nest (Chapter 8)? Often the answer to such questions about behavior is yes, and the behavioral interpretations of traces must take into account that tracemakers may have had multiple and inseparable reasons for forming any given trace.

While keeping in mind the three pillars of wisdom in ichnology—substrate, anatomy, and behavior—two other principles of ichnology will help with forming, testing, and revising ichnologically based hypotheses. These principles are (1) different species of tracemakers can make the same type of trace; and (2) the same species of tracemaker can make a wide variety of traces. An understanding of both of these ideas lends well to knowing that ichnodiversity (the amount of trace variety) should never be confused with biodiversity (Chapter 10).

With regard to the first principle—different tracemakers can make the same type of trace—think of how many animals on the Georgia coast can make a simple burrow: worms, spiders, ghost crabs, fiddler crabs, bees, wasps, armadillos, gopher tortoises, alligators, swallows, moles, gophers, chipmunks, and beavers (Chapters 5–8). How, then, to distinguish the tracemakers? As mentioned previously, size-appropriate distinctions certainly can be applied. For example, the preceding list of tracemakers can be arranged in order of body mass to narrow down the possible originator of any given burrow. In other words, a gopher tortoise did not dig a 2.5-cm (1-in) wide burrow; this is too small for even a newly hatched juvenile (Chapter 7). A second method is to ask about anatomical signatures or fingerprints, sometimes termed bioglyphs (Mikuláš 1998) that may have added to a trace. These clues include body impressions, claw marks, or other impressions that can help to identify a tracemaker, or at least narrow down its identity (Figure 1.7). Yet another way to figure out who made a particular burrow, and perhaps the most basic, is to simply ask the question, "Who lives here?" For example, an armadillo probably did not dig a burrow in the intertidal zone of a beach, because they live in maritime forests. Similarly, a ghost crab almost surely did not make a burrow in a maritime forest, because they live in back-dune meadows, dunes, and the upper parts of beaches. Consideration of ecology and tracemaker habitat is paramount when examining any trace (Chapter 3), although a theme we will commonly revisit is how tracemakers will sometimes surprise ichnologists with their extensive forays into habitats seemingly unfriendly to their modes of life.

The second ichnological principle—the same species of tracemaker can make a wide variety of traces—is demonstrated easily by following the life history of any given individual animal and thinking about the traces it makes during its lifetime. Naturally, the number and variety of traces will depend on many factors; for example, appropriate substrates encountered, anatomical attributes for making traces, types of distinctive behaviors, and length of the tracemaker's life. Nonetheless, one short-lived individual animal, representing one species, can make a huge variety and number of traces during its life, covering a disproportionately wide

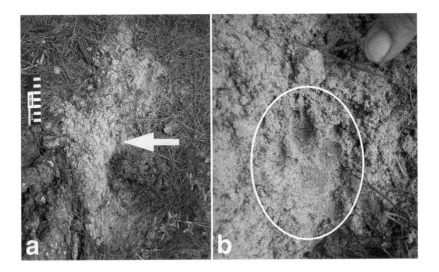

Figure 1.7. (a) Shallow excavation made by an armadillo (*Dasypus novemcinctus*) in a maritime forest as it looked for infaunal insects at the base of a loblolly pine (*Pinus taeda*). Scale in centimeters. (b) Close-up of the armadillo track (right rear foot) indicated the by arrow in (a), confirming the identity of the burrow maker: Sapelo Island.

volume of substrates. Perhaps my favorite example of an animal that embodies a trace-filled life is the ghost crab (*Ocypode quadrata*), which is also ubiquitous on Georgia beaches (Chapter 6). Individual ghost crabs live for about three years, and if they live this long, they make (at a minimum) the following traces (Figure 1.8): (1) near-vertical burrows, about a meter (3.3 ft) deep, often with two entrance–exit holes; (2) their distinctive trackways, made with eight walking legs, which lead to and from burrows, and affect areas of more than 100 m^2 (1,090 ft^2) per day; (3) large piles of sand balls made from digging out their burrows, but also used to visually mark them; (4) scrape marks from pulling their claws across beach surfaces to feed on algae in the sand; (5) shallow excavations from which they have extracted small bivalves from the surf zone and eaten them; and (6) resting traces made in wet beach sands, showing where they have rehydrated after extended times outside of their burrows (Frey et al. 1984; Frey and Pemberton 1987; Martin 2006a). Multiply these behaviors, any of which can happen on a given day in a ghost crab's life over the course of three years, and hundreds of thousands of traces will have been made by a single animal.

This potential for trace production and diversity by an individual animal also points toward the huge advantage enjoyed by neoichnologists over most other naturalists. After all, which has a better chance of being noticed: an individual tracemaker, or hundreds of thousands of its traces? Additionally, these records of unobserved behaviors far outnumber observed behaviors, meaning that neoichnologists have far more data to study than most animal behaviorists who rely on direct observation of their subjects. Moreover, these observations are done with much less effort. (It doesn't quite seem fair, does it?) Paleoichnologists might also have a problem of abundance, in which they can only study a small sample of thousands of trace fossils at a given geologic site. In contrast, other paleontologists may search for days (or weeks) for a single elusive body fossil at the same site.

Nonetheless, a major challenge presents itself when geologists and paleontologists encounter trace fossils and ask the question, "Who made this trace?"—an inquiry made when the original environments are long gone and contain only these remnants of past behavior in sediments or rock. Furthermore, despite their great numbers, not all traces make it into the geologic record, or only parts of the original traces are preserved (Chapter 10). Connecting traces with tracemakers is difficult enough, but it becomes nearly insoluble when confronted with mere shadows of the original traces from millions of years ago. A corporeal analogy is expressed in the John Godfrey Saxe poem, *The Blind Men and the Elephant,* in which each of six blind men metaphorically (and mistakenly) interprets the animal nature of an elephant on the basis of feeling only one part (a flank, tusk, trunk, leg, ear, and tail). Similarly, ichnologists can easily have their vision taken away unless provided with search images based on entire and interconnected modern examples. This is where neoichnology can serve as one of the most powerful tools in a paleontologist's kit, requiring mostly eyes, knowledge, and imagination, rather than relying primarily on technological fixes (Chapter 11). Consequently, those who become skilled in neoichnology often justify why they are sought to identify odd structures in ancient sedimentary rocks that do not seem to fit any other known category (Chapter 10).

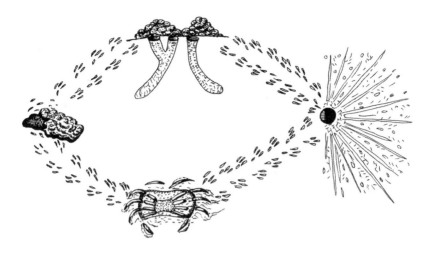

Figure 1.8. The wonderfully diverse traces made by ghost crabs (*Ocypode quadrata*): burrows (Y shaped and J shaped) and sand piles outside of burrows; surface scrapings and sand flung outside their burrows; body outlines left while resting; and excavations made while looking for edible clams, all connected by their tracks.

ICHNOLOGY AND UNIFORMITARIANISM

Uniformitarianism, a long-standing principle applied to some aspects of geology and evolutionary biology, is a mouthful of a term but simple to explain. The concept of uniformitarianism is often summarized through the phrase "the present is the key to the past"; its less syllabic synonym is *actualism*. In other words, actual geologic processes we observe today, such as tides, waves, wind, erosion, and deposition, are assumed to have also happened in the geologic past. Likewise, biological processes happening today, such as reproduction, growth, movement, herbivory, and predation, are rightfully assumed to have taken place long before humans ever thought about these processes. By using uniformitarianism as a guide, then, one can also surmise that plants and animals behaved in ways that caused their anatomies to interact with substrates in the geologic past, and they accordingly left traces comparable to those observed today.

Sure enough, over the past 200 years or so, people have documented trace fossils that were clearly identifiable as the former products of organismal behavior. For example, college student Pliny Moody found dinosaur tracks in rocks of Massachusetts in 1802, decades before the word *dinosaur* had even been invented; yet he and others instantly recognized these and similar trace fossils from the surrounding area as animal tracks (Steinbock 1989; Pemberton et al. 2007a). In an instance where uniformitarianism was more explicit, fossil vertebrate tracks found in Scotland in 1824 inspired William Buckland to use live tortoises to make trackways in various substrates, such as clay, mud, and unbaked pie crusts. (And no, I did not make up the latter: Pemberton et al. 2008a.) Through this method, Buckland compared the fossil tracks to modern ones and noted their similarities and differences. In contrast, many invertebrate trace fossils initially provoked confusion in the early to late nineteenth century, as many scientists accepted these as fossilized algae, or fucoids. These, however, were eventually linked with invertebrate tracemakers once the trace fossils were more carefully described (Osgood 1975; Rindsberg and Martin 2003; Pemberton et al. 2007b).

Ichnology, however, is a science that actually can flip the phrase of actualism to "the past is the key to the present," in that trace fossils found in rocks from specific environments provided incentive for finding modern analogues. Indeed, this reversal of actualism happened where structures and other clues preserved in some sedimentary rocks showed that they were formed in deep-ocean basins, yet these rocks also contained numerous complex trace fossils. Because modern deep-sea traces are thousands of meters below the ocean surface, these are necessarily the most difficult traces for humans to observe directly. As a result, the hypothesis that these trace fossils were originally formed in deep-sea environments remained untested for decades. Nonetheless, when scientists looked at sediment samples from the deep-ocean seafloor or visited these areas in submersibles, they found modern traces that matched trace fossils interpreted as coming from deep-sea environments (Ekdale 1980; Wetzel 1983; Uchman 2007; Rona et al. 2009). In other words, trace fossils were used to predict what modern traces we should find in analogous environments, and many modern-day animals make traces identical—or at least similar—to some ancient traces. In turn, this perspective helped to establish a foundation for the further application of neoichnology to paleoichnology.

Uniformitarianism in its original sense, however, is epitomized by neoichnology, particularly through the study of invertebrate and vertebrate traces in modern marginal marine environments. The nascence of modern neoichnology was in the 1920s, begun by German geologists on the tidal flats of the Wadden Sea along the coast of the Netherlands, Germany, and Denmark (Schäefer 1972; Osgood 1975; Cadée and Goldring 2007). Having recognized the importance of investigating modern environments for similar traits discernible in the geologic record, in 1928 these geologists founded the Forschungsstation Senckenberg am Meer (the Senckenberg Field Station on the Sea, also known as the SAM) in Wilhelmshaven, Germany (Cadée and Goldring 2007). Among the fields of study at the SAM was neoichnology, which was deemed as an integral component of actuogeology or actuopaleontology; the prefix *actuo* refers to what is observable in modern environments. The neoichnological research that came out of the SAM in the 1920s through the 1950s was classic and influential in modern ichnology. It also had far-reaching effects in a related field, taphonomy (from *taphos,* "burial," and *nomos,* "law"), which is the study of how body and trace fossils are preserved in the geologic record (Shipman 1981; Cadée 1991; Bromley 1996, chap. 10).

In fact, the effects of this research reached so far that it moved across the Atlantic Ocean to Georgia. As explained later (Chapter 2), collaborations between a few of these same German geologists with U.S. geologists in the 1960s inspired similar research on the Georgia barrier islands. This legacy is thus one of the reasons why the Georgia coast deserves greater public recognition for its scientific importance as a place where geology, paleontology, and ecology came together to form a holistic approach to the study of organismal traces.

WHY THE GEORGIA COAST AS A PLACE TO STUDY ICHNOLOGY?

Because traces can be found wherever life leaves marks of its behavior in substrates, ichnology can be studied nearly everywhere on the surface of the earth. Yet the Georgia coast is one of the best places to study neoichnology—a bold statement that will be justified by the rest of this book. I further submit that most neoichnological research on the Georgia coast does not require massive grants, research vessels, submersibles, or other expensive equipment. Using our three pillars of ichnological wisdom for guidance, we can understand why such audacious claims can be made:

Substrates—The Georgia coast, with its barrier islands and back-barrier environments, contains abundant and diverse substrates that easily record billions of organismal traces on any given day (Chapter 3). Although other barrier island systems may have similarly abundant sandy beaches, muddy marshes, and other environments, the Georgia barrier islands are unique in also having consolidated relict marshes and Pleistocene deposits. Some of the relict marshes are hundreds or thousands of years old, and the oldest Pleistocene sediments are at least 40,000 years old. These ancient landscapes are now emerging from erosion largely caused by rising sea level (Hoyt et al. 1964; Hoyt and Hails 1967; Frey and Basan 1981; Frey and Howard 1986). With the unveiling of these older and firmer substrates, modern tracemakers, such as fiddler crabs, insects, and trees, take residence in them, and thus merge new and old traces. Even a Pleistocene gastropod shell exhumed on a Georgia shore can become a new substrate once occupied by a modern hermit crab (Chapter 6), just as humans willingly move into old homes and leave their own distinctive and more recent marks in these preexisting structures.

Maritime forests, often overlooked in neoichnological studies, also supply important substrates—wood and other plant tissues—which preserve traces of wood-boring insects and birds, as well as feeding marks made by herbivores (Chapters 4, 5, 8). Moreover, once wood from this maritime forest has washed out to sea and drifted long enough, it can also preserve traces of wood-boring marine invertebrates (Chapters 2, 6), combining older terrestrial with newer marine traces in the same substrate. With a similar perspective, we can look for ichnologic clues in the geologic record, where tens, hundreds, or thousands of years may have passed in an area before a prevailing ecosystem was replaced by a new one, as indicated by distinctive suites of tracemakers (Chapter 10). The Georgia coast is one of the few places where comprehensive studies of substrate-dependent traces have been conducted, and all within a small area that is readily accessible to scientific researchers and the general public alike.

Anatomies—Although the Georgia barrier islands cannot rival more tropical areas of the world in its biodiversity, the number of species native to these islands is nonetheless impressive. This biodiversity is directly attributable to the large number of ecosystems in this relatively limited space, as well as recent geologic history, in which plants and animals of the mainland part of North America blend with those endemic to the islands. This bio-

Figure 1.9. The tracemakers involved in the murder mystery (at the start of the chapter) revealed: (a) Giant Atlantic cockle (*Dinocardium robustum*); (b) Laughing gull (*Larus altricilla*).

diversity also translates into a wide variety of tracemaker anatomies. The study of how organismal anatomies mirror abilities to perform tasks (or, simply put, how "form reflects function") is called functional morphology. Implicit to functional morphology is how any given species' anatomies, and even those of fossil species, represent evolutionary adaptations to environmental parameters (Seilacher 1973; Savazzi 1999).

For example, the thick shell of the Atlantic cockle (*Dinocardium robustum*), mentioned at the beginning of this chapter, serves it well for protecting it against powerful shoreline waves and most predators. Its muscular foot helps to conceal it in wet sand during low tides. When placed on an exposed sand flat, this foot, like a wagging tongue, will pop out of the shell to rapidly (and comically) push the animal into a position more amenable to burying the shell. The cockle's shell is also variegated, with a pinkish and dark brown mottling that tends to blend into the sand surface, providing some camouflage. Pronounced ribbing on the outside of the shell increases its surface area and correspondingly strengthens the shell: think of corrugated cardboard as an analogue. Last, this ribbing accentuates surface tension of the saturated sand surrounding the cockle, which better holds it in place when buried. All in all, these anatomical traits add up to a formidable defense against most predators (Figure 1.9a).

Nonetheless, the visual acuity and nervous system of a laughing gull (*Larus altricilla*), combined with its stout beak, arms and shoulder girdles modified for flight, and flight feathers, all constitute anatomical traits that allow it to successfully prey on cockles, no matter how well protected the latter might seem (Figure 1.9b). The gull's nervous system in particular gives it an advantage in sensory and cognitive abilities as a predator; it is sophisticated enough for gulls to watch and learn from other gulls. In contrast, no amount of a bivalve watching another bivalve will result in it learning how to avoid gull predation. Bivalves, after all, are not just slow learners, they do not learn at all. Similarly, vultures have enlarged olfactory lobes that allow these scavengers to detect decaying carrion from kilometers away, making the flight from a maritime forest (where they normally live) to a beach with dead bounty a rewarding exercise (Figure 1.4a; Chapter 8). From a vegetative perspective, the deeply penetrating and tapering (pointy) vertical roots of loblolly pines (*Pinus taeda*) are good at finding freshwater and other nutrients, as well as avoiding salt water (Chapter 4). In other words, anatomy matters, and the effects of organismal behavior on substrates will not be imparted without the use of an organism's given set of body parts.

Behaviors—Just as anatomies reflect evolutionarily selected adaptations, the myriad of behaviors displayed by organisms along the Georgia coast is the result of ecological diversity. Ecosystems and their changes through time are the result of natural selection, which favors or disfavors adaptations over time, including genetically linked behaviors. The study of how behavior often represents the outcomes of evolutionary processes in the context of ecological factors is termed behavioral ecology (Danchin et al. 2008). In this sense, ichnology is easily linked with behavioral ecology. For example, during April–May of every year, horseshoe crabs, also known as limulids and represented in Georgia by the species *Limulus polyphemus*, emerge from the ocean to lay and fertilize eggs on Georgia beaches (Shuster et al. 2003; Martin and Rindsberg 2007). Limulids are the only marine invertebrates that must leave the sea to reproduce, and the only arthropods that fertilize eggs externally (Chapter 6). Both adaptations, then, provoke a simple scientific question: Why? The hypothesis currently favored by evolutionary biologists and paleontologists is that limulids represent an ancient lineage of marine invertebrates. Somewhere in geologic time during the history of this lineage—which is at least 450 million years old—marginal-marine environments, such as beaches and tidal flats, were widespread. Natural selection somehow favored individuals that went onto those beaches and tidal flats to lay and fertilize eggs. The ancestors of modern limulids must have had genes that enabled this behavior to be passed on to future generations. Given enough selection pressures exerted over time, it eventually became the normal behavior that we witness today.

Interestingly, natural selection also resulted in coevolved behaviors for shorebird predators of limulid eggs and juveniles. Each year along the east coast of the United States, in Delaware and New Jersey in particular, ruddy turnstones (*Arenaria interpres*) and red knots (*Calidris canutus*) time their lengthy spring migrations so they arrive in the midst of limulid mating. This punctuality is not accidental, as these shorebirds can then consume massive numbers of limulid eggs and newly hatched limulids (Castro and Meyers 1993; Walls 2002; Botton et al. 2003a). How long did it take for this behavior to evolve in both birds? Good question, as shorebirds in one form or another have been around for only 100 million years (or only about 20 percent of limulid history), and the lineages for ruddy turnstones and red knots are considerably shorter than that. Did this same behavior evolve along different evolutionary pathways of shorebirds in the past? Ah, that is an even better question, and one that could be answered by ichnology. Limulids make diagnostic traces (especially when mating), and co-associated avian tracks and feeding traces could be described carefully too. These traces then could be used as search images in the rock record, thus developing hypotheses that can better answer such evolutionarily based inquiries.

Nonetheless, probably the most important aspect of behavior to remember is its malleability. In other words, knowledge of genes and evolutionary heritage does not always ensure 100 percent reliability when predicting how a plant or animal will behave at any given moment. Genes are not destiny, and most complex organisms are not machines that blindly follow predetermined programs, no matter what the arguments to the contrary. Instead, organisms will initiate behaviors according to ever-changing circumstances in their environments, sometimes in ways that contradict previous perceptions and amaze the observer. For example, when I encountered tracks of an adult alligator (*Alligator mississippiensis*) coming out of the surf and onto a Georgia beach one morning, I had to revise my expectations of alligator physiology that confined these animals to freshwater environments (Chapters 7, 9). Similarly, trampled areas and trails leading into salt marshes associated with tracks of feral cattle (*Bos taurus*) and cropped tops of smooth cordgrass (*Spartina alterniflora*), persuaded me to reevaluate the habitat range and feeding preferences of these animals on Georgia barrier islands (Chapters 8, 9). Yet other ichnological oddities I have encountered on the Georgia coast were long, shallow tunnels of moles that traveled from behind coastal dunes and stopped at the high-tide mark of the beach (Chapters 8). In comparison, a discovery of dozens of resting traces made by ghost crabs on a Georgia beach just after a spring tide seemed conventional; after all, ghost crabs live abundantly in these environments. Yet these traces had not been described in this otherwise well-studied species (Martin 2006a). In essence, ichnology expands the world of nature observation to include the vast realm of unobserved behaviors embodied in traces, which can then be applied to the interpretation of similar trace fossils. A sentiment often shared by experienced trackers is that traces frequently hold surprises that go against the accepted conventional wisdom for the natural history of some species (Rezendes 1999; Elbroch and Marks 2001; Elbroch 2003; Young and Morgan 2007).

The preceding reasons might seem sufficient for making the point about the aptness of the Georgia coast for studying ichnology, but fortunately the case was already made long before this present sales pitch. Geologists and paleontologists have studied organismal traces on the Georgia coast, and particularly those of the barrier islands, since the 1960s (Weimer and Hoyt 1964; Frey and Howard 1969; Frey and Mayou 1971; Hertweck 1972; Basan and Frey 1977; and many more). This longevity and concentration of study is admittedly related to some appealing aspects of the Georgia barrier islands—yes, geologists, biologists, and even paleontologists also like long stretches of beautiful beaches—but is also connected to the history of science, and the development of modern ecology in particular (Chapter 2). Granted, organismal traces of some other coastal systems have been studied longer, such as those of the Wadden Sea tidal flats. Nevertheless, the latitudinal advantage enjoyed by Georgia coast tracemakers provides for a deciding factor: plants and animals are both more diverse and more active in a warmer climate, and for more months of the year (Chapter 3). Ichnologists can accordingly be more assured of making year-round observations of plant and animal traces when visiting the Georgia coast, while also seeing seasonal variations in behavior that enter into ichnological assessments. Given these factors, as well as the plethora of environments and substrates available for study, these life traces of the Georgia coast create ideal outdoor laboratories for exploration, discovery, and insights into how modern traces can be our guides to better understanding behavior in the geologic past.

2

HISTORY OF THE GEORGIA COAST AND ITS ICHNOLOGY

A RISING SEA, AND LIFE TRACES ON THE MOVE

About 55,000 years ago, as the earth warmed and glaciers melted, the barrier island moved across the landscape in synch with the rising sea. Its oceanward side shifted to the west toward the coastal plain, and its resident biota adjusted and adapted to new locales. Laterally adjacent environments began to succeed one another vertically, and in this instance, the sands of offshore environments piled onto new surfaces scoured out of formerly onshore environments. Salt marsh mud also swelled in area and volume, spreading throughout low-lying areas, as forests and other terrestrial environments were displaced and replaced. Life in all of these environments—maritime forests, dunes, back-dune meadows, freshwater swamps, salt marshes, tidal creeks, beaches, and shallow subtidal sandy bottoms—flowed and ebbed like the tides, and all attendant traces of plant and animal behavior were transplanted with their substrates.

On what would some day be the modern Georgia coast, this linkage between sea level and life was apart from and unaffected by humans. It was a time when the world operated on its own unhampered schedule. The still-new *Homo sapiens* would not arrive in this part of North America until about 40,000 years later; small populations of this species were applying their own version of ichnology to tracking game animals in Africa and Australia. In the mainland part of North America, and just west of the barrier islands, giant ground sloths, mammoths, mastodons, bison, and glyptodonts—armadillo-like animals the size of a compact car—were among the most visible animals. Nonetheless, each spring and fall, even these huge mammals were rendered insignificant by flocks of migrating birds so numerous they blotted out the sun, avian clouds that cast vast shadows across landscapes and seascapes. Their massed calls, used to keep flocks together, were deafening, temporarily overwhelming any auditory communications between land-dwelling animals. A fine, white mist of feces was left in their wake, peppered with seeds from berries consumed by these birds as fuel for their journeys. Well-worn trails made by mammals large and small crisscrossed coastal plain forests. These anastomosing, low-lying pathways were better revealed after frequent forest fires, ignited by lightning and fueled by fire-adapted pine and wiregrass communities. Plants living in these ecosystems used these blazes as opportunities to disperse their seeds and grew faster in the absence of competing vegetation. During such firestorms, many smaller forest-dwelling animals would seek refuge in deep gopher tortoise burrows. After a fire, perhaps they would remain in these burrows for a while longer, as predators—saber-toothed cats, panthers, wolves, and bears—took advantage of the lack of protective cover following a burn and easily found live prey, or at least barbecued remains of animals that had failed to avoid the flames.

Megafaunal trails tended to connect waterways and served as inadvertent canals, especially wherever they intersected the banks of rivers, lakes, and marshes. Lakes were largely the handiwork of those great prehuman engineers, beavers (*Castoroides ohioensis*), which, like modern beavers (*C. canadensis*) chopped down vegetation and dammed streams, thus encouraging the expansion of freshwater wetlands wherever they lived. Such rodents were nearly unrecognizable in size and activity, though; these upscaled versions of modern beavers were half the size of black bears. In more swampy environments between coastal plain forests, alligators (*Alligator mississippiensis*) similarly widened their more modest waterholes by undulating their tails and bodies in saturated sands and muds, activities that coincided with the annual arrival of spring and summer rains. The largest of these alligators, as well as newly arrived crocodiles (*Crocodylus acutus*) that had moved north with the warmer climate, walked across salt marsh surfaces at low tide, and through maritime forests when looking for food and mates. Through these habitual and seasonal movements, their feet, bellies, and heavy tails plowed through mud and sand to make new trails that, like those of the megafauna, linked previously separate water bodies. In this way, the mammals, reptiles, and other animals were parts of the landscape, but their traces were also responsible for its character and appearance.

As sea level rose, mute and immobile maritime forests on the barrier islands, which had taken thousands of years to develop as ecosystems during a sea-level lowstand, stood helpless against the once distant but now encroaching shore. Within just a few decades, the arrival of daily tides and waves erased the lives of trees and undergrowth alike, replacing them with plants better adapted to the change in salinity. On the seaward side, massive pines (*Pinus* spp.) and live oaks (*Quercus virginiana*) that had survived fires and hurricanes for a hundred years or more foundered and toppled. Sands that once held them securely were instead undermined by ocean water, whisking away their former stability. Their root systems, sheared from above as the trees fell, were still vertical in the surf and their buried, distal parts marked exactly where the trees once lived. Birds nesting in trees proximal to their fishy foodstuff, such as ospreys (*Pandion haliaetus*) and bald eagles (*Haliaeetus leucocephalus*), abandoned these places as soon as they became too wobbly. The trees, however, started dying long before they tilted toward the sea. Their fatal illness began as salt water filled the pores between sand grains with each high tide, twice daily. These intrusions killed symbiotic fungi surrounding the roots, thus depriving the trees of important nutrients. The salt content of the water also sickened the trees that could only handle occasional influxes of such salinity. Likewise, cabbage palm trees (*Sabal palmetto*), normally tolerant of salt spray in coastal environments, succumbed to the heightened sea level on forest margins adjacent to salt marshes. These trees left behind beheaded upright trunks bearing nest holes of red-bellied woodpeckers (*Melanerpes carolinus*).

Yet all of these dead trees were not wasted, ecologically speaking. As they progressed from upright to recumbent in the surf zone, they became places for the attachment of sedentary marine animals, such as barnacles and oysters. Pieces of wood exposed on the shore or that floated out to sea became new homes for marine invertebrates that settled onto the woody surfaces, such as the wedge piddock (*Martesia cuneiformis*) and marine isopods (*Limnoria* spp.). Throughout their growth to adulthood, these bivalves and arthropods bored deeply into the wood, making tight compartments for homes as they rafted in a heaving sea. In some instances, their borings cut across those formed previously by wood-eating animals that lived in the forest, such as carpenter ants, bees, beetles, and termites. The traces of the forest insects in turn had been crosscut by the holes of ivory-billed and pileated woodpeckers (*Campephilus principalis* and *Dryocopus pileatus*, respectively), drilled into the wood as these birds ate wood-boring insects. A few of the trees bobbing on the ocean surface even contained their nest holes, which differed from those made by red-bellied woodpeckers. As more tree branches and trunks rafted into the Atlantic Ocean, a symptom of rapidly rising sea level, resident bivalves and their traces accordingly increased in number. Once any of these pieces of wood became waterlogged, they sank into greater oceanic depths, or they drifted onto newly made shorelines, carrying the ecologically mixed marks of their bioeroders. Some trees also contained the empty shells of dead piddocks, still ensconced in their former domiciles.

The first arrival of seawater on the edges of maritime forests also triggered a sort of subterranean migration. Cicada nymphs (species of *Magicicada* or *Tibicen*), some of which had lived on and near tree roots for more than a decade, burrowed away from the first, sudden incursions of salty water, as did other insects that were feeding, growing, and producing offspring in the sandy soil. Entire ant colonies picked up and went to high ground wherever they could find it and constructed new, complex multichambered nests. Ant lions, the carnivorous larvae of winged insects, also had to move with their intended food. Their shallow horizontal burrows, trails, and funnel-like vertical burrows hence appeared on the surfaces of sandy substrates wherever ants might be roaming. In these same environments, particularly in the maritime forests, eastern moles (*Scalopus aquaticus*) in search of their infaunal invertebrate sustenance—earthworms, nematodes, insects, and other such treats—also had to burrow away from the seawater. Some of their shallow tunnels briefly intersected beaches and other intertidal zones. A distant relative of the eastern mole, the star-nosed mole (*Condylura cristata*), was less affected by rising water levels because of its proclivity for burrowing alongside water bodies. These semiaquatic moles accomplished much of their hunting underwater. Nonetheless, these burrowing animals likewise could not stay in or near salty water, as most of their prey consisted of freshwater worms (especially leeches) and aquatic insect larvae. So these species and their immediate descendants moved too. Freshwater crayfish (*Procambarus* spp.) emerged from their deep vertical burrows and similarly chambered quarters below the water table, and over generations swam or walked to areas that still retained freshwater. Crayfish descended from marine lobsters nearly 250 million years ago, but they found their niche early on in freshwater environments, and none of their descendants were going back into the ocean any time soon. Those invertebrates that did not move far enough, or could not move at all, died immediately. These animals included the larvae and pupae of ground-nesting bees (*Agapostemon virescens*) and wasps (*Sphecius speciosus* and *Stictia carolina*), still encased in their underground brooding chambers as they drowned. Insects, despite their 400-million-year history, vast numbers, and ubiquity in modern terrestrial environments, had evolved little tolerance for living in salt water, let alone during sedentary growth cycles.

Soon this mixture of terrestrial invertebrate burrows and nests was crosscut by the wider, deeper, and distinctively shaped burrows of newly arrived and more ecologically appropriate fiddler crabs and ghost crabs in salt marshes and coastal dunes, respectively. Where moles used to burrow, female sea turtles, such as loggerheads (*Caretta caretta*) and leatherbacks (*Dermochelys coriacea*), arrived nightly in large numbers from late spring through late summer. These turtles dug deep, bowl-like nests in the same dunes inhabited by ghost crabs, deposited clutches of 100–150 eggs, buried them, and returned to sea. Expectant mothers repeated this egg laying as many as five times in a season, meaning that 1,000 mother sea turtles were capable of producing 5,000 nests per year on these newly formed barrier islands. Overlapping in time with this annual reptilian nesting was an orgiastic invertebrate event, which took place with each high tide and just slightly downslope on beaches and tidal runnels. Millions of female horseshoe crabs (*Limulus polyphemus*), or limulids, crawled out of the surf to lay their eggs in the beach sands. Even greater numbers of male limulids competed to fertilize these eggs, pushing one another out of the way and latching onto the backs of females to ensure they would be in the right place to let loose their gametes once eggs were deposited. Their trackways and resting traces, some doubled by female–male pairs, churned the uppermost part of the beach sand. These sediments were then mixed even more by the feet and beaks of shorebirds feasting on the eggs and newly hatched juveniles. Other flocks of shorebirds stopped to feed on burrowing bivalves and gastropods in the surf zone, leaving countless tracks, probe marks, cracked and empty shells, feces, and regurgitants filled with finely comminuted shell and crab parts in their wake. Some birds were year-round residents, whereas others were there seasonally for nesting and the bountiful provisions, which expanded considerably during times of warmer temperatures and sea-level rises. A few species of shorebirds tended to communally nest in the same places each year, a site fidelity shared with sea turtles. Nonetheless, as the sea advanced on nest sites in the upper shore and dunes, these too had to move more inland.

Then, as sea level went up just a bit more, the deep, reinforced burrows of callianassid shrimp—species of *Callichirus* or *Biffarius*, more commonly known as ghost shrimp—intersected underlying sediments of former dunes, slicing through remnants of ghost crab burrows and sea turtle nests. Some of these turtle nests had successfully hatched their clutches, whereas others were connected with ghost crab burrows, a sign of successful egg predation. Dark brown fecal pellets of mud, packaged at the size of large sand grains, were pumped out of deep ghost shrimp burrows and deposited near the tops of their small, volcano-like openings. These pellets rolled into nearby ripple troughs and accumulated to form thin, dark muddy streaks interspersed with whitish quartz beach sand. In the previous site of dunes, sand was pounded, rolled, eroded, and deposited daily by waves and tides. Drier, emergent spots had their less dense quartz sand removed, reworked, and redeposited by near-constant winds from the east. At low tide, regular echinoids, also known as sand dollars (*Mellita isometra*), coordinated hundreds of their tube feet to rapidly burrow into the surf zone in attempts to stay hydrated in the wet saline sand. Some clams in the same surf zones, such as species of *Donax, Mulinia,* and *Ensis,* likewise burrowed quickly when exhumed by waves, trying to escape predation from shorebirds or anything else that found them delectable, such as southern stingrays (*Dasyatis americana*). During high tides and in shallow water, stingrays preyed on buried bivalves and other invertebrates by settling onto sandy surfaces, shooting high-pressure streams of water into the sand and chomping on nakedly exposed molluscans and crustaceans. Sea stars, distant relatives of the echinoids, were also sometimes exposed on sand flats at low tide. These animals were poor burrowers and instead made complex surface trails while seeking life-sustaining moisture before the next high tide, or they hid under thin covers of sand. Amphipods, small crustaceans known today as beach fleas, roiled the sand with their sheer numbers and rapid, shallow burrowing. Juvenile limulids, ranging from less than the size of a capital letter in this sentence to nearly half the width of this page, made looping and criss-crossing bilobed trails as they crawled expansively throughout the tidal flats and in runnels. During their furrowing journeys, they grasped and ate tiny bivalves, worms, and other morsels that might have been in their paths. The top predator of these sand flats, however, was the seemingly inconspicuous moon snail (*Neverita duplicata*). Its silent calling card was a single small, beveled hole drilled through the shells of its molluscan victims; some of these holes were even left in shells of their own species, traces of Pleistocene cannibalism.

Some of these same empty gastropod shells, though, took on new lives as hermit crabs slipped inside them and traveled much farther and faster on sandy and muddy surfaces than their corporeal owners ever would have experienced. These hermit crabs not only left distinctive trackways of their four walking legs on the sand flats, but also wore down shell bottoms by dragging them across the harder quartz sand. These polished spots would later identify the former presence of their secondary tenants. Gastropod rivals of moon snails in the predatory realm of this environment were whelks, consisting of various species of *Busycon*. Each whelk employed its outermost lip as a wedge to pry open obstinate clams, and such attempts were recorded on shells as chippings of their edges, barely noticeable, yet evidence of life-and-death struggles. Some of these whelks, however, wore their own scars from battles they fought and won against shell-crushing portunid crabs. Preserved in every whelk that successfully escaped a lip-peeling crab was a ragged but healed tear, visible as a jagged line running the length of an otherwise smooth shell. In other words, some tracemaking predators themselves carried the traces of other predators, and secondary occupiers (hermit crabs) left their own traces on gastropod shells, imparting detailed tales of life, death, and lives after death.

Beneath these molluscans and crabs, and sharing the upper layers of sand flats with callianassid shrimp, were burrowing polychaete worms and acorn worms. The homes of the polychaete worms, consisting of species of *Diopatra, Chaetopterus, Onuphis,* and many others, protruded from sandy surfaces as parchment-like vertical tubes. If washed out by waves, some of these tubes accumulated in ripple troughs like thousands of collapsed paper straws. Tubes of *Diopatra*, however, were reinforced on their top ends by small bivalve shells that had been pasted into tube walls. These were the behavioral equivalent of *Diopatra* adding bricks to a chimney, a way of ensuring greater longevity for its burrow in the face of constant shoreline erosion. Tops of acorn worm burrows were identifiable by fecal castings, which were pyramidal and stacked piles of sand resembling soft-serve ice cream that would not be invented for more than 50,000 years.

Within a hundred years or so, a mere jot in the Pleistocene, a sea-level rise of about 50 cm (20 in) had completely replaced the maritime forest with offshore sand and mud. The buried remnants of the former ecosystem included numerous traces of its former inhabitants, but few remains of the tracemakers themselves. Moreover, although daily erosion of sand related to this rise erased most burrows, tracks, trails, and other superficial traces, the deeper parts of burrows made by ghost crabs and callianassid shrimp were preserved. So were many bivalve and gastropod shells, pieces of driftwood, and other hard substrates bearing traces made by other organisms. Thus the sediments preserved a detailed record of successive generations of traces, telling a story of ecological change that happened tens of thousands of years before the words "global warming" (in any language) were uttered by humans.

Meanwhile, on the western (landward) side of the islands, salt marshes expanded their boundaries into the lower edges and interiors of former maritime forests and dunes. This progression was aided by more frequent summer storms, some of which were full-blown hurricanes. These storms generated surges that breached dunes and washed sand over and into back-dune meadows and forests. As with the rising sea, the salt water that poured into these environments with each storm instantly killed plants and subsurface insects unused to having more sodium and potassium introduced so suddenly to their environments. However, the thick sand layers of these storm-washover deposits and newly saline groundwater provided fresh real estate for sand fiddler crabs (*Uca pugilator*), which quickly colonized these surfaces. Within only a few weeks, thousands of fiddler crabs were digging burrows for protection and mating, and were scraping algae off the surface for food, rolling little balls of sand from these scrapings. Males made territorial displays around their burrows and worked hard at impressing potential mates by seemingly shouting, "Look at my claw!" while waving their large appendages, effective come-hither moves in their world. Within just a few years, the edges of these storm-washover deposits gained a distinctive suite of salt-tolerant plants (halophytes), which started to spread across the sandy expanses. In instances where storm-washover sands had covered muddy salt marshes, the most common species of halophyte, the smooth cordgrass (*Spartina alterniflora*), poked through the sand and continued to grow as long as it received sunlight, nutrients, and salty water. Marsh periwinkles (*Littoraria irrorata*), small snails that grazed on the stalks of the *Spartina*, quickly became abundant, as did several species of insects that stayed above the ground, such as the marsh grasshopper (*Hesperotettix floridensis*). Soon the fiddler crabs, with more plants overhead, had even more protection, and were thus encouraged to further colonize the marsh surface, which increased the flow of nutrients within the top layer of the young salt marsh.

With further inundation by the heightened sea, mud began to deposit on top of these sandy surfaces, the result of muddy feces pumped out by ribbed mussels (*Geukensia demissa*) and eastern oysters (*Crassostrea virginica*). With more mud, the marsh changed. Smooth cordgrass grew far taller, reaching nearly 2 m (6.6 ft). Mud fiddlers (*Uca pugnax*) replaced their close relatives, the sand fiddlers, excavating their own burrows in the newly deposited muddy substrates. Clumps of ribbed mussels attached below stalks of smooth cordgrass and added more mud; plant stalks also slowed tidal waters so that much of this mud was retained. Thus muddy marshes returned to some of the same places where sand had been so suddenly introduced by storms.

The added defense afforded by towering stands of *Spartina* in either sandy or muddy marshes did not help fiddler crabs much against some predators, though. With such large numbers of crabs and their readily available protein, great blue herons (*Ardea herodias*), great egrets (*Ardea alba*), sandhill cranes (*Grus canadensis*), and other birds could fly in and easily find quick snacks of stragglers caught outside of their burrows. Raccoons (*Procyon lotor*), which had moved landward into the shrinking maritime forest as sea level rose, soon found these recently expanded ecosystems served up nightly all-you-can-eat crab feasts. Fiddler crabs that retreated into burrows, a defense that worked well against most predatory birds, were summarily dug out and eaten by raccoons. Conical pits with claw marks on their walls, along with displaced sand or mud heaped outside of these holes, told how raccoons procured their meals. Frequent raccoon forays into these new marshes wore down trails through the marsh vegetation that were just slightly wider than the average raccoon. Raccoon feces, filled with fiddler crab body parts, punctuated open areas on the higher parts of the marshes, including in the middle of trails, serving as territorial markers for other raccoons to take heed.

As ebb and flood tides began scouring new tidal creeks and channels through these sandy and muddy flats, oysters settled onto any firm surfaces and stabilized creek banks. At low tide, these banks would show the occasional playful four-by-four loping trackway patterns and belly-slide marks of river otters (*Lutra canadensis*). These traces were accompanied by impressions of large, sunning alligators and crocodiles that tolerated the salinities of tidal creeks to find a greater variety of food. Occasionally, a mammoth or ground sloth lost its way while walking through a maritime forest and stepped into the edge of a salt marsh. As these huge mammals extracted their feet from the black muck, the rotten-egg smell of hydrogen sulfide (H_2S) gas was released, a by-product of anaerobic bacteria breaking down organics in the marsh mud. Herons, egrets, and cranes, all with large feet bearing four long, thin digits, also changed marsh surfaces. Whenever their feet punched through algal films on high-marsh substrates, such intrusion initiated drying of the underlying mud. Summer sunshine then encouraged the growth of mud cracks from each toe impression; these cracks spread in fractal-like patterns throughout the marsh surface until the original footprints became unrecognizable.

Like all cycles, though, everything reversed. After about 10,000 years, the global temperature went down, and the global water budget flipped toward less liquid and more solid, locking up nearly 95 percent of all freshwater into alpine and continental glaciers. Accordingly, sea level lowered, and the barrier islands dutifully followed the shoreline, but eastward this time. Coastal plain ecosystems, such as pine and oak forests, replaced areas formerly subjected to daily tides. Wherever these islands once were, all that was left were slight, linear, topographic highs of sandy sediments on an otherwise flat coastal plain covered mostly by pines, grasses, and freshwater wetlands that closely followed river valleys.

This sea level low was responsible for forming the foundations of some present-day Georgia barrier islands, composite islands made of Pleistocene and Holocene sediments, the latter of which were deposited in the past 10,000 years or so. Dunes from tens of thousands of years ago are now far inland and marked by northeast–southwest trending ridges of sand that would be completely unnoticed if driving through on a paved highway, and changes in elevation would barely break a sweat in a person walking up their slopes. Yet each is the vestige of a former barrier island, and within each ridge are the trace fossils of ghost crabs, ghost shrimp, and other animals whose burrows are so similar to those of modern examples that they are diagnostic of their original ancient environments. These trace fossils tell us where the islands once resided and how the back-and-forth, up-and-down motions of the sea altered coastal ecosystems and the behaviors of its flora and fauna, all of which took place long before humans became aware of the history of this landscape.

RIVERS AND SEAS, MUDS AND SANDS: GEOLOGIC HISTORY OF THE GEORGIA COAST

The preceding narrative about traces, their tracemakers, and environments changing through time is intended to show how tracemakers and their traces are related to a larger tableau, which is the geologic history of the Georgia barrier islands. This history, in turn, imparted the traits we now observe in the ecosystems and sedimentary environments of the islands, which affect where organisms live and how they behave. These fluctuations of sea level culminated in the contemporary situation of the Georgia barrier islands where we have always known them to be. Nonetheless, these and all of the islands that preceded them were always in flux. Thus what seems static now from our perspective—such as the present-day islands and their positions—is not, and never will be. The tracemakers that interacted with environments on and around these islands never had knowledge of this dynamism; they simply lived where it was appropriate.

Geologists mapped the coastal plain ridges, recognized their original paleoenvironmental identities from their trace fossils, and later interpreted a series of six linear, parallel but discontinuous landforms that spoke of past sea-level changes and former islands. These same geologists then gave colorful and historically apt names to these extinct Pleistocene barrier islands, mixing Native American and European sources: Wicomico, Penholoway, Talbot, Pamlico, Princess Anne, and Silver Bluff (Figure 2.1), listed in order of height above sea level and distance from the present-day shoreline (Hoyt and Hails 1967; Henry et al. 1993). For example, the Wicomico ridge system represents the highest reach of the sea up the coastal plain; it has the highest elevation of any ancient barrier island system along the Georgia coast; and it is the oldest. In contrast, the remains of the Silver Bluff barrier island system are near present-day sea level, although the sediments were deposited at a minimum of 1.5 m (5 ft) above present-day sea level. Silver Bluff deposits are being actively eroded today, but they are also covered in part by modern Holocene sand and mud associated with the most recent oceanic rise, which began about 11,000 years ago at the end of the Pleistocene Epoch. All barrier island systems in between the Wicomico and Silver Bluff ones represent varying ages and sea levels, recording the fluctuations of worldwide temperature, as well as melting or freezing of glacial ice. The Penholoway barrier islands were formed when sea level was about 23 m (75 ft) higher than today; the Talbot at about 14 m (45 ft); the Pamlico at 8 m (25 ft); and the Princess Anne at slightly more than 4 m (15 ft) above modern sea level (Hoyt and Hails 1967; Henry et al. 1993; Bishop et al. 2007).

Interestingly, at least one ancient barrier island system from a sea-level low, correlated with a widespread glaciation, also formed about 32 km (20 mi) offshore from the present-day Georgia coast, and at about 22 m (72 ft) below present-day sea level. This sea-level low happened in the Pliocene Epoch, which was just before the start of the Pleistocene, about 2–3 million years ago. This former barrier island was then later alternately exposed and drowned during the next few million years. During the most recent post-Pleistocene sea-level rises, however, and starting about 7,000 years ago, this higher area on the seafloor made an ideal place for reef-building animals, such as sponges and corals. Today the site of this former barrier island is occupied by Gray's Reef, one of the best-developed temperate reefs in the eastern United States (Hyland et al. 2006; Garrison et al. 2008). How do we know this was a barrier island, more connected with the land than the sea after the Pliocene? Paleontological and archaeological investigations of bedrock composing the core of Gray's Reef revealed fossil bones of *Bison*

Figure 2.1. *Facing.* Surface expressions of ancient barrier islands on the Georgia coastal plain: the Wicomico (Wi), Penholoway (Pe), Talbot (Ta), Pamlico (Pa), Princess Anne (PA), and Silver Bluff (SB) islands. Modern composite (Pleistocene–Holocene) islands indicated (Cumberland → Ossabaw); see Figure 2.2 for Pleistocene and Holocene components. Rivers are denoted as "Savannah R.," and so on; note how these either flow between ancient island chains, or cut directly across them.

HISTORY OF THE GEORGIA COAST AND ITS ICHNOLOGY 39

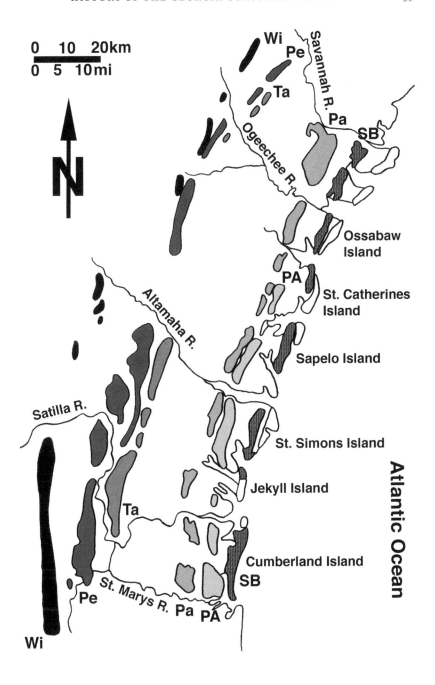

(buffaloes), *Equus* (horses), and *Mammuthus* (mammoths), as well as shallow-water invertebrates (oysters) and Native American artifacts (Garrison et al. 2003). Thus the reef, now under more than 20 m (67 ft) of water, may have been a place where people gathered and terrestrial wildlife thrived, and it owes its present existence to an extended period of global cooling in the past.

On the coastal plain of Georgia, old barrier island ridges also affect the courses of freshwater rivers of the Georgia coastal plain. These rivers flow in between ridges in some places; elsewhere, they cut directly through them (Figure 2.1). The present-day St. Marys and Satilla rivers, for example, make right-angle turns where their flow is restricted to low areas between the Talbot and Penholoway ridges. In contrast, the Altamaha and Ogeechee rivers dissect any ridges they encounter and move in a more or less southeastern direction and directly to the coast (Rich 2007). These differences in drainage of coastal plain rivers result in interesting and unique ecosystems along the way. One of these is probably the best-known freshwater swamp in the world, the Okefenokee in southeastern Georgia. The Okefenokee owes its fame to the Suwanee River, which originates there, and songwriter Stephen Foster, although he never saw the river of which he rhapsodized in "Old Folks at Home." More recently, this swamp served as the setting for cartoonist Walt Kelly's comic strip, *Pogo*. The Okefenokee is bordered on the south by the Wicomico barrier island system, which formed the topographic high of Trail Ridge, a former beach dune complex. Salt marshes that formed during the Miocene Epoch, about 5–23 million years ago, compose clay layers underneath. These impermeable clays prevent water from draining downward, and the ridge stops the lateral flow of water so that it accumulates on the westward side of the ridge. Thus both Trail Ridge and these clays, vestiges of ancient barrier island ecosystems, effectively block the drainage of the Suwanee and nearby St. Marys rivers, and helped to form the extensive present-day cypress forests and bogs of the Okefenokee.

As coastal plain rivers flow downslope and empty into the Georgia coast, they effectively feed the present-day barrier islands and their associated environments with sand and mud. Sand is to barrier islands what cells are to bodies, providing firmament for their ecosystems and substrates for tracemakers. Accordingly, the youngest of the Georgia barrier islands, like many such barrier islands worldwide, would simply erode away and vanish if not for this constant supply of sand. The sand and mud carried by the rivers is ultimately derived from the Appalachian Mountains: pick up a handful of sand on most Georgia barrier island beaches, and you are also holding pieces of the Appalachians. The Appalachians weathered with each cycle of uplift caused by the crashing of continental and oceanic plates during the Paleozoic Era, were reworked along the edges during sea-level fluctuations of the Mesozoic Era, and were (and are) having their sediments moved back to the sea during the Cenozoic Era. Because the Appalachians are oriented northeast–southwest, surface water flows either to the west or east of this continental divide. This eastern continental divide is more modest in height than the one farther west in the Rocky Mountains, and hence most rivers in Georgia—particularly in the coastal plain—lack the whitewater rapids craved by adrenaline-addicted rafters, and instead attract more sensible and sedate canoers and kayakers.

The demarcation between the harder igneous and metamorphic rocks of the Blue Ridge and Piedmont Provinces and the softer sedimentary rocks of the Coastal Plain Province is called the fall line, in recognition of how the abrupt change in slope caused many waterfalls at this geologic boundary. Many Georgia towns were founded along the fall line, partially because its faster-moving water was useful for mills and agriculture: Columbus, Macon, Milledgeville, and Augusta, for example, all line up on this geologic boundary. Moreover, boats traveling upstream from the coast had to stop at the fall line, which contributed to the growth of these towns as important places for shipping commerce (Coleman 1991; Stewart 1996). With softer rock to cut through, rivers also tend to meander more on the coastal plain, making broad floodplains that resulted in flat, relatively fertile areas, explaining why southern Georgia became the agricultural nexus for the state. The fall line also represents the furthest known extent of the Cretaceous seaway from slightly more than 70 million years ago. Dinosaur bones are still recovered from some coastal plain deposits in the southwestern part of Georgia, where their bodies must have floated out to sea before final burial of these rare bits and pieces (Schwimmer et al. 1993; Schwimmer 2004). Because the coastal plain is downslope and its rocks are poorly consolidated, it is threaded with rivers that flow mostly southeast toward the Georgia coast. These rivers carry a load of sand and mud that is ultimately deposited and reworked on the coast by waves, tides, and organisms. Hence any dams or other human structures built on these waterways would interrupt their flow and cut off that sediment supply of sand, adversely affecting the islands.

A close look at the sand composing most Georgia beaches and dunes reveals its provenance from the Appalachians. Look at it—perhaps the same handful I previously told you to hold—by examining it with a hand lens or magnifying glass, and you will find that most of it is composed of clear, well-rounded grains of quartz (SiO_2). (Incidentally, *sand* is a size term, denoting any sediment $1/16$ to 2 mm in diameter, regardless of mineral composition.) Quartz is extraordinarily stable chemically. On the Georgia coast, quartz sand not only survives constant exposure to wind, surf, and tides, but also passes through the guts of suspension- and deposit-feeding animals without appreciable change to its composition. Other than quartz, the remainder of the sand is dark, some of it black. These grains are heavy minerals, such as rutile (titanium oxide, or TiO_2), ilmenite ($FeTiO_3$), zircon ($ZrSiO_4$), magnetite (Fe_3O_4), and various types of garnets, such as almandine ($Fe_3Al_2(SiO_4)_3$). Such minerals are termed heavy because they are denser than quartz: quartz has a density of about 2.7 g/cm³, whereas heavy minerals have densities ranging 4.2–5.5 g/cm³ (with water as a standard, at 1.0 g/cm³). Heavy minerals are concentrated by winds affecting Georgia beaches and dunes, which blow away the lighter (less dense) quartz grains and leave behind the dark mineral layers. In some Pleistocene barrier islands, heavy minerals accumulated into economically valuable deposits that were—and still are—mined in places (Pirkle and Pirkle 2007; Vance and Pirkle 2007). For example, rutile is an important industrial pigment and is used as a whitener in plastics, foods, and paper; think of how a small part of this page may be composed of eroded mountains. All in all, this sand was broken down mechanically by hundreds of kilometers of movement down rivers flowing from the Piedmont and Blue Ridge provinces. As the sand is recycled at and near the shore, its grains are continually diminished ever so slightly by grain-to-grain impacts and other alterations imparted by winds, tides, waves, and organisms.

Clays deposited and reworked on the Georgia coast have the same place of origin as the sand, but a different history of weathering and erosion. Clay, like sand, is a term for sediment size, referring to particles less than $1/256$ mm in diameter. Nonetheless, clay also has a compositional connotation. Often simply called clay minerals, these fall naturally into the clay size range and are the result of chemical weathering of feldspar minerals from the Appalachians or coastal plain. These by-products of weathering then travel downstream and are more often deposited in salt marshes, rather than on beaches. Clay minerals typically have complex chemical formulas, but the one with the simplest formula is also the most economically lucrative, kaolinite—$Al_2Si_2O_5(OH)_4$. Kaolinite, also called kaolin, inspired the commercial name for the antidiarrhea medication Kaopectate. Its consumption for digestive ailments, called geophagy, actually originates from western Africa. Slaves coming from that region carried on this tradition in the southeastern United States, which had geologically similar soils to western Africa (Hunter 1973; Burrison 2007). This coincidence of geology and culture is thus related to both plate tectonics and slavery, as displaced people more than 300 years ago easily recognized the same types of sediments and their uses on both sides of the Atlantic Ocean.

Kaolinite is also used in paper production as an emulsifier and filler in manufactured products, which, combined with abundant pine forests on the coastal plain, led to a significant paper industry in Georgia that continues to use both resources today (Kogel and Shelobolina 2007). Again, look at this page and ponder how you are also likely looking at parts of the Georgia coastal plain. Georgia clay deposits represent a combination of deposition during the Cretaceous Period, from about 70 million years ago, and the Eocene Epoch, from about 45–50 million years ago. These thick, profitable clay layers probably formed in vast salt marshes and other embayments associated with coastal environments during those times. How did this clay get deposited in the first place, though? That will be discussed in an ichnological context in Chapter 3.

Given all of these geological considerations of sediments, the primary substrates for preserving most traces made on the Georgia coast, the uniqueness of the present-day Georgia barrier islands and their scientific significance are better realized. The Georgia barrier islands, along with one island to the south in Florida (Amelia) and one to the north in South Carolina (Hilton Head), comprise what are also known as the Sea Islands. This chain of barrier islands is different from all others in the world, in that both Pleistocene and Holocene processes and sediments, deposited at separate times, formed six of the Georgia islands: Cumberland, Jekyll, St. Simons, Sapelo, St. Catherines, and Ossabaw (Figure 2.2; Pilkey and Fraser 2003). As a result of their diachronous and multiple origins, these six islands are also among the largest of the Georgia barrier islands. In fact, these islands are simply large: for example, Cumberland Island is bigger than Manhattan in New York City. Barrier islands are actually rare on marine coasts worldwide, numbering only a little more than 2,000 (Pilkey and Fraser 2003). Moreover, most of these islands are composed of sand and other sediments deposited by the most recent Holocene sea-level rise during the past 5,000 years or so. Thus, what happened to cause the composite nature of the Georgia barrier islands was a unique circumstance of birth and rebirth made possible through a happenstance of geography, sediment movement, and sea-level change.

The sea-level high associated with the Silver Bluff barrier island system, which is what makes up the Pleistocene cores of the composite islands, took place about 40,000 years ago, an interpretation based on studies of rock outcrops on Sapelo, St. Catherines, and a few other Georgia islands. The Silver Bluff sea-level rise, or transgression, peaked at about 1.5 m (5 ft) higher than the present one (Hoyt and Hails 1967; Linsley et al. 2008). This means that modern intertidal sediments are being deposited on top of what were originally subtidal sediments, which also results in present-day tides and waves eroding older sediments. Accordingly, these older sediments represent environments that were originally farther west (inland) than their modern counterparts. The outcome is that former marsh deposits, still containing burrows and other traces made by their original occupants, are now exposed on barrier island beaches and being reworked by tracemakers of those environments. Likewise, former coastal dunes and beaches with old ghost crab burrows are now in maritime forests with their extensive tree roots, insect burrows, and armadillo burrows.

As a modern barrier island migrates with rising or falling seas, sometimes at rates of about 10 m (33 ft) per year, it is exposing and then covering the previously existing sediments of a much older island. For example, if you see black mud associated with disarticulated oysters on a Georgia beach, you are probably on a composite island, in which modern sediments are actively migrating across Pleistocene or older Holocene sediments. How do we tell the modern Holocene islands from the Pleistocene ones? The key distinction is that Holocene islands are sometimes discernible as companions to the eastward sides of Pleistocene islands. Sapelo Island, for example, has the Holocene islands of Cabretta and Blackbeard to its east, separated from the main part of a Pleistocene island by tidal channels. Cumberland Island has Little Cumberland as a Holocene companion to

HISTORY OF THE GEORGIA COAST AND ITS ICHNOLOGY

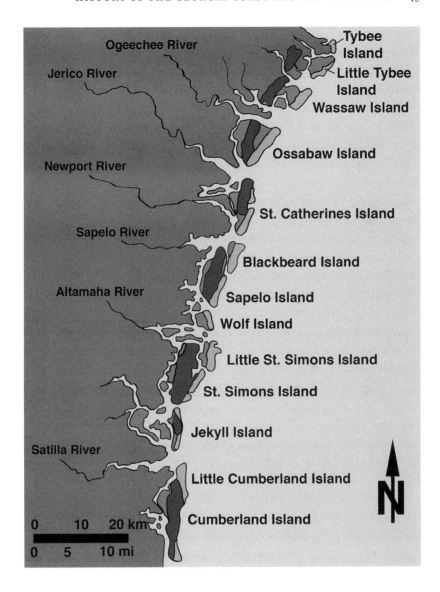

Figure 2.2. The composite (Pleistocene–Holocene) and Holocene barrier islands of the Georgia coast, with darker shade indicating Pleistocene (Silver Bluff islands, to the west) and lighter shade indicating Holocene (east). Notice how the outcrop pattern of the Pleistocene splits and veers to the west in the northernmost composite island, Ossabaw.

its northeast, and St. Simons Island has Little St. Simons to its northeast as another recent newcomer. Pleistocene foundations of composite islands are also easy to spot from the ground and from aerial photos. Maritime forests cover much of these areas, and these were the preferred sites of human settlement and agriculture (Figure 2.3). Last, the composite Georgia barrier islands are in the southern part of the chain, whereas shorelines of the more northern Holocene islands veer off to the east of the former Silver Bluff shoreline. For example, the relatively new island of Wassaw, formed only about 1,000 years ago, is immediately east of its Pleistocene neighbor, Skidaway Island (Alexander and Henry 2007). Just south of Wassaw, Ossabaw Island shows a split between its Pleistocene and Holocene parts that widens to the north. This situation is because of increased sand input by the Savannah River to the north, which pushes sand further out from the shoreline than other rivers emptying into the coast to the south (Pilkey and Fraser 2003). Longshore drift, or the movement of sand by currents along a shoreline, also tends to transport sand to the south, which affects the erosion and deposition of sand on barrier islands.

As a result of this concurrence of ancient products and modern processes, body and trace fossils made during Silver Bluff times are mixed freely with the remains and traces of contemporary organisms. Consequently, it is completely normal on the Georgia coast to see, for example, a modern sea anemone or barnacle attached to a fossil oyster; a modern oyster on a fossil ground sloth bone; a hermit crab carrying a shell provided by a Pleistocene whelk; and fiddler crabs living in (and modifying) burrows made by their ancestors in former marsh mud deposits from hundreds or thousands of years ago. This recycling and blending of new and old makes the composite Georgia barrier islands more challenging to interpret than those that appeared more recently and are simply composed of unconsolidated sand and mud. For example, in a moment of scientific humility, famed coastal geologist Orrin Pilkey, after just a short time of study on

Figure 2.3. *Facing.* Satellite photo of St. Catherines Island, Georgia. The darker areas in the western part of the island represent maritime forest and coincide with Pleistocene bedrock. Salt marshes mostly compose the lighter areas on the east side of the island and the back-barrier (landward) side. Also note the numerous tidal channels and creeks on the island and landward, a direct result of a mesotidal regime. From Google Earth (source U.S. Geological Survey), image taken May 31, 2008.

HISTORY OF THE GEORGIA COAST AND ITS ICHNOLOGY 47

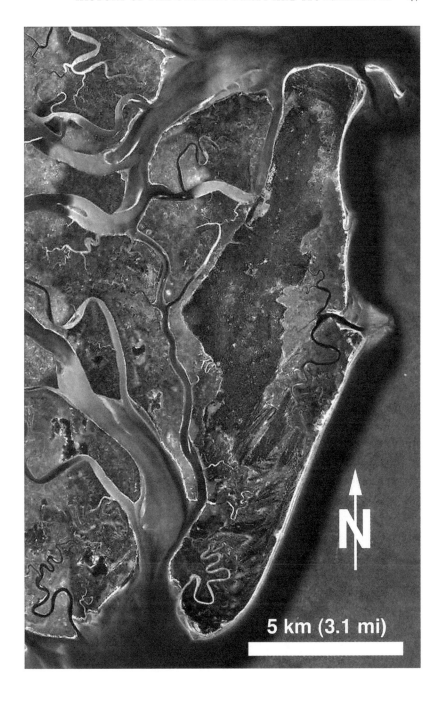

Sapelo Island in the 1960s, realized that he was wrong about nearly everything he previously took for granted about barrier islands (Pilkey and Fraser 2003). Properly warned, coastal geologists who study the barrier islands of the Georgia coast do not hold too tightly onto their hypotheses, and they remain aware that oft-told generalities become much more interesting by searching for their exceptions.

With each transgression, wherever the rivers of the past flowed downslope on the coastal plain and met the rising seas, estuaries formed. Most basically, estuaries are embayments in which freshwater and marine water mix, but these mean much more to the organisms that live in these special ecosystems. Estuaries on the Georgia coast can be divided into two broad categories, riverine and salt marsh (Frey and Howard 1986), both of which are to the west of the barrier islands. Barrier islands effectively reduce wave energy from the westward-trending winds and storms on their eastern sides (hence their "barrier" label), thereby creating low-energy zones on their western sides. These quiescent places are where fine-grained sediments, composed mostly of mud, are deposited and marine organisms have their nurseries.

Riverine estuaries are broad-mouthed rivers where water of marine salinity intrudes and mixes. This happens on the Georgia coast twice every 24 hours with each flood tide and is exacerbated by droughts, which cause saltwater wedges to travel far inland on what otherwise looks like normal freshwater rivers (Frey and Howard 1986). Higher-salinity water is also denser than freshwater because of the added dissolved load. For instance, water of normal marine salinity (35 parts per thousand, or ppt) has a density of 1.02 g/cm^3, whereas distilled freshwater is 1.0 g/cm^3. This density difference means that higher-salinity water will hug the bottom of a riverine estuary, rendering a fine habitat for bottom-dwelling (benthic) organisms adapted to such conditions, while also excluding freshwater-only organisms. Of course, some mixing of the water does occur, especially if inland rainstorms contribute large amounts of runoff. This mixing causes salinities that are quite variable, from 5 to 25 ppt (Frey and Howard 1986; Albert and Sheldon 1999). Higher salinities on average, though, mean a riverine estuary is more likely to have euryhaline (tolerating a wide range of salinities) animals, or fully marine animals, such as barnacles and burrowing mud shrimp (*Upogebia*). In contrast, freshwater clams, aquatic insects, crayfish, and some fish are excluded by small concentrations of salt in the water. Accordingly, burrows and other traces made by a benthic fauna are more like an impoverished marine assemblage, rather than a strictly freshwater or fully marine one (Chapter 10). Examples of riverine estuaries on the Georgia coast include the lower reaches of the Altamaha and Ogeechee rivers, which send freshwater into salt marsh estuaries or the open ocean.

Salt marsh estuaries are environments on the landward side of barrier islands in which formerly low-lying coastal plain areas are drowned by transgressing seas. These marshes are also considered as lagoonal, although they are quickly filled with mud held in place by abundant *Spartina alterniflora*, oysters, and compaction. So why are salt marsh estuaries places for mud deposition? Mud is trapped in these environments through a confluence of factors. Most important, the tidal range affecting estuaries of the Georgia coast is rather high compared to most barrier islands, at about 2.5–3.5 m (8–11 ft), classified as mesotidal (Hayes 1975; Frey and Howard 1986). This slight variation in tidal range, or the range within the range, is attributed to the difference between neap tides, coinciding with quarter moons, and spring tides that come with a new and full moon, which happens once every two weeks for each. Regardless of these variations, a mesotidal regime contributes to a huge volume of water that carries mud out of riverine estuaries and into salt marsh estuaries with each ebb tide. Mud, composed largely of clay minerals and some silt-sized particles, is normally kept in suspension by this constant motion of water. Nevertheless, friction caused by dense growths of *Spartina alterniflora* slows down the water flowing through a marsh, which encourages some mud deposition. Suspension-feeding organisms, such as oysters and mussels (*Crassostrea virginica* and *Geukensia demissa*, respectively), complete the rest of the job. These bivalves filter mud out of the water column and deposit it as feces (Pryor 1975; Smith and Frey 1985). This deposition in turn enables colonization of muddy substrates by *Spartina alterniflora*, which further stabilizes the mud.

Accompanying the mud in these salt marshes is a phenomenal amount of finely suspended organic matter. How much organic matter? When compared to the most fertilized of human agricultural systems, Georgia salt marshes are nearly 20 times more productive in nutrients (Teal 1962; Teal and Teal 1964). Organics produced in the marshes from *Spartina alterniflora* and other organisms are cycled through the complex food web of salt marsh ecosystems, and high tides also result in massive amounts of water flooding into and draining from Georgia salt marshes. Such mass movements put minute particles of organic matter into the water column with the clay particles, imparting an impenetrably dark mocha to water of a typical tidal creek. Some clay minerals will also flocculate with organic compounds while in the water, although clays and organics often come together more easily via the deposition of feces. Much of this organic material is then incorporated with the clays to form the thick, gooey, richly black mud so emblematic of most Georgia salt marshes.

A sort of ecological positive feedback happens in salt marshes because of so many nutrients in such relatively small areas. Suspension- and deposit-feeding animals increase dramatically in numbers, their predators multiply, and many marine animals lay their eggs in the marsh because of the plentiful food and quiet-water conditions for their young. Furthermore, flood tides bring in floating and swimming larvae of marine invertebrates, which settle in the marsh; if not eaten (which most of them are), they grow up to become larger invertebrates. As can be imagined, the sheer number of traces produced by the usual denizens of salt marshes is overwhelming. As in riverine estuaries, traces made by salt marsh organisms are more typical of animals that tolerate salinities above 10 ppt (Basan and Frey 1977; Howard and Frey 1985). Nonetheless, emergent parts of a marsh can also include traces of terrestrial animals, such as insects, birds, and mammals, as well as a few terrestrial plants that accept a little salt in the soil, such as red cedars (*Juniperus virginiana*).

Overall, the relatively high tides of the Georgia coast affect many aspects of the Georgia barrier island ecosystems. Georgia coastal tides influence the following coastal features: (1) the number of tidal creeks that drain the marshes; (2) the number and spacing of inlets between the barrier islands; (3) the lengths of the islands; and (4) the sizes of ebb-tidal deltas (Pilkey and Fraser 2003). The sheer amount of water flowing in and out of the marshes with each tidal cycle erodes salt marsh mud, regardless of the stabilizing effects of *Spartina* roots, oysters, and mussels. This erosion results in many beautifully intricate, branching, and sinuous tributaries (tidal creeks) cutting through the marsh, a pattern nearly invisible to an observer within the marsh but readily viewed from above. These creeks then empty into larger tidal channels, which in turn are often connected to riverine estuaries. Tidal creeks or channels separate the Georgia barrier islands from one another, either cutting through the older Pleistocene cores of islands or effectively preventing the deposition of Holocene sediment. Accordingly, higher tides result in more islands, and islands are shorter than those that would form under a lower tidal range. For a simple analogy, take one long loaf of bread, cut it in three places, and you will instantly have four pieces of bread, albeit all shorter than the original single loaf. Now do the same with one long island, using the tides as the knife.

Inlets are where these creeks or channels cut through islands and connect with the open sea to the east of the barrier islands; they are places where sediment is actively redistributed. Ebb tides are stronger than flood tides because the water is traveling downhill with each ebb, and hence its higher flow rate. As a result, more sandy sediment is flushed out of a riverine or salt marsh estuary on the seaward side of a barrier island than transported into the estuaries (Figure 2.4). These sandy deposits at the mouth of a creek or channel form ebb-tidal deltas, which are essential sources of sand for subtidal sandy bottoms, beaches, and dunes inhabited by a significant number and variety of tracemakers (Frey and Howard 1986, 1988).

Aside from the Pleistocene foundations of some islands, connections between riverine and salt marsh estuaries, unusually high nutrient contents, and relatively high tides, the Georgia coast has yet another geological situation that distinguishes it from the rest of the eastern U.S. coast: lower wave energy and a wide shelf. Waves are weaker on the Georgia coast than most of the eastern seaboard because of a broad expanse of shelf just offshore, called the Georgia Bight. This shelf slows down wind-driven waves through friction against the shallow bottom (Frey and Howard 1986). In coastal systems, lower wave energy tends to correlate with finer-grained sands, which are deposited more easily with lower flow regimes.

Wind-driven waves impinging on barrier island shorelines also contribute to the everyday movement of sand along beaches, or longshore drift. This southward transport of sand from ebb-tidal deltas causes the Georgia barrier islands to assume a shape like a chicken drumstick, in which sand deposition results in islands thicker to the north and thinner to the south. Some sand accretion, though, does take place on the southern ends of islands, also through longshore drift, where currents pull sand around the southern end of an island to form recurved spits. As a result, accreting sand can cause beachfront property on the southern end of a barrier island to become further estranged from the shoreline as sand continues to pile up offshore. For example, homeowners in the latter half of the twentieth century on the southern end of Jekyll Island and eastern St. Simons Island have

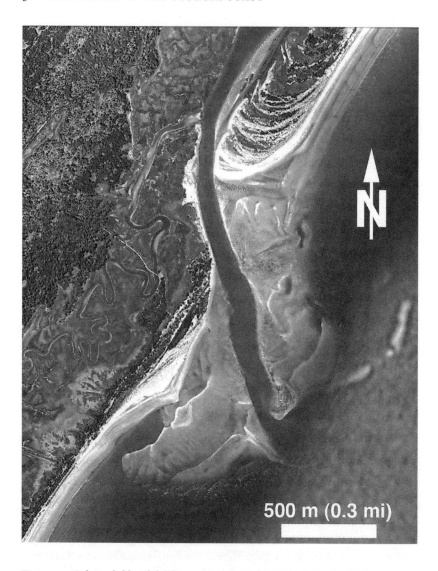

Figure 2.4. Inlet and ebb-tidal delta associated with Blackbeard Creek, which separates Blackbeard Island (*right*), composed of Holocene sediments, from the main part of Sapelo Island (*left*), which is largely Pleistocene. The sand from this ebb-tidal delta results in much building of Cabretta Beach and its coastal dunes, which lends well to these environments hosting many tracemakers and their traces. From Google Earth (source U.S. Geological Survey), image taken May 31, 2008.

certainly witnessed this phenomenon; they now have to walk much farther to the beach than in previous decades. Longshore drift and other processes that redistribute sediments can accordingly make an island migrate up or down a coastline (Bush et al. 1996; Pilkey and Fraser 2003).

Hurricanes also redistribute prodigious amounts of sand very quickly, interrupting the predictability of daily sand transport. Nonetheless, the Georgia Bight decreases the likelihood of hurricanes making it to the coast with their full fury intact, as its breadth lessens wave energy from tropical storms once these come into contact with shallow-shelf environments. Indeed, powerful hurricanes hitting the Georgia barrier islands are relatively rare compared to neighboring states. Nonetheless, when hurricanes do strike the Georgia coast, they instantly change where tracemakers live (or die) and inflict major damage to ecosystems and property (Fraser 2006).

Last, and most importantly from an ichnological perspective, the broad shelf of the Georgia Bight also has an effect on organisms that live there. This may explain why limulids of the Georgia coast are the largest in the world, hence making the most prodigious of modern limulid traces (Chapter 6). Generally speaking, finer-grained sands also mean easier transport by waves and wind, which corresponds to the formation of wider beaches and dunes. This situation is a lucky one for neoichnologists, who are then afforded great expanses of easily accessible and varied substrates in which organisms can leave abundant and varied marks of their behaviors (Chapters 6, 8).

RECENT ARRIVALS: HUMAN TRACES OF THE GEORGIA COAST

Traces of a human presence along the Georgia coast, which of course continue to be made today, are relatively young compared to other such evidence from the Americas, beginning only about 7,000 b.p. (before present). For example, the oldest known human trace fossils in North America are coprolites (fossil feces) dated at about 14,000 b.p. from the Pacific Northwest (Gilbert et al. 2008). Nonetheless, human traces in Georgia include the oldest known use of clay for pottery in North America, which is abundantly represented by artifacts from St. Catherines Island, as well as some of the oldest and most impressive shell rings of the eastern United States. What is a shell ring? Well, it is certainly not worn on a finger, and will be explained next.

Native American Tracemakers

Native Americans who lived on the Georgia barrier islands probably moved there from the mainland just after the post-Pleistocene sea-level rise slowed to a standstill, starting about 7,000 b.p. Sea level after the end of the Pleistocene Epoch, at about 10,000 b.p., had been going up at a steady rate, then stabilized at about 4,500 b.p., before global warming of the last 200 years or so accelerated it (DePratter and Howard 1981; Colquhoun et al. 1995). This sea-level standstill resulted in some of the Georgia barrier islands staying more or less in one place, and the composite ones are close to their positions from 5,000 b.p., although with differently configured shorelines.

As a result, these composite islands provided relatively secure places to live, as did some of the islands formed later. But why else did people move from the mainland and establish semipermanent settlements on the islands? The reasons are clear from the great abundance of natural resources on and around the islands, especially near salt marshes, where many Native American communities were situated. Widespread salt marshes with their tidal creeks held bivalves, fish, and other plentiful sources of easily acquirable animal protein, which were indeed procured in abundance, as indicated by human traces, or artifacts. Artifacts include shell rings, burial mounds, ceramics, and tools, collectively inspiring detailed scenarios of life and behaviors before the arrival of Europeans (Thomas 2008a). In contrast, and with only minor exceptions, relatively few skeletal remains of indigenous peoples from this time are preserved at or near their former homes.

Of the traces attributed to Native Americans, the oldest, most prominent, and intriguing on the Georgia barrier islands are shell rings. Shell rings, which also occur in South Carolina and Florida, are, appropriately and simply, ring-like structures composed largely of molluscan shells (Figure 2.5). Moreover, Georgia shell rings are prominent, semicircular, and upraised, with the largest 3–4 m (10–13 ft) high and 60–75 m (197–246 ft) wide (Russo 2006; Andrus and Thompson 2008). The molluscan framework of the shell rings is represented primarily by oysters (*Crassostrea virginica*), whelks (various species of *Busycon*), southern quahogs (*Mercenaria mercenaria*), and ribbed mussels (*Geukensia demissa*), but also includes parts of other edible invertebrates, such as blue crabs (*Callinectes sapidus*). More-

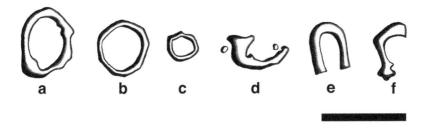

Figure 2.5. Outlines of shellrings on the Georgia barrier islands, traces of Native America habitations. (a–c) Sapelo Island. (d–e) St. Simons Island. (f) Skidaway Island. Based on schematics provided by Russo (2006). Scale = 100 m (330 ft).

over, rings hold minor amounts of vertebrate bones, such as fish, deer, and alligators (Waring and Larson 1968; Crook 1980). The interlocking of so many bivalve shells, like spoons in a drawer, results in dense, robust structures resistant to weathering. Soils even developed on the tops of some of these structures and support abundant vegetation, including mature live oaks. As a testament to the resilience of these shell rings, some continue to stand, despite the weathering and erosive effects of a semitropical climate and many hurricanes and human alterations of the surrounding landscapes over the past few thousand years.

Archaeologists have proposed three main hypotheses for the formation of these remarkable remnants of past indigenous communities. The first is that they represent monuments of ceremonial (religious) significance, particularly for rituals that involved feasting on the foodstuffs indicated by cast-off shells and bones (Waring and Larson 1968; Russo 2004). (Incidentally, an oft-told joke in anthropology is that whenever any prehistoric artifact defies a facile interpretation, the default explanation is that the object must hold some religious significance.) The second hypothesis is that the rings are refuse piles, or middens, accumulated from everyday life, with a seasonal or permanent village in the middle. The ring then would have formed by gradual deposition of refuse out the back of each dwelling place over time (Waring and Larson 1968). The third hypothesis is a synthesis of the preceding two: these structures were the result of ceremonies and relative status within a community, as well as marking the former places of villages (Russo 2004; Thompson 2007). Regardless, all three explanations presuppose that villages were arranged with circular outlines. This scenario is likely, considering that French explorers reported that villages of one tribe they encountered in the sixteenth century, the Timucuan, were enclosed in circular fortifications. The oldest, best-preserved, and most easily visited shell ring on the Georgia coast is on Sapelo Island; according to radiocarbon dating, it is about 4,200 years old (Simpkins 1975; Thompson 2007). Another impressive—but mostly buried—ring of about the same age is on St. Catherines Island.

Shell rings of the Georgia barrier islands are scientifically significant for several reasons. Although similar rings have been found in coastal areas north and south of the Georgia coast (Dame 2009), the Georgia ones are the largest known on the east coast of the United States. Consequently, these shell rings provide better evidence of human interactions with Atlantic coastal ecosystems than most other archaeological sites. The Georgia shell rings also point toward thriving civilizations in these areas beginning about 4,000–5,000 years ago, in which people settled at one place instead of subsistence hunting, which required following game animals. People staying put, then, very likely was related to abundant resources provided by nearby marshes and the open ocean. A sea-level standstill would have caused a maximum extent of salt marshes—providing many molluscans, especially oysters—while also retaining terrestrial ecosystems on island interiors with their plant-based resources and game animals.

Other large traces that are more overtly related to Native American spirituality are burial mounds, which contained the skeletal remains and various personal items of former inhabitants of the islands. Such mounds were more commonly constructed on the mainland, and the best known (and most prominent) of these structures in the southeastern United States are the Etowah mounds in northwest Georgia, some of which are nearly 20 m (67 ft) tall (King 2004). The best-studied mounds on the Georgia coast were on Ossabaw and St. Catherines Island, with at least one burial site yielding graves of more than 75 people (Thomas 2008a). Unfortunately, primary archaeological research done at a few sites in the 1890s destroyed the original structure of the mounds, so these cannot be observed today. Modern archaeological research, in contrast, is much more carefully conducted, and its researchers are keenly conscious and respectful of these sites as places where people lived and worshipped. Indeed, the investigation of a recently discovered shell ring on St. Catherines is being conducted through mostly noninvasive techniques such as ground-penetrating radar and other remote sensing methods (Royce Hayes, personal communication 2010).

Associated with shell rings, burial mounds, and other archaeological sites on the Georgia barrier islands are pottery sherds, which are unique among early ceramics in the Americas. This pottery is among the oldest known in North America, dating from more than 4,000 years ago (Bullen 1961; Sassaman 1993; Russo 2006). Furthermore, it is the oldest to demonstrate a technique called fiber tempering, in which plant fibers—in this instance, from Spanish moss (*Tillandsia usneoides*)—were mixed with the clay before firing (Bullen 1961; Simpkins and Allard 1986). Tempering, which is the addition of materials other than clay to an intended ceramic item, is done mainly to prevent the high water content in pure clay from exploding pieces when fired. Other tempering materials included quartz sand (recycled from Pleistocene bedrock in inland environments), shell fragments, and limestone; the latter was sometimes available through trade with mainland tribes.

Much of this pottery, while functionally having served as vessels for food and water, certainly represents traces of human behavior in and of themselves, but some pieces also contain traces of aesthetic or artistic expression. Such vestiges began with wooden or clay paddles, which contained inscribed designs and were slapped against the side of still-moist vessels before firing (Sassaman 1993; Williams and Thompson 1999). In these examples, the designs in each paddle and their transference to vessels were traces of human creativity, superimposed on the traces represented by the ceramic containers themselves (Figure 2.6). Similarly, tools made of local materials (bone, wood, and shells) were also used to etch lines onto unfired surfaces; both decorative techniques were used for more than 2,000 years in the pottery of the Georgia coast (Williams and Thompson 1999). Motifs imparted by these paddles and other instruments are often abstract and intricate concentric spirals, or parallel wave-like patterns, which do not easily correspond to known natural objects on the barrier islands and thus tempt anthropologists to invoke spiritually oriented interpretations. Pottery was further decorated with vegetable dyes and clay-based pigments, rendering colorful variations on otherwise similar vessels. The presence of abundant clays for pottery was, of course, also a function of their deposition in local sedimentary environments, particularly the

Figure 2.6. Example of stamped pottery from St. Catherines Island, Georgia, attributed to the Guale (Native American tribe) and dated to AD 1300–1580. Specimen is about 10 cm (4 in) wide, and is on display at Fernbank Museum of Natural History, Atlanta, Georgia.

salt marshes, and ultimately their origins from weathered feldspars of the Appalachian Mountains.

Other traces made by Native Americans on the Georgia barrier islands relate to their development of subsistence agriculture, but these are either too subtle or no longer noticeable on the islands themselves. Rather, the indirect evidence of food choices shows up in their bones and teeth. Answers come from paleopathologists who study the effects of disease and other ailments in prehistoric or more recent humans. For example, studies conducted on relatively recent (< 1,000 years old) skeletal remains from the Georgia coast reveal that tooth decay and bone disorders increased with greater use of maize in their diets; the maize was probably cultivated (Ezzo et al. 2005). This shift in health became more acute after first contact with the Spanish in the sixteenth century, resulting in more emphasis on growing food instead of harvesting from the bountiful goods of their surroundings, such as live-oak acorns, oysters, and deer.

Perhaps the subtlest mark left by the presence of the first Americans is an absence—namely their apparent impact on the largest terrestrial mammals in North America. Mammals of unusual size represented the so-called North American megafauna, including the previously mentioned mammoths (*Mammuthus*), mastodons (*Mammut*), giant ground sloths (*Megalonyx*), glyptodonts (*Glyptodon*), giant beavers (*Castoroides ohioensis*), dire wolves (*Canis dirus*), saber-toothed cats (*Smilodon*), and short-faced bears (*Arctodus simus*). All of these animals dwarfed any comparable species still around today, and this megafauna was widespread: most of the species listed previously were in the southeastern part of North America, on and near the present-day Georgia barrier islands (Kurtén and Anderson 1980; Hulbert and Pratt 1998; Laerm et al. 1999). Nonetheless, they all vanished in a wave of extinctions that nearly coincided with humans arriving in most parts of the Americas. After the elimination of these species, the largest mammal left in the southeastern part of North America was the bison (*Bison bison;* that's its species name, not a stutter). This simple pattern—people show up, animals go extinct—has happened repeatedly throughout human history and implies a probable causal relationship (Martin 2005). Ichnologically speaking, this sudden extinction of significantly abundant and huge tracemakers means their traces were no longer made after the Pleistocene and the arrival of humans. As a result, if paleontologists find, say, a massive fossil beaver lodge, mammoth tracks, or bones with dire wolf tooth marks in sediments of the Georgia coastal plain, they can say with confidence that these trace fossils were very likely made more than 10,000 years ago.

Unfortunately, no information is available about whether the indigenous people of the Georgia coast had developed any form of ichnology, such as tracking or other forms of awareness about animal traces. Nevertheless, the availability of abundant sandy and muddy substrates, game animals, and edible intertidal molluscans on the islands all imply that ichnological methods were almost certainly applied for subsistence. In fact, archaeologists have attempted to replicate foraging techniques and caloric productivity of the surrounding ecosystems with Native American ichnological practices in mind, such as the identification of sea turtle trackways leading to clutches of edible eggs (Thomas 2008b).

European, African, and American Tracemakers

European peoples—the Spanish, French, and British—arrived on the Georgia coast beginning in the sixteenth century. They found that indigenous people on the barrier islands belonged to two tribes: the Timucuan, who were on what is now called Cumberland Island, and the Guale, who lived on islands north of Cumberland (Thomas 1996; Milanich 1999). With the Georgia barrier islands, as elsewhere, the Europeans followed a pattern sadly familiar to anyone who has studied the post-Colombian history of North America: repeated attempts (some gentle, some forceful) to convert the natives to dominant religions; military conquest and colonization; forced labor; and genocide of the native people. Granted, most of the latter was unintended, caused by the transmission of European diseases. Still, the Guale and Timucuan peoples, as distinctive ethnic communities, were extinct by the late seventeenth century. Hence much of what we know about Native American lives and cultures on the Georgia coast, particularly from before European contact, is inferred from their just-described traces. Additionally, auditory signs of their ancestry are discernible from transliterated accounts by the Europeans of the Timucuan and Guale languages, which resembled other Native American language groups on the mainland. For example, Guale was linguistically similar enough to Muskhogean that some anthropologists are confident about its original provenance from further inland (Thomas 1996; Milanich 1999).

Evidence of a Spanish presence still haunts many places on the Georgia coast, and none so much as St. Catherines Island. St. Catherines is the former site of Santa Catalina de Guale, one of the oldest Spanish missions in North America (Thomas 1987, 1996). Historical records mentioned the mission, which was founded in the 1590s and abandoned about 90 years later after Spanish defeat by the British. Its former site, however, was not detected until after an intensive search that began in the late 1970s. Archaeologists associated with the American Museum of Natural History found the mission in 1981 and documented it thoroughly over the next 15 years. Upon investigating, they were astonished to discover that the ground below one of the churches contained the remains of more than 400 people—apparently all

Figure 2.7. Site of the Santa Catalina de Guale, one of the oldest Spanish missions in North America, on St. Catherines Island. Cabbage palms (*Sabal palmetto*) outline the footprint of the church, and a stump (foreground) marks the former entrance.

Guale, and converted to Christianity—thus constituting the only known Spanish cemetery of the Georgia barrier islands (Larsen 1990; Thomas 1996). The mission site also yielded many artifacts of everyday religious and secular utility, thus helping to reconstruct daily life for the people who lived there. Cabbage palms (*Sabal palmetto*) outline the church foundation today, but otherwise no surface features indicate its former presence (Figure 2.7). Once the archaeological investigation was completed, all of the bodies were reburied there; this place is thus still a graveyard.

Another lingering sign of the Spanish occupation of the Georgia coast is evident in the name of Sapelo Island, as "Sapelo" is probably a corruption of the Guale name for a village on the island. Spanish missionaries, who were on Sapelo for more than a hundred years (1573–1686), modified the Guale name of "Sapalaw" to "Zapala," which later was doubly corrupted to its present-day Anglicized "Sapelo" (Swanton 1922). Nearly all traces of this mission (San Jose de Zapala) are gone except for a few pieces of Spanish ceramics found near the Sapelo shell ring in the northern part of the island. Nonetheless, its ruins were identified in a few photographs from 1927 to 1928, and they were mentioned several years later in a book (Cate 1930), which lends to speculation about its actual whereabouts today. More likely, though, these "mission ruins" were actually the degraded remains of English, French, or American plantations, and thus were misidentified traces, a dilemma shared by many ichnologists.

After territorial tussling over this part of North America among the Spanish, French, British, and various buccaneers—including the famed pirate Blackbeard, who even had a Georgia barrier island named after him—the British eventually gained control of the Sea Islands by the middle of the eighteenth century. As a result, island landscapes were changed radically. Once in place as occupiers, the British promoted an African slave trade as a brutally economical means for ensuring profitable, large-scale agriculture in the colonies. The Americans then continued slavery after the British until emancipation in 1863. Plantations resulted in the dramatic and widespread removal of maritime forests on the islands. Island soils, which owed whatever productivity they might have had to their Pleistocene and Holocene geological and ecological histories, were then used for growing crops for food, fibers, and dyes. Likewise, salt marshes were drained for rice production, a practice that was developed most extensively in South Carolina, but with some in Georgia. Much of the rice growing and processing in Georgia took place near and at the Wormsloe Plantation south of Savannah; the cultivation of rice, though, persisted on Sapelo into the early part of the twentieth century. Most of the expertise in rice cultivation came from the African slaves themselves, who were already experienced at growing rice in coastal areas of western Africa (Carney 2001; Morgan 2010). Historical accounts from the plantation era also refer to a common practice of nutrient enrichment provided by the salt marshes, in which the marsh mud, chock full of organics aided by mulch from *Spartina*, was exhumed (temporarily depriving fiddler crabs of their homes) and spread over the fields as a natural fertilizer (Stewart 2002; Reitz et al. 2008). The slight acidity of the marsh mud was buffered by the addition of crushed oyster shells, which increased its pH and thus better encouraged the growth of cotton, sugar cane, and other crops.

Ruins of buildings made during this time of plantations, which were used for dwelling and storage of slaves, domestic animals, and agricultural goods, still stand on a few islands, a tribute to their tabby construction

(Figure 2.8). Tabby (from Spanish *tapia*, "mud wall"; Manucy 1952) is a construction method that took full advantage of abundant calcium carbonate resources in and around the nearby salt marshes. Tabby required using a cement mixture of lime ($CaCO_3$ that had been powdered and baked), water, quartz sand, and ash, and molluscan shells (especially oysters) as the bricks or framework (Sickles-Taves 1997; Sickles-Taves and Sheehan 2002). Much of the molluscan component for the tabby on a few islands was mined from the nearest available concentrations of shells: the Native American shell rings. For example, the shell ring on the northwestern cor-

Figure 2.8. Tabby construction, using mostly molluscan shells (primarily eastern oysters, *Crassostrea virginica*) and held together by lime cement, a building technique used by Europeans and Americans on the Georgia barrier islands from the 17th through 19th centuries: Sapelo Island.

ner of Sapelo, dated at more than 4,000 years old, still bears a large gouge on one side where, in the early nineteenth century, shells were dug out of the structure, carted to a nearby plantation, and used for tabby (Johnson 2009b).

Major landowners, such as James Oglethorpe, Thomas Spalding, and Spalding's son, Randolph, were responsible for encouraging the widespread use of tabby for buildings on the Georgia barrier islands through the eighteenth and early nineteenth centuries, which in some instances depleted shell middens. Other, perhaps unexpected, materials were often added to the tabby to strengthen it, and a close examination of these walls will sometimes reveal horse or hog hair amid the cement. The development of a more modern formula for Portland cement in the late nineteenth century, however, decreased the need for tabby construction (Sickles-Taves and Sheehan 2002). This situation, when coupled with the end of slave-based agriculture, meant that tabby buildings mostly became a thing of the past, although many of their walls (and a few reconstructed buildings) still remain. Nonetheless, this architectural method was imitated in new structures, in which their builders used Portland cement with molluscan shells to make a sort of Tabby Nuevo.

With the end of the American Civil War in 1865, the intensive agriculture on the Georgia barrier islands, which depended on slave labor, was mostly curtailed. Nonetheless, another form of indentured servitude, sharecropping, persisted on some of the Sea Islands and the nearby mainland into the early twentieth century. On at least one island (Sapelo), though, freed slaves stayed where many generations of their mixed ethnicities had lived, a legacy still evident in the community of Hog Hammock (Johnson 2009b). An auditory trace in Hog Hammock and other places on and near the Sea Islands is the persistence of a distinctive dialect directly reflecting west African heritage, despite the passage of more than 300 years since the ancestral speakers arrived in Georgia. This dialect is often referred to as Geechee (or Gullah), synonymous with the name given to the enduring, distinctive cultures in Georgia and South Carolina (Morgan 2010). Geechee consists mostly of easily recognizable English, but also mixes in words from the western African dialects of Ewe, Hausa, Ibo, Kimono, Mende, Twi, and Yoruba (Pollitzer 1999). Other reminders of African heritage include weaving techniques that use grasses (Figure 2.9), as well as the making of nets, both of which link

Figure 2.9. Weaving of sweetgrass (*Muhlenbergia filipes*) into distinctive patterns, a cultural trace of African traditions and associated with African American Gullah communities of the Georgia coast, demonstrated by Yvonne Grovner of Sapelo Island.

shared practices of coastal communities from either side of the Atlantic to their local ecosystems. Perhaps the most widely recognized African tradition still present on the Georgia coast is fine storytelling, which is linked in particular with people on Sapelo Island (Bailey and Bledsoe 2000; Jones 2000; Crook et al. 2003).

All human alterations of the islands aside, the most significant meteorological and sedimentological event to affect the Georgia barrier islands during the past 500 years or so was the Sea Islands hurricane of 1893. On the basis of historical descriptions of its landfall (near Savannah, Georgia, on August 27, 1893) and its aftermath, this hurricane was likely a category 3 on the Safer-Simpson scale (Marscher and Marscher 2004; Fraser 2006). Storm surges from the hurricane were estimated at heights of nearly 5 m (16 ft), which would have easily immersed single-story buildings and washed through second-story windows. The effects of this hurricane on the human population were horrific, killing more than 2,000 people and destroying most homes along the Georgia and South Carolina coasts, leaving about 30,000 people homeless. Conditions were made much worse less than 50 days later, when another category 3 hurricane hit just north of the Georgia barrier islands near Charleston, South Carolina, on October 13, 1893. Like most hurricanes affecting barrier islands, it also permanently changed them geologically and ecologically. For example, sandy hillocks in the middle of Wassaw Island are remnants of sand dumped by this hurricane, but they are now covered by maritime forest and burrowed by modern terrestrial tracemakers. As might be imagined, poststorm colonization of sedimentary surfaces by plants and animals is an important subject within ichnology, one that will be explored in more detail later (Chapters 4, 6, 9).

During calm times between storms, wind- and steam-powered European and American ships glided into and out of the Georgia coast, and traces of their former movements are visually obvious in the form of ballast islands (Figure 2.10). These small, isolated islands typically bear a few pines, cedars, and oaks, but they are otherwise on the edges of tidal channels and surrounded by salt marsh. The islands mark where crews of European and American ships dumped rocks, soil, and other forms of ballast (Emery et al. 1968; Stewart 2002) and are distinct from small, natural back-barrier islands, called hammocks, which owe their existence to prehuman processes. With regard to the formation of ballast islands, ships typically came from various ports of Europe with light cargoes; hence their ballast was used to maintain an even keel in the water while sailing. Once ships were loaded with cotton, lumber, and other agricultural goods, their ballast was no longer needed. As a result, heaps of rocks and soil were dumped

Figure 2.10. Ballast islands, artificially high and vegetated areas in back-barrier environments, traces of 19th- and 20th-century ships that dumped their ballast. (a) Overhead view of ballast islands (between arrows) forming a string along a tidal-channel margin in Sapelo Sound, between mainland and Sapelo Island. From Google Earth (source, U.S. Geological Survey), image taken May 31, 2008. (b) View from the water in Sapelo Sound, between mainland and Sapelo Island.

from the ships into the marshes. These piles then effectively slowed down water flow and increased local sedimentation, thus quickly becoming small patches of terrestrial ecosystems in back-barrier environments. Considering how these rocks and soils came from Europe, they also provided for exotic contributions to the local geology and ecology, including colonization by invasive species of plants, such as salt cedar (*Tamarix gallica*). Soils used as ship ballast in particular are also notorious for carrying seeds and egg-laying insects of invasive species into new territories. Perhaps the most infamous of invasive species in the southeastern United States is the imported red fire ant (*Solenopsis invicta*; Chapter 5), which arrived in Alabama ports in the early twentieth century via ships from South America (Wilson 1951; Buhs 2004).

Some ballast removal from ships, however, was also done in response to disease prevention. Ships arriving in ports on the Georgia coast in the nineteenth century often carried ballast water containing yellow fever. Ballast was then removed and treated to mitigate this disease (Lenz 2002; Sullivan 2008). This practice stopped in the early twentieth century, after the widespread use of yellow fever vaccines eliminated the need for quarantine. Hence the deposition of marine ballast and the formation of ballast islands along the Georgia coast ended with the development of vaccines. Accordingly, ballast islands are vestiges of times before such vaccinations.

Modern Human Traces, Connected with the Past

Today, human traces are all too obvious in the terrestrial ecosystems of developed Georgia barrier islands (e.g., Tybee, St. Simons, and Jekyll) as buildings, roads, golf courses, and so forth. More subtle yet lasting traces of human behavior are also noticeable on those islands often labeled as pristine, advertised in glossy brochures containing refined Georgia rutile and kaolinite, and made of wood pulp processed on the Georgia coast. In actuality, these islands are palimpsests, bearing the repeated, overlain, and interwoven effects of human interactions with their landscapes throughout the past 6,000 years. In some situations, the traces become iterative. For example, the raw material for tabby buildings was derived from the mining of nearby Native American shell rings. Straight, unpaved, and sandy roads cutting through now-thick maritime forests, used by modern vehicles or hikers, are on former routes of horses and wagons that carried cotton and other crops grown and harvested by slaves. The maritime forests themselves are mostly secondary growth, products of ecological succession that followed logging of original forests for intensive agriculture or the use of the wood for shipbuilding. For instance, the dense wood of Georgia live oaks was prized for its durability, and was used in the U.S.S. *Constitution,* more popularly known as "Old Ironsides" because British cannonballs would bounce off its sides, leaving only dents instead of punctures. The failure of most agriculture on the Georgia barrier islands coincided with the freeing of slaves. Yet single, massive, and very old live oaks are still seen in places, marking former sites of subjection and deforestation. Before the American Civil War, these trees were left in otherwise open areas to provide shade for slaves during hot Georgia summers. Thus these "slave oaks" serve as eerie living reminders of that recent human past (Reitz et al. 2008).

Even some of the modern biota of the barrier islands and their traces, or lack thereof, speak of recent human history and its effects on the ecosystems. For example, people transported the much-despised red fire ant to the southeastern United States; its extensive and complex ground nests are now omnipresent in terrestrial environments throughout these states (Chapter 5) (Buhs 2004; Tschinkel 2006). Furthermore, feral hogs (*Sus scrofa*), cattle (*Bos taurus*), and horses (*Equus cabullus*) now living on some Georgia barrier islands are invasive species introduced, reintroduced, and otherwise encouraged by European people and their descendants. This was done partially for animal husbandry, but also to change the barrier islands into places more familiar to those people. These species have since radically changed terrestrial habitats and negatively impacted native species of plants and animals on the Georgia barrier islands (Chapters 8, 9, 11). Consequently, whenever I track these animals in maritime forests and on dunes and beaches of "pristine" Sapelo, Cumberland, St. Catherines, or Ossabaw Islands, I also think of these mammals as traces of human hubris.

Nearly coinciding with the arrival of these exotic tracemakers to the islands were local extinctions of a few large mammalian carnivores native to the islands and the nearby coastal mainland, such as pumas (*Puma concolor*), wolves (*Canis lupus*), and black bears (*Ursus americanus*). Because all three carnivores were perceived as threats to the imported and domesticated species, they were hunted relentlessly in southeastern states (Bowers et al. 2001). These extirpations were also likely hastened by widespread habitat alterations that accompanied European-style agriculture.

Coyotes (*Canis latrans*), however, persisted in mainland environments; they probably increased greatly in numbers after the local extinction of wolves, which were their primary competitors. As of this writing, coyotes are on nearly every Georgia barrier island, even the bridgeless ones, as verified by both sightings and their traces. (True to their trickster reputation in Native American lore, no one is quite sure how these wild canids get onto the islands.) Bobcats (*Felis rufus*) are now the largest nonhuman mammalian carnivores on any of the Georgia barrier islands, but they are only on Cumberland, where they were brought in to decrease the deer populations there (Baker et al. 2001). The Pleistocene and Holocene fossil record, however, does not show any evidence of bobcats living on these islands (Laerm et al. 1999), so their reintroduction may actually be a mere introduction. Regardless, the disappearance of so many native predators, which included Native Americans, meant that their prey, such as white-tailed deer (*Odocoileus virginiana*) and raccoons, became much more numerous on the islands. This circumstance resulted in a concomitantly sharp increase of deer and raccoon traces in terrestrial ecosystems soon after the extinction of their predators, an increase in traces that was inversely proportional to a decrease in traces of the native megafauna once people first spread through the Americas. Now deer and raccoons are mostly held in check by alligators, which will snatch an occasional unwary individual stopping for a drink at a freshwater pond (Shoop and Ruckdeschel 1990).

Among the most important alterations to the landscapes of the Georgia barrier islands wrought by humans, yet one that probably would be unnoticed without a historical perspective, was the dramatic decrease in available freshwater and freshwater ecosystems on island interiors. Native Americans, who were entirely dependent on freshwater supplies on the islands while surrounded by saline waters, clearly thrived over thousands of years on the islands. The abundance of freshwater is verified by geological investigations confirming the long-term presence of freshwater wetlands (Reitz et al. 2008). In historical accounts, Spanish and English explorers also remarked about the bountiful freshwater wetlands of the Georgia islands. Much of this water was supplied by natural artesian seeps, where water held under pressure in underlying aquifers—particularly from the Upper Floridan Aquifer—flowed easily to the surface in springs, causing wetlands in low-lying areas on island interiors. These wetlands, in turn, were expanded or otherwise modified by alligators (explained further in

Chapters 3) and seasonal rainfall, especially during hurricane season. One such wetland, reported from St. Catherines Island in the eighteenth century, was described as a freshwater lagoon that covered an estimated 15–20 km² of the island interior (Bryan et al. 1996). Widespread agriculture during the eighteenth through the nineteenth centuries, however, diminished such ecosystems through a combination of draining water for irrigating fields of cotton and other crops, and the overuse of groundwater wells that tapped into shallow aquifers. Overtapped artesian wells also lost their pressure and eventually ceased bringing water to the surface, corresponding to fewer natural springs. Hence the only rare (and smaller) freshwater wetlands left behind are now replenished almost entirely by seasonal rainfall. This loss was exacerbated by increased coastal development of the twentieth century, which pumped large amounts of water from wells for use in the paper processing plants that still dominate parts of the Georgia coast (and often can be smelled before being sighted). As a result, freshwater lakes are now nearly absent on the Georgia barrier islands, resulting in fewer modern analogues for ancient freshwater wetland environments and their tracemakers.

Of course, the most significant human trace currently affecting the present and near future of the Georgia barrier islands is global climate change. This trace is atmospheric, composed of greenhouse gases (in particular carbon dioxide and methane), which insulate infrared radiation near the earth's surface and increase average sea and air temperatures. The elevated temperatures are consequently linked to the rise in sea level that accompanies the input of liquid water from the melting of alpine and continental ice. For the Georgia barrier islands, this rapid rise in sea level means that (1) coastal erosion rates are increasing; (2) formerly terrestrial and marginally marine environments are changing to marine-dominated ones—for example, salt marshes will drown under too much water and turn into open, muddy lagoons; (3) intrusion of saline water into groundwater in coastal aquifers will become more common; and (4) the risk of intense tropical storms hitting the coast, which is currently low compared to most parts of the southeastern United States, will become more likely (Michener et al. 1997; Greenland et al. 2003). Observations of such overall changes have already been made within the last few generations. Moreover, coastal geologists and ecologists predict that they will be able to continue measuring how this human imprint on the integrated ocean, atmosphere, and land will affect the coasts and the movements of tracemakers with the shifting of their habitats (Chapter 10).

In summary, any observations of modern organismal traces on the Georgia coast must be viewed through a filtered lens that takes into account human impacts on the environments hosting the traces, whether these influences were caused by peoples who called themselves Guale, Timucuan, Spanish, French, English, African, African American, European American, or simply *Homo sapiens*. These traces of human behavior are suffused throughout the sand, mud, plant tissues, animal bodies, and other substrates of the barrier islands in ways that are both covert and obvious, but that should be kept in mind whenever comparing modern traces to analogous trace fossils. Uniformitarianism is a broken mirror that still provides for reflections on the fossil record, but ichnologists cannot ignore the cracks that interrupt our projection of these modern-day pictures into the ancient past.

THE ECOLOGISTS AND THE BIRTH OF A MODERN SCIENCE

The Georgia barrier islands have been the setting for many fascinating tales about prominent Americans who have lived on, visited, or otherwise interacted with these places during the past 100 years or so. These people include aviator Charles Lindberg; various scions of the Carnegie and Rockefeller families; tobacco magnate R. J. Reynolds Jr.; United States presidents Calvin Coolidge, Herbert Hoover, Lyndon Johnson, and Jimmy Carter; and the son of a former president, John F. Kennedy Jr. In recognition of the volumes already written about these people and their connections with the Georgia barrier islands, nearly nothing will be retold about them here. Instead, we will explore the worldwide importance of these islands to the natural sciences of ecology and geology, as well as tell the neglected stories of the scientists who contributed to the development of neoichnology. In a sense, scientists and their works made the Georgia coast more internationally renowned than the "lifestyles of the rich and famous," which are more of a provincial concern.

For example, any discussion of the history of ecology would be remiss without mentioning Eugene "Gene" Odum, often termed the father of modern ecology, but who also had important ties to the Georgia barrier islands (Craige 2002). In a broad sense, Odum and his younger brother, Howard T. Odum, were responsible for the development of ecology—the study of interactions between organisms and their environments—as a mainstream science. In turn, Odum would have been negligent if he did not credit the lush, natural laboratories of the Georgia coast for providing the raw materials that inspired many of his eclectic hypotheses about the flow and exchange of matter and energy in ecosystems. In particular, Odum noted that salt marshes are ecosystems in which the organisms are constantly producing and recycling massive amounts of nutrients (Odum and Smalley 1959; Odum 1968, 2000). Much of this organic flux can be attributed to ichnologic processes too (Chapter 3).

The beginning of intensive ecological research on the Georgia barrier islands is credited to the generosity of millionaire philanthropist R. J. Reynolds Jr., coupled with the encouragement of Reynolds by Eugene Odum in the 1950s. Reynolds, who inherited lots of money from tobacco profits, bought Sapelo Island from the estate of Howard Coffin Jr. (the founder of the Hudson Motor Company) in 1933, during the nadir of the Great Depression (Sullivan 2000; Craige 2002). After a chance meeting between Odum and Reynolds on Sapelo Island in 1948, Reynolds carried out the idea of donating his extensive land holdings and some buildings on the south end of Sapelo to the University of Georgia, Athens (UGA). This proposal came to fruition in 1953 with the founding of the Sapelo Island Research Foundation, which led to the formation of the University of Georgia (Athens) Marine Institute, or UGAMI (Figure 2.11). More than 50 years later, the UGAMI is still coordinating ecological research and education with its outdoor laboratories and classrooms—namely the marshes, beaches, maritime forests, and other ecosystems there. Preservation of these ecosystems as places of learning was ensured with Georgia state ownership of the south end of the island in 1976, the establishment of the Duplin River National Estuarine Sanctuary to the west of Sapelo, and state control of the northern half of the island (Sullivan 2000, 2008). The few privately owned parts of Sapelo are associated mostly with the Hog Hammock community.

Through their research, Odum and his colleagues found that the Georgia salt marshes represented ecosystems of the highest known productivity, which they attributed to several factors: (1) enormous amounts of biomass (and hence nutrients) provided by three major producers (photosynthesizers), consisting of floating algae (phytoplankton), algae on the marsh surface, and *Spartina alterniflora;* (2) a mild climate that enabled high rates of productivity throughout any given year; (3) and movement of nutrients

Figure 2.11. University of Georgia (Athens) Marine Institute, also known as UGAMI; the birthplace of modern ecology, and host for much of the neoichnology done in the 1960s–1980s. Note how its architecture reveals its former use as a domicile for domesticated, hoofed mammals. Photo by Kathryn Henderson.

into and out of the ecosystem via the relatively high tides (Teal 1962; Teal and Teal 1964; Chalmers 1997). As a result of such ecological studies, the Georgia coast became better known scientifically, and it later attracted the attention of researchers from other scientific fields, such as archaeology and geology.

Immediately north of Sapelo is another composite island, St. Catherines Island, which, like Sapelo, became a locus for scientific research because of the generosity and scientific curiosity of a millionaire philanthropist, Edward John Noble Jr. Noble, who became rich from people excessively consuming small, ring-shaped candies known as Life Savers, bought the island in 1943 (Sullivan 2000). After his death in 1958, the Edward John Noble Jr. Foundation continued to fund scientific research there. The island also became connected with the Wildlife Conservation Society and the Bronx Zoo, which have used it as a sort of Noah's ark for exotic species. For example, semiwild ringtail lemurs (*Lemur catta*) have been living and breeding on the island since 1985, making interesting traces in maritime forests that otherwise never experienced the behaviors of nonhuman primates (Parga and Lessnau 2008). In other scientific endeavors, the American Museum of Natural History (AMNH) began intensive and long-running archaeological research on St. Catherines Island. This research by AMNH personnel on traces left by the island's former Guale and Spanish inhabitants firmly established St. Catherines as one of the most important archaeological sites in North America (Thomas 2008a).

North of St. Catherines is a Pleistocene island—Skidaway—that has hosted the Skidaway Institute of Oceanography (also known as SKIO), associated with the university system of Georgia. SKIO was founded in 1968 after Robert and Dorothy Roebling donated land on Skidaway to the state of Georgia for the expressed purpose of creating a new marine institute. From its inception, neoichnology was one of the primary areas of study for at least one scientist at SKIO, James Howard. Research at SKIO is broad, covering plankton ecology, marine biochemistry, marine pollutants, coastal sedimentology, and oceanography. Gray's Reef National Marine Sanctuary likewise falls under the responsibilities of SKIO; it is managed through the National Oceanographic and Atmospheric Administration (NOAA), with offices housed there. Skidaway, unlike the Pleistocene islands south of it, has no active beach and is located west of Wassaw, a Holocene island that bears the brunt of waves from the Atlantic Ocean. Nonetheless, salt marshes and tidal creeks surround Skidaway, which, like Sapelo, provide immediate access for scientists studying these environments.

Other Georgia barrier islands, especially composites ones like Cumberland, Jekyll, St. Simons, and Ossabaw, have been the sites of previous and ongoing research projects in a variety of fields. Nonetheless, geologists originally conducted the most comprehensive studies of the origins of these islands, and it was geologists who, along with the Odums, helped to put the Georgia coast in the minds of the scientific community worldwide.

THE GEOLOGISTS, OR WERE THEY ECOLOGISTS TOO?

The arrival of geologists to the Georgia coast in the 1960s was slightly behind ecologists who conducted field research there, yet when all was said and done, they likewise had a far-reaching and influential scientific impact. The beginnings of geologists' research was inspired by uniformitarianism, in which they surmised that present-day processes of sediment erosion and deposition in environments of the Georgia coast might very well provide models for interpreting the products of ancient environments. Indeed, they were correct in this respect, but even more so when they combined the long-term perspectives so inherent to geologists with the methods of behavioral ecologists. This synergism was pioneered by the German sedimentologists of the Senckenberg am Meer (SAM) in the 1920s–1930s (Chapter 1), and its rebirth on the Georgia coast was linked to a few European scientists who visited and studied the islands there in the 1960s and 1970s (Dörjes 1972; Hertweck 1972; Howard et al. 1972; Howard and Reineck 1972a, 1972b; Reineck and Howard 1978). American sedimentologists (people who study sediment) also encouraged this renaissance through examining the resultant traces imparted by animal behavior on the substrates of the Georgia coast (Howard 1968; Frey and Howard 1969; Frey and Mayou 1971; Howard and Elders 1970; Howard and Frey 1973, 1975). So were these scientists most properly called geologists, ecologists, or a combination of the two? Both labels are certainly applicable, but in terms of their overarching synthesis of these fields, they were greater than the sum of their parts: they were ichnologists.

The UGAMI on Sapelo Island was the primary academic home for most geologically inclined scientists who studied the Georgia coast, its biota, and wherever those two intersected with relation to ichnology, although

SKIO has also hosted some outstanding scientists. Among the first of the geologists who came to the UGAMI in the 1960s were John H. Hoyt and Orrin H. Pilkey Jr. Hoyt began his studies there in 1960, and in the ensuing 10 years, he and other scientists were responsible for making many of the connections between the neoichnology of the nearshore environments and the paleoichnology of Pleistocene outcrops in the Sea Islands and coastal plain (Weimer and Hoyt 1964; Hoyt et al. 1964; Hoyt and Hails 1967). Unfortunately, Hoyt died in a light plane accident in 1970, but his prolific output during the 1960s was influential on geologists who overlapped with his subjects of study. Pilkey was on Sapelo for a three-year appointment (1962–1965) with the UGAMI, where he soon realized that the geologic history of the Georgia islands was exceptional in comparison to most previously accepted norms for the formation of barrier islands (Pilkey and Fraser 2003). V. J. "Jim" Henry Jr. of UGAMI aided Hoyt in his investigations of the physical sedimentary processes of the Georgia barrier islands, and the two coauthored several scientific papers before Hoyt's death (Hoyt et al. 1964; Hoyt and Henry 1967). Henry went on to have the proverbial long and successful career with the University of Georgia, Georgia State University, and Georgia Southern University (affiliated with SKIO). Despite his official retirement, he remained active and did fieldwork on the Georgia islands until his death (at the age of 81) in 2010.

Without a doubt, though, the most ichnologically significant partnership on the Georgia coast was that of James "Jim" D. Howard and Robert "Bob" W. Frey (Figure 2.12). Howard, a sedimentologist, began his academic career at the University of Georgia in 1965, but he moved much closer to his research area with an appointment at SKIO that started in 1968. Frey began his fruitful career with the University of Georgia in 1972. He was a paleontologist and geologist by training, but he was also among the first academic ichnologists in the United States, having edited one of the classic tomes of ichnology, *The Study of Trace Fossils* (Frey 1975). His ichnologic interests were developed at Indiana University, where he completed a dissertation on the trace fossils of the Late Cretaceous Niobrara Chalk in Kansas. This formation also contained abundant remains of marine reptiles, ammonites, and other denizens of the Cretaceous seaways from about 85 million years ago (Frey 1969, 1970a; Everhart 2005). While studying German for his foreign language requirement at IU, Frey also became aware of the pioneering neoichnological works of the German sedimentologists,

Figure 2.12. Robert W. Frey, who was an important tracemaker, both literal and figurative, in the neoichnology of the Georgia coast, particularly for Sapelo Island (where he is pictured). Photo by Stephen Henderson.

and he translated some of their articles into English. He later corresponded with some of these same sedimentologists and ichnologists, such as Hans Reineck, Walter Häntzschel, and Dolf Seilacher. In 1972 and 1974, he and Howard then coedited comprehensive volumes on the neoichnology and sedimentology of the Georgia coast in the German journal *Senckenbergiana Maritima*, named after the SAM.

Appropriately enough, Frey's new home of Georgia and its barrier island systems, along with Howard's overlapping ichnological interests, encouraged their collaboration at a time when ichnology was still a relatively unexplored discipline in North America. Interestingly, Howard had also attended Indiana University as an undergraduate student, which meant that he and Frey had connections by chance with both Indiana and Georgia. Frey and Howard's work concentrated on organism–sediment relationships, answering questions such as the following: (1) Which organisms lived in which sediments? (2) What traces did their behaviors impart on those sediments? and (3) What was the preservation potential for such traces in the fossil record? These works emphasized invertebrate tracemakers in shallow marine and marine-influenced environments (intertidal, beach, dune, marsh), although they occasionally studied vertebrate traces in these environments too (Howard et al. 1977; Reineck and Howard 1978; Frey and Pemberton 1986). Some of the research approaches they used were innovative. For example, box cores and x-radiography were applied as tools for better describing distinctive traces and the mass effects of bioturbation (organismal mixing of sediments) in Georgia riverine and salt marsh estuaries (Howard 1968, 1969). They also successfully applied much of what they learned about modern traces of Georgia to Pleistocene trace fossils in Georgia and Florida (Howard and Scott 1983) and Cretaceous trace fossils in the western United States (Howard and Frey 1984; Frey and Howard 1982, 1985, 1990).

Frey was joined briefly at the University of Georgia by George Pemberton, who likewise was interested in using neoichnology as a way to better understand trace fossils, particularly through studying marginal-marine invertebrate traces. Although Pemberton only stayed at the University of Georgia for a few years (1978–1981) before heading back to his original home of Alberta (Canada), he and Frey continued to work together on neoichnological research on the Georgia coast throughout the 1980s (Frey et al. 1984; Pemberton and Frey 1985; Frey and Pemberton 1986, 1987). They also cofounded the first scientific journal started at the University of Georgia, *Ichnos*. *Ichnos* is an international journal about plant and animal traces, and as of this writing, it is still publishing a wide variety of peer-reviewed articles dealing with plant and animal traces, both modern and fossil.

After this frenzied output, Howard retired from SKIO in 1988 and promptly began sailing solo around the world. In one such voyage, he even met with Frey in Inchon (Republic of Korea) to study the neoichnology of the tidal flats there. Oddly enough, three of their last papers together were about modern traces on the other side of the world from Georgia (Frey et al. 1986, 1987a, 1987b). Frey, who was ill throughout much of the late 1980s, died of brain cancer in 1992, abruptly ending his cooperative ichnological endeavors with Howard and Pemberton. Furthermore, Howard ceased publishing or otherwise conducting research after 1992. Regardless, Howard, Frey, and Pemberton, along with their graduate students from the late 1960s through the 1980s, jointly left behind a formidable body of literature on the neoichnology of the Georgia coast that has not been matched since. Furthering this legacy, though, is the continued ichnological work of Pemberton, who is now one of the more famous ichnologists in the world, for reasons related to his studies of the Georgia coast. He is probably best known among other ichnologists for his adherence to the ichnofacies paradigm (Chapter 10), but he and his graduate students have also studied the influence of bioturbation on porosity and permeability in sediments (Gingras et al. 2004a; Pemberton and Gingras 2005; Pemberton et al. 2008b). The latter research resulted in his attracting attention from oil companies, who were interested in learning how, as Pemberton puts it in his public talks on the subject, "burrows can hold billions of barrels of oil" (Chapter 10).

What happened to the neoichnology of the Georgia coast after 1992? It languished considerably, and only recently has been picked up again in part by Frey's successor at UGA, Sally Walker; Murray Gregory of the University of Auckland in New Zealand; Andrew Rindsberg of the University of West Alabama; and me (Gregory et al. 2004, 2006; Martin 2006a; Martin and Rindsberg 2007, 2011). Most importantly, I have taken Emory University undergraduate students on weekend field trips to various Georgia barrier islands every year since 1998. Teaching in the field with new, fresh ichnological examples with each visit increased my familiarity with the bountiful ichnological resources of the Georgia islands while convincing me further of their general educational importance.

The story of the neoichnology of the Georgia coast, as the cliché goes, is a work in progress. Nevertheless, it will continue to be told against a backdrop of the geologic and human histories of the Georgia barrier islands.

These left interpretable imprints of one another that, like Schrödinger's cat, offer various states of interpretation that may not become apparent until we actually observe present-day traces and their tracemakers. The constant change epitomized by the Georgia barrier islands, although having been grounded originally in the Pleistocene, also provides linkages between past and present unlike those of any other barrier islands in the world. This situation means their traces and trace fossils present unique opportunities for testing the basic precepts of uniformitarianism, which is intimately connected to changes in ecosystems through time.

3

TRACEMAKER HABITATS AND SUBSTRATES

CHANGE IN A FEW EASY STEPS

I nearly stepped on the snake, mistaking it for a stick on the trail leading into maritime forest. My oversight was understandable, as the trail was cluttered with branches, leaves, and pine needles, and the snake was relatively small, only about 50 cm (20 in) long. It was also stiff and torpid from the cool shade cast by the long shadows of late afternoon, helping it to blend in with other nearby twigs. Sensing a learning opportunity that could not be ignored, I reached down to pick it up, and it responded by writhing slowly and gently along my forearm. Nonetheless, a few students behind me screamed when they realized what I held. I was amused that my Moses impression had provoked such a response, and I waited a few seconds until their anxiety lessened before talking about snakes, their tracemaking, their habitats, and the substrates of those habitats.

The trail was well known to me because of many field trips led along it, and that day was another one. I was accompanied on this trip by a large group of students and my friend and colleague, Steve Henderson from Oxford College of Emory. The University of Georgia (Athens) Marine Institute (Chapter 2) on Sapelo Island officially names this path the Nature

Trail and it is within easy walking distance of their facilities. The route is short and flat by Appalachian Trail standards, only about 3 km (1.8 mi) in length, with no noticeable changes in elevation along the way other than a footbridge crossing a tidal creek. Yet it offers a sample of nearly every major ecosystem of the Georgia barrier islands along the way: maritime forest, high salt marsh, low salt marsh, tidal creeks and channels, well-vegetated relict dunes, back-dune meadows, modern (active) dunes, beach, intertidal sand flats, and lower shore face. The trail is mostly grassy and firm, but in the forests, pine straw and cones crunch underfoot. Later, closer to the seashore, the footing becomes increasingly soft and silenced. Despite its short length, an attentive stroll along the entire trail should take at least three or four hours. As an educational resource used to introduce students to these ecosystems, it is invaluable, a textbook filled with life, substrates, and the life traces imparted on those substrates. Moreover, every time I place my feet on this trail, I encounter something new and interesting, such as the nearly trod-upon snake, an animal that would never have to worry about stepping on another.

This expectation of discovery also reinforced one of the many reasons why I insist on my being in front whenever taking students along this path. In my experience, first-time visitors to the Georgia coast fail to notice much of what is around them, hurrying along such routes in a needless race against time. As a result, small and otherwise subtle traces are missed: narrow deer trails cutting across our path, sprinkled with their pelletal feces; shallow armadillo dig marks in the forest floor; variously sized woodpecker drill holes in trees; the upraised tumulus of a bee or wasp burrow; fiddler crab scrapings on salt marsh surfaces at low tide; and mouse tracks dotting a dune face, to name a few. All of these traces could be overlooked as soon as student conversations turn toward favorite topics that deal with places and situations other than the here and now. Safety was another consideration, as I wanted to make sure that my novice students did not accidentally stumble onto something that presented them with unforeseen challenges. For example, although I had readily identified this snake as a harmless corn or rat snake—some species of *Elaphe*—I also knew that the much larger and venomous eastern diamondback rattlesnake (*Crotalus adamanteus*) was common along parts of the Nature Trail. Although I had not yet seen a rattlesnake on Sapelo, I had observed their trails many times along this same pathway. These traces were discernible as wide, meandering displacements of leaves and pine needles on the forest floor, sometimes tunneling underneath the debris, which neatly distinguished them from similar disturbances made by armadillos. These wakes not only indicated their former presence, but also the huge sizes of some individuals, some of which were more than 20 cm (8 in) wide. These traces also spoke of how they moved in and through the leaf litter of the forest floor while hunting for small rodents. Other snake trails were also noticeable on sandy surfaces of well-vegetated relict dunes and coastal dunes. In fact, during a solo visit to Sapelo the following year, I saw such a sinuous pattern on top of a footprint I had left earlier in the afternoon.

Field trips to the Georgia coast with students typically take a minimum of four to five hours of driving in a van, a ride that starts from our Atlanta campus on a Friday afternoon, just when the infamous and horrendous Atlanta traffic solidifies into a slightly moving parking lot. The juxtaposition of seemingly endless concrete, steel, and burning petrol with the vast greenish-gold expanses of coastal salt marsh at either end of this trip thus becomes all the more jarring for both first-time and veteran visitors of the coast. If we go to Sapelo Island for our field trip, then an overnight stay at a motel on the mainland follows this Friday afternoon drive. An early Saturday morning jaunt to a nearby dock for ferry transport to the island also means disrupting the long-traditional and cherished student pleasure of sleeping in on a weekend morning. This situation then presents challenges in making sure that everyone is up and on time to catch the boat to the island. Why even take students on such field trips, considering all of the potential difficulties in planning and execution? The main reason is simple: so that ecological and ichnological concepts are taught using actual ecology and ichnology, and in a setting where there is little else to distract. I like to think those slithering ribbons of interstate highways, filled with roaring motorcycles, cars, and trucks, become a faded memory of someone else's dream (or nightmare) once students are immersed in the Georgia coastal environments.

Modern ecology, founded on Sapelo Island (Chapter 2), generated a term I like to use in class when testing my students on basic ecological concepts: ecotone. If they have never before heard this word or cannot remember its meaning, the confusion manifests instantly on their faces, and I imagine their reaching for technological explanations, such as nonpolluting cell-phone ring tones. Typically, though, at least one student has heard

Figure 3.1. Sharply defined ecotones observable on a Georgia barrier island, in which salinity differences determine where different tracemakers live. In the foreground is high marsh, dominated by glasswort (*Salicornia virginica*); the ecotone (indicated by arrow) is the high marsh with black needle rush (*Juncus roemerianus*); and in the background is maritime forest dominated by live oaks (*Quercus virginiana*): Cumberland Island.

this word and can give a reasonable explanation of its meaning: it is the transitional area between one distinct ecosystem and another, each with its own set of physical, chemical, and biological parameters. The follow-up topic to this recitation is always related to reality. How do we actually see these transitions? For example, the phasing of one terrestrial ecosystem into another is normally a subtle one that requires expert identification of plant and animal species. Furthermore, in today's feral world, knowledge of invasive or otherwise exotic species overprinting environments is also needed. Nonetheless, the Sapelo Nature Trail contains easily observable ecotones, their visibility enabled by differences in salinity of subsurface waters between the maritime forests and adjacent marshes, as well as in dunes and beaches (Figure 3.1). Like all things, ecotones are ephemeral, especially when viewed in the context of geologic time, but normally they connote a sort of reliability that lasts more than, say, the span of a day. For example, the snake I was holding had been occupying a sort of smaller and more temporary transition zone, not an ecotone. It had been lying in a patch of the forest floor that fell into darkness as trees just to the south of our position had blocked the sunlight. This interruption dropped temperatures enough that the tracemaking activity of the snake was slowed to a halt.

As the snake warmed up and my students calmed down, it moved around my arm while I remarked on its beauty and talked about the functional role of snakes in terrestrial ecosystems. Not wanting to bother it too much, and the educational lesson imparted (mostly, "Pay attention!"), I set it down, bid it adieu, and wished it happy hunting. As we traveled further along the trail, within about a hundred meters (330 ft) or so, we encountered an ecotone between maritime forest and marsh, a distinction that could be smelled if observers remembered to use more than one sense when observing.

The maritime forest, with its live oaks, Spanish moss, pines, and associated fauna, contrasted easily with the salt marsh next to it. The marsh had no trees and was composed of only a few species of halophytes. The smooth cordgrass *Spartina alterniflora* dominated the lower part of the marsh, whereas the salt-meadow cordgrass (*Spartina patens*), black needle rush (*Juncus roemerianus*), glasswort (*Salicornia virginica*), and salt grass (*Distichlis spicata*) were in the high marsh. Along the fringes of the high marsh were sea-oxeye daisies (*Borrichia frutescens*), which hold beautiful yellow blooms throughout much of the year. Directly adjacent to, and barely upslope of these plants, were small trees, such as red cedars (*Juniperus virginiana*), which, as one can tell from its genus name, is not a cedar at all, but a juniper. If my students were persuaded to draw a line between the ecosystems of maritime forest and high marsh, the demarcation could be made between the red cedars and the plant assemblage of the high marsh.

A distinctive suite of animals, but dominated by only a few species, accompanied the marsh plants. These were at first invisible, but once eyes and ears were tuned into this ecosystem, the small, turret-like burrows, scrape marks on marsh surfaces, and feeding pellets of sand fiddler crabs (*Uca pugilator*) came into focus. The subdued clicking and rustling of fiddler crab activity also could be perceived before spotting their movement between the clumps of salt-meadow cordgrass and glasswort. Hence we were looking at the high marsh. A further look revealed salt marsh periwinkles (*Littoraria irrorata*), small gastropods looking like beads pasted onto the stalks of *Spartina*; but if one were patient enough and watched closely, these were crawling along the stalks, grazing on algae growing on these surfaces (Figure 3.2). Trackways of raccoons (*Procyon lotor*), denoted

Figure 3.2. Salt marsh periwinkles (*Littoraria irrorata*), small but ecologically significant tracemakers in salt marshes of the Georgia coast for their contributions to organic detritus through their grazing on smooth cordgrass (*Spartina alterniflora*): Sapelo Island.

by side-by-side pairs of rear and front feet, ambled across barren areas of the salt marsh. Their five digits looked very much like little human feet and hands, although with sharp claw marks at the ends. In the scatological realm, the fecal pellets of deer in the maritime forest, filled with digested terrestrial grasses, were now replaced by much larger, cylindrical raccoon scat containing the remains of hapless fiddler crabs. Competitors to the raccoons were noticeable through the numerous large (12–15 cm, or 5–6 in long), four-toed, spindly tracks of predatory great blue herons (*Ardea herodias*) and great egrets (*Ardea alba*) that had landed on the marsh surface for early-morning crab snacks. The sandy areas of the maritime forest mixed with organic layers of decomposing leaf litter—scented by pine and oak—also differed considerably from the sulfur-odored muddy sand before us. Hence with the added sensory information provided by the traces of animals, combined with the visible biota and substrates, the ecotone between the maritime forest and the marsh became all the more obvious.

Nonetheless, it was time for a quiz on ecotones. Steve and I, through the time-tested application of the Socratic method—"Where is the marsh?," "Where is the forest?," "Where is the low marsh and the high marsh?," "What plants distinguish each division?," "What traces are diagnostic of the maritime forests and the salt marshes?," "Which traces can be found in both ecosystems, and why?"—were then able to guide the students to a place where they could point to the ecotones (low marsh → high marsh → maritime forest). This gave the students just enough confidence for Steve and me to shatter it only a few minutes later.

The unsettling of their newfound abilities took place when we pointed out another and more subtle ecotone they had not noticed within the low marsh. Near this point, the Nature Trail continues along an earthen causeway just wide enough for a single car to drive along it. As we walked along the grassy causeway toward the east, low marsh was on our left and high marsh on our right, each bordered by rows of red cedars. Branching to the right from the causeway was a short wooden boardwalk leading to a pavilion, mounted above the middle of what seemingly was still high marsh. The stroll to the pavilion did indeed place us in the middle of the marsh, but in an area containing no vegetation, not even halophytes. Owing to its lack of plant cover, the ichnology of the surface was nakedly obvious, revealing fiddler crab burrows, feeding pellets, scrape marks, and numerous tracks left by raccoons, herons, and the occasional errant armadillo (Figure 3.3). So we asked the students a simple question with a complicated answer: "Why no plants?" The students then proposed hypotheses for this oddity; tested these on the basis of the available evidence, which grew with the quantity and quality of their observations; rejected a few of the explanations proffered; and finally one was conditionally accepted, temporarily satisfying their instructors.

Their hypothesis was like so. This part of the high marsh had more salt content in its sandy mud than the surrounding area, meaning that the substrate excluded the growth of halophytes, as none had adapted to handle such high salinities. The high salt content was probably attributable to high evaporation rates accelerated by the lack of shade normally provided by vegetation, a form of positive feedback. The absence of plants also limited fiddler crabs from colonizing the surface, as these crustaceans tend

Figure 3.3. High salt marsh with patchy vegetation dominated by the burrows, tracks, and feeding traces of numerous fiddler crabs (*Uca* spp.), as well as the traces of terrestrial and marginal-marine vertebrate tracemakers hunting these crabs: Sapelo Island.

to burrow underneath *Spartina* and other halophytes for protection from predators. Moreover, vegetation provides readily available detritus for their feeding. Some of this detritus falls to the marsh surface from above as a result of the grazing periwinkles and other herbivores, which break down *Spartina* while moving across the leaves and stalks. Fiddler crab burrowing and feeding is also an essential facet of the nutrient flux in a salt marsh; once this activity stopped, salt marsh substrates undergo rapid change.

Of course, there was more to the story than that, but this hypothesis was a good start. We then explained how these areas in high salt marshes are appropriately called salt barrens because of their hypersaline soils and absence of foliage. We also discussed briefly how the predominant grain sizes of the low marsh (mostly mud) and high marsh (relatively more sandy) are additional factors in ecological partitioning of fiddler crab species: *Uca pugilator* is known colloquially as the sand fiddler, whereas *Uca pugnax* is the mud fiddler, and each prefers to burrow in substrates dominated by those grain sizes (Chapter 6). We did not confuse them any further, though, by discussing a third and less common fiddler crab in the area: *Uca minax*. This species prefers to live in tidal-creek banks in organics-rich mud, and it better tolerates low oxygen and freshwater. Dichotomies provide starting points for discussion, and can always unfold into other dimensions later.

Our students, now more attentive to the concept of ecotones, the importance of substrates as ecological factors in ecosystems, adaptations of their biota, and ichnology as an essential tool for discerning such factors, were then more easily led, step by step, into changed perspectives along the rest of the trail. Exposed banks of tidal creeks at low tide hosted the open burrows of the squareback marsh crab (*Armases cinereum*) and other mud-loving crabs, such as *U. minax*. Encrusting oysters (*Crassostrea virginica*) could be seen stabilizing meander bends of these creeks, while also providing hard surfaces for the attachment of future generations of oysters. Rattlesnake trails and armadillo excavations in the pine needles and sandy soils were clearly evident on the floor of a maritime forest dominated by needles shed by loblolly pine (*Pinus taeda*). Relict dunes from a coastal system of the recent past were occupied by mature stands of wax myrtle (*Myrica cerifera*) and yaupon holly (*Ilex vomitoria*). Ant nests, ant-lion burrows, deer trackways, and mole burrows were all apparent in more open sandy patches between plants (Figure 3.4). Modern coastal sand dunes held in place by sea oats (*Uniola paniculata*) were inscribed and punctuated by burrows and trackways of ghost crabs (*Ocypode quadrata*), tiger beetle burrows, and vertebrate tracks made by mice, mourning doves, and marsh rabbits. The relatively dry, windswept, and fine-grained sandy berm, just below the dunes but above the high tide mark, was imprinted with more ghost crab burrows and their trackways, as well as limulid tracks. A few of these limulid trackways ended with a dead limulid, providing good examples for our students of a dying trace, also called a taphichnion (Chapter 6). Had we been there during May through August, we would have also seen the heavy, wide trackways of nesting sea turtles connecting the sea with the dunes. The lower, rippled part of the beach was peppered with sand-

Figure 3.4. Relict dunes, formed earlier in the Holocene Epoch but now hosting back-dune meadows and maritime forests: Sapelo Island. These areas are inhabited and shaped by a wide variety of tracemakers, including terrestrial plants, insects, arachnids, amphibians, reptiles, birds, and mammals.

sized but mud-filled fecal pellets of ghost shrimp (*Callichirus carolina* and *Biffarius biformis*), which accumulated in ripple troughs. Along the edges of runnels were the burrows of shallowly buried sand dollars (*Mellita isometra*), whelks (various species of *Busycon*), olive shells (*Oliva sayana*), and other invertebrates, either keeping themselves moist or hunting for food in the saturated sand between high tides. Also visible at low tide were feeding pits left by southern stingrays (*Dasyatis americana*), accompanied by an overwhelming number of shorebird tracks and beak prod marks, feces, and regurgitants.

Thus our noting of these traces, who made them, and where they occurred helped us to mark the present-day locations of ecotones and to keep in mind how such traces can be diagnostic of past ecotones. This is where ichnology, especially when united with uniformitarianism, becomes an essential facet of both ecology and paleoecology. Just as ecologists look for ecotones as markers for boundaries between ecosystems, paleoecologists look for clues that similarly show where different facies may abut or succeed one another in the geologic record. Facies ("face") are sedimentary rocks containing characteristic traits that help to identify their original environments; as a result, many of the substrates and traces we saw during our brief tour could be related directly to facies. Thus this jaunt with our students was short in distance but lengthy in knowledge, leading directly to a more enlightened view of how traces inform us about present and past habitats, the transitions between those habitats, and the substrates that preserve the traces.

SALT MARSHES: *SPARTINA* ABOVE, AND MOSTLY MUD BELOW

Salt marshes are among the best studied of ecosystems, and the Georgia salt marshes in particular have long been the focus of salt marsh studies. Early ecological studies on Sapelo Island by Eugene Odum and others starting in the 1950s focused on the flux of matter and energy in marsh ecosystems (Teal 1958, 1962; Odum and Smalley 1959; Odum 1968), but coastal geologists and ichnologists in the 1960s onward were also intrigued with the organismal traces made in such environments and the biological processes behind mud deposition (Frey and Howard 1969; Frey and Mayou 1971; Frey et al. 1973; Howard and Frey 1973; Basan and Frey 1977; Edwards and Frey 1977; Letzsch and Frey 1980; Smith and Frey 1985). Indeed, to anyone who has walked into a Georgia salt marsh, perhaps their most memorable experience has been that sinking feeling, perceived literally once feet, ankles, and sometimes knees disappear under the soft, sulfurous-smelling black mud (Figure 3.5a).

What causes sedimentation of this mud, and how does it relate to the organisms that live above, on, and in it? As mentioned earlier (Chapter 2), organisms—particularly filter feeders—are responsible for depositing most mud in salt marshes and many other coastal environments (Pryor 1975; Kraeuter 1976). As a case study for this general principle, one species of filter-feeding molluscan, the ribbed mussel (*Geukensia demissa*), was studied specifically for its role in dumping mud in Georgia salt marshes (Smith and Frey 1985). In such studies, researchers measured amounts of feces produced by mussels in an aquarium under controlled conditions, and they varied temperature to see how it affected mussel physiology and mud production. They also measured this mud production in the marsh itself. These mussels, along with oysters, accomplish this feat by sucking water in through siphons (Chapter 6). This ingested water contains suspended clays and organics, which are separated internally so that the mussel only uses the nutritious organics. Meanwhile, unused clays and other waste products are packaged into sand-sized particles of mud, shrink-wrapped in mucus, and ejected from the animal as feces (Figure 3.5b). This biode-

94 LIFE TRACES OF THE GEORGIA COAST

Figure 3.5. Forms of feces deposited by ribbed mussels (*Geukensia demissa*) in Georgia salt marshes, showing changes from those deposited in water (*left*) to those that become slightly dehydrated with exposure (*right*). Modified from an illustration by Smith and Frey (1985). Scale = 1 mm (0.04 in).

position results in mounds of mud pellets around a cluster of mussels and oysters. Tidal currents may then transport these pellets, or their mucoidal packaging may break down and liberate once-bound clays, which become suspended again in the water column. The net effect, though, is the accumulation of fine-grained particles that become the foundation for benthic animals that live in or on a salt marsh; they thus depend on bivalve feces to compose their habitat. At least some of these feces are used as food by other invertebrates too (Frankenberg and Smith 1967). Along those lines, my students are either amused or appalled when I tell them, in the middle of a good wade through marsh mud, just how that mud got there. Nonetheless, some students will also humorously connect this information metaphorically with what teachers sometimes put them through for a good grade.

Mud fiddler crabs (*Uca pugnax*) and other mud-loving crabs are also responsible for accumulating mud in sand- or pebble-sized (>2 mm diameter) packets in salt marshes (Frey and Mayou 1971; Frey et al. 1984). *U. pugnax* make such pellets by scraping the marsh surface at low tide with one of their claws. Their claws and mouthparts meticulously manipulate each small bolus of mud; the mouthparts separate the organics—such as algae growing on the muddy surface—from the inorganic particles. Once this deposit feeding is done, crabs will have made many small (2–4 mm wide) balls of sediment, dropped unceremoniously on the marsh surface (Chapter 6). Such pellets are sometimes termed pseudofeces because of their resemblance to fecal pellets, but with a different origin, coming out of the other end of the animal. Furthermore, such pellets are not even the direct result of ingestion; hence these traces are not classified as regurgitants either. The results of this behavior are dozens of pseudofeces made by each crab per day, multiplied by thousands of *U. pugnax* in any given marsh, ultimately causing an enormous amount of mud to be deposited as large particles. This mud is also weathered and eroded at the marsh surface, which makes it easier to break up and become suspended again with each tidal cycle.

With such intensive biological activity combined with tidal cycles, one might ask what holds the mud in place once it is deposited. The short answer is *Spartina alterniflora* (which as a species name is not so short) and its root systems. *S. alterniflora*, a type of grass, is the most abundant and tallest plant in the low salt marsh, and its ubiquity in that environment is related to evolutionary hardiness. In the geologic past, its terrestrial-grass ancestors must have lived near salt marshes in ecotones that underwent strong ecological pressures from saltwater intrusion in the surrounding soils, which in turn would have favorably selected grasses with the right genes for handling higher salinities. This, however, is only part of the remarkable story of *S. alterniflora* as the ruler of the salt marsh. Its present-day root systems were also selected because of their ability to keep plants entrenched in a gooey mud that is frequently liquefied or otherwise made unstable by the flooding and draining of tidal cycles. To do this, *Spartina* roots have vertical and horizontal components, as well as fine rootlets, that spread throughout the marsh mud (Chapter 4). Colms (horizontal shoots from the base of the stem) also enable clonal growth, where each individual plant may have more than one stem (Proffitt et al. 2003). The net effect of these roots and colms is a broad meshwork that binds together and helps to compact the sediment, making a relatively stable surface out of what should be a huge, soupy mess of mud. These root systems also help to preserve ancient marsh deposits.

Another factor in holding together mud, and a significant one, is the binding of muddy surfaces by algal films. The uppermost surface of a salt marsh is the part that receives direct sunlight, which naturally attracts photosynthetic organisms, such as halophytes and various algae, the latter scraped and eaten by mud fiddler crabs. Among the algae living in this illuminated zone are diatoms (a type of brown algae) and green algae (Holland et al. 1974; Williams 2007). These and other types of algae form mucoidal mats that stick to and contain the uppermost layer of mud, much like how a thin plastic wrap placed over pudding keeps it somewhat in place, however precariously. The effects of such biofilms also influence the preservation of invertebrate and vertebrate traces that punch through them, particularly when these surfaces become dehydrated along their broken edges. For example, I have often seen tracks of deer or large birds (herons or egrets) that cut through these films and into the underlying soft, moist mud in high marshes. Once exposed to direct sunlight, drying begins along the edges of the tracks, which is soon followed by the development of mud cracks that radiate outward from each footprint and intersect with one another. This phenomenon of tracks and burrows initiating mud cracks has been documented in many instances where a combination of avian trackmakers, mud, and drying happen, even in Arctic environments (Mason and Bruu 1978; Wallace and Wallace 1992; Martin 2009a). Nonetheless, it is often related to the presence of a thin algal film binding the top surface of the mud.

Salt marshes are not all alike, although these environments can be subdivided on the basis of slight differences in topography, which in turn control the distribution of different tracemakers (Edwards and Frey 1977). If one had a boat and could start a cross-sectional jaunt from a tidal creek, a typical sequence of low-marsh to high-marsh subenvironments might include creek bank; streamside and levee marsh; low (meadow) marsh; salt barren; meadow marsh (again); the ecotone between the low marsh and high marsh; high marsh; and the ecotone between the high marsh and maritime forest (Figure 3.6). Subenvironments within these are often distinguished on the basis of their dominant plant species. The default dominant species in the low marsh, though, is *S. alterniflora*. As a result, root traces made by plants and invertebrates, if identifiable to species, would be diagnostic of specific ecological conditions within a salt marsh complex preserved in the geologic record (Chapter 4). Accordingly, invertebrates and vertebrates living in or on—or simply visiting—salt marshes will frequently have their behaviors affected by whatever plants and substrates might be there. Hence the locations and forms of their traces will also reflect these ecological factors.

TRACEMAKER HABITATS AND SUBSTRATES   97

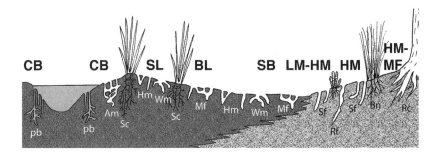

Figure 3.6. Idealized cross section of a Georgia salt marsh, with differences defined by its substrates, relative elevation, plant communities, invertebrate burrows, and other factors: creek bank (CB); streamside levee marsh (SL); back levee (meadow) marsh (BL); salt barren (SB); ecotone between the low marsh and high marsh (LM-HM); high marsh (HM); and the ecotone between the high marsh and maritime forest (HM-MF). Also note a transition of muddy (left) to sandy substrates (right) toward the maritime forest. Key of traces (from left to right): pb = polychaete burrows; Am = Atlantic mudcrab (*Panopeus herbstii*); Sc = smooth cordgrass (*Spartina alterniflora*); Hm = heavy mud crab (*Sesarma reticulatum*); Wm = white-clawed mud crab (*Eurytium limosum*); Mf = mud fiddler (*Uca pugnax*); Sf = sand fiddler (*Uca pugilator*); Rf = red-jointed fiddler (*Uca minax*); Bn = black-needle rush (*Juncus roemerianus*); Rc = red cedar (*Juniperus virginiana*). Modified after figures and descriptions by Basan and Frey (1977) and Howard and Frey (1985).

Despite the seemingly large number of genus and species names just mentioned, salt marshes are tough environments that limit most life. Hence salt marsh tracemakers are either adapted as a result of severe selection pressures, or are just passing quickly through these environments. For instance, twice-daily high tides with their influx of saline water, separated by periods of exposure, will exert natural selection on any species that intend to live and reproduce there. Just think of the effects of a hot Georgia sun on black mud during the summer and how long you would last without protection, while not factoring in predators looking for an easy meal. Yet plants living in marshes manage to pull out nutrients from low-marsh or high-marsh sediments while also tolerating salt water. Most animals in marshes are either descended from fully marine ancestors and adapted to extended times out of the water (during low tides), or they are terrestrial animals that flee the rising water of high tides by leaving the marsh, or climbing the stalks of *S. alterniflora*. Along these lines, perhaps the most unexpected inhabitant of a salt marsh is a species of ant, *Crematogaster clara*, which lives inside the stems of *S. alterniflora* (Teal 1962; Teal and Teal 1964). Members of the colony feed on the outside of the plant during low tide, but with the onset of high tide, they retreat inside through a single hole that serves as both the entrance and exit. Once inside, a special large-headed individual in the colony blocks the hole during the tidal cycle—a finger in the dike that saves the colony twice every 24 hours.

Of course, for most mobile, benthic, and marine-adapted invertebrates, the best and time-tested adaptation for dealing with the stresses of tidal fluctuations is burrowing. Consequently, a marsh surface at low tide is peppered with thousands of burrows made by fiddler crabs and their kin, polychaete worms, and many other species of invertebrates. As explained in detail later, the primary purposes of these burrows may vary (Chapter 6), but such structures are united behaviorally by their common use for protection against the environmental extremes inherent to salt marshes.

FRESHWATER PONDS AND MARSHES: ALLIGATOR-MADE HOMES

Surface expressions of freshwater, such as ponds and marshes, are rare on the present-day Georgia barrier islands, limited in area, and seasonally dependent. For example, the largest standing freshwater body on any of the islands is Lake Whitney of Cumberland Island, composed of only 4–5 hectares (about 10 acres) of open water (Frick et al. 2002). As a result of the geologic framework of the Georgia islands, constant saltwater influx via tidal creeks, and depletion of shallow aquifers by previous and current human populations (Chapter 2), free-flowing freshwater streams are nearly unknown today, despite their common presence on the mainland Georgia coastal plain. Nonetheless, currently existing small bodies of freshwater are sites of deposition for mud (with minor influxes of sand) and are ichnologically important ecosystems to study. This circumstance is largely because most small ponds owe their origin to a single tracemaker (*Alligator mississippiensis*), and then are modified by many other tracemakers.

Alligators are a keystone species of the Georgia barrier islands and much of the lower coastal plain. If alligators are reduced in number or completely removed, changes in those ecosystems are notable and swift. As the top predators in most areas, their preferred prey, such as raccoons, deer, and various fish (Shoop and Ruckdeschel 1990), would in their absence also become much more abundant in closely associated ecosystems. This experiment has been unwittingly performed on developed Georgia barrier islands, such as Jekyll and St. Simons, where the filling in of alligator-made ponds for housing and golf courses led to explosions in local deer and raccoon populations (Chapters 7, 8). Habitat alteration is responsible for decreasing areas where alligators can stay and breed, as alligator ponds are not just used for ambush predation but also as places for protecting adults, nesting, and raising young.

Alligators of sufficient size, capable of walking over extended distances, occupy nearly any water body large enough to conceal them and keep their skin moist, but they can also act directly to expand their habitats. With the arrival of more rain in spring and summer, low-lying areas in terrestrial environments, such as between relict dunes, may fill with standing pools of water. This situation is most likely to happen wherever impermeable sediments, such as layers of compacted mud, underlie sandy, porous soils. These muddy layers may have originated in previous freshwater or salt marshes, but they now confine the downward flow of water underneath sandy soils, much like how a plastic liner underneath a sandbox will hold standing water after a hard rain. Alligators, if not already living in such pools, will seek out these places with ponded water and then make them bigger. They enlarge their homes by wallowing in the muddy sand, swirling their thick, ponderous tails, and undulating their bodies through the moist sediment, which breaks up (and spreads out) the banks of the water body. These "gator holes" then become low-lying areas for further accumulation of water, which encourages colonization by wetland-adapted plants, forming the foundation for other trophic levels in these ecosystems (Figure 3.7). Alligators also further modify these ponds by digging burrows on their peripheries; these burrows are dens that allow for protection during times of torpor prompted by wintertime temperatures, as well as for brooding of hatchling alligators. These marvelously large and complex burrows, along with some of the other organisms they affect, will be examined in considerable detail later (Chapter 7).

Other tracemakers that affect freshwater ponds and marshes include plants, invertebrates, and small vertebrates, but not many animals would live there without plants. The largest and most obvious plants in freshwater wetlands are trees adapted for wet conditions, such as pond cypresses (*Taxodium ascendus*) and red maples (*Acer rubrum*). Cypresses are dis-

Figure 3.7. A gator hole, a freshwater pond surrounded by maritime forest that was formed and expanded through bioturbation by generations of alligators (*Alligator mississippiensis*). Wassaw Island.

tinctive for their wide, buttressed bottoms and root systems that project above the ground (or water) surface as knees. Abundant narrow-leafed cattails (*Typha angustifolia*) might co-occur with a bevy of other water-loving plants, such as scouring rushes (*Equisetum* spp.), pickerel weed (*Pontedaria cordata*), water lilies (*Nymphaea stellata*), bulrushes (*Scirpus* spp.), arrowhead (*Sagittaria latifolia*), and green-arrow arum (*Peltandra virginica*). Boggy parts of a wetland also might have those trophic-level iconoclasts, pitcher plants (*Sarracenia* spp.), which augment their nitrogen and phosphorous uptake by luring insect victims with nectar and digesting them in enzyme-filled pools within their cupped leaves (Slake and Gate 2000; Schnell 2002). On the banks of these environments are the appropriately named pond pines (*Pinus serotina*) and other trees that can handle moist soils for much of the year. Away from the islands and more inland along freshwater riverbanks are giant sawgrass (*Cladium jamaicense*), cutgrass (*Zizaniopsis miliacea*), and wild rice (*Zizania aquatica*).

Aquatic insects are the most common invertebrate tracemakers in freshwater ecosystems of the Georgia barrier islands, composing about 90 percent of known invertebrates in those environments (Frick et al. 2002). Insects include the usual suspects of any temperate freshwater assemblage, such as water bugs (hemipterans), water beetles (coleopterans), dragonflies and damselflies (odonates), and mayflies (ephemeropterans). Beetles are represented by scavenger and predaceous species, and one genus (*Hydrocanthus* spp.) is a member of the group called burrowing water beetles (Noteridae). Most traces collectively made by insects in freshwater environments may be through interactions with aquatic plants (herbivory) or burrows on submerged or emergent muddy surfaces. However, only a few species burrow in either their larval, nymph, or adult phases (Chapter 5). Probably the most ichnologically significant freshwater invertebrate tracemakers on the Georgia barrier islands are burrowing crayfish (*Procambarus* spp). These crustaceans make visually prominent towers at the tops of their burrows adjacent to water bodies and can burrow several meters down to local water tables (Chapter 5). A few species of gastropods, nicknamed pond snails, also live in freshwater environments, but they are not nearly as numerous as their marine relatives. Their grazing trails are evident on either vegetation or sedimentary surfaces in these ecosystems.

Vertebrates much smaller than the average alligator are important members of freshwater ecosystems, including various bony fish (teleosts), frogs, salamanders, water snakes, turtles, birds, mink (*Mustela vison*), river otters (*Lutra canadensis*), and (very rarely) beavers (*Castor canadensis*). Some fish species, especially bottom feeders, may form incised grooves with their ventral fins or marks made by sucker-like mouths on soft, muddy bottoms, and of course they are always providing a gentle rain of heavenly fish feces, a major source of organics. A few fish species, such as bluegill and warmouth (*Lepomis macrochirus* and *L. gulosus*, respectively), also may make semicircular depressions on the sediment surface, which serve as places for females to lay eggs for fertilization (Gross and MacMillan 1981; Neff et al. 2004). All amphibians of the Georgia barrier islands are dependent on either standing or seasonal pools of freshwater for reproduction, and thus tracks of nearly every species might be found in fine-grained sediments on the banks of these environments. Some amphibians also are burrowers; they burrow for overwintering or for other forms of protection. Aquatic snakes, such as the cottonmouth (*Agkistrodon piscivorus*), peninsula ribbon snake (*Thamnophis sauritus*), and Florida water snake (*Nerodia fasciata*), do just the opposite of amphibians with their reproduction: they come up on land, either to lay eggs (cottonmouths) or bear live young (ribbon and water snakes). Hence their sinuous trails may be seen emerging from water bodies. Likewise, water turtles, including the snapping turtle (*Chelydra serpentina*) and mud turtles (*Kinosternon* spp.), may make distinctive trackways and shell drag marks during their quests to lay eggs on land, and later dig small depressions above the waterline for their egg clutches (Jensen et al. 2008; Ernst and Lovich 2009). Birds that make a living in freshwater ponds and marshes, such as various waterfowl, herons, egrets, and bitterns, trample or otherwise leave tracks on the banks of these water bodies, but they also construct nests in these areas. High above the water, old live or dead cypresses, pines, oaks, and other tall trees become sites for rookeries of many larger birds, and their rickety stick nests are common sights in both freshwater and saltwater wetlands (Chapter 8).

Perhaps surprisingly, the list of potential freshwater tracemakers on the Georgia barrier islands included beavers (*Castor canadensis*). Although beavers were once native to the Georgia barrier islands, these prolific tracemakers (and ecosystem engineers) are apparently absent now. The last live one seen on any of the islands—Cumberland—was in the 1970s, although dead ones, presumably flushed out to sea by coastal rivers, will also occasionally wash up on a barrier island beach (Carol Ruckdeschel, personal communication, 2008). Nonetheless, the well-documented and considerable traces made by beavers in nearby coastal plain freshwater ecosystems can be applied as search images for detecting their former presence in freshwater environments of the Georgia barrier islands. Their major impact on freshwater ecosystems is their construction of dams that slow water flow, which increases sedimentation rates while also expanding wetland areas, thus creating ideal habitats for wetland-loving plants. As a result, beavers are similar to alligators in their large-scale effects on freshwater environments. Accordingly, though, their absence on the Georgia barrier islands means that their previous habitats are also absent or otherwise unrecognizably modified. For example, I am often surprised to find decades-old beaver chew marks on trees and their eroded bank trails in low-lying places near stream drainages in the middle of urban Atlanta. Whenever I find such traces, a quick survey of the surrounding area reveals that I am standing in a grass- or building-covered floodplain that is actually the former site of a beaver pond. In most instances, the makers of that pond are long gone, but their effects on the surrounding landscape remain.

For whatever reasons, freshwater environments of the Georgia barrier islands and their biota have not attracted nearly as much attention from barrier island researchers and ichnologists as salt marshes, beaches, and maritime forests. This neglect, however, is probably related to the relative rarity of these ecosystems. Nevertheless, freshwater plants and animals comprise an important suite of tracemakers worthy of further study, especially for better defining their traces and contrasting these with their better-known cohorts in marine-influenced environments (Chapters 5, 7, 8).

MARITIME FORESTS: LIVE OAKS AND SPANISH MOSS (OF WHICH IT IS NEITHER)

Take a quick look at a bookshelf devoted to the Georgia coast, a Web site advertising a vacation on the coast, or other communiqués about what is there, and their photographs will center around two ecologically oriented themes: beaches and maritime forests. The maritime forest photos are, invariably, beautifully clichéd. Spanish moss will hang languidly from branches of live oaks, and the forest floor is limned with a jagged mass of saw palmetto, all bathed with golden sunlight, or, if slightly more creatively shot, with diffused light shining through the Spanish moss (Figure 3.8). (What is not shown in the photos are traces of excessive trampling and rooting wrought by feral hogs in these same forests, but never mind.) Once an ichnologic eye is cast upon these forests, however, they become far more interesting than the superficial take presented by marketing artifice. Part of their scientific quality is owed to the wide variety of substrates available to tracemakers in maritime forests, as well as their diversity of plant and animal tracemakers. Yet another aspect of maritime forests that makes these ichnologically distinct from most other ecosystems on the Georgia coast are their vertical zones of substrates. This situation is mostly a function of large, tall trees and their root systems in a maritime forest, which provide plentiful plant tissues as substrates for tracemakers. These substrates, coupled with sandy or muddy soils, such as ultisols, podosols, histosols, and entisols (Chalmers 1997), thus lend to a bonanza of trace-preserving media in forest ecosystems.

Maritime forests can be divided into three types on the basis of their surrounding topography and proximity to the sea: low hammock, upland

Figure 3.8. A photograph of maritime forest depicting all of the popular subjects of Georgia coast photographers—live oak (*Quercus virginiana*) with Spanish moss (*Tillandsia usneoides*) hanging from its branches and saw palmetto (*Serenoa repens*) in the forest understory—but omitting traces made by its numerous ground- and tree-dwelling animals: Jekyll Island.

hammock, and strand (Bellis and Keough 1995; Albers and Albers 2003). More inland forests—low and upland hammock—will be less affected by sea spray and other limiting factors associated with higher-salinity water, and consequently they develop into diverse communities. Low hammocks occupy slight depressions in a barrier island, perhaps in the swales of former dunes and near salt marshes. These areas tend to collect more moisture than the well-drained soils of the upland hammocks; hence their plant assemblages differ. In low hammocks, sweet bay (*Magnolia virginiana*), loblolly bay (*Gordonia lasianthus*), and red bay (*Persea borbonia*) are abundant, whereas upland hammocks more often have live oaks (*Quercus virginiana*), southern magnolia (*Magnolia grandiflora*), laurel oak (*Q. laurifolia*), and American holly (*Ilex opaca*). Strand forests are those that abut beaches and salt marshes and normally contain only cabbage palmettos (*Sabal palmetto*) and red cedars (*Juniperus virginiana*) as dominant trees. Such forests I always think of as marked for death because rising sea levels will surely take out the living in the next decade or so (Chapter 2). Of these three, upland forests are the ones normally featured in tourism advertising, but all forest ecosystems are ichnologically important too.

Numerous tracemakers and their traces exist above the maritime forest floor and in the vegetation. Plant tissues make excellent substrates for preserving organismal traces, but plant parts themselves are often used to make other traces, such as bird and squirrel nests (Chapter 7). The most common organismal traces left in leaves and wood, though, are associated with feeding and reproduction, such as those formed from insect–plant interactions (Chapter 5). Examples of such traces in leaves include marginal and nonmarginal incisions on leaves made by feeding insects; leaf mines, in which the insect meanders beneath the outer covering (cuticle) of a leaf and chows down on its soft, green layer of mesophyll (in between the cuticles); minute holes made by piercing mouthparts; and galls, which are swellings made by a plant as wound reactions after insects lay eggs in living plant tissue (Chapter 4). Some insects also bore into wood to make feeding galleries, brooding chambers, and communal nests, represented by beetles, carpenter bees, carpenter ants, and, of course, termites (Chapter 5). The colonization of woody substrates by insects also illustrates important ecological relationships between herbivores, detritivores, and insectivores. Think of how many woodpeckers and other insect-eating birds that drill into wood or pull off bark for their food would be hampered by a lack of insects in dead and dying wood. Once trees are hammered enough by a woodpecker, excavations in the wood that began as a few drill holes can also turn into nest holes, which are distinctive enough to identify woodpecker species (Chapter 7). Other bird nests of varying shapes and materials may also be lodged in trees, and frequent roost spots of predatory birds may have cough pellets and droppings located directly below their perches.

Various fungi and the resurrection fern (*Polypodium polypodioides*) also leave traces (scars) on the outer surfaces of trees. The resurrection fern, which reproduces by windblown spores and commonly attaches to branches and trunks of live oaks, derives its common name from looking brown and wilted on a live oak, but it is brought back into full green-leafed glory by a soaking rain. Fungi are abundant and can become quite large in a humid maritime forest. These colonies, however, are not just on the outside of a tree above the ground surface (evident as mushrooms) or exist as thick, horizontal, wedge-like growths on tree trunks. Extensive fungal colonies also surround the roots of most trees and other plants, making visually obvious traces of their former presence (Chapter 4).

Spanish moss (*Tillandsia usneoides*), so beloved to nature photographers on the Georgia coast, imparts no apparent traces on the live oaks, a tree it prefers because of its laterally extensive branches. This plant is actually not a moss at all but a type of bromeliad, a flowering plant closely related to pineapples. It also is native to the Georgia coast, so it is not Spanish either. Bromeliads are sometimes nicknamed "air plants" because they derive all of their nutrients from the surrounding atmosphere, rather than sinking roots into the ground. Spanish moss is not a parasitic plant, unlike mistletoe (*Phoradendron* spp.), which will burden and eventually kill an oak tree if left unchecked, thus delivering a kiss of death.

Interacting with the forest floor are its normally observed denizens, and any saunter through the woods will result in a long list of animal traces caused by these animals. A few of these traces include shallow, horizontal burrows of the eastern mole (*Scalopus aquaticus*); trackways of white-tailed deer (*Odocoileus virginianus*), raccoons (*Procyon lotor*), and opossums (*Didelphis virginiana*); armadillo tracks (*Dasypus novemcinctus*) and their copious excavation marks, whether as shallow food-seeking pits or deep dens; snake trails; broken husks of acorns and other nuts cracked by eastern gray squirrels (*Sciurus carolinensis*), southern flying squirrels (*Glaucomys volans*), eastern chipmunks (*Tamias striatus*), or, more rarely, southern fox squirrels (*Sciurus niger*); piles of pelletal feces left by deer and cord-like scat (often filled with the remains of berries) dropped by raccoons; and rodent tooth marks on white, exposed bones of deer, raccoons, and armadillos, where rodents wore down their rapidly growing teeth while gaining some calcium in their diets (Chapter 7). A closer look at the bones might reveal pockmarks caused by dermestid beetles, which stripped flesh from the bones when the animal was still freshly dead (Chapter 5).

Subsurface tracemakers that modify substrates of maritime forests include the following: moles and their insectivore competitors, shrews; halictid bees, which burrow to make single (but usually clustered) hole nests; burrowing sand wasps and their brooding chambers, denoted by small piles of excavated sand covering burrow tops; cicada nymphs and their burrows; burrowing beetles; earthworms; and ants (Chapter 5). Because maritime forests are often composed of well-drained, sandy soils, these are excellent places for insect nests and earthworms. Accordingly, for moles and shrews that prey on subterranean insects, these substrates also make for stupendous hunting grounds. Insects and earthworms, however, through their sheer numbers are the true rulers of the underground in a maritime forest. Consequently, their numerous, multigenerational traces will outnumber and overwhelm all traces formed by small fossorial vertebrates.

Of course, the root systems of trees and other plants are also prolific subsurface tracemakers in forests and attract burrowing insects. The growth and movement of roots through soils pushes sediments out of the way and locally compacts it; naturally, roots also help to keep sediment in place, preventing erosion. As mentioned before, fungal colonies are directly associated with most roots as a form of symbiosis (Smith and Read 1997). Cicada nymphs are also gratified by the presence of so many plant roots in maritime forests. These insects, which spend 13–17 years underground before developing into adults, are sap suckers, piercing the roots with their mouthparts and living off sap given by tree roots. This is one of the reasons why cicada–nymph husks are often found on tree trunks: the emergent nymphs just climbed up onto the same tree they most recently fed upon. Cicada nymphs are marvelous burrowers, making meandering and back-filled tunnels, and with more than a decade to burrow, each nymph can make a huge number of traces near tree roots (Chapter 5).

The largest subsurface traces made by vertebrates in maritime forests are the burrows of armadillos and gopher tortoises (*Gopherus polyphemus*). Armadillos will often seek out a sandy area under a tree to begin digging a den; these vary considerably in size, but they can be as much as 5 m (16 ft) long and more than 1 m (3.3 ft) deep. Gopher tortoises, which are more typical of cleared areas in and around maritime forests, will dig long—as much as 8 m (26 ft)—and 2 m (6.7 ft) deep, slightly spiraled burrows that are used for protection, nesting, and overall environmental stability (Doonan and Stout 1994; Jensen et al. 2008). Gopher tortoise burrows are extremely important ecologically, as they also can serve as homes for more than 200 species of other animals (Chapter 7).

In short, maritime forests are excellent places for preserving fungal, plant, and animal traces made by a wide variety of taxa using a vast array of behaviors. Although much of the neoichnology conducted on the Georgia barrier islands has emphasized coastal environments, maritime forests are certainly worth a close look for the clues they can provide of these complex ecosystems (Chapters 4, 5, 7, 8).

DUNES AND BACK-DUNE MEADOWS AS THE TRANSITION BETWEEN LAND AND SEA

Dunes and back-dune meadows serve a utilitarian purpose to people who live in coastal communities, acting as slight (yet important) barriers to storm surges or spring tides. These environments, however, are also homes for a rich diversity of tracemakers reflecting ecotones between the terrestrial and marine realms (Figure 3.9). Dunes and back-dune meadows are mostly composed of very fine- to fine-grained quartz sand, with some heavy minerals in the mix, but a few areas also will accumulate mud. On the more landward side of the coastal dunes and in back-dune meadows, low areas, known as swales, tend to collect freshwater and hence encourage plant growth, accumulation of organics, and formation of more humus-rich soils. The plants and animals in a back-dune meadow are more typical of fully terrestrial conditions, although they are occasionally affected by salt spray or storm surges coming from the nearby sea. Of course, dunes have more marine-related tracemakers, but they also include a mix of terrestrial plants and animals.

The most prominent plant tracemakers in back-dune meadows are wax myrtle (*Myrica cerifera*) and youpon holly (*Ilex vomitoria*), which are ac-

Figure 3.9. Back-dune meadows, terrestrial environments between maritime forests and coastal dunes, which contain many traces made by an interesting mixture of terrestrial and marginal-marine tracemakers: Jekyll Island.

companied by many understory plants, including other flowering plants and grasses. Infaunal insects that also live in maritime forests, such as burrowing ants, wasps, bees, ant lions, cicadas, and beetles, are common tracemakers in back-dune meadows and the lee sides of coastal dunes, making nests and burrows used for dwelling and feeding (Chapter 5). Seemingly out of place in these trace assemblages are burrows of large adult ghost crabs, which will dig their obliquely oriented and cylindrical shafts in these same areas. Invertebrate trackways are also quite common, as are the more subtle trails of pulmonate gastropods. Vertebrate tracks from amphibians,

110 LIFE TRACES OF THE GEORGIA COAST

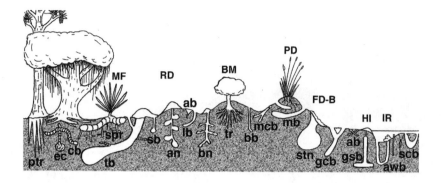

Figure 3.10. Idealized cross section of a Georgia barrier island from a maritime forest to the beach, showing its divisions defined by plant communities, vertebrate and invertebrate burrows, relative elevation, exposure to the tide, and other factors: maritime forest (MF); relict dunes (RD); back-dune meadows (BM); primary dunes (PD); foredune berm (FD-B); high intertidal (HI); and intertidal runnel (IR). Some tracemakers are restricted to these zones, whereas others cross them. Key to subsurface traces (from left to right): ptr = pine tree roots; ec = earthworm chamber; cb = cicada burrow; spr = saw palmetto roots; tb = gopher tortoise burrow; sb = spider burrow; an = ant nest; ab = antlion burrow; lb = lizard burrow; bn = bee nest; tr = tree roots; bb = beetle burrow; mcb = mole cricket burrow; mb = mouse burrow; stn = sea turtle nest; gcb = ghost crab burrow; ab = amphipod burrows; gsb = ghost shrimp burrow; awb = acorn worm burrow; scb = sea cucumber burrow.

reptiles, birds, and mammals tend to shout out the invertebrate traces, especially if an errant alligator, prowling bobcat, or other large predator moves along dune swales. Less noticed in these same areas are snake trails and shallow mole tunnels. Moles constantly dig and consume infaunal insects at alarming rates, and upraised linear zones of disturbance through sandy areas denote their progress.

More seaward, coastal dunes will contain a suite of plant tracemakers well suited to the challenging conditions imposed by the salinity of sea spray, storm surges, spring tides, and other insults to terrestrial plant physiology. The most common species of plant in the dunes, and perhaps the most famous for its role in publicity photos of the Georgia coast, is sea oats (*Uniola paniculata*), a tall grass nearly by itself on the crests of the most shoreward dunes (Bellis and Keough 1995). Roots of *U. paniculata* are complex and dispersed through a broad amount of the subsurface, anchoring these plants firmly in the fine sand and thus becoming guardians of dune integrity (Chapter 4). Near sea oats are creeping vines and other low-lying plants with tightly bound and mostly horizontal stem and vertical root systems that resist dislodging from wind and water. These plants include beach elder (*Iva imbricata*), fiddle-leaf morning glory (*Ipomoea stolonifera*), railroad vine (*Ipomoea pes-caprae*), and pennywort (*Hydrocotyle bonariensis*). The most visually prominent of dune plants is the spiky-leafed Spanish bayonet (*Yucca alifolia*), from which arises a single stalk bearing beautiful white flowers.

Infaunal traces in primary dunes closer to shore are more dominated by ghost crab (*O. quadrata*) burrows and trackways. Insect traces are still here, although fewer and less diverse. These consist of open, near-vertical brooding burrows of wasps; slightly upraised, horizontal tunnels of beetle larvae; occasional ant nest mounds; and adult beetle trackways mixing with those of ghost crabs (Chapter 5). Surfaces just behind the dunes are also places for bird nests, such as the shallowly scraped bowl-like depressions of American oystercatchers (*Haematopus palliatus*) and Wilson's plovers (*Charadrius wilsonia*). Subsurface nests, however, are made by one of the ichnological stars of the Georgia coast, the loggerhead sea turtle (*Caretta caretta*), and two other sea turtle species. These nests may also be marked by trackways of both adults and juveniles, assuming the latter successfully make it through incubation, hatching, exiting nest structures, and escaping predators (Chapter 8).

BEACHES, DOWNWARD AND UPWARD

Georgia coast beaches, which are promoted as among the most beautiful on the eastern coast of the United States (and rightfully so), are also awe-inspiring playgrounds for ichnologists because of their fine-grained quartz sand, intermittent muddy areas, molluscan shells, newly dead animals washed up on shore, and driftwood, all of which preserve traces of organisms living there or elsewhere. The beach proper is often defined in terms of its slope and where the tidal range affects it: a steeper slope translates into a narrower intertidal zone and smaller beach, whereas a low-angle (nearly flat) beach imparts an intertidal zone that may extend far out from shore. Of course, sand supply will also affect how much beach might exist in any given place. Hence beaches down current from tidal inlets are often rather extensive, whereas those that are sand starved will undergo net erosion and expose older, underlying sediments. Just above the average high-tide mark on a beach is a strip of sand deposited by wind, spring tides, and storms, called a berm. Berms are ecotones representing transitions between intertidal beaches and coastal dunes.

The upper parts of sandy beaches, including berms, are places traversed by sea turtles, ghost crabs (*O. quadrata*), and adult limulids (*Limulus polyphemus*), but they are also where one can find a few species of amphipods and insects feeding on detritus deposited at the high-tide mark. This linear accumulation of material, composed mostly of dead *Spartina* flushed out of nearby marshes with ebb tides, is called a rack line. Beachcombers are familiar with the rack line as a rich source of flotsam and jetsam, human made and otherwise, brought in with the waves and tides. Sufficiently large dead animals, however, will also attract vultures, seagulls, and other avian scavengers, which leave their numerous tracks and beak marks near these bodies (Chapters 5, 8). Typical examples of scavenged animals may include sea turtles, sharks, and dolphins, but I have also seen traces of vultures scavenging on limulids (Figure 1.4a). However, macroscopic traces in this part of the beach are dominated by ghost crabs, which leave an astonishing number and variety of signs throughout much of the year: burrows, mounds of sand outside of burrows, radial patterns of flung sand (yes, ghost crabs fling sand; Chapter 6), trackways, resting traces, and much more.

Although the beach above and below sand surfaces truly seems governed by ghost crabs, amphipod tracemakers actually dominate inside the sand. These minute crustaceans feed on the microscopic algae that grow on the surfaces of sand grains and detritus (Croker 1968; Howard and Elders 1970), causing subtly mixed sediment, a process termed cryptobioturbation (literally, "hidden life mixing"). Amphipods probably have the volumetrically greatest biological effect on beach sediments, analogous to the effects of earthworms on terrestrial soils (Chapters 5, 6), although the original planar bedding of the berm and lower beach formed by waves and wind may still be preserved in its original form.

Burrow tops and feces of burrowing callianassid shrimp, such as the Carolina ghost shrimp (*Callichirus major*) and the smaller Georgia ghost shrimp (*Biffarius biformis*), dominate the lower portions of beaches, as defined by the intertidal zone. These traces are evident as numerous small, 5–10 cm (2–4 in) wide, low-profile mounds with narrow apertures at their tops, and are typically adorned by dark brown, cylindrical fecal pellets that look very much like chocolate sprinkles, but are most assuredly less tasty. Actively occupied burrows are easily spotted as erupting volcanoes, in which pulses of water flow out of the aperture, some of it including pellets. These muddy pellets are exceedingly abundant; they are transported as sand-sized grains and collect in the troughs of sandy ripples (Figure 3.11a). Similar to the processes of mussels, oysters, and fiddler crabs in salt marshes, this is how most mud gets deposited in the lower intertidal zone of a beach (Pryor 1975; Bishop and Bishop 1992). These thin streaks of dark mud are then surrounded by quartz and heavy-mineral sand, lending to a type of stratification observed in the geologic record called flaser bedding.

The burrows of shallowly buried bivalves, gastropods, holothurians (sea cucumbers), echinoids (sand dollars), and stelleroids (sea stars) are also common in lower intertidal sands of Georgia beaches, especially during the summer. A few burrows by polychaete worms may be in situ in the same areas, vertically oriented and with their tops exposed on the beach, but more often they are transported out of their original settings and are seen recumbent on beach surfaces. Some burrows, like those by the parchment worm (*Onuphis microcephala*), are lined by mucus and papery in texture. These frequently wash out of the sand and accumulate in ripple troughs like soaked confetti. Other polychaete worm burrows, such as those made by the tube worm (*Diopatra cuprea*), are more robust, strengthened by bivalve shells the worm pasted into the burrow lining. Still, these are also frequently moved from the original places where the worms lived, providing rare examples of infaunal traces that are not in their initial locations (Chapter 1). Hermit crab traces, observable as tightly stitched trackways with an occasional drag mark of their borrowed gastropod shells, will run into and out of shallow water that might pool in the troughs of ripples. In the spring and early summer, ripple troughs also will be decorated by many meandering, looping trail-like trackways and shallow horizontal burrows of juvenile limulids, which are feeding on amphipods and other food items much smaller than themselves (Martin and Rindsberg 2007). Similarly, the mud snail (*Ilyanassa obsoleta*) makes complicated and intersecting grazing trails in ripple troughs, especially those with lots of mud provided by disaggregated callianassid fecal pellets. Soft, stacked piles of cord-like sand on intertidal surfaces are fecal castings that mark the U-shaped burrows of balanoglossids, or acorn worms. These animals are only worm-like in outer body plan and are actually hemichordates, sharing a common ancestor with vertebrates (Chapter 6).

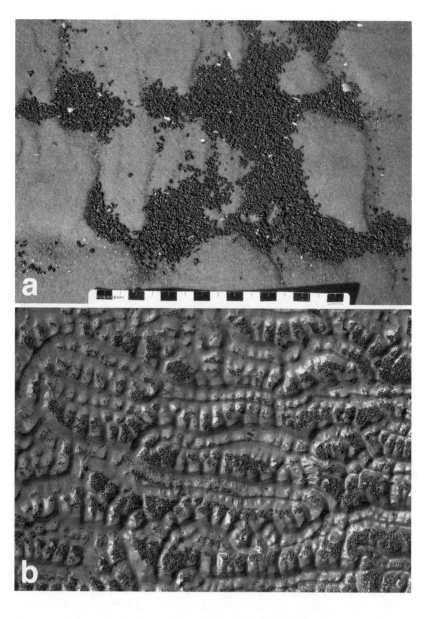

Figure 3.11. Physical sedimentary structures and their effect on the deposition of biogenic material in intertidal sand flats. (a) Accumulation of ghost shrimp (callianassid) fecal pellets, composed of mud packets, in sandy ripple troughs: Sapelo Island. Scale in centimeters; (b) Interference ripples formed by waves that moved sand in two different directions and thus interfered with one another, making small depressions and peaks; the former were sites for deposition of finely ground plant debris: Ossabaw Island.

A keen observer on these Georgia sand flats might notice that many tracemakers are restricted to ripple troughs and other small dips on beach surfaces. These checks on lateral movement point toward the importance of physical sedimentary structures in affecting tracemaker behaviors. Active transport of beach sand and mud by waves and tides causes the physical sedimentary structures, which then imbue a sort of microtopography to the lower beach (Figure 3.11a,b). From the perspective of a small invertebrate tracemaker, these structures are the equivalent of hillocks, mountains, valleys, ridges, and river systems. Examples of physical sedimentary structures common to the intertidal part of a Georgia beach are wave ripples, current ripples, runnels, rill marks, and underwater dunes. Although the word *dune* is often associated with wind deposition, it actually is a specific term for a sand body of a particular size, and thus this name is not contingent on its origin (Allen 1982; Reading 1996). Appropriately, wave ripples are formed by waves, and current ripples by currents, but the latter are specifically made by the more bidirectional flows of ebb- and flood-tide currents. Runnels, which are wide, sometimes deep, meandering ebb-tidal drainage channels that cut through an intertidal zone at low tide, will frequently yield a variety of invertebrate traces on their edges and within channels. In most instances, these tracemakers are burrowing to escape desiccating conditions imposed by the low tide, seeking moist sand that can keep them alive until the next high tide. On the tops and sides of runnels, dendritic drainage patterns—rill marks—run down the sloping surfaces, creating miniature deltas with distributary channels. In some instances, rill marks also emanate from ghost shrimp burrows, in which water actively pumped out of burrows runs down the slopes of the burrow mound. The largest physical sedimentary structures, however, are dunes, which look like gigantic ripples, and actually, they are. In a fractal-like iteration of forms, dunes will have smaller wave and current ripples on their surfaces, although these surfaces are often punctuated by wide, shallow depressions. These structures are feeding pits, made by the southern stingray (*Dasyatis americana*) when the surface was submerged during high tide (Chapter 8). As a result, physical and biogenic sedimentary structures can merge, blend, and influence one another in ways both obvious and subtle.

Borings and fractures are also common traces on sandy beaches, wherever hard substrates occur. For example, driftwood deposited on beaches often contains borings from the marine bivalves, such as the wedge piddock

(*Martesia cuneiformis*). Empty gastropod and bivalve shells will also bear small beveled holes near their apices, and others show signs of chipping along their edges. Both types of traces are made by predatory gastropods (Chapter 6). Other gastropods, including the larger predatory ones, will bear their own scars: ragged, healed lines on their shells, marking an attack by a crab and successful escape of the gastropod (Chapter 6). Shorebirds visit these environments often, particularly at low tide when their favorite prey items are most vulnerable, leaving copious tracks, broken shells, regurgitants, feces, and beak-prod marks in the sand (Chapter 9).

Owing to the amenability of most Georgia beaches to unencumbered movement, clear lines of vision unobstructed by vegetation or buildings, and holding many varied traces, these are understandably popular places for neoichnologists to spend their time, and not just for gathering data, but simply looking. An enormous number of ecological and natural history lessons can be gleaned from several square meters of beach, communicating insights that cannot be taught in books (even this one) or gleaned from reading research papers. Unsurprisingly, these are also popular environments for teaching neoichnology, especially for students who may love beaches for reasons other than their organismal traces.

BELOW THE TIDES AND OFF THE SHORE

Subtidal environments contain a wide variety and number of traces in their layers of mud and sand, and marginal-marine invertebrate burrows dominate these trace assemblages, although vertebrates occasionally interact with these sediments too. Most traces made in subtidal environments, however, cannot be observed without collecting samples from a boat. As a result, for the purposes of our mostly land-bound readers, these environments, their substrates, and traces will be described vicariously by relating the results of research largely done by Robert Frey and James Howard in the 1970s–1980s (Chapter 2). In these ichnological studies, many sediment samples were taken from submerged environments next to the Georgia barrier islands, such as in riverine estuaries, salt marsh estuaries, sounds (open-water areas between the mainland and barrier islands), foreshores, and the shallow, extensive shelf of the Georgia Bight.

In their early studies, Frey, Howard, and other researchers looked at subtidal sediments taken from box cores and applied radiography to these, a technique that revealed the patterns of individual traces, as well as more general textures or fabrics imparted by the repeated mixing of sediments by benthic animals (Howard 1968, 1969; Frey and Howard 1972). What is a box core? As might be expected from their name, these are box-like in shape, but more specifically they are devices used to take manageable samples of sediment from subtidal areas. The core, which is about the size of a microwave oven, is dropped from a boat onto the seafloor with one end open so that it penetrates bottom sediments. Once in the sediment, a handle on the top of the core is turned so that it moves small doors on its bottom, securing the sediment sample. The core is then hauled to the surface, and once on the boat, its top hatch is disassembled (four wing nuts hold it in place) and the sediment sample is extracted. The resultant rectangular chunk of sediment provides two-dimensional views of vertical sediment sequences (cross sections of bedding) and horizontal bedding planes. Researchers then can slice through the sediment to get a sense of the three-dimensional architecture of physical and biogenic sedimentary structures. Occasionally, the sediment in a box core will also include the tracemakers within their burrows, an added scientific bonus for neoichnologists attempting to link specific traces with species of benthic organisms. For example, during the course of one study, 73 species of live animals were recovered from box cores (Howard and Frey 1975). In instances where researchers did not want to distort or otherwise disturb sediment layers or traces by cutting through it, they used the noninvasive technique of radiography. Core samples were placed into an x-ray machine, and the transmission of x-rays provided black-and-white images of bedding and bioturbation in their entirety.

The substrates of subtidal environments are varied, ranging from muddy sand to sand lacking mud (so-called clean sand), mud, and coarse-grained sediments composed of molluscan shell debris, pebbles, mud clasts, and other pieces of solid material eroded from underlying bedrock. These sediments in turn compose a profusion of physical sedimentary structures, such as large-scale cross-bedding, ripples, and varying proportions of sandy and muddy layers, forming continuums from interbedded mud and sand, to flaser mud beds enclosed in sand, to lenticular (lens-like) sand beds enclosed in mud.

These textural contrasts in sediment help considerably in discerning organismal traces. For instance, as a tracemaker burrows from one type of sediment to another, its burrow may become filled with sediment differing from that around the burrow. The net effect of continuous burrowing, though, is to take formerly discrete, differently colored and textured layers of sediment and stir them so that these layers became blurred or mixed completely. In situations where the sediment is blended, which organisms did it, and how many times did this happen? These questions are analogous to asking how many times a cup of coffee with cream (or soy milk, if you prefer) was stirred, and whether the stirring was done by a spoon, knife, fork, pencil, finger, or—well, use your imagination. In many instances, ichnologists have no way of telling exactly how these sediments were homogenized, but we can say this process does happen, and its products are also recognizable in the geologic record whenever we find similar disruption of original sedimentary bedding (Chapter 9).

As can be envisaged from their abundance in the intertidal zone, ghost shrimp also burrow extensively in the saturated sands of subtidal environments. Above this callianassid underworld are the surface traces—shallow burrows, trackways, trails—of horseshoe crabs, hermit crabs, true crabs, echinoderms (sea stars and sand dollars), southern stingrays, and other epifaunal animals that may interact with the sediment–water interface (Chapters 6, 9). These traces, unfortunately, are almost never observed directly. The murky, mud-laden waters of the estuaries and open ocean near the Georgia barrier islands are not ideal for directly looking at surface traces; in other places, such as the Bahamas, this could be accomplished by snorkeling or scuba diving. Hence, the bulk of studies in these environments have concentrated on the infauna and their traces, much of which was derived from box cores.

In such research, one of the more interesting outcomes was how much of the infauna was ecologically partitioned on the basis of substrate. Similar to how fiddler crabs in salt marshes are split into mud-adapted and sand-adapted species, numerous burrowing invertebrates of the subtidal environments and their traces reliably show up either in muddy or sandy sediments. For example, upogebiid shrimp, such as the mud shrimp (*Upogebia affinis*), are burrowing crustacean compatriots of callianassid shrimp, but their common name alludes to their restriction to burrowing in muddy substrates (Dörjes 1977; Prezant et al. 2002). Mantis shrimp (*Squilla empusa*), despite their "shrimp" appellation, are not directly related to upogebiid and callianassid shrimp, but they are also impressive burrowers in muddy subtidal environments, such as the lower submerged parts of tidal creeks in salt marsh estuaries. In contrast, callianassid shrimp of the Georgia coast are exclusively sand dwellers (Bishop and Bishop 1992; Prezant et al. 2002).

Polychaete worms are among the more common tracemakers in subtidal environments, such as the previously mentioned onuphid worms, but also include capitellid worms (*Notomastus, Heteromastus*). Because some of these tracemakers are passive carnivores, lying in wait for their victims to pass by their burrows, their traces are typically simple open burrows, vertically oriented, and thin. If preserved in the fossil record, these traces would be easily recognized as probable former polychaete burrows, or at least would constitute a starting hypothesis (Chapter 10). Potentially more confusing are the thin, branching feeding burrows of polychaetes that resemble plant-root traces. Other infaunal invertebrate tracemakers in subtidal environments include anemones, bivalves, gastropods, amphipods, as well as echinoids (heart urchins), sea stars, and ophiuroids (brittle stars). Ophiuroids are especially interesting ichnologically, in that their morphology, consisting of five thin arms and a small, disk-like central body, does not exactly scream "burrower!" to a budding neoichnologist. Nonetheless, they make distinctive burrows in subtidal environments that are easily spotted on radiographs of box cores (Chapter 6).

Bioerosion also takes place in the muddiest and sandiest of subtidal environments, as long as hard substrates—shells, clasts of eroded bedrock, driftwood, and bones—are there to be bored. Bioerosion traces often show up as the minute, complicated branching borings of the sponge *Cliona celata* on dead oyster shells, as well as the U-shaped borings of the polychaete worm *Polydora* sp. on chunks of rock (Chapter 6). Bivalve borings are also documented from subtidal environments, made in rocks by pholad clams, such as the false angelwing (*Petricola pholadiformis*) and scorched mussel (*Brachidontes exustus*), but some wood-boring species also leave their marks. As mentioned earlier, one species of bivalve, the wedge piddock, bores into driftwood, or even bones that might sink to subtidal bottoms (Frey et al. 1975; Mann and Gallagher 1984). Of course, any dead bivalve or gastropod shell should be examined for evidence of predation by hunting gastropods, showing up as holes made in their shells. The common moon snail (*Neverita duplicata*), the thick-lipped oyster drill (*Eupleura caudata*), and the eastern oyster drill (*Urosalpinx cinerea*) make such traces (Chapter 6).

Vertebrate traces are comparatively rare in subtidal sediments, but a few can be recognized against all of the background noise caused by invertebrate burrowing. Among the more interesting vertebrate traces are those made by snake eels (*Ophichthus* spp.), which are simple, open, and semihelical burrows (Howard and Frey 1975). Southern stingray feeding pits also extend well into the subtidal realm, wherever delicious bivalves and other infauna might be hiding in the sediment (Chapter 8). Other vertebrate traces in subtidal environments also include tooth marks on bones. These traces are most likely made by sharks, and normally they are formed by their scavenging on floating corpses. However, such traces will not be preserved unless the bones bearing these marks are buried quickly in subtidal sediments.

Ichnologists and sedimentologists who have studied the subtidal ichnology around the Georgia coast, however, have not just focused on individual species and their respective traces. Other lines of interest include the cumulative effect of organisms on substrates (how much is sediment mixed?); the effects substrates have on organismal behavior (who likes to burrow in mud, who likes to burrow in sand?); and the spatial distribution of amounts of bioturbation within subtidal environments (is a tidal-channel bottom more or less burrowed than its bank?). In other words, a holistic approach is taken: assemblages of traces and the overall effects of organismal behavior in host substrates take precedence over just listing individual traces or tracemakers. With this perspective in mind, several observations and generalities have emerged from box-core studies of subtidal sediments around the Georgia coast (Howard and Frey 1973 1975, 1985; Dörjes and Howard 1975; Frey and Howard 1986; Gingras et al. 2008a):

1) Wherever freshwater rivers empty into estuaries (salt marsh or riverine), the diversity of traces and amount of bioturbation increase further out toward the open sea.
2) Shallower margins of estuaries, such as tidal-channel banks, contain a greater diversity of traces than the deeper parts of estuaries, such as in the lower parts of channels.
3) Muddy sediments have discrete, identifiable burrows, but muddy sand or sandy mud may simply be mixed up such that individual burrows are difficult to discern.
4) Traces are rare in sediments with grain sizes larger than mud and sand, such as pebbles and gravel.
5) Hard substrates, such as bedrock, shells, bones, and wood, will bear numerous borings by a well-defined suite of bioeroders.

The traces also indicate organismal responses to sedimentation and erosion, which along with physical sedimentary structures reveal the shifting energies of waves and currents on estuarine bottoms. Additionally, assemblages of traces often show a dominance of either horizontally or vertically oriented burrows, in which tracemaker behavior is influenced by grain size (mud or sand) and energy (high- or low-velocity currents) in their given environments (Howard and Frey 1973, 1975). Last, because of the inherent mixed signals given off by an estuarine environment (freshwater, salt water; high tide, low tide; high energy, low energy), trace assemblages of subtidal environments will reflect ecological blends, in which organisms find what works best for them in relatively small areas within substrates.

Exceptions to these ichnological generalities have been found, of course —science thrives on finding nits and oddities—but nonetheless these and other aspects of Georgia neoichnology have held up well as a model for trace assemblages in temperate to subtropical estuarine environments for the past 40+ years. Important details added since the pioneering work of Frey and Howard in the 1970s and 1980s have also helped to clarify how these modern traces and their substrates are comparable to trace fossil assemblages and ancient sedimentary rocks made in similar environments (Chapter 10).

OLD LANDSCAPES REBORN: RELICT MARSHES

One of the more interesting ichnological discoveries associated with Georgia coast research in the 1960s and 1970s was that of relict salt marshes, formerly buried but newly exhumed, and how substrates of these ancient facies provide present-day habitats for tracemakers (Figure 3.12). By geological standards, relict marshes are rather young: the few dates calculated for some suggest most are only 500–1,000 years old (Frey and Howard 1969). Thus these were thriving ecosystems during times of Native American occupation of the islands (Chapter 2). So far, well-exposed relict marshes have been found on Cumberland, St. Catherines, Sapelo, and Ossabaw islands. Originally buried underneath thick layers of dune and beach sand, their unveiling now is a result of either of two factors: rising sea level, which has accelerated coastal erosion and worn away their sandy covers; or longshore drift, which erodes coastlines regardless of whether sea level is going up or staying steady (Chapter 2). Once exposed at the shore and inundated twice daily by tides, relict marshes, which are much firmer than their modern counterparts, become sites for colonization by many modern marine invertebrates, some of which specialize in inhabiting more solid substrates (Morris and Rollins 1977; Frey and Basan 1981; Pemberton and Frey 1985). Among these secondary colonizers are bivalves, some of which bore into the old marsh sediments (*Petricola pholadiformis, Barnea truncata*) and others that attach to its surface (*Brachidontes recurvus*). Perhaps surprisingly, fiddler crabs (*Uca* spp.) are known to reoccupy the substrates of their ancestors, although they normally need a layer of softer mud or sand on top of the relict marsh to entice them to move back into such an aged neighborhood.

The relatively recent history of these relict marshes means that body parts and some traces of original inhabitants in the marsh are still there, making for ghostly shadows of these formerly thriving environments. Root systems and cropped stems of smooth cordgrass are there, with nearly all their original color gone, and yet are otherwise recognizable. Clusters of ribbed mussels that must have aggregated under stands of cordgrass in the low marsh are also there. You can also see scattered, whitened beads of the marsh periwinkles, their shells eroding out of the stiff mud. Eroded, circular pockmarks indicate the former burrows of mud fiddlers and other mud-loving crabs. Thick layers of imbricated oysters indicate former positions of tidal creek banks, where they prefer to live. Tidal creeks that cut through the original marsh are not only marked by clumps of oyster shells, but also by what is not there. Beach sand fills the low areas of these creeks, making for wide, linear patches of white sand that cut across the dark-brown former marsh. Overall, the resemblance of these relict marshes to modern ones is eerie for some, similar to walking into a house that had been abandoned for hundreds of years.

However ethereal these biological traits might seem, one big difference can be felt in the substrate that brings you back to the present: compaction and loss of water from the original mud formed a solid, easily walked-upon foundation. In contrast, a similar amble in a modern salt marsh would result in full-body immersion in the mud, and shoes surrendered to the geologic record. Appropriately enough then, these relict marshes are examples of what ichnologists and paleontologists call firmgrounds. Firmgrounds are substrates intermediate in consistency between softgrounds (unconsolidated sediment), which yield considerably to any applied pressure—to the point of enveloping whatever is causing the pressure—and

Figure 3.12. Relict marsh on Cabretta Beach of Sapelo Island, an excellent example of a firmground environment with a distinctive mix of traces made by previous and more recent inhabitants. (a) Overall view of eroding outcrop of relict marsh, providing 1-m (3.3-ft) tall vertical section of ancient marsh and containing original remnants of smooth cordgrass (*Spartina alterniflora*). (b) View of top surface, showing nubs of old smooth cordgrass and clumps of ribbed mussels (*Geukensia demissa*); surface also contains holes that are old burrows from the original marsh, newly modified old burrows, or new borings. Sandal (*left*) for scale.

hardgrounds (wood, bone, shells, or rock), which for all practical purposes have no compressibility (Bromley 1996; MacEachern et al. 2007a, 2007b). Firmgrounds will be examined later in the context of how organisms use these substrates, their relation to ecological niches, and what their trace assemblages tell us about changing environments through time (Chapter 10).

THE ICHNOLOGICALLY FORMED LANDSCAPE (OR SEASCAPE)

An attentive reader might have already discerned that the habitats and substrates of the Georgia barrier islands are not blank slates, waiting for organisms to leave their marks, nor are their plants and animals passive receivers of environmental factors, waiting for these to shape and mold their behaviors. Instead, these habitats and substrates owe their very existence to the tracemakers. Through their behaviors and effects on substrates, tracemakers render entire ecosystems as composite traces, ichnologically created landscapes and seascapes upon which other plants and animals depend.

Only a few species on the Georgia coast are responsible for altering habitats through their tracemaking activity, but their strength is in behavior, numbers, and (sometimes) size. For example, experiments done on one species of burrowing callianassid shrimp (*Glypterus acanthochirus*) in the Bahamas revealed that if one 10-cm-long (4 in) shrimp lives to be eight years old, it will have processed a cubic meter of sediment (Curran and Martin 2003). Scaled up, this is analogous to a human digging more than 1,500 tons of soil during the same time, and without the aid of a shovel or backhoe. Although similar experiments have not yet been conducted on Georgia callianassid shrimp, their burrowing activity and behaviors are certainly comparable. Thus when these abilities are combined with sheer abundance, ghost shrimp are assuredly churning through an enormous volume of sand, while also depositing a prodigious number of sand-sized mud fecal pellets. For example, on St. Catherines Island, more than 400 individuals of *Biffarius biformis* per square meter (10.8 ft^2) were counted in the lower intertidal zone (Bishop and Bishop 1992; Bishop and Brannen 1993). Slightly higher on the beach, ghost crabs are similarly modifying sandy substrates in a major way, with their 1–1.5 m (3.3–5 ft) long burrows, pyramidal mounds of sand balls, and millions of tracks across beaches and dunes. Interspersed throughout part of the ghost crab range are sea oats, which hold dunes in place with their root systems. These dune plants in turn are aided by sea turtle nests, which increase the nitrogen content of dune soils and help vegetation to grow there, a sort of ichnological and ecological "you scratch my back, I'll scratch yours" synergism (Hannan et al. 2007).

Meanwhile, over in the salt marshes, the ichnological dynamic duo of ribbed mussels and mud fiddler crabs are depositing large volumes of mud and overturning the top 10–15 cm (4–6 in) of sediment through burrowing, respectively. One study showed that only three species of molluscans (*Geukensia demissa, Polymesoda caroliniana, Littoraria irrorata*) and two decapods (*Uca pugnax, U. pugilator*) were responsible for processing nearly 50 percent of the organic productivity in a salt marsh (Kraeuter 1976). In those same marshes, *Spartina alterniflora* and other halophytes perform a function similar to that of sea oats, binding together sediment so that it stays mostly in one place.

More landward, alligators (*Alligator mississippiensis*) create waterholes, which are expanded with multiple seasons and generations of alligators, then turned into freshwater marshes. In wiregrass communities between patches of maritime forest, gopher tortoises (*Gopherus polyphemus*), with their deep, extensive burrows, form subterranean microhabitats that become homes for hundreds of other animal species. In the maritime forest, earthworms, ants, and cicada nymphs all transform the soils into completely bioturbated blends of sand, mud, and organics. Trees in those forests, whether living, dying, or dead, are filled with wood-boring tracemakers, which in turn provide food for wood-boring predators of those insects—woodpeckers—which leave their beak marks and hole nests throughout the forest.

In short, one cannot go to the Georgia coast and say, "I don't see any traces," unless that person has his or her eyes closed, or is otherwise sight deprived. Traces are, in fact, everywhere to be seen. This is where one might reach for the idiom, "You can't see the forest for the trees," but the perceptual perspective is actually reversed: one should instead pay attention to the parts to better understand the whole. The Georgia barrier islands and their surrounding habitats, with their sand, mud, wood, shells, bones, and other substrates, can be viewed as massive complex and composite traces, serving as the preserving media for evidence of nearly all life activities that take place on or near the islands. As a result, much of the remainder of this book is about those parts—life traces—that compose the ichnologic landscapes and seascapes of the Georgia barrier islands, and how these traces help us to interpret the geologic past.

4

MARGINAL-MARINE AND TERRESTRIAL PLANTS

THE PRESERVATION OF ROOTEDNESS:
A TALE OF TWO ISLANDS

It was July 2001, and two full days of fieldwork on Sapelo Island had nearly exhausted our small band of five ichnologists and three eager undergraduate assistants, but in a good way. During those days, we had perused two extensive beaches—Nannygoat and Cabretta, both among my favorite beaches in the world—and looked at these and their dunes for invertebrate and vertebrate traces. So far, we had been rewarded with many ichnological observations and insights. Moreover, the second day was highlighted by a visit to a relict marsh at the north end of Cabretta Beach. Relict marshes, mentioned earlier (Chapter 3; Figure 3.12), are neither fossil nor modern marshes, inhabiting a nether world in between. Because of their relatively recent entry into the geologic record, they present rare opportunities for ichnologists to witness an ecosystem shift from one fully breathing and recycling its nutrients to an inert, stony vestige of its former self. A few of my ichnologist colleagues had read about this one on Sapelo but had never seen it, and they eagerly anticipated the enhanced learning that inevitably stems from direct experience.

The modern marshes of the Georgia barrier islands are nearly unavoidable for any visitor there, particularly in the back-barrier areas near the mainland, where they form expansive greenish-golden waves of tall grass composed largely of smooth cordgrass (*Spartina alterniflora*). This monotypical assemblage is an example of natural selection, in which this one species has adapted so well to this ecosystem that it has either crowded out nearly all other plants, or is the only one adapted for such harsh conditions. Smooth cordgrass can become rather tall, nearly 2 m (6.6 ft), and might as well be a sequoia forest from the perspective of fiddler crabs (*Uca* spp.) living in burrows between their stalks. Its roots extend straight downward as much as a meter (3.3 ft) and branch into fine rootlets that push through the mud and gather nutrients from its organic richness. Individual plants will also send shoots (colms) into adjacent areas to form new stalks, thus propagating clonally and neatly avoiding any messy problems associated with sexual reproduction. Their stalks host a small gastropod, the marsh periwinkle (*Littoraria irrorata*), which continually grazes on fungal films that grow on individual leaves (Fierstien and Rollins 1987; Figure 3.2). This grazing inevitably damages the plants, though, just as rubbing any leaf inflicts collateral damage, thus encouraging further fungal growth. As a result, the overall upward growth of the cordgrass is markedly inhibited, and periwinkle-removal experiments have revealed this mode of snail-induced height suppression (Silliman and Bertness 2002; Silliman et al. 2004). Periwinkle cohorts in herbivory include grasshoppers, which reproduce and go through their life cycles in terrestrial forests, but can also easily fly into nearby salt marshes for an endless feast of grass that is never mowed. Gastropod- and grasshopper-provided pieces of cordgrass raining from above form an important component of the nutrient cycling in the marsh mud. Fiddler crabs consume these bits along with algae growing on the marsh surface, then produce organics-rich pseudofeces and real feces (Chapter 3). Ribbed mussels (*Geukensia demissa*) also cling to marsh surfaces with secreted threads (byssae) and tend to clump near and around cloned plants. As mentioned before, these prolific producers of pseudofeces are responsible for most of the mud deposition in a salt marsh (Chapter 3).

Imagining that this idealized marsh existed only 500 years ago, and what would be there now, is readily testable with the actual item. The Sapelo relict marsh includes nearly every macroscopic part of the original ecosystem, except for the tall, thin green leaves that projected above the surface. Most of the below-surface and near-surface portions of the marsh were preserved: fiddler crab burrows; colms and roots of smooth cordgrass, its closely cropped stalks seen only as hollowed, pencil-like stubs on the surface; partially bleached shells of mussels, some still attached by their threads, even in death; and periwinkle shells. These shells occasionally take on second lives and can be seen crawling across the ancient marsh surface courtesy of modern hermit crabs, unmindful that their new homes are hundreds of years old. The same mud that would have engulfed entire lower halves of human travelers several hundred years in the past was now firm enough to hold all of us, not even mussing our shoes. Larger hermit crabs scuttled across the surface with us, carrying whelk shells that more properly belonged to the sea only a few meters away. The beach, which once covered the relict marsh, had eroded as the shoreline shifted during the past few hundred years, thus unveiling the muddy facies, incongruous with so much clean, whitish beach and dune sands nearby.

All of these traits were fascinating in themselves, but the cross section of the relict marsh on the oceanward side, eroded and forming a vertical face, comprised its most beguiling feature. Here was a dense mass of interpenetrating roots and colms, telling us about life underground in this marsh from about the time of Spanish colonization of what is now called Georgia (Chapter 2). Although I had seen and studied this cross section before, the company of ichnological colleagues and students, along with our ensuing discussion, greatly enriched its meaning. We mostly discoursed on how the vertical, oblique, and horizontal patterns wrought by the yellowed and mostly decayed roots and colms might translate into the geologic record as root traces. Once all of the organic matter of the original plant tracemakers has disappeared, the evidence of their growth and movement will remain as disruptions of layers, locally visible compaction of surrounding sediment, and differences in color imparted by root respiration (during life) and root degradation (after death). Interestingly, root trace fossils in much older sediments or sedimentary rocks are often noticeable as variegated haloes around the former sites of roots. In some instances, I have even seen the carbonized remnants of roots in the center of these chemically altered zones: Carboniferous rocks of West Virginia, from more than 300 million years ago; Jurassic rocks of the Yorkshire coast in the U.K., from about 180 million years ago; and Cretaceous rocks of Alaska, from 70 million years ago. Often in paleontology we are told that these are rare instances of tracemakers being identified by direct association with their former traces, but the immobility of plants with roots makes this comingling all the more likely.

With this salt marsh model of root traces carried in our minds, a new search image and way of thinking had surreptitiously seeped into our worldview of traces. Indeed, the next day of fieldwork then demonstrated how such musings and remembrances could collide with previously misunderstood phenomena. On the third full day of the field trip we went to Raccoon Bluff, an outcrop on the east side of the island with the only major exposure of Pleistocene rock. Now we were going back nearly 40,000 years into the Pleistocene Epoch, or about two orders of magnitude older than the relict marsh. Raccoon Bluff is a dark brown to brownish-yellow outcrop forming a nearly 400 m (1,300 ft) long cliff, although embarrassingly inadequate by western U.S. standards, standing only about 3 m (10 ft) tall. The poorly consolidated rocks in the outcrop had been eroding rapidly in recent years, retreating further into the upland maritime forest just above it. Its position on the edge of a major tidal channel (Blackbeard Creek) meant that it not only endured daily tidal fluctuations, but frequent waves caused by wakes of speeding boats that passed through the channel.

The main reason why we were there was because I wanted a mystery to be solved. In a previous visit to the outcrop, I had identified large, enigmatic structures there that did not fit the search images of any trace fossils I had ever recognized (Figure 4.1a). These structures were filled with white, fine-grained sand reminiscent of the same white sand of coastal dunes we had seen in the preceding few days, giving them an appearance of being illuminated from within. Two-dimensional outlines expressed by what used to be three-dimensional structures were oval, circular, or conical; tips of these cones pointed down. Cones also bifurcated in places, or otherwise had their own small conical branches that tapered downward. These structures were big, with the cone-shaped ones nearly 2 m (6.7 ft) in vertical extent, and the circular or oval ones as much as a meter (3.3 ft) wide, although most were about 20–70 cm (8–28 in). All had dark brown or nearly black sandy zones surrounding the white sand, highlighting their color differences. Nonetheless, boundaries between the two were blurred as a result of much smaller cylindrical structures filled with half-moon-

MARGINAL-MARINE AND TERRESTRIAL PLANTS   131

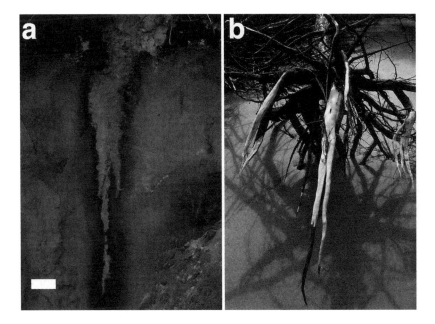

Figure 4.1. The once-mysterious structures of Raccoon Bluff on Sapelo Island, formed during the Pleistocene Epoch about 40,000 years ago, and fortuitously revealed as tree-root trace fossils. (a) Outcrop view of white, sand-filled, tree-root trace fossil. This structure also has fossil invertebrate burrows in its sand fill, and the dark geochemical halo is attributed to fungal cohorts, hence it is more properly a composite trace fossil. Scale = 10 cm. (b) Root architecture of loblolly pine (*Pinus taeda*), exhumed in eroded shoreline less than 2 km (1.2 mi) from Raccoon Bluff on Sapelo Island. Note the resemblance of its main taproot to the structure in (a).

shaped lines, also known as meniscae. These structures I had recognized as invertebrate burrows, in which the tracemakers had actively and progressively filled their burrows as they moved through the sediment. Otherwise I had no clue about the origin of the larger structures, or how the small burrows related to them. One half-witted hypothesis, based solely on thinking big, was that they were fossil burrows and dens of gopher tortoises (*Gopherus polyphemus*). But even this explanation did not match the criteria of modern gopher tortoises, as their burrows are much smaller and differently shaped (Chapter 7). So I was left flummoxed, yet hopeful that my ichnological colleagues, with their considerable cumulative experience, could shed some light on their origin of these odd features.

The easiest way to get to the outcrop, requiring the least amount of bushwhacking through the maritime forest, was to drive our vehicle to an open field near its northern end, then hike down and around that end. Low tide afforded us space for walking on the muddy areas just underneath the outcrop. Our shoes, which had stayed so clean on the relict marsh the previous day, soon collected a weighty amount of muck that we accepted with cheery resignation. Our group, other than the three student assistants (Nick, Molly, and Chris), consisted of my coworker and friend from Emory University, Steve Henderson (Chapter 3), and ichnologists Richard Bromley, Alfred Uchman, and Murray Gregory from Denmark, Poland, and New Zealand, respectively. (Yes, I know, it sounds like the start of a joke, as in "A Dane, a Pole, and a Kiwi walked up to an outcrop one day...").

I had primed (and baited) my colleagues about the outcrop, saying that it contained something I could not figure out, but I had no idea of the serendipitous science that would happen next. As soon as we turned a corner and were in front of the outcrop, Murray Gregory (the Kiwi) stopped in his tracks. I looked askance at him, wondering what he was thinking as he stared at the outcrop and his mouth moved soundlessly. Finally the words came out of him: "We've got exactly the same thing in New Zealand!" he exclaimed. Once he became a little less excited, the story was pieced together for our benefit.

Just a few months before, in Holocene rocks of the North Island of New Zealand, he had seen structures remarkably similar to the ones in front of us. The rocks resembled the Raccoon Bluff ones before us in that they were poorly consolidated and not too old (geologically speaking), estimated at less than 10,000 years old. After much detailed analysis of the structures and their host sediments, Murray and his colleagues had concluded that these were tree-root trace fossils. How and why? The main piece of evidence supporting the hypothesis was how the overall shapes, sizes, and arrangement of the structures matched those of mature tree roots in that part of New Zealand. Moreover, these structures were in rocks that likely formed in coastal dunes and near upland forests; the outcrops were only a few kilometers from the present-day east coast of New Zealand. The trees, once they died, had rotted in place and left hollow spaces in the soil, rendering local depressions that accumulated windblown sand. Once buried, these sand-filled structures became places where moisture accumulated locally in the soil, which, along with abundant organic matter, attracted burrowing insects, especially cicada nymphs and beetle larvae (Chapter 5). These insects burrowed along the interfaces of the white sand fill and the dark boundaries caused by fungi, and effectively broke down the previously sharp demarcation between the two so that these sediments melded.

Part of this story we put together at the outcrop as we listened and commented on Murray's proposed ichnological solution to the mystery structures, and the rest fell in place rather easily later. For example, he and I figured out that the dark haloes around the structures were probably the result of fungal colonies that once lived around the roots, called mycorrhizae (more about those later). The conical, downwardly pointing structures also matched the shapes and sizes of roots for pine trees native to the Georgia barrier islands, such as *Pinus taeda* (Figure 4.1b). The white sand that filled the root structures was originally windblown, which we suspected happened with our Raccoon Bluff structures on the basis of its resemblance to Sapelo beach and dune sands. While at the outcrop, we further examined the structures and their other associated trace fossils and enjoyed some mild debate about the details of the hypothesis, although everyone seemed generally in agreement with its main tenets. This sort of accord is rather unusual in science, but sometimes it happens.

What we had on opposing sides of the world, and in rocks close to the same age, were composite trace fossils. These are trace fossils made of closely associated but different parts made by more than one species of tracemaker. In these instances, trees, fungi, and insects had all contributed to the formation of some beautifully complex structures, no longer mysteries and compelling in their tales of changing life, behavior, and taphonomy. End of the story, right? No, not really. Like any scientific hypothesis, it required more testing from us, and eventually peer review from others of our ilk. For that reason, motivated by both science and the desire to take a vacation in a very nice place, I traveled to New Zealand with my wife Ruth. While there, we met with Murray and he took us to the outcrops containing the structures near the north end of the North Island. Our tandem examination in the field, with Murray and I having visited these strikingly comparable structures on the two islands, further affirmed the hypothesis, while also filling in a few details we had missed previously. We then put together our descriptions of the structures from each place with the assistance of one of Murray's colleagues at the University of Auckland,

Kathleen Campbell, who had jointly described the New Zealand structures with Murray. We summarized the information and wrote a paper for submission to a scientific journal.

But before the manuscript was ready to submit for peer review, we performed one last test of the hypothesis, although at the temporary expense of Kathleen. One of the figures in the paper was of side-by-side photographs of the large, vertically oriented, conical tree-root structures: the one from Raccoon Bluff of Sapelo Island was on the left, and the other from New Zealand was on the right. In this figure, though, Murray and I had done something sneaky. The Sapelo photograph contained Murray's photo scale, with "New Zealand Geological Survey" clearly marked on it, whereas the New Zealand photograph had my distinctive yellow-and-black photo scale in it. Upon seeing the photographs, Kathleen immediately informed us that we had mixed up the captions to our figure: surely the one on the left was from New Zealand and the one on the right was from Georgia. When I gently informed her via e-mail that no, the captions were correct and our scales had fooled her, another point was made. The two structures were so similar that the only way to distinguish them was to place a recognizable humanmade object in front of them. With all three authors now convinced, the paper was submitted, successfully passed peer review, was published, and now serves as an example of how tree-root traces can form templates for complex subsurface ecological relationships (Gregory et al. 2004). I also presented our preliminary results at two paleontological meetings in 2002 and 2003, and Murray, Kathleen, and I co-led a field trip to the New Zealand outcrop in 2005. All of this public exposure resulted in more scrutiny and further confirmation of our hypothesis by our peers. So far, so good.

What were the ichnological lessons we learned from the Cabretta relict marsh and Raccoon Bluff? It was that plants and their traces matter. The overwhelming majority of ichnological literature and study is devoted to invertebrate and vertebrate traces, and most ichnologists readily divide themselves into invertebrate and vertebrate ichnologists. In contrast, few ichnologists will say they study plant traces, and in fact they may be cowed into inaction whenever they encounter what they think are root trace fossils in the geologic record. Nonetheless, this not should be the case because of a lack of material, as plants have made trace fossils by sending their roots into sediments for more than 400 million years (Driese et al. 1997; Brundrett 2002; Hillier et al. 2008). In terms of neoichnology, and as might have been noticed previously, nearly every habitat of the Georgia coast, with the notable exception of intertidal and subtidal environments, contains plants and is sometimes most easily defined by its plant communities (Chapter 3). Furthermore, nearly all of these plants impart visible traces of their behaviors on their host substrates.

However fixed plants may be, they nonetheless live and behave. Thus, a look at their neoichnology provides important clues about their former habitats and ecological connections with their better-studied invertebrate and vertebrate tracemaker companions. Accordingly, the confidence to identify and describe plant trace fossils in the geologic record can be boosted by a thorough treatment of their modern equivalents. Hence this chapter is about recognizing plant traces and how these can be used to interpret plant trace fossils.

BUT PLANTS DON'T REALLY BEHAVE AND MAKE TRACES, DO THEY?

The title of this section is a question I have provoked whenever I discuss plant traces with others, and it comes not just from nonscientists but also from zoologists and other ichnologists. Botanists, on the other hand, just grin, nod their heads vigorously, and thank me for recognizing a truism they regard as settled yet often have to defend. Granted, plants do not behave much like animals; all plants are producers, or photoautotrophs, using photosynthesis for their matter–energy conversions and taking up water and nutrients via roots or other anatomical and physiological attributes. Yet some plants are also passive carnivores, augmenting their photosynthetic production of food with a diet of insects. This adaptation shatters every stereotype of linear, producer–consumer relationships in terrestrial ecosystems glibly labeled as food chains, and is most commonly expressed by plants that live in nitrogen- or phosphorus-poor soils. Moreover, carnivorous plants are sometimes coupled with low-oxygen aquatic environments such as freshwater swamps or bogs (Schnell 2002). For example, the Okefenokee Swamp in southeast Georgia, which has such conditions (Chapter 2), is filled with pitcher plants and sundews. These plants either compete with some of their predaceous insect cohorts for their meals, or eliminate the competition by trapping and digesting them too. Other than the supplementary acquisition of insect protein by these plants, though, plants produce their own food. They also do not normally pick up and move from one place to another while living. Plants that are firmly rooted in a substrate, however, still move in some ways and will take action or otherwise react to stimuli as quickly (or more quickly) than some animals living in the same habitats.

Plants exhibit a variety of behaviors but can be interpreted broadly as dwelling (staying in one place), "feeding" (but not really, as will be explained soon), reproduction, and movement, the last of which can be very much like locomotion. For the latter, think of creeping vines, especially the notorious invasive plant of the southeastern United States, *Pueraria lobata*, otherwise known as kudzu. Plant movement is demonstrated best by tropism, which is intrinsically related to dwelling and feeding (Capon 2005). Of course, except for the aforementioned carnivory, most plants do not feed like animals; nonetheless, they need to acquire nutrients to better facilitate photosynthesis and production of food. Plant movement is thus affected by light, soil, the nearby presence of other plants, and ever-present gravity, among other factors. Once in a while, we can actually observe plant movement without the aid of time-lapse photography, such as in the "sensitive plant," *Mimosa pudica*, which curls its leaves within seconds as a response to touch; this movement is called thigmotropism (Fasano et al. 2002; Capon 2005). Plant movement is perhaps most dramatically demonstrated by stimulus–response reactions of some carnivorous plants, which trap insects by either using changes in water pressure, seen in species of *Dionaea* (Venus flytraps), or rapid cell growth, which happens in *Drosera* (sundews).

Other forms of plant behavior are covert and remarkably sophisticated, including protective measures taken in response to attacks by animals—usually insects—that find plants desirable to eat. Among the most common reactions by plants are wound responses, which change tissues or other body parts to prevent water loss, or otherwise toughen their tissues to prevent further attack (de Bruxelles and Roberts 2001; Cornilessen and Fernandes 2001). For example, galls, which are swellings on plant stems or leaves, are visible examples of wound responses.

If this sort of passive, after-the-fact protection against herbivory does not impress, perhaps the concept of plants warning others of their own or other species will elicit some grudging murmurs of admiration. Yes, plants can communicate with one another, but using chemotransmitters (analogous to pheromones in animals) instead of sound or body language. Plants thus cautioned then trigger chemical defenses that actively prevent insect herbivory (Dolch and Tscharntke 2000; de Bruxelles and Roberts 2001). Best of all, some plants will call in their animal friends, using chemotransmitters to attract predatory insects, which then attack and eat the offending herbivores (Havill and Raffa 2000; Karban and Agrawal 2002). In other words, these plants are doing the vegetative equivalent of saying, "You mess with me, you mess with my friends."

Now that the gentle reader is convinced that plants do indeed behave, perhaps thoughts will wander back to the most important consequences of those behaviors: traces. Reaching back to the foundation of ichnology, the holy trinity of substrate, anatomy, and behavior (Chapter 1), we then ask ourselves about parts of plant anatomy that can interact with substrates so that traces are left for ichnologists to interpret. Unfortunately, chemotransmitters have poor preservation potential, and their effects would be difficult to pick out using an ichnological approach. Partially digested insect parts left by carnivorous plants are also unlikely candidates for fossilization, especially if taken out of the context of their plant tracemakers. Fortunately, most plants have anatomical parts that interact with substrates in ways that leave visible effects of behavior that also have good preservation potential: roots and root traces, respectively. Consequently, most of this discussion of plant ichnology will center on root behavior and its traces, but with brief coverage of wound responses in plants too.

Roots move mostly downward because of positive gravitropism, also called geotropism, which is facilitated by a semisolid or solid substrate, such as sediment (Capon 2005). Some bromeliads, such as Spanish moss (*Tillandsia usneoides*), and a few ferns that grow on trees, like resurrection ferns (*Polypodium polypodioides*), are exceptions to this generality because they do not require soils. The effects of roots may be preserved in the geologic record more readily than most other traces of plant behavior because these are below the sediment surface and thus are already buried. Moreover, roots commonly extend to depths that may rival (or exceed) the height of the plant. Very simply, the deeper the roots, the better the fossilization potential. Some roots will also take hold in consolidated substrates, such as rocks and bones, and make traces in these (Mikuláš 2001; Montalvo 2002), although most roots invade a medium that can be more easily moved apart.

Roots make traces of plant behavior in substrates through various means. They can shift or compact sediment grains with their growth and movement, and as they increase in girth and length, they push aside whatever might be impeding their progress. Through this slow but powerful process, roots can even initiate fractures in consolidated substrates. For example, anyone who has lived in a city with old sidewalks adjacent to mature oaks has witnessed the buckling strength of slow growth and movement of roots. Besides local compaction and movement of sediments around roots, geochemical reactions in the soil can also result in easily observable changes. The results of geochemical alterations are often seen as differently colored haloes surrounding roots or, in fossil examples, the former locations of roots. These haloes are caused by root respiration, water flow, and exchanges of nutrients between the roots and the soil, as well as the effects of other organisms (Gregory et al. 2004). Organisms that interact with roots include fungi; soil bacteria, such as nitrogen-fixing bacteria; and insects that prey on the roots, such as termites, beetle larvae, and cicada nymphs (Chapter 5).

Of these, one of the most important symbiotic relationships in terrestrial ecosystems is what evolved between plant roots and fungi. Fungi, when directly associated with plant roots, are called mycorrhizae (mentioned earlier), and when fungi surround roots, they are termed ectomycorrhizae (literally, "outside fungi roots"; Smith and Read 1997). Some plants would not grow as well as they do without ectomycorrhizae helping them. These fungal colonies envelop roots and enhance the uptake of phosphorus and nitrogen in plants by sending fine tendrils (hyphae) into the surrounding soil, thus dramatically increasing the absorptive capacity of the roots (Figure 4.2). Ectomycorrhizae are not just helping plants out of fungal altruism, though; they gain carbohydrates from the roots as food for their continued growth and reproduction. Fungi are normally classified ecologically as saprophytes, meaning they mostly fill niches that deal with the decomposition of dead organic material and other media. (Although, frighteningly enough, some fungi are also classified as carnivorous [Schmidt et al. 2007], a thought deserving less exploration for now.) Fungi, like plants, also behave: anyone who has watched time-lapse photography of a moving slime mold will attest to its tactile mobility as a colony, as well as understand how they have inspired creepy science fiction stories of horrible alien species. As a result, fungi also fit the holy trinity of ichnol-

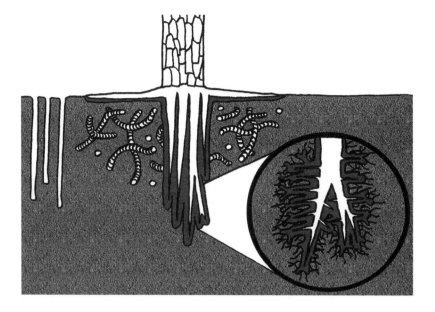

Figure 4.2. Fungi (mycorrhizae) surrounding the roots of a pine tree root system, sending its hyphae into the surrounding soil (inset) and thus making composite traces with the tree roots. Associated with the roots are the backfilled burrows of cicada nymphs and other invertebrates; to the far left are vertical escape burrows made by emerging cicada nymphs (Chapter 5).

ogy (substrate, anatomy, and behavior; Chapter 1) and leave traces of their symbiotic behavior with plant roots.

Perhaps surprisingly, mycorrhizae are documented in the fossil record from well before the Pleistocene Epoch. Two examples of fossilized fungal colonies surrounding root structures have been found in rocks from the Paleogene Period (Eocene Epoch) and are about 50 million years old. These colonies were associated with conifers such as *Pinus* spp. and *Metasequoia milleri* (LePage et al. 1997; Stockey et al. 2001). If that sounds impressive, it is, but not as much as ectomycorrhizae that have been identified from the Paleozoic Era. At least one example is from the Devonian Period and another is from the Silurian, about 350–400 million years ago (Brundrett 2002). These occurrences imply that the symbiotic relationship between terrestrial plants and fungi has not only been around a long time, but also probably represents a mutualism that coevolved early in the history of terrestrial ecosystems. Traces of this type of symbiosis are then noticeable as

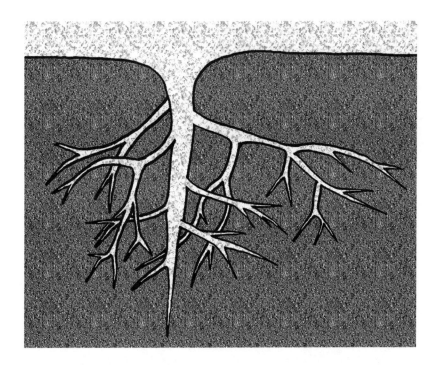

Figure 4.3. An idealized search image for root traces in the fossil record based on traits of modern plant roots (mostly downward orientation, distal tapering, lack of swelling at branches, and so on), also showing how such trace fossils can be preserved by passive filling of sediments from above. System is based on a plant that had a central taproot.

differently colored zones around the former locations of living roots (the previously mentioned haloes). Additionally, these zones may or may not hold the carbonized remains of the plant roots. What should be kept in mind, however, is the possibility that roots decayed after a plant died. This postmortem process also may have contributed to geochemical changes in the nearby sediment and further enhanced color differences seen in a paleosol (fossilized soil) where roots may have lived.

DESCRIBING ROOT TRACES

Now that we all agree that plant roots and their fungal friends make traces and trace fossils, what to do next? Well, when geologists or paleontologists encounter what they think might be root-like structures in the geologic record, they may do one of two actions: write down the word *roots* in their field notebooks and be done with it; or make a careful, detailed description of the structures to better ascertain their origin and potential behavior. As one might imagine, I recommend doing the latter. After all, such structures may not be root related at all and instead may be unrecognized invertebrate or vertebrate burrows; or worse, physical sedimentary structures that have no relation to life traces whatsoever. To better facilitate such descriptions, the following checklist can be used for documenting root-like structures, which will also help to test their initial identification. Ask yourself these questions (adapted from Gregory et al. 2004) and keep in mind some corresponding search images (Figure 4.3):

1) Do they show secondary and tertiary branching (do branches also branch)?
2) Do the structures have unequal diameters throughout their lengths?
3) If they do have uneven diameters, do they become especially narrower (taper) farther away from branching points?
4) Do they show Y- or T-shaped branching with junctions that maintain a normal width (i.e., do not show enlargement at junctions)?
5) Do they indicate near-vertical, downward pointing orientations? (This comes with some exceptions, though, based on how individual plants may have responded to a substrate and other environmental cues.)
6) Do the structures show any evidence favoring a passive fill of sediments from overlying layers? (Plants can do many things, but they cannot actively fill their former sites of roots with sediment.)
7) Are any preserved plant remains left in the structures?

A "yes" response to nearly all of these questions probably means that you have root traces. The one most likely to result in a "no" answer is the last question, relating to partial preservation of the original tracemakers. As is typical in ichnology, most trace fossils do not have any bodily remains of the tracemaker directly associated with their traces, with only a few notable exceptions (Chapter 6). As is likewise typical in ichnology, however, evidence for body parts of the plant tracemaker is among the least important of interpretative criteria for plant-root traces. The most important diagnostic standards for recognizing root traces are their narrowing downward (tapering) and branching while also showing no enlargements at branch junctions; in roots, these are known as nodes. So what is the big deal about enlarged junctions between branches? These wider parts are normally present in branching animal burrows, particularly those made by crustaceans. Enlarged junctions between burrow branches allow sufficient room for a stiff-bodied animal to turn around; hence they are also called (simplistically enough) turnaround junctions (Chapter 6). Plant roots, on the other hand, have no need for turning around within themselves, as they mostly grow outward and downward, away from the stem.

Descriptions of root architecture appropriately require knowledge of terms applied to plants in general and roots specifically, which can be a bit challenging but is easily surmounted with a top-down approach. In this case, however, "top" refers to the ground surface, where the main stem meets the roots and other subsurface plant parts. The stem, though, can have horizontal parts underground (rhizomes) that have other parts coming off these, such as shoots (these grow upward) and roots (these grow downward). Rhizomes, however, are not to be confused with stolons, horizontal stems that grow outward on a soil surface or just below it, sending roots from their nodes or shoots of new plants from their ends (Mauseth 2008). Wherever roots originate, their ends grow from an apical meristem, which uses three types of cells: meristem cells, which produce more such cells (thus encouraging growth of the root); root cap cells, which cover and protect a root, analogous to an epidermis; and undifferentiated root cells (Nobel 2009; Mauseth 2008). Undifferentiated root cells compose the majority of any given root, and their formation is what causes the end of a root to push through whatever substrate might be holding it. Roots, regardless of whether they are oriented horizontally, obliquely, or vertically, then branch from nodes. Primary roots are just above where they branch, and secondary roots are below the nodes. Of course, as anyone who has weeded a garden knows, root systems can keep branching into third-, fourth-, and fifth-order roots, getting ever finer with further distance from the stem.

Interestingly, the root systems of flowering plants can be predicted on the basis of how many petals are in their flowers. This is a nice piece of information that prolongs the life of the plant if a curious investigator is trying to decide whether to pull it up from the ground, yet is not weeding a garden. Flowers with petals in multiples of three indicate that the plants are monocots, a group of flowering plants (angiosperms) so named on the basis of their having only one seed leaf (cotyledon) after seed germination (Walters and Keil 1996; Mauseth 2008). Monocots, which include all true grasses, orchids, and a host of other flowering plants, typically have diffuse root systems that primarily hold a plant in place, rather than just absorb water or other nutrition from the soil. These root systems are often dense and finely fibrous, and secondary roots are difficult to pick out from the primary roots unless they are followed carefully from the stems.

Contrasting with monocots, but not quite their opposites, are dicots, which (you guessed it) have two seed leaves after germination. Dicot flowers also have petals in multiples of four or five (Mauseth 2008). Dicots normally have taproots, which are single, central roots with a well-defined direction (usually straight down) and some secondary roots coming off the main shaft. These roots, although they also help to anchor a dicot in one place, have two main purposes: food storage, which is similar to, but not quite like, tubers; and finding groundwater that might be far below the surface. Taproots can penetrate to astounding depths, in some trees to more than 60 m (200 ft; Stone and Kalisz 1991). Monocots and dicots also differ in their leaf patterns. Monocots have parallel veins; whereas dicots have veins with more complicated, reticulate patterns.

In other words, a quick count of the petals on a flower and a glance at the leaves on the same plant can lead to an initial hypothesis about its root systems. Conversely, if a diffuse root or taproot system is identified via root traces or trace fossils, the probable number of petals in the flowers (or at least its multiples) of the plant tracemaker can be surmised, assuming that the tracemaker is identified as an angiosperm. As a result, this method only works for root traces from Cretaceous and younger rocks, as angiosperms are only documented from the fossil record of those times (Magallón and Sanderson 2005).

Plants adapted for storing large amounts of food produced by photosynthesis have their roots swell into tasty, carbohydrate-laden treats for herbivores that dig them up, also known as tubers. Many well-known tubers, such as potatoes and sweet potatoes, are native to the Americas and were cultivated by Native Americans for thousands of years, meaning that this selective breeding led to the disproportionately large tubers we enjoy eating today (Brown 1993; Sleper and Poehlman 2006). As a result, root traces left by prehuman versions of these plants may not be as obvious as those of their inbred descendants. Nonetheless, plants with tubers would have formed noticeable swellings along the course of a root trace that, if observed in isolation, could be mistaken for either large invertebrate or small vertebrate burrows (Chapters 5, 6). These swellings should not be confused with enlarged junctions in branching animal burrows, however, because the swellings are along the root, rather than at nodes.

Once root traces or trace fossils are identified, some further challenges might arise in what to call them. Once encountered in the field and recognized, I simply call them "root traces" and their fossil equivalents "root trace fossils." Once labeled, I describe them in loving detail so that adequate interpretations can be made later of their taphonomy, behaviors, paleoecological relevance, and tracemakers. Nonetheless, the scientific literature is replete with many terms for root trace fossils and root body fossils—for example, rhizolith, rhizomorph, rhizocretion, and ichnorhizomorph. The term "rhizomorph" is probably the most confusing of these because botanists apply it to describe ectomycorrhizae that make imitative forms of the roots they surround (Kwansa 2002; Mihail et al. 2002). In contrast, geologists have used rhizomorph as a catchall term for any rootlike structures they encounter in the geologic record; after all, it literally means "root form." Pointing out this discrepancy is not meant to bash geologists, though, considering how most botanists have not been inclined to study plant-root trace fossils, or even to describe modern plant roots for us nonbotanically oriented barbarians. After all, look in a plant identification guide, and nearly every description of any given plant deals with its above-surface appearance, with scarce mention of its root system or how those roots may affect their host substrates. Granted, most botanists do not necessarily pull plants out of the ground to identify them, which is especially beneficial for the continued health of the plant. Also, if botanists know their monocots from their dicots—which they most assuredly do—there is no need for them to look at a plant's roots.

In short, although the study of root traces can make for perplexing reading, knowledge of their terminology and diagnosis can help these to become a source of important information related to plant behavior, habitats, and subsurface interactions between plants and animals. Accordingly, an inquisitive attitude should be adopted when exploring the neoichnology of root traces on the Georgia coast, a subject that can easily rival or surpass the neoichnology of their animal companions. Granted, this is a relatively new field of study, begging for more attention from botanically trained researchers with some geological savvy. So the hope here is to provide an introduction to the subject of root traces through an examination of the ichnological standouts in various Georgia coast habitats and substrates, rather than an exhaustive overview.

A LITTLE SALT IS FINE: HALOPHYTES

Every discussion about salt marshes of the Georgia coast mentions its most visible and ecologically important plant, smooth cordgrass (*Spartina alterniflora*), and how this species is singly responsible for both the beauty and functioning of a salt marsh (Chapters 2, 3). The subsurface parts of smooth cordgrass may garner less attention than its magnificent stalks and leaves, but these roots and stolons are also essential for making the salt marsh what it is. It is a prairie, submerged twice daily by marine-flavored water, which keeps its soil in place through an intricate subsurface network. Although a large amount of mud, sand, and organic matter is flushed out of these marshes by tides, much of it remains. Consequently, the sedimentary surface of the marsh stays at about sea level. Lower this sedimentary surface, though, either through accelerated erosion or heightened sea level, and all of the cordgrass would drown (Proffitt et al. 2003). Dead plants result in fewer roots, which means less sediment staying put, translating into yet more erosion—a positive feedback with an overall negative impact on the original ecosystem. In other words, if not for roots and their cohesive effects, these environments would become lagoons. The opposite effect, a lowering of sea level, would also inflict death on a salt marsh, as the prodigious flux of nutrients and organic productivity facilitated by tidal exchanges would mostly shut down, meaning no more nutritive soils for the cordgrass.

The roots and stolons of smooth cordgrass have diagnostic traits that help to distinguish these from those of similar halophytes and could likewise be applied in the event of finding their root trace fossils in ancient marsh deposits. Because cordgrass is indeed a grass, it is also a monocot, which means its roots are adventitious. Adventitious roots start from the stem, not the primary root, and then project downward from the base of each stolon or nodes of rhizomes (Figure 4.4). Such root systems are correspondingly diffuse and form a network meant to increase absorption, but while also holding a plant in place (Mauseth 2008). Above-ground portions of the plant (leaves and stems) similarly originate at these nodes, but grow upward. Hence roots and foliage each exhibit the classic plant behaviors of gravitropism and phototropism, respectively. Smooth cordgrass, like many plants, can reproduce clonally, with stolons making a lateral false root that joins two or more separate vertical stems. Rhizomes are typically 4–8 mm (0.15–0.3 in) wide, circular to oval in cross section, and smooth on their exteriors, except for where fine rootlets (< 1 mm wide) might be present, giving the root a hairy appearance. These rootlets are massed and interwoven in the upper 20–30 cm (8–12 in) of the marsh, composing a meshwork that secures and consolidates marsh mud around each plant. Cordgrass roots also go deep. They can be as much as a meter (3.3 ft) long and are oriented vertically, but with horizontal jointing along the way. The joints are quite regular, at 2–3 cm (1 in) intervals along the length of a given root (inset of Figure 4.4).

This is the situation for the low marsh, though. In contrast, the high marsh involves numerous other plants, such as the close relative of smooth cordgrass, salt-meadow cordgrass (*S. patens*); black needle rush (*Juncus roemerianus*); sea-oxeye daisy (*Borrichia frutescens*); glasswort (*Salicornia virginica*); saltwort (*Batis maritima*); and marsh elder (*Iva frutescens*), to name a few (Bellis and Keough 1995; Reitz et al. 2008). Perhaps surprisingly, all of these plants—except for sea-oxeye daisy—are monocots. To nonbotanists, the thought that magnolia trees and glasswort belong to the same taxonomic division of flowering plants may seem unbelievable, but that is evolution for you: deal with it. As a result, these plants' root systems, like those of smooth cordgrass, are also diffuse. The big difference between these and the roots of smooth cordgrass, however, is in their limited vertical and horizontal extent.

One potential ichnological dilemma posed by cordgrass occurs where older (relict) marshes are buried and remnants of the plants' vertical stems are buried by eolian sand from above (Figure 4.5). Stem impressions imparted on the sand could then look much like vertical burrows made by an intertidal invertebrate, such as a polychaete worm (Chapter 6) and thus make false "trace fossils" (Gregory et al. 2006). For example, unsuspecting geologists encountering their equivalent in the geologic record might not notice fine rootlets connecting to these "burrows," or the vertically oriented ridges associated with the stems. As a result, they may simply label these structures in their field notes as fossilized vertical burrows. Such an

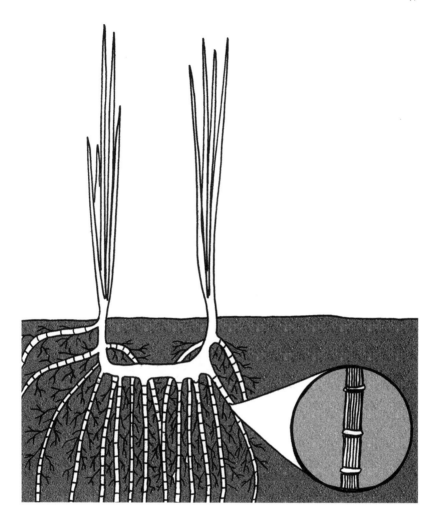

Figure 4.4. The adventitious roots systems of *Spartina alterniflora* and their details. (a) Overall form, showing where roots can begin at the stem and project downward from the stolon to form rhizomes (with rootlets), forming a dense network in marsh sediment in the upper part of a marsh mud. Inset shows root details, with jointing and longitudinal striations, helping to distinguish these from invertebrate burrows.

interpretation, however, would be doubly wrong. Not only would these "traces" be from plants, they would not be traces: after all, dead bodies do not behave (Chapter 1). As a result, geologists documenting what they think might be trace fossils should keep in mind the tricks that plants of the past might play on them.

Figure 4.5. Stems of 500-year-old smooth cordgrass (*Spartina alterniflora*) on a sandy beach, projecting from an underlying relict marsh (see Figure 3.12), and doing their best to imitate vertical burrows made by invertebrates in such sandy environments: Sapelo Island.

BESTOWING STABILITY: DUNE GRASSES AND BACK-DUNE MEADOW PLANTS

Just as salt marshes depend on smooth cordgrass for their existence, dunes along the extensive sandy beaches of the Georgia coast mostly owe their presence to one species of grass, sea oats (*Uniola paniculata;* Figure 4.6). The analogy is further extended by the role of roots in both instances: marsh mud would not be held in place if not for the roots of smooth cordgrass, and dune sand would be blown away in the absence of roots from sea oats. Because *U. paniculata* is a grass, it is also a monocot, and accordingly

MARGINAL-MARINE AND TERRESTRIAL PLANTS   149

Figure 4.6. The main stabilizer (and plant tracemaker) in coastal dunes of the Georgia barrier islands, sea oats (*Uniola paniculata*). (a) Overall view of its dominant role on coastal dunes, which is to stabilize sand and retard sand transport (background), in contrast to where a lack of sea oats corresponds to a lack of dunes (foreground): Jekyll Island. (b) Close-up of its roots, showing extensive, fine meshworks throughout sandy substrate: Wassaw Island. Scale in centimeters.

Figure 4.7. The initial formation of dunes from dead smooth cordgrass (*Spartina alterniflora*) deposited at the rack line, in which accumulations of cordgrass stems and roots slow the wind and help to deposit sand. This is not a trace per se, but nonetheless serves as an example of how plants, even when dead, can affect the development of dune environments: Wassaw Island. Boot (lower left) for scale.

has dense, finely diffuse root systems, concentrated in the uppermost 30 cm (12 in) of a soil. Unlike most monocots, it can extend taproots several meters downward (Hester and Mendelssohn 1989, 1991). Although sea oats derives its common name from prominent clusters of seeds, looking much like the "amber waves of grain" of the American Midwest, it propagates more often via its horizontal rhizomes. Through their lateral spread, these rhizomes also tend to keep large amounts of sand in place.

Interestingly, dunes often get their start through smooth cordgrass, although not by its behavior. Dead stalks of *Spartina* are flushed out of marshes with each ebb tide, which are then incorporated into the flotsam and jetsam of the open sea just offshore of Georgia beaches. Wind-driven waves and flood tides then bring this *Spartina*-dominated detritus to the shoreline, where it accumulates abundantly in the rack line. These clumps of dead plants then become windbreaks, slowing the wind as it blows across the beach surface and consequently piling the sand. These protodunes then become small hillocks on the berms and upper parts of beaches that are more amenable to colonization by sea oats (Figure 4.7). Once these small areas are raised above the high-tide mark of the beach, windblown seeds of sea oats will accumulate on them, and seedlings will germinate and take hold (Bellis and Keough 1995; Hayes and Michel 2008). Soon afterward, individual plants will spread out their rhizomes and roots with growth. This extension is then what starts the positive feedback of plant growth corresponding to dune growth: more sand held in place by rhizomes and roots imparts a larger barrier to wind movement. This is turn slows down the wind so that more sand accumulates on the downwind sides of plants. Strength in numbers also becomes important, as closely spaced plants with overlapping root systems then cause broader swaths of bound sand, which then stack above the berm surface and eventually create the familiar hill-and-dale topography of dunes. Thus, in a sense, smooth cordgrass is ultimately responsible for the start of most dunes on the Georgia coast, although not in its living role (and, sadly for ichnologists, not through the making of traces while living). Nonetheless, sea oats in life finish the job with their roots.

Other plants that take root in dunes next to the beach, and sometimes in the foredunes, include a few flowering vines, which make distinctive traces as a result of their extensive horizontal movement; a single plant can extend as long as 30 m (100 ft). Two closely related species of creeping vines on Georgia coastal dunes are railroad vine (*Ipomoea pes-caprae*) and fiddle-leaf morning glory (*Ipomoea stolonifera*). Each of these is well known for its beautiful flowers, and such plants are more popularly lumped into the category of morning glories (Kaplan 1999). As a bit of taxonomic trivia, *I. pes-caprae* derives it species name from the resemblance of its petals to a goat's foot, with *caprae* meaning "goat's" and *pes*, "foot." As hinted by the species name of *I. stolonifera,* these vines bear extensive stolons that allow this plant to spread quickly in horizontal directions. Roots originate at nodes and grow straight down from the horizontal stems into the sand, anchoring the plant in shifting dune surfaces, or even in upper intertidal areas (Figure 4.8). Think of taking a rope and pinning it down every 20 cm

Figure 4.8. Railroad vine (*Ipomoea pes-caprae*), forming extensive, horizontal traces from stems and short, vertical traces from roots in the foredune–upper intertidal zone: Sapelo Island. Boat-tailed grackle (*Quiscalus major*) tracks and ghost crab (*Ocypode quadrata*) burrow (lower right) for scale; burrow is about 5 cm (2 in) wide.

(8 in); then the rope becomes difficult to pick up. Railroad vine is not only tolerant of salt water (it has to be, living next to and on the beach), but its seeds can float in salt water and are often dispersed by oceanic currents. Hence this species inhabits all of the Georgia barrier islands and is found throughout the Bahamas and much of the Caribbean. This plant, along with sea oats, also helps to stabilize foredune areas, and its runners sometimes stray down to the upper part of the intertidal zone.

Although trace fossils from creeping vines might seem unlikely, they have been interpreted from the geologic record, and not too far from the Georgia coast. Horizontal stolon and vertical root impressions attributed to railroad vine, or a closely related species, are preserved in relatively young Holocene rocks (about 5,000 years old) in the Bahamas and are particularly well expressed on bedding planes of ancient dune deposits of San

Salvador Island (White and Curran 1997; Martin 2006c). External molds of these plants are interpreted as trace fossils because modern, living plants show cementation of sand grains around their stolons. This is seemingly a result of plant respiration and would not happen if the plant had not moved onto and into these sediments. So why would this fossilization of creeping vine traces also not happen on the Georgia barrier islands? Bahamian sand has a huge advantage over Georgian sand in its preservation potential for plant and invertebrate trace fossils, namely its calcareous composition, consisting of aragonite and calcite (both of which are $CaCO_3$). Plant respiration then easily causes local geochemical change in calcareous sands surrounding a root, stolon, or other body part, encouraging early cementation. This phenomenon, however, is not mirrored in the relatively inert, quartz-rich sands of the Georgia coast. Still, uniformitarianism cannot always predict what happened in the past, so I would not be surprised if trace fossils of creeping vines are someday found and reported from quartz-rich sandstones of former coastal dune deposits.

Three other species of plants on and near primary dunes are beach elder (*Iva imbricata*), bitter panic grass (*Panicum amarum*), and pennywort (*Hydrocotyle bonariensis*). Beach elder is a dicot, so its roots have different traits from most of its monocot associates near a beach, such as having a well-defined taproot. Moreover, this plant, like sea oats, is a perennial (Franks 2003). Hence it may stay in the same place for years, increasing its likelihood for tracemaking activity. Bitter panic grass will make root traces similar to those of sea oats, but of course on a much more modest scale. Pennywort is a dicot as well, with a small but prominent taproot and limited horizontal components. Nevertheless, it is abundant enough in coastal dunes that its traces may be noticeable too.

If dunes are not immediately in front of a maritime forest or salt marsh, the back-dune area, largely consisting of old dunes no longer directly next to the coast, may be quite broad, comprising areas hundreds of meters wide. This ecosystem typically contains the following plants, taller and more deeply rooted compared to those on beach dunes: Spanish bayonet (*Yucca alifolia*), wax myrtle (*Myrica cerifera*), and yaupon holly (*Ilex vomitoria*). All of these plants can merge with an ecotone shared with a maritime forest, but they are not so common along the edge of a salt marsh, which is more likely to have red cedar (*Juniperus virginiana*). One of the more important of these plants as root tracemakers is *Y. alifolia*, which, as a plant adapted to dry conditions, can extend its roots horizontally over a wide range.

Trace fossils of coastal dune plants have been interpreted in the geologic record and are especially well defined in ancient dune deposits composed of cemented calcareous sand (limestones), such as those in the Bahamas (Curran 1992; Martin 2006c). In fact, degrees of phytoturbation—bioturbation caused by plants—in these dune deposits are used as indicators of the original maturity of the dunes, and they are correlated with sea-level lows and highs. For example, during sea-level lows, lots of formerly shallow-marine sand is exposed and made available on the surrounding sand flats. This sand is transported by offshore winds and piled up into tall dunes. In turn, windblown seeds and other plant parts, especially those that can survive a sea journey or propagate by vegetative growth, then colonize sandy surfaces. The role of bird excrement in starting new plant communities on newly formed dunes cannot be underestimated, either. Any migratory birds that eat fruits—many of which contain indigestible seeds—can fly to other islands and bomb them with seeds surrounded by fecal fertilizer. The effects of this sort of evolutionary bribery by plants (or, "I give you food, you carry my child") are especially evident wherever birds nest on dunes; plant colonization and primary growth will naturally take place where any large groups of fruit- and seed-eating birds pass through an area or stay for a while (Wilkinson 1997; Cain et al. 1998). Once plants colonize a dune, dune sands shift much less, which enables more plants to take hold, while also allowing the original pioneers to grow for longer times. This all has a net effect of more and deeper roots from these plants.

During sea-level highs, the opposite happens, as coastal dune communities are drowned by waves and tides, made sick by constant sea spray, or turned into tossed salad by hurricanes. Accordingly, thin dune deposits with a paucity of root trace fossils are considered as immature and more likely formed during sea-level highs. Root trace fossils, or their lack thereof, thus can be used as clues for transgressions or regressions in the geologic record wherever coastal dunes and root traces are preserved (White and Curran 1997; Curran and White 1999).

In modern dunes of the Georgia barrier islands, removal of rooted native plants and subsequent human development can also result in a lack of dunes, the first line of defense against storm waves or rising sea level (Rodgers 2002). Although the Georgia coast is not as impacted by tropical storms as other parts of the eastern United States, such as Florida or North Carolina, it had some horrific hurricanes in the past (Fraser 2006). Such storms are also more likely to arise in the near future from global climate change (Chapters 2, 11). The importance of sea oats and other dune plants in stabilizing dunes has resulted in protection of these plants, and signs are often posted near dunes to dissuade people from walking across these environments and damaging the vegetation. Nonetheless, illiterate vertebrates, such as feral horses (on Cumberland Island) and hogs (on nearly every Georgia island) flout these warnings by grazing, trampling, and uprooting dune plants. As a result, the ichnological impact of invasive species can act as ecological agents of change, suppressing the ecological and sedimentological effects of native plant species in these environments.

DEEP ROOTS: TERRESTRIAL SHRUBS AND TREES

The most defining traits of a maritime forest are, of course, its trees, especially its mix of live oak (*Quercus virginiana*), various pines (*Pinus* spp.), and its prominent understory of saw palmetto (*Serenoa repens*). Fortunately for neoichnologists, all of these trees have prominent root systems, which hold the potential for making spectacular root traces. Even better, neoichnologists do not have to hire backhoe operators to pull up these plants for a peek at their roots, because nearly every beach on a Georgia barrier island has a so-called tree "boneyard," filled with the skeletal remains of dead trees (Figure 4.9). These fallen trees, marking the sites of former maritime forests killed by changing shorelines and rising sea level (Chapter 2), are often lying on their sides or otherwise in nonvertical positions. As a result, they afford the opportunity for up-close study of their roots. Root architecture and dimensions then can be translated into models for expected forms of trace fossils in the geologic record. In fact, this is exactly how the mystery structures of Raccoon Bluff, mentioned earlier in the chapter, were identified specifically as the trace fossils of a species of *Pinus*, and not those of *Quercus* or other trees (Gregory et al. 2004).

Speaking of pines, the Georgia coast has a good number of species: slash (*Pinus elliottii*), loblolly (*P. taeda*), long leaf (*P. palustris*), short leaf (*P. echinata*), and pond (*P. serotina*). In terms of their geologic record on the Georgia coast, pine pollen is abundant in 47,000 and 37,000-year-old (Pleistocene) rocks on St. Catherines Island and Skidaway Island, respectively (Booth et al. 1999, 2003). Of course, we now know that pine-root traces were

Figure 4.9. A so-called tree boneyard on St. Catherines Island. These are common sights along Georgia beaches, where a combination of erosion from longshore drift and saltwater intrusion from sea-level rise has killed former maritime forests and exposed their root systems, better helping ichnologists to understand their root architecture.

interpreted from 40,000-year-old rocks on Sapelo Island too (Gregory et al. 2004). Modern assemblages of pines, however, may not represent their coastal ancestors very well, largely because of human alterations of maritime forests, which include the growing and logging of pines (Chapter 2). Consequently, relative proportions of pine species in modern maritime forests may be viewed as a recent anomaly. Nonetheless, because pines have a geologic history on the Georgia coast, as well as stretching back into the fossil record, a description of their probable root trace fossils is justifiable. Pines have a network of near-horizontal roots, but most are vertical or nearly vertical ($\geq 80°$), tapering downward into sharp points (Figures 4.1b). Some bifurcation is evident on the way down, which can be as much as 2–3 m (6.5–10 ft) deep, depending on the average depth of the local water table. Roots are circular to oval in cross section, with smooth exteriors, and they are thickest proximal to the trunk, reaching as much as 30–40 cm (1–1.3 ft) in diameter.

Of course, maritime forests on the Georgia barrier islands are epitomized by live oaks, and thus any consideration of plant traces on the islands must take these into account. Oaks, all of which are in the genus *Quercus*, also have a long geologic history on the Georgia coast, as their pollen shows up frequently in Pleistocene sediments, although not as commonly as pine pollen (Booth et al. 1999, 2003). Despite the overlap in root sizes, live-oak root systems are distinct from those of pines in both orientation and overall form. Like pines, oaks tend to have near-surface and horizontal root systems that are far more abundant than their vertical components. Apparently, their taproots are surprisingly shallow, with most vertical components extending down less than a meter (3.3 ft). Added vertical parts of roots may, however, grow in the opposite direction (up) as a result of clonal sprouting that sends shoots above the ground surface. Root forms also differ from those of pines in having more or less constant diameters over their lengths until abruptly tapering into relatively blunt ends (Figure 4.10a). Oak roots can also be extremely dense, forming complex and interlocking networks that cover 10–20 m^2 (33–66 ft^2), which would result in complicated trace fossils (Figure 4.10b). A native species that may have its root traces confused with those of live oaks, though, would be the Southern magnolia (*Magnolia grandiflora*), which likewise grows thick and mostly horizontal roots connecting to a robust trunk.

Other traces that may accompany live-oak roots include borings, made (appropriately) by the larvae of the live-oak root borer (*Archodontes melanopus*), a species of beetle (Solomon 1995). As mentioned previously, root traces may also have insect burrows in surrounding or nearby sediments, and root trace fossils can be intersected by backfilled burrows of infaunal insects, such as those of cicada nymphs and beetle larvae (Chapter 5). On branches well above the ground, resurrection ferns (*Polypodium polypodioides*) leave minute scars on live oak bark, whereas epiphytes, such as Spanish moss (*Tillandsia usneoides*), leave no visible traces whatsoever (Callaway et al. 2002). The preservation potential of their traces seems unlikely but may be worth studying for better defining commensal relationships in maritime forests.

As an interesting behavioral aside, live oaks, more than any other trees of maritime forests, show visible adjustments to the local weather patterns in

Figure 4.10. Root architecture of live oaks (*Quercus virginiana*), helpful models for similar trace fossils. (a) Horizontal view of exhumed tree, showing nearly equal-diameter roots with blunt ends staying in a relatively narrow, near-surface zone; total horizontal width about 5 m: Sapelo Island. (b) Dense horizontal network of roots viewed from below, providing a hint of the complexity of any similar trace fossil: St. Catherines Island.

their above-ground parts. Look at any collection of live oaks near a shoreline, and the branches are clearly oriented in the direction of prevailing winds coming from offshore. This orientation also takes on a noticeably upward slant, with the lowest part of the branches nearest the shoreline. This "go with the flow" form is a response by the tree growing less on the side associated with environmental stress, such as wind and its attendant sea spray. Is this ichnological, and would it count as a trace fossil? The answers are yes to the former, as it fulfills the guideline of anatomy responding behaviorally, and the effects recorded by the tree itself (the substrate in this case); and yes to the latter, but only if enough of the tree were preserved in the geologic record and this preferred orientation of its branches were somehow discerned.

Saw palmettos (*Serenoa repens*) are without a doubt the most ichnologically important understory trees in Georgia maritime forests, and they compete with live oaks and Spanish moss for the attention of cliché-driven nature photographers in those environments (Figure 3.8). Fortunately, their root systems have been described in detail (Fisher and Jayachandran 1999), which correspondingly can be applied to visualizing their traces. Their stems (trunks), which bear many leaf scars (the former sites of palmetto fronds), are prostrate—that is, growing mostly horizontally as they move sinuously through the forest floor. The compound leaves of the palmetto fronds actually grow from the vertical and much shorter part of the trunk, which performs an abrupt upward twist from just below the ground surface. As a result, saw palmettos are among the few maritime-forest trees that can make prominent horizontal traces with their stems. Their trunks, which are about 10–25 cm (4–10 in) wide, are clearly evident after forest fires take out surrounding brush, or wherever stems may cross well-worn trails (Figure 4.11a). Two of the most striking traits of saw palmetto stems are (1) the corrugated distal parts of the trunk; and (2) thin (1–5 mm) adventitious rootlets, both vertical and horizontal, that mostly extend from the bottommost part of the trunk, downward to as much as 50 cm (20 in), and laterally for as much as 5 m (16 ft; Figure 4.11b). Saw palmetto roots can also get quite complicated, having second- and third-order branching, in which these might turn down, up, or sideways (Fisher and Jayachandran 1999). If structures made by both the stem and roots were preserved in the geologic record, they would look much like large-diameter backfilled bur-

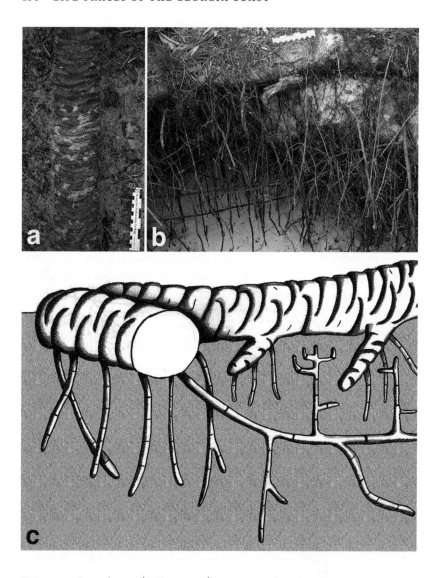

Figure 4.11. Saw palmetto (*Serenoa repens*) stems can make some interesting traces, or mistaken traces. (a) Top view of a trace of a horizontal trunk of saw palmetto with false meniscate structure: Jekyll Island. Scale in centimeters. (b) Abundant and closely spaced rootlets extending down from a trunk, imitating vertical burrows: Jekyll Island. Scale in centimeters. (c) Idealized sketch of the overall near-surface and subsurface form of a saw palmetto, with its smaller-diameter roots branching downward and upward from a main trunk (curving toward the viewer); based on a figure by Fisher and Jayachandran (1999).

rows with numerous, extensive, smaller commensal burrows emanating from the main burrow tunnel (Figure 4.11c).

Could such a mistake be made, confusing tree trace fossils with animal burrows? In a word, yes. Murray Gregory and Kathleen Campbell, mentioned at the beginning of the chapter, noted a tree-root trace fossil in New Zealand that mimicked the trace fossil *Phoebichnus* (Gregory and Campbell 2003). *Phoebichnus* is an invertebrate-made trace fossil consisting of a wide vertical shaft, from which radiate smaller-diameter horizontal burrows containing backfilled (meniscate) structures. The tracemaker presumably made each burrow as feeding probes from a main burrow shaft. In contrast, the vertical taproot and horizontal roots of a nikau palm tree (*Rhopalostylis sapida*) were responsible for making its imitator in New Zealand. "So what?" says the casual armchair ichnologist: if such an error is made, no big deal, because *Phoebichnus* is a relatively uncommon trace fossil anyway. Nevertheless, the problem with accepting such a mistake is twofold and compound. First of all, *Phoebichnus* is a trace fossil that only occurs in rocks of formerly marine environments (Bromley 1996), whereas nikau palms only grow on land. Secondly, *Phoebichnus* is apparently restricted to the middle of the Mesozoic Era, meaning its tracemaker went extinct more than 100 million years ago. Nikau palms, on the other hand, are relatively recent tracemakers, having left their marks in Pleistocene and Holocene sediments, and are still common in modern terrestrial environments of New Zealand. In other words, "So what?" becomes "So wrong!," with an incorrect ichnological diagnosis that misses the correct age, environment, and tracemaker. With the preceding description and cautionary tale in mind then, traces made by saw palmetto roots and stems should be unmistakable if their geometry, interconnections, and paleoenvironmental context are carefully noted and identified. Nonetheless, if stem and root traces are found separately, or otherwise treated as individual and unrelated traces, these parts could be confused with invertebrate and vertebrate burrows.

Cabbage palms (*Sabal palmetto*) are probably underestimated ichnologically, and modern examples are models for how these plants were likely tracemakers in ancient terrestrial environments. In some instances, these trees comprise the only upper-story foliage along a Georgia coastline; for example, the long beaches of St. Catherines Island are adorned with these beautiful, tropical-looking trees. The pencil-thin roots of cabbage palms

Figure 4.12. The impressive root architecture of a cabbage palm (*Sabal palmetto*) exposed in a 3 m (10 ft) tall vertical section of the Pleistocene Silver Bluff Formation, with thick clusters of vertical to near-vertical roots, some of which extend 2–3 m (6.7–10 ft) below the surface; arrow points to the base of the trunk: St. Catherines Island.

grow in amazingly dense arrangements, with hundreds of roots projecting from trunks that are only about 50 cm (20 in) wide (Figure 4.12). Their roots radiate downward and outward at vertical and near-vertical angles (about 70°) and can extend below for nearly 5 m (16 ft). These trees, like red cedars, serve as ecotone markers: they tolerate some brackish water along the edges of beaches or marshes but succumb once the salinity gets too high. Consequently, root traces of either cabbage palms or red cedars, along with traces indicating a mixture of terrestrial and marginal-marine invertebrates—such as insect traces overlapping with those of ghost crabs—could be diagnostic of these transitions between environments if preserved in the fossil record. Prudence is warranted, though: I have seen roots of modern, living cabbage palms extending well into Pleistocene sediments, mixing their traces with those left by ancient tracemakers, including root trace fossils of long-gone plants. This means that a hasty description of traces in an outcrop could miss such anachronisms, resulting in a botched interpretation of the paleoenvironments.

Root traces from these trees on the Georgia barrier islands are nearly always formed in sandy substrates; soils in which they grow in the southeastern United States include ultisols, spodosols, histosols, and entisols (Chapter 3). A few quick distinctions between these soils are as follows: ultisols are mineral-rich soils lacking calcareous material; podosols are well-drained sandy soils that typically form in conifer forests; histosols are mostly composed of organics; and entisols are soils barely weathered from their parent material (Messina and Conner 1996). Thus, the observation and classification of tree-root trace fossils in the geologic record may depend on a sediment fill that contrasts with that of original soil, consisting of a different color, grain size, or both.

WOUND REPONSES: HEALED INCISIONS, GALLS, AND OTHER VISIBLE TOUGHENING OF PLANT TISSUES

When insects attack plants, the resulting traces are actually two-for-one records of behavior: that of the insect and that of the plant. Usually these traces are placed solely under the category of insect traces, assigned from the viewpoint that insects initiated their making. Nonetheless, as we discovered earlier, living plants are not inert and passive receivers of these attacks, but have evolved defenses and other responses, which in many instances leave visible marks documenting these behaviors. Thus, this section will be told from the perspective of the plants, not of their herbivores. Besides, insect tracemakers will receive their due in great detail later (Chapter 5).

Pick up any given deciduous leaf that may have fallen from a tree, look carefully at it, and your eyes may soon pick up all sorts of flaws: breaks in the continuity of the outer margin of the leaf; open holes inside the margin; minute raised bumps; tiny punctures; or convoluted mazes between the thicker, outer layers (cuticles) of the leaves. If you are having problems seeing these features, hold up the leaf so it is backlit, and all such traces should become clearer. If the leaf is thin enough, you may even see your fingerprints where you handled the leaf. Once you have seen these traces, close your eyes and feel the surface to detect the smaller nicks, dimples, and pimples you may have missed by relying on just one sense. (Of course, make sure the leaf is not from poison ivy or other plants that chemically defend themselves against your well-meaning attempts to learn more about them.) Insects initiated most of these marks, but look a bit closer and you will see how at least some will be discolored, showing up as brown edges or toughening of the plant tissues. These are traces made by the plant, living reactions like the hardening of a scab or the annealing of a broken bone. If fossilized, how many of these would be overlooked completely with the excitement of finding a fossil leaf? Yet fossilized they are, with some occurring in plants from the early part of the Paleozoic Era, from more than 400 million years ago (Labandeira et al. 2007; Scott 2008).

Such traces are described on the basis of both where they occur on (or in) a plant, as well as how they affect it. The most basic categories of traces are marginal, nonmarginal (or hole), leaf mines, galls, and piercings (Scott 1992; Labandeira et al. 2007: Figure 4.13a). Marginal traces are incisions or other cuts made along the outer edge of a leaf, some looking like a cookie cutter or scissors took a chunk out of them. These insect traces are then further outlined by the plant's reactions to the attack, marked by a brownish coloring and closing of the cuticle over the soft, inner tissues of the leaf (the mesophyll). These are similar to a scab but really are more like a scar because the leaf never reforms its previous shape. Nonmarginal traces are also incisions but consist of relatively large circular to oval holes within the main body of the leaf, like the mark a hole-punching tool might leave in paper (Figure 4.13b). Insects, such as the unimaginatively named leaf-cutter bees (*Osmia* spp.); some beetles; butterfly and moth larvae (caterpil-

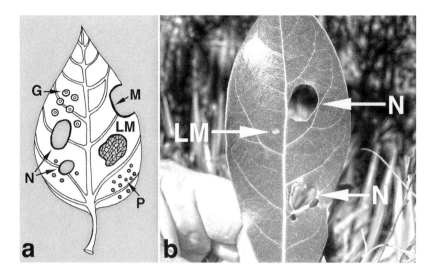

Figure 4.13. General categories of insect-made traces in plant leaves and examples. (a) Schematic sketch of traces in a typical deciduous leaf, including galls (G); leaf mine (LM), marginal (M), and nonmarginal (N) incisions; and piercings (P), all with darkened rims a result of wound reactions by the plant. (b) Insect traces in leaf of red bay (*Persea borbonia*), including small leaf mine (LM) and nonmarginal incisions (N), but also showing wound reactions by the plant, making for a composite trace: Little St. Simons Island.

lars); and others with shearing mandibles are responsible for marginal and nonmarginal traces (Chapter 5).

Leaf mines seem like the most insidious of attacks, taking place between the cuticles. In such traces, the offending insect—usually in its larval form—follows a tightly meandering path between the upper and lower cuticles of the leaf, feeding on the mesophyll. Sometimes this part of the leaf is not enough, though, and the same insect or others will start eating the cuticle too. Hence these traces may not be obvious until a leaf is nearly skeletonized, showing only bits of the original cuticle clinging to secondary and tertiary leaf veins. Judging from the amount of damage caused by insects feeding in such a manner, plants seemingly have few defenses against these attacks. More subtle traces are piercings, which are pinpoint-like holes caused by an insect with stabbing mouthparts. These traces mark where the insect poked into the cuticle of a leaf to suck out some of the softer and more nutritious material within the mesophyll.

Among the more obvious traces of combined insect and plant behavior are galls. These are visible swellings of plant tissue made in response to insect infestations, usually caused by the laying of eggs and hatching of hungry larvae. Galls can be found on leaves, stems, and branches; leaf galls normally are numerous and smaller, whereas stem or branch galls are few and larger. Branch galls are the most ecologically important, as they become sites that attract other insects, such as those that want to use the new, expanded plant tissue as a new home, or those that prey on the gall insects (McGavin 2001). Many galls are caused by cynipid wasps, which probably have a long history of coevolution with host plants. A plant's forming of a gall as a wound reaction, though, often helps to better protect the insects living inside it, which means it mostly benefits the insects. Other instances of unequal relationships in coevolution are common whenever discussing insects, and insect tracemakers in particular provide rich examples of these seeming imbalances (Chapter 5).

Occasionally, fungal attacks can also discolor a leaf so that it mimics the effects of insects, such as piercings, but these are usually less precise in their targeting. As a result, a plant may react by forming an entire reaction zone (or "front") around the general site of a fungal infection, such as around the leaf margin (Labandeira et al. 2007). On a much larger scale, however, are fungal infections that affect interiors of stems, trunks, or branches of trees, such as live oaks. These assaults provoke prominent abscesses, looking much like the tree has an out-of-control tumor. Fungal infections of trees are often started and then encouraged by wood-boring insects, such as certain beetles, which will then sometimes harvest the fungi as food (Chapter 5). Consequently, some plants have evolved at least two levels of defense: producing sap when insects attack plant tissues, followed by a hardening of the tissue ahead of a fungal colony to slow or stop its advance (Wagner 2002). Trace fossils of fungal infections and corresponding wound reactions in trees have been interpreted from the geologic record (Taylor and Krings 2008), although multiple phases of defense in plants to both insect and fungal attacks remain undiagnosed. This is where trace fossils could help considerably, through the identification of coinciding marks left by fungi and arthropods in plant tissues, as well as evidence of wound reactions from the plants. Where could such search images originate? Why, from modern examples, of course. In other words, a neoichnological approach could aid in unraveling how fungus–arthropod–plant interactions evolved together in terrestrial ecosystems, a process that began nearly 400 million years ago and continues today.

PLANT ICHNOLOGY: QUO VADIS?

As noted earlier, the study of root traces, as well as other aspects of plants imparting visible changes to substrates through their behavior, including alterations to their body parts, is a neglected subset of ichnology, particularly in neoichnology. Fortunately, this disregard is expected to change, and the preceding examples from the Georgia coast should supply sufficient reasons for more intensive follow-up research on this fascinating subject. Suggested lines of growth in research that may blossom and bear fruitful results include the following:

1) Discerning details about the geochemical changes imparted on sediments by the symbiotic relationship between plant roots and ectomycorrhizae (e.g., how much do certain elements have to be concentrated around roots for them to become visible to the naked eyes of ichnologists, versus those changes that might be more cryptic?).
2) Better quantification of root architectures and how the geometry of monocots, dicots, and other types of plants may be more easily identified from their root traces.
3) The ecological effects of roots on subterranean animals, particularly insects, and how roots may influence their behaviors.
4) How root traces might be used further for interpreting sea-level changes along shorelines, in which shorelines either lost or gained sediment wherever plant communities were adjacent to the sea.
5) Integrating modern studies of leaf-cutter bees, beetles, gall wasps, and other insects that are well known for attacking plants with analogous trace fossils in fossil plant material.

No doubt these suggested avenues of research may also lead to other, new pursuits of knowledge about plants and their traces in substrates of the Georgia coast.

5

TERRESTRIAL INVERTEBRATES

THE CRAYFISH OF JEKYLL ISLAND

I was surprised, yet not surprised, to find out I was wrong. This feeling of humility is a common one among scientists, especially natural scientists who go outside to test what they learned indoors, whether this knowledge was gained through books, journal articles, Web sites, hearsay, or Web sites repeating hearsay. The habit of correction becomes even more acute among ichnologists, particularly when we practice actualism: What are the traces being made by modern organisms, and where are these traces? Traces, as the products of behavior that form whether humans witness it or not, also often point toward an animal presence in places that defies our expectations. In this particular instance, the traces were burrows, the animals were freshwater crayfish, and the place was Jekyll Island on the Georgia coast.

Although Web pages, just like books (yes, this one too), are never to be trusted completely, the brief item I encountered while reading one—stating that Jekyll Island had freshwater crayfish—was intriguing news. For the previous several years, I had assumed that none of the Georgia barrier islands should have these animals. My reasoning was based largely on knowledge of sea-level rise, salinity, and the biology of crayfish. Once the Pleistocene Epoch ended about 11,000 years ago and the last of the great North American continental glaciers melted, sea level went up and filled the low areas between the Pleistocene versions of the Georgia barrier islands and the mainland (Chapter 2). This transgression formed the back-barrier marshes we see today, but also effectively dug a moat that prevented many salt-intolerant animals from migrating over from the mainland. Additionally, the islands were small enough that resident animal populations originating from the mainland were left behind. These animals then must have undergone severe selection pressures, such as shortages in habitat, resources, and mates. Accordingly, their numbers may have decreased as no new individuals (or genes) were provided over generations; inbreeding then would have inflicted genetic bottlenecks and made them more vulnerable to disease and other environmental stresses. For freshwater species, one major pressure was being surrounded by the ocean and salt marshes. An occasional hurricane was all that was needed to soak a low-lying island with salt-infused water, a situation in which few crayfish or other freshwater animals could survive. So in a relatively short amount of geologic time, some freshwater animals even may have become locally extinct. With all of these factors in mind, crayfish were presumed to have been part of the original resident fauna on former barrier islands connected to the mainland during sea-level lows. Nonetheless, crayfish were on my list of animals least likely to have survived on the islands since the Pleistocene. This assumption seemed affirmed by scarce mention of them in the voluminous peer-reviewed literature written by Frey, Howard, and other ichnologists about the biota of the Georgia barrier islands (Chapter 2).

Otherwise, crayfish (or crawfish, in Southern vernacular) are quite common in Georgia, and the southeastern United States is one of two notable places in the world for crayfish biodiversity; the other is in southeastern Australia (Hobbs 1988). Among the 67 known species of crayfish in Georgia are some ichnologically important ones: burrowers that form 2–3 m (6.5–10 ft) deep, laterally extensive, and complex burrow systems in river floodplains, freshwater swamps, and on the banks of rivers. Crayfish, which are decapod crustaceans, either inherited this burrowing ability from lobster-like marine ancestors or evolved it independently in their various lineages during the Mesozoic Era (Bedatou et al. 2008; Martin et al. 2008).

In their burrowing, they use their prominent front claws (chelae or chelipeds) for pushing and removing sediment, and their carapaces and tails for packing the walls of the burrow interior. Tops of crayfish burrows are readily identifiable by pyramidal, 10–20 cm (4–8 in) tall towers on floodplain surfaces. Look closely at these piles and you will see they are stacked balls of mud, sandy mud, or muddy sand, rolled for ease of transport out of the burrow. In some instances, crayfish burrows are evident as circular holes, typically about 1–5 cm (0.4–2 in) wide and each set dead center in a tower. Together, these holes and towers make them look like small, lumpy volcanoes. The overall geometry of the burrow in the subsurface can be complicated, but normally descends as a vertical shaft until the crayfish tracemaker encounters the water table. From there, the burrow system spreads horizontally into branching tunnels and interconnecting chambers. This is a world we humans rarely see, but these are the places where crayfish communities live, eat, mate, fight, and die. Burrowing crayfish are rarely seen above ground, perhaps only briefly glimpsed at the tops of burrows.

These crayfish, having rudimentary gills, spend much of their time in the water-filled horizontal components of the burrow system, but also move up the vertical shafts to avoid drowning when the water table rises. This adjustment to rising subsurface water is why crayfish towers will suddenly appear overnight on floodplains after a heavy rain. As rainwater infiltrates the ground, the water table goes up, and burrowing crayfish move up with it. In fact, the presence of crayfish towers on a ground surface is a sure indicator that the water table is just below your feet. People who have lived in rural Georgia know this well, whereas most city dwellers are often surprised to find such burrows in their backyards and gardens after a rainstorm, and may not even know what made them. Floodplain maps could be drawn and revised on the basis of crayfish burrows; hence homeowners in the suburbs of Atlanta and other cities in Georgia would benefit greatly by identifying them. I also recall how crayfish burrows provoked eerie feelings while I attended a geology field trip to southwestern Georgia in 1994. Only a few months after severe floods in the Flint River had receded from that area, I was astonished to see months-old crayfish towers next to a paved road and high above a stream valley nearly 200 m (660 ft) away and well below us. The burrows comprised a vestige of former floodwaters in that area, and in my imagination these traces conjured visions of furious, orange-tinged currents filled with debris and bodies, inundating all of the low areas in the now-dry landscape.

With regard to Jekyll Island, and according to the Web site that mentioned the crayfish burrows, the traces were there in a freshwater wetland, next to the main paved road and a shopping center on the eastern side of the island. I took a chance that the Web site was fairly reliable, based on several premises. It was written and posted by environmental activists opposed to further development of Jekyll, and emotional appeals aside, my experience has been that specific statements about resident biota in affected areas are often factually based. In this instance, the presence of crayfish burrows in a definite place also was an easy one to check out, whether to disprove or confirm. Moreover, the activists included a few naturalists and other scientists, who were more likely to have some experience in identifying crayfish burrows, versus, say, random piles of mud. Last, these people used their real names accompanied by contact information, meaning I could talk to them about what they found.

Despite these reasons for me to be optimistic, though, I was still doubtful of their claims. For one, Jekyll is more altered by an ongoing human presence than most other Georgia barrier islands, and thus less likely to contain much of a native fauna. Unlike most of these islands, a causeway connects it to the mainland, and it hosts just less than a thousand permanent residents, as well as hundreds of thousands of visitors each year. The small shopping center near the freshwater wetland area was also accompanied by nearby golf courses—both miniature and normal sized—a recreational water park, hotels, restaurants, paved bike paths, boardwalks, neighborhoods with permanent residents and rental properties, and various other amenities that made it a easy place to accommodate city dwellers looking for a coastal getaway. Parts of the beach, however appealing it might seem to most tourists, look odd to a coastal geologist or ichnologist, having been changed considerably by beach erosion structures in the northeastern part of the island, such as jetties, seawalls, and riprap. These alterations then caused imbalances in the sediment supply to the beach, resulting in unwanted erosion, which in turn required some beach renourishment along the shoreline. "Beach renourishment" is an oxymoron if there ever was one, in that sand from somewhere else—usually from the mainland, such as the Georgia coastal plain—is dumped suddenly into areas where beach sand may have been eroded (Pilkey and Fraser 2003).

The net effect of such a sudden influx of sand, though, is to suffocate whatever suspension-feeding infauna might have been living in the subtidal and intertidal areas. Consequently, organismal traces in exposed sand flats at low tide on the northern end of Jekyll are not as common or varied as on the other Georgia barrier islands, although the southern end of the island is a bit better, neoichnologically speaking.

These differences in beaches, even on the same island, are stark reminders of the unintended consequences of environmental restoration efforts. Fortunately, pockets of maritime forest and freshwater wetlands were still extant on parts of Jekyll, and these were among the ecosystems the environmental activists were trying to preserve in perpetuity. Of particular interest in their preservation efforts was the wetland area containing the purported crayfish burrows, an area that also had large, mature red maples (*Acer rubrum*), cypresses (*Taxodium* spp.), and a unique species of water-adapted fern, the Virginia chain fern (*Woodwardia virginica*). In other words, this wetland sounded as if it was a special place to visit, both ecologically and ichnologically.

As a result of this preliminary information, followed up by some e-mail exchanges with the activists, I arranged for a brief visit to Jekyll with my wife, Ruth, in December 2008 during a three-week field trip to the islands. Honestly, I might have otherwise passed it by in favor of other islands. After all, its heavily altered terrains and beaches meant that its traces and habitats comprised poor models for comparison to what might be preserved in the fossil record (Chapter 2). Jekyll is a nice place to visit for many people, but not for ichnologists who, of course, are not most people.

Did I mention previously that I was wrong? The wetland area, located just west of the shopping center—exactly where the activists said it was located—did indeed contain crayfish burrows (Figure 5.1a). One member of their group, Steve Newell, a retired biologist from the UGAMI who had recently moved to Jekyll, met Ruth and me in the parking lot of the shopping center, and he led us into the site. I immediately spotted a conical tower, and once our eyes and minds were quickly trained by the next few, we soon realized that hundreds of burrow mounds were scattered throughout the visible area of the wetland. Circular cross sections of burrow openings ranged in diameter from 1.5–3 cm (0.6–1.2 in). Centimeter-wide pellets accumulated on the flanks of the burrows, forming the familiar small, lumpy volcano shape of a crayfish tower. Furthermore, the vertical shafts

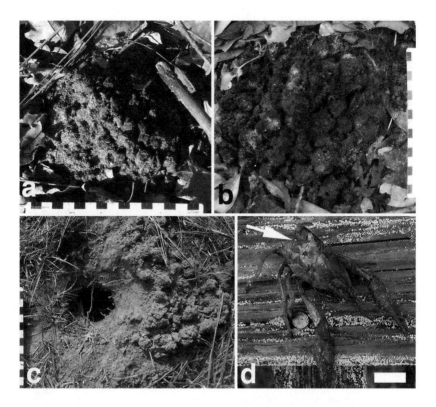

Figure 5.1. Trace and body evidence of freshwater crayfish on the Georgia barrier islands. (a) Burrow mound on Jekyll Island in freshwater wetland; scale in centimeters. (b) Burrow mound on Cumberland Island in freshwater wetland; same scale as in (a). (c) Probable burrow on Sapelo Island adjacent to freshwater wetland, with mound eroded (pellets to right) and open burrow shaft exposed; same scale as in (a) and (b). (d) Partially eaten crayfish on wooden handrail above freshwater wetland, Sapelo Island. Arrow points to beak marks on carapace, presumably made by predatory bird; scale bar = 1 cm (0.4 in).

descended directly downward for 20–30 cm (8–12 in), which I measured using a handy scientific instrument known colloquially as a stick. Some burrow mounds did not show obvious openings, but gentle removal of their tops revealed open, circular burrow cross sections below. This trait was also ichnologically consistent, as crayfish often plug the tops of their burrows. Most importantly, the burrows were in the right habitat: a lowland area containing plants that were specially adapted for saturated conditions—

red maples, cypresses, and ferns—rather than plants of a typical upland, well-drained maritime forest. In short, although we had not seen a crayfish, I was convinced, then and there, that these animals were present, active, and abundant on Jekyll Island. This was an astonishing find, made even more incongruous by walking less than two minutes from the parking lot of a shopping center to see what I had thought was absent from the Georgia barrier islands.

Of course, hypotheses can become more robust with further testing, especially if the same results are repeated independently. Only a few days before our visit to Jekyll, we had been in a very different setting, on Cumberland Island. While there, the eminently experienced naturalist Carol Ruckdeschel, who has lived on Cumberland and studied its biota for more than 30 years, led us to a freshwater wetland there that was also dotted with crayfish towers (Figure 5.1b). These, like the Jekyll ones, had probably been constructed about two weeks previously, bearing signs of recent weathering, but their identity was again undoubted. Years beforehand, Carol had distinguished two species of crayfish on the island, with body parts of one species contained in river otter feces. (Scatology, the study of feces, is probably Carol's favorite method of ichnological divining.) Something I did not know at the time was that one of the world's experts on crayfish, Horton Hobbs Jr., had listed a species of crayfish on Cumberland, *Procambarus lunzi*, in his comprehensive volume *The Crayfishes of Georgia* (Hobbs 1981). Although the book was published 27 years previously, the possibility loomed that this was one of the two species of crayfish on Cumberland. The other species was *Procambarus talpoides*, identified by Carol from the otter scat.

So thanks to the power of neoichnology, the claim of two species of crayfish on Cumberland was not so outlandish, either. Our cursory examination of the burrows in the Cumberland wetland suggested that these fell into two size categories based on burrow diameters, as one was noticeably smaller than the other. At the time, I wondered aloud if this was a possible species-linked difference in size. Moreover, Cumberland is a relatively undeveloped barrier island, which implied that it was more likely to retain native species than the heavily altered Jekyll. Of course, the hypothesis needed a bit more investigating to fill in the details, but as far as my mind's-eye view of Georgia barrier island crayfish was concerned, where I previously had none, I now had two.

The list of crayfish-bearing islands grew as the three-week-long trip in December 2008 progressed. Royce Hayes, the superintendent of St. Catherines Island who, like Carol, had lived on "his" island for more than 30 years, first reacted with wide-eyed surprise at my asking such a basic question about crayfish there, but he said yes too. He did not know the species of crayfish, but stated confidently how he had seen both living ones and their burrows, and knew exactly where they could be found. The next week, Jon Garbisch, the educational program director of the UGAMI on Sapelo Island, also affirmed the presence of crayfish and their burrows on Sapelo. During that same visit to Sapelo, James Nifong, a graduate student from the University of Florida conducting fieldwork on alligator ecology there, answered my inquiry (with a smile), "I have one [a crayfish specimen] dried out and in my freezer right now." Although he had not yet identified it, he said crayfish were common components in the stomach contents of juvenile alligators he had been studying in freshwater areas of Sapelo.

This all sounds believable, right? Nevertheless, anecdotes, even those coming from scientists I trusted, were not enough to satisfy my curiosity and sense of discovery, so Ruth and I sought out some field examples while still on Sapelo. Out we went, and a brief roadside stop next to a freshwater slough later revealed a few likely candidates for old, weathered crayfish burrows (Figure 5.1c). Strong circumstantial evidence of a crayfish presence in the same area came in the form of a chicken turtle (*Deirochelys reticularia*) next to the burrows; these turtles eat crayfish as a regular part of their diets (Jensen et al. 2008). Even so, I was not completely convinced of this evidence, and would have been more satisfied by newer burrow mounds with their characteristic knobby exteriors, or a chicken turtle actively munching on a crayfish. This skepticism, however, was erased by a revelation from my own past experiences on Sapelo. While reviewing a folder of digital photographs taken there in June–August 2004, I came across several pictures of a partially eaten crayfish, left on the handrail of a boardwalk above a freshwater pond (Figure 5.1d). A bird—probably a small heron—must have caught and eaten part of the crayfish just before I arrived. The proximity of the crayfish to the pond just below it was damning. I had completely forgotten about these photographs and their evidence of a crayfish presence on Sapelo, only to be reminded by the crayfish burrows of Jekyll Island four years later.

All of this information, some admittedly subjective and needing more rigorous and independent corroboration, led to new, exciting research questions. What crayfish species were on all of these islands? How did they get there? Were they relicts of the Pleistocene, trapped by rising sea level from the previous worldwide melting of ice-age glaciers? Could their larvae possibly have been carried on the feet of wading birds that flew from wetlands on the mainland to the islands, an aerially aided transport? Or, the most likely, and hence the most boring explanation (I hated to admit it): Did people bring them to the islands in only the past couple of hundred years, because these crayfish were good to eat? (As a big fan of crayfish étoufée, I understood how this could happen.) The problem with this scenario, though, was that open-water, bottom-dwelling crayfish are much easier to harvest than burrowing ones. After all, people do not just go out and collect burrowing crayfish for a meal unless they want to get a lot of exercise before dinner. So regardless of their origin, how did these crayfish persist while freshwater habitats shrank on every island, particularly Jekyll? All of these questions added up to a multifaceted scientific puzzle looking for a solution.

Thus it was that at the start of 2009 I undertook a more intensive review of the scientific literature, including Hobbs's (1981) book, which describes the two species of crayfish, *P. lunzi* and *P. talpoides*. A tip on the first species (*P. lunzi*) came from Carol, who told me in an e-mail message it was reported on Cumberland, St. Simons, and Sapelo Islands. In fact, she related in the same message how she sent a specimen to Hobbs himself in 1983, who readily identified it. The odds that Hobbs correctly identified this species were rather high, seeing that he had originally described and named it in 1940. The geographic distribution of *P. lunzi* included McIntosh County of Georgia, which includes Sapelo Island, and Hobbs recounted how one specimen was recorded from Sapelo and another from St. Simons Island. The other species, *P. talpoides,* was likewise named by Hobbs, although more than 40 years after *P. lunzi,* in 1981; it was listed in an online taxonomic list that also indicated its presence on Cumberland. Then there was the clincher for me. Hobbs had described both *P. lunzi* and *P. talpoides* as burrowers (Hobbs 1981). Thus the two differently sized burrows on Cumberland might be explained by size differences between adults of those two species. In short, I now had independent verification in the literature of at least two species of burrowing crayfish from the Georgia barrier islands. All thanks to a potentially dodgy Web page report by environmental activists of crayfish burrows on Jekyll Island, a place I would have least expected to have them.

Wrong, wrong, wrong. How good it was to be wrong! Now I considered the likelihood of four, possibly five Georgia barrier islands—Cumberland, Jekyll, St. Simons, Sapelo, and St. Catherines—with burrowing crayfish, when only a few months before I had assumed these places were devoid of such tracemakers. Moreover, what was most exciting about their presence on these islands was their Pleistocene origin (Chapter 2). This coincidence suggested that these burrowing crayfish might represent an element of a relict fauna, isolated from the mainland when sea level rose during the past 10,000 years. Of course, in science we do not prove, but disprove. As a result, the next steps will be to test the species identifications, find more burrows, carefully study the burrows in the context of their ecosystems, collect specimens of the tracemakers (but not too many, just in case these are rare species), and ponder how these burrowing crayfish relate to the broader geologic history of the Georgia barrier islands.

Regardless of whatever science will emerge from these new revelations, a lesson was learned about my prejudicial assumption that a developed barrier island could not be a potential source of ichnological discoveries. As the cliché goes, you never know until you ask, but just to modify that adage, you never know until you ask about—and look for—traces.

TEN LEGS GOOD: DECAPOD TRACEMAKERS

Our discussion of terrestrial and freshwater invertebrate tracemakers could start with any one of the major taxonomic groups, but for the sake of simplicity will descend in order of number of legs possessed by different invertebrate taxa (ten to zero: decapods → arachnids → insects → worms) before increasing abruptly once we consider traces made by the many-legged centipedes and millipedes. As will be discussed, the presence, absence, and number of legs are important anatomical traits to consider when evaluating the tracemaking abilities and subsequent traces made by terrestrial and freshwater invertebrates.

The most popularly known decapod representatives are crabs, crayfish, shrimp, and lobsters, some of which are best known for their edibility, but are also well loved among biologists and paleontologists for their scientific worth. Along those lines, the ecological and behavioral diversity of decapods is nothing short of remarkable. Although the greatest variety and numbers of decapods are associated directly with marine environments, a few have managed to adapt to freshwater conditions, whereas others live on the fringes of terrestrial environments, Among the freshwater dwellers are burrowing crayfish, which have diversified quite well in those environments. For decapods living on the edge (salinity-wise) on the Georgia coast, some crabs also burrow, but only in areas affected by tidal influxes of saline water. Nonetheless, because some of these crabs often make forays into terrestrial environments, they are mentioned briefly in this section on terrestrial decapod tracemakers.

A few species of burrowing crayfish occasionally go for strolls outside their burrows and water bodies, and hence leave distinctive trackways, described later. These traces are nowhere nearly as common, nor as likely to be preserved, as their deep burrow systems. As a result, ichnological studies have concentrated on their burrows, which have received much attention from crayfish biologists and ichnologists alike, especially because their fossil burrows have been interpreted from Mesozoic and Cenozoic rocks (Horwitz and Richardson 1986; Hogger 1988; Hasiotis and Mitchell 1993; Kowalewski et al. 1998; Bedatou et al. 2008; Martin et al. 2008). The description of crayfish burrows in the preceding section, however, may also cause an attentive reader to wonder, "How do you tell these apart from any other decapod burrow?" Indeed, this is a concern for paleontologists attempting to distinguish trace fossils made by freshwater decapods—probably crayfish—versus those made by decapods in marginal-marine and fully marine environments—probably crabs, shrimp, or lobsters. Other extant burrowing terrestrial decapods that might have their burrows confused with those of crayfish include land crabs, such as species of *Gecarcinus* and *Cardisoma* (Bright and Hogue 1972; O'Mahoney and Full 1984). These crabs, though, are more typical of tropical areas and are quite common in the Bahamas and parts of the Caribbean, far south of the Georgia barrier islands. Appropriately then, a brief comparison will be made here between the decapod burrows of crayfish, fiddler crabs, and ghost shrimp, with details about the marginal-marine and marine decapod burrow systems provided subsequently (Chapter 6).

As mentioned before, crayfish burrows are easily recognized at the surface by relatively tall, 10–20 cm (4–8 in) conical towers composed of muddy

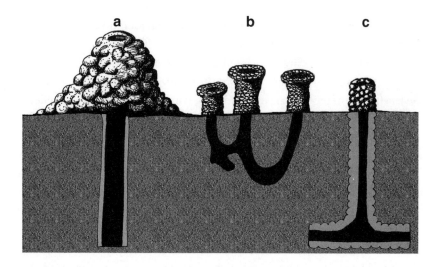

Figure 5.2. Crayfish burrow tower and its differences from other decapod-made towers with pelletal linings, in vertical view with subsurface portions of burrow and drawn to the same scale. (a) Crayfish tower. (b) Mud fiddler crab (*Uca pugnax*) turret. (c) Eroded part of callianassid shrimp burrow. Full vertical and horizontal dimensions of crayfish and callianassid burrows are abbreviated.

pellets; these pellets are about 5–15 mm (0.2–0.6 in) wide. Moreover, some of these structures are rather broad, about 15–30 cm (6–12 in) at their bases, making them look much like iconic images of stratovolcanoes (Figure 5.2a). Mud fiddler crabs (*Uca pugnax*) also make towers above their burrow shafts, but these are relatively small, cylindrical, wider (flared) at their tops, and composed of much smaller pellets (Figure 5.2b; Chapter 6). Hence these look more like a castle turret, rather than a volcano. Superficially similar to both of these burrows are the eroded lower parts of callianassid shrimp burrows, which are cylindrical and have knobby exteriors made of pellets (Figure 5.2c). These, however, have uniform widths and, if followed downward, may show extensive connections with horizontal and vertical components, forming complex branching networks (Chapter 6). In their overall geometry, callianassid burrows somewhat resemble crayfish burrows, but the crayfish burrow walls lack the knobby exteriors of callianassid burrows. Crayfish towers on a ground surface are actually more likely to be mistaken for molehills made by the eastern mole (*Scalopus aquaticus*), which are similarly sized and overlap ecologically with crayfish in terrestrial environments (Chapter 8). This ichnological confusion is especially understandable if the crayfish tracemaker plugged its burrow so that the opening to the vertical shaft is hidden, turning it into a conical mound of sand or mud. The key diagnostic trait of a crayfish tower in such instances, however, will be its ball-like pellets, which will last for weeks in the absence of rain. In contrast, mole hills, which are formed by a mole pushing up sediment from an underlying horizontal tunnel, will be composed of more irregularly arranged and sized plugs of sediment.

A commonality of burrow networks made by decapods and most other burrowing crustaceans, regardless of habitat, is a slight widening at each branch junction. This extra space allows for the burrow inhabitant to turn around or move around corners in the burrow. Unlike worms, decapods' stiff bodies need extra accommodation to make sharp turns, and they cannot easily reverse course once in a burrow. Geometry, nevertheless, is not all that matters in identification. Diameters of crayfish, callianassid shrimp, and fiddler crab burrows overlap only in the lower range of the first two groups; fiddler crab burrows, even those made by full-sized adults, are typically less than 2.5 cm (1 in) wide, whereas the interiors of callianassid shrimp burrows are 2–3 cm (0.8–1.2 in) wide. In contrast, Georgia crayfish burrow interiors can be more than 5 cm (2 in) wide.

Of course, with modern burrows, ecological context is paramount for proper identification. For example, decapod burrows on the edge of a freshwater pond on a Georgia barrier island clearly belong to crayfish, whereas those on salt marsh mudflats are from fiddler crabs. Where such simplistic identifications fail, however, is when a change in the salinity of standing water and sediments after a storm makes a formerly freshwater pond a more attractive environment for fiddler crabs (Chapter 6). Or, alternatively, a tidal channel could shift its position, abandon a salt marsh, and fill with freshwater, which may then make its mud more amenable for burrowing crayfish. A basic principle in ichnology is as environments shift and otherwise alter, tracemaking organisms will move with them (Chapters 2, 3).

Just in case anyone is doubtful that a pile of lumpy mud with a hole in its middle might be a crayfish burrow, these structures may possess a few more details that further confirm their tracemakers. For example, the top of the burrow may have a beveled and sculpted appearance, where the crayfish swept a cheliped across the muddy surface like a bricklayer smoothing mortar. The burrow interior, if composed of a firm mud, may show linear scratch marks from its legs, as well as longitudinal grooves caused by movement of the carapace and tail up and down the burrow. Repeated vertical movement in the burrow also results in a compacted inner liner, several centimeters thick and distinct from the surrounding soil or other sediments. Last, tracemakers with a motive more basic than intellectual curiosity—such as raccoons (*Procyon lotor*)—can leave no doubt about the identity of the animal that constructed the burrow. Raccoons will eagerly excavate any crayfish burrow in which the crayfish is near its top, meaning their dig marks and tracks will be directly associated with suspected crayfish burrows. Of course, raccoon feces that contain crayfish body parts will also attest to nearby crayfish.

Trackways of modern crayfish are rarely observed on land and fossil trackways of crayfish are unknown. This situation may be partially attributed to their not being recognized, but is more likely because crayfish simply do not travel far on dry land. The few crayfish trackways I have seen in parts of Georgia—but not yet on the barrier islands—show the four-by-four pattern so typical of a decapod, made by eight walking legs and two chelipeds held off the surface, but are also accompanied by a central, tripartite drag mark. These three longitudinal grooves are caused by the tail fan of a crayfish, composed of three body parts: a central telson and two lateral uropods. These trackways invariably lead out of and into open water for only short distances, and affirm that these crayfish are not well adapted for long forays away from water or their burrows.

The only other decapod tracemakers that one might encounter in or near fully terrestrial or freshwater ecosystems of the Georgia barrier islands are wharf crabs (*Armases cinereum*), marsh crabs (*Sesarma reticulatum*), and ghost crabs (*Ocypode quadrata*), although the last of these are much more common in foredunes and back-dune environments (Chapter 6). Wharf crabs are not so good at burrowing, bested in that respect by marsh crabs, whose burrows occur in salt marshes and are especially common along the banks of tidal creeks (Basan and Frey 1977). Nonetheless, both species tolerate low salinities—as little as 2 ppt, with normal marine salinity being 35 ppt—and consequently can be found quite close to maritime forests, where they sometimes forage (Zimmerman and Felder 1991). Accordingly, their tracks may show up in sandy and muddy substrates of maritime forests and potentially could be mistaken for trackways of large insects, or, if in back-dune environments, ghost crabs. All three decapod tracemakers—wharf, marsh, and ghost crabs—are thus what ichnologists might term as facies crossing, in that they normally live in and make traces in one ecosystem but can cross over to another and make traces in it too. These tracemakers will be discussed in more detail with other marginal-marine invertebrates (Chapter 6). In the meantime, make a mental note that animals on the move will make traces outside of their normal, expected ecosystems, a theme that will be revisited subsequently (Chapters 6–9).

EIGHT LEGS BETTER: ARACHNID TRACEMAKERS

Arachnid tracemakers of the Georgia barrier islands include both scorpions and spiders. Other than lumping these two groups of animals together under the rubric of "dangerous multilegged creatures that either bite or sting you," arachnids have anatomical traits, like their eight walking legs, that point to their shared genetic heritage. Arachnids originated from marine ancestors more than 450 million years ago during the Paleozoic Era, then some of these diverged and adapted to marginal-marine, freshwater, and fully terrestrial environments (Pisani et al. 2004). The most famous, awe-inspiring, and frightening of arachnid relatives from the Paleozoic Era were eurypterids, which were the largest arthropods that ever lived. Some eurypterids were more than 3 m (10 ft) long, and at least one left a fossil trackway nearly a meter wide (Whyte 2005). These were among the top predators of the seas until vertebrates underwent their own evolutionary increases in body size later in the Paleozoic Era. Modern-day but distant relatives of eurypterids and arachnids living in marine environments are limulids, which not only represent a long-lived lineage of arthropods but also are impressive tracemakers in their own right (Chapters 2, 6).

Unlike spiders, which are diverse and common throughout Georgia, including near and on the barrier islands, only three species of scorpions are known in all of Georgia: one with a religiously inspired name, the Southern devil scorpion (*Vejovis carolinianus*), and two species of the striped scorpion (*Centruroides vittatus* and *C. hentzi*). Of these three, only *C. hentzi* is known in southern Georgia and the barrier islands. In 1821, one species of scorpion was named from the Georgia islands, *"Buthus vittatus."* Unfortunately, the original specimen was lost and is presumed to have belonged to the same genus *Centruroides*; it was probably *C. hentzi*, which lives on Cumberland Island (Shelley and Sissom 1995). *C. hentzi* is a gorgeous scorpion, measuring about 7–8 cm (2.8–3.1 in) long, and derives its common name from three longitudinally arranged, muted orange stripes alternating with dark brown stripes, giving it some camouflage in maritime forests and other upland environments. Despite its menacing appearance, this scorpion's sting is only temporarily painful and decidedly nonfatal. Hence neither panic nor death-dealing blows with shoes, books (including this one), wine bottles, and other objects are necessary if one is spotted.

Many scorpions in the western United States are burrowers (Williams 1987; Brown et al. 2002; Bost and Gaffin 2004), but the striped scorpion is not known to make such traces, preferring to live in rotting vegetation, under rocks, and generally in nooks and crevices. As a result, its most probable traces will be tracks, which would likely be spotted on the fine-grained sand of dunes and back-dune meadows near its preferred habitats, maritime forests, and particularly in pine-dominated portions of forests. Scorpion trackways are characterized by rows of three to four tracks slightly offset on each side of the trackway (Figure 5.3a). Why only three leg impressions on each side? Although scorpions have four pairs of walking legs, only the last three pairs typically register in their trackways (Davis et al. 2007). Their large pincers, or pedipalps, are ahead of and above the walking legs, and normally do not leave any marks in a trackway. Regardless of the number of tracks left, though, nearly all scorpion trackways have a simple, slightly undulating medial groove, caused by dragging a heavy abdomen, which carries its infamous stinger. Scorpion trackways showing such traits have been interpreted from the fossil record and are assigned to the ichnogenus *Palaeohelcura* (Sadler 1993; Braddy 1998). Surprisingly, experiments conducted on *C. hentzi* running speeds show it is capable of fast bursts for as much as 20 seconds and top speeds of 9 body lengths per second, although it normally slows considerably and reaches near-total

Figure 5.3. *Facing.* Arachnid traces. (a) Scorpion trackway pattern, with arrow indicating direction of movement. Size depends on species of scorpion, so no scale given. (b) Longitudinal (vertical) section of wolf spider burrow (*Geolycosa* sp.), based on descriptions and figures by Ratcliffe and Fagerstorm (1980) and Platt et al. (2010). Keep in mind how burrows of *Geolycosa* can be about half the size of *Hogna* burrows. Scale = 10 cm (4 in). (c) Entrance to wolf spider burrow (*Geolycosa* sp.) lined with silk and plant debris: St. Catherines Island. Scale (left) in millimeters. (d) Typical spider trackway pattern, with arrow indicating movement; compare to scorpion trackway.

TERRESTRIAL INVERTEBRATES 185

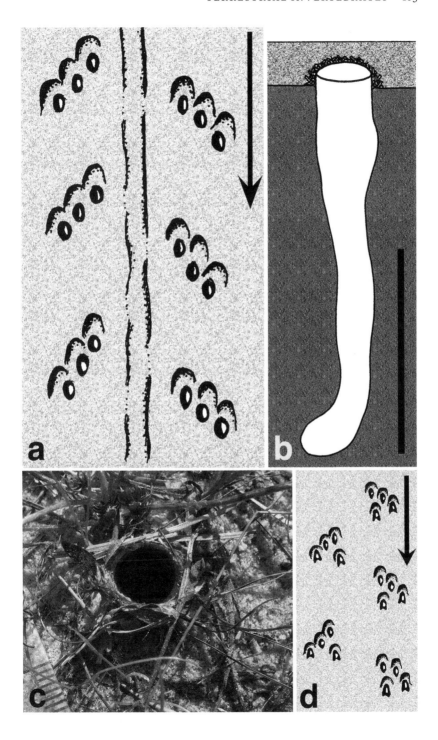

fatigue after 30 seconds (Prestwich 2006). These results imply that scorpion trackways indicating running may be rare compared to their normal walking or resting traces.

Perhaps the mere mention of spiders conjures thoughts of webs, along with all of the metaphors they have inspired, which can be readily researched on the World Wide Web. Indeed, webs are the most easily observed of arachnid traces in terrestrial ecosystems of the Georgia barrier islands, although their preservation potential in the fossil record is extremely low. Nonetheless, the former presence of spider webs from the fossil record is inferred indirectly from unusual assemblages of macerated insect parts in a small volume of rock, or rare instances of spider silk preserved in amber, in some cases with associated prey (Peñalver et al. 2006; Vollrath and Selden 2007).

Direct evidence of spiders in the fossil record is normally discerned from body fossils, but many were also capable of making traces with a better preservational potential than webs, such as burrows. Most spider burrows are simple, vertical to near-vertical shafts that are occasionally lined with webbing, which helps to waterproof the burrows against moisture intruding from the surrounding sediment. Top and bottom ends of burrows may be slightly bulbous, which allows enough room for the spider to turn around during its various trips to and from the top of the burrow (Figure 5.3b). Some burrow tops are concealed by small lids hinged with spider silk; these are the work of so-called trapdoor spiders. Other burrows, such as those made by burrowing wolf spiders—various species of *Geolycosa* or *Hogna*—are discernible as open circular outlines and slightly raised turrets supported by sediment or plant debris, and bound together by silk (Wallace 1942; McCrone 1963; Shook 1978; Marshall et al. 2000; Figure 5.3c). These turrets probably serve as home alarm systems, which transmit vibrations from animals outside of the burrow, as well as waterproofed dikes that prevent water from entering the burrow. Although tarantulas are not native to Georgia, their burrows lack these turrets, and thus are distinguishable from wolf spider burrows by this trait alone.

Burrows of *Geolycosa* and *Hogna* may be distinguished from one another by size. For example, *Geolycosa* burrows are often only about 1 cm (0.4 in) wide and 5–9 cm (2–3.5 in) deep (Halloran et al. 2000). In contrast, *Hogna* burrows can be about twice as large, measuring as much as 2 cm (0.8 in) wide and 20 cm (8 in) deep, such as those of the giant wolf spider,

Hogna carolinensis (Shook 1978). Bigger sizes, however, also may be a result of spiders taking over and modifying another animal's burrow, and thus they may not necessarily reflect spider size. Burrowing wolf spiders are active predators and use their burrows mainly for protection, either waiting for potential prey to wander nearby, or going out and nabbing it. As a result, these burrowing spiders normally dig their burrows in open, sandy areas such as back-dune meadows. Perhaps surprisingly, fossil burrows from the Miocene Epoch (24–5 million years ago) of Namibia have been preserved with webbing impressions similar to those of an extant genus of burrowing spider in that area today (Pickford 2000). Such finds hold promise that the search images provided by modern spider burrows, like those of the Georgia coast, might likewise help in discovering more such trace fossils.

Spider trackways are other recognizable arachnid traces if made in substrates that can record such light footfalls. Spider trackway patterns often reveal eight-legged tracemakers (Figure 5.3d), although these trackways are potentially confused with those made by similarly sized scorpions and decapods. Once studied more closely, however, the leg movements of both spiders and scorpions are quite different from those of decapods, and an understanding of these motions helps to prevent their misidentification. Spiders have a four-by-four pattern in their trackways, with alternating groups of four impressions on either side, similar to those of ghost crabs. Nonetheless, even the largest of Georgia spiders never approach the size of adult ghost crabs (for which many people are thankful), and spider tracks are quite small and difficult to spot, even in fine-grained sand. Yet another distinctive trait of spider tracks is that the foremost two of the four tracks on either side of a trackway sometimes split into two points (didactyl) and the pair in the rear have a single point (unidactyl). As mentioned earlier, scorpion trackways are also noticeable because of a tail drag mark in the middle of the trackway, whereas spider trackways will lack these. Moreover, trackways made by scorpion species native to Georgia are often slightly wider than those of spiders, thus further narrowing in on their identification.

Although the neoichnology of arachnids on the Georgia barrier islands might seem straightforward, little has been done on this subject. Most of the preceding ichnological information is based on known behaviors of selected species of scorpions and spiders combined with their geographic ranges. Furthermore, few arachnids are reported directly from the islands, and none has had its traces studied. Consequently, much more can be learned about arachnid traces, their distinguishing traits, and fossilization potential, especially for researchers who can overcome social fears to appreciate the ichnological potential of these eight-legged beauties.

SIX LEGS BEST: INSECT TRACEMAKERS

Likewise, the ichnology of insects is surprisingly underdeveloped in comparison to the study of marginal-marine invertebrate traces on the Georgia coast (Chapter 6). Seeing that insects compose the bulk of animal biodiversity, and live in modern terrestrial environments from pole to pole, one might expect that their traces would have received much attention from ichnologists. Alas, they have not. Part of this bias may originate from the pioneering neoichnological work by German sedimentologists on the North Sea tidal flats, which emphasized intertidal and subtidal marine invertebrates, followed by German and American scientists and their similar focus with Georgia coast tracemakers (Chapter 2). In both instances, only snippets of these voluminous works mentioned insect traces. Nonetheless, easy access to terrestrial and freshwater environments of the Georgia barrier islands should enable much future research on insect tracemakers. Furthermore, ichnological investigations of arachnid and insect traces might be combined to better establish predator–prey relations in those ecosystems. Overall, insect ichnology of the Georgia barrier islands would constitute an entire book unto itself, but the gentle reader will be spared such pedantry for now, and instead will only be given an overview of their ichnologic standouts.

Before covering insect tracemakers in general, however, a brief review of insect life cycles is needed to provide insights on how the ontogeny (growth history) of a tracemaker can result in quite different traces during the lifetime of a single insect. Insects have two main types of life cycles, holometabolous and hemimetabolous (McGavin 2001). Holometabolous insects have complete metamorphosis, transforming from egg to larva, pupa, and adult. Probably the most familiar of the holometabolous insects are lepidopterans (butterflies and moths), but these also include coleopterans (beetles), dipterans (flies and mosquitoes), and hymenopterans (wasps, bees, and ants). Hemimetabolous insects, in contrast, have incomplete metamorphosis, which is characterized by a nymph stage before the adult and no pupation; thus the sequence is egg, nymph, and adult. This development is not so simple, though, as nymphs undergo their own morphologically distinctive stages (instars) that more closely resemble an adult as they progress in their growth. The end of one instar, and the start of a new one, is marked by the shedding of an outgrown exoskeleton (ecdysis). Instars are labeled from youngest (first instar) to oldest (as many as 30 in some species), and are often defined visually on the basis of wing growth (Merritt and Cummins 1996). Examples of hemimetabolous insects are orthopterans (grasshoppers and crickets), blattoideans (cockroaches), odonates (dragonflies and damselflies), and homopterans (cicadas and others). Hemimetabolous insects can be further subdivided on the basis of whether the nymph and adult live in the same habitat (paurometabolous) or different ones (heterometabolous; Capinera 2008). For example, grasshoppers develop on and then above ground surfaces (subaerial), whereas dragonfly nymphs grow up in fully aquatic environments, yet the adults later become masters of the air.

Differing stages of development and habitats mean that a single insect, depending on its species, may make different traces during its lifetime in a variety of ecosystems and substrates. This situation is especially true for holometabolous insects, in which larvae spend much of their time and energy eating, hence causing many feeding traces. In contrast, adults of the same species may be moving or brooding, making numerous locomotion and reproduction traces, respectively. Also, think of how cicada nymphs may live 13–17 years burrowing underground, and then emerge to live only a brief aerial and arboreal existence. As a result, the vast majority of cicada traces will have been made while it was a nymph, whereas traces from its adulthood will be as fleeting as its last year of life. Thus these developmental stages in insects and their associated traces will be noted in their descriptions, while also taking into account how some traces are more likely to be preserved in the fossil record.

Ants

With this review of life cycles in mind, a discussion of insect tracemakers can commence, and is best started with ants. Why? Famed ecologist and ant entomologist E. O. Wilson probably said it best: when once asked, "What's the deal with ants?" he simply replied, "Because ants are cool." Indeed, facts attesting to the coolness of ants are overwhelming. Ants are hymenopterans, an evolutionarily related group of insects that include bees and wasps (Grimaldi and Engel 2005). Although ants only consist of about 12,000 species, they compose about 15 percent of terrestrial animal biomass worldwide, outweighing humanity in tonnage (Hölldobler and Wilson 1990). If all continents north of Antarctica vanished tomorrow but left behind ants, they (along with worms) would comprise ghostly outlines of those same continents. Accompanying their sheer numbers are their biological diversity and a bewildering variety of behaviors. As might be expected, they are also ichnologically significant, causing massive, wholesale changes in sedimentary and woody substrates of terrestrial environments. Ants, in many ways, rule the near-surface world of terrestrial environments. Cool, indeed.

How do ants burrow? Such a question may be based on a mistaken premise: ants do not actually dig like some mammals or even like crayfish, both of which use specially adapted appendages for shoveling, compacting, and pushing aside sediment. Instead, they employ a blend of rugged individualism and collective action, a seeming dissonance that might never be reconcilable in human societies. Ants are prime examples of eusocial insects that are polymorphic, having morphologically distinctive castes—workers, soldiers, and queen—and with each caste serving a function in a colony (McGavin 2001; Capinera 2008). For example, in nest making, a worker ant may grasp a single grain of sand, seed, pebble, or other particle with its mouthparts, and carry it out of the region of the intended nest. This ant will then repeat this action, perhaps hundreds of times that same day. Multiply this activity by hundreds of thousands (or millions) of worker ants in a colony, along with the application of a little saliva as cement, and *voilá*, a nest structure is made, complete with horizontal tunnels, vertical shafts, and expansive chambers. The amount of sediment moved through such activity is staggering. A single ant colony can displace many tens of cubic meters of sediment, and when scaling for size, construct subterranean structures more massive than the largest human modifications of landscapes, including the Great Wall of China or the latest megamall. Ants use saliva to harden the walls of their nests; this helps to keep the integrity of the nest intact, but also increases its likelihood of fossil preservation. Thus between their depth, size, and robustness, ant nests have a good chance of becoming trace fossils, albeit exceedingly complex ones that represent the combined, unified behaviors of millions of individual tracemakers.

Consequently, they may be difficult to identify properly, especially if a paleontologist sees only a small part of what might have been an originally huge nest.

Although little research has been conducted on ground ant nests of the Georgia barrier islands specifically, studies have been done on modern ants burrowing into Pleistocene sand dunes from former barrier islands on the mainland part of Georgia. Otherwise, at least three species of ground-dwelling ants are reported from Sapelo Island (Ipser et al. 2004): the Florida harvester ant (*Pogonomyrmex badius*), the slave raider ant (*Polyergus lucidus*), and the funnel ant (*Aphaenogaster miamiana*). Plant-dwelling species include the acrobat ant (*Crematogaster clara*), which lives in the stalks of smooth cordgrass in the salt marsh (Chapter 3), and the common Florida carpenter ant (*Camponotus floridanus*), a frequent modifier of living and nonliving trees. Many entomologists also acknowledge the presence of the much-loathed imported red fire ant (*Solenopsis invicta*), but try to ignore it in favor of the native fauna, perhaps in the vain hope that it will go away.

One of the more ichnologically important species of ant on the Georgia barrier islands is the Florida harvester ant, which makes spectacular and complex nest structures (Tschinkel 2003, 2004). Their nest architecture was studied through casting nests with plaster and molten zinc, which made three-dimensional facsimiles of the originals. Yes, you read that right: molten zinc was poured into the nests, and yes, more than a few ants were likely harmed by this technique. Harvester ants are so called because they gather seeds for food. Their nests can be more than 3 m (10 ft) deep and more than a meter (3.3 ft) wide, depending on the nature of the substrate (they prefer burrowing in fine-grained sand) and the depth of the water table (most ants do not like being fully immersed in water). Nests are composed of thin, vertical to near-vertical and spiraling shafts that connect with horizontal, disk-like chambers (Figure 5.4a). The density of these shafts and chambers is greater in the uppermost one-third of the nest. This is a result of older worker ants concentrating their efforts in this region of the nest, whereas younger workers construct the deeper parts of the nest (Tschinkel 2003, 2004). The preservation potential of these nests as trace fossils is high, considering their depth and the amount of sediment disturbance.

Ants that nest in environments of the Georgia coastal plain composed of former barrier island sediments (Pleistocene coastal dunes) include the

Figure 5.4. Nesting traces of the Florida harvester ant (*Pogonomyrmex badius*). (a) Subsurface architecture of a small- to medium-sized nest; based on a cast made and figured by Tschinkel (2004). Scale = 25 cm (10 in). (b) Surface expression of the nest entrance and sediment mound: Cumberland Island. Scale = 10 cm (4 in).

pyramid ants (*Dorymyrmex bureni*), lawn ants (*Forelius pruinosus*), and leaf-cutting ants (*Trachymyrmex septentrionalis*; Isper et al. 2004). The last of these, the leaf-cutting ants, are little farmers, growing their own fungi for food. Workers cut and gather pieces of leaves and install these in modestly sized, 5–10 cm (2–4 in) wide subterranean chambers, connected by thin vertical shafts. Fungi then develop on these leaf parts, and are consumed by the colony. The agricultural prowess of this species, however, is humbled by that of leaf-cutting ants of the subtropical genus *Atta*, which make huge, semispherical fungal chambers in their colonies (Hölldobler and Wilson 1990; Moreira et al. 2004). All three species of ants—pyramid, lawn, and leaf-cutting ants—are great examples of modern tracemakers overprinting substrates of older environments, which potentially can lead to complex and composite traces. This situation also shows how the products of past barrier island environments, including the traces of biota from their original ecosystems, can still influence the behaviors of modern tracemakers (Chapter 10).

To identify the maker of an ant nest, one does not necessarily have to capture specimens or cast their nest interiors, as the outward appearance of a nest can sometimes be allied with a species of ant. For example, piles of sand and other fine-grained material ejected by the Florida harvester ant make a broad, low-relief depression with a central hole at the nest entrance (Tschinkel 2004; MacGown et al. 2008; Figure 5.4b). In contrast, conical, steeply heaped mounds of ejecta, looking like little craters, denote nest entrances of pyramid and lawn ants (*D. bureni* and *F. pruinosus*, respectively). Even more distinctive are the crescentic mounds of leaf-cutting ants (*T. septentrionalis*), which look much like miniature parabolic dunes on one side of the nest entrance hole. Surface features of an ant nest, though, are often eroded first by wind, rain, trampling, and so on. In contrast, the deeper parts of the nest structure may become filled with overlying sediments, making a natural cast of the original structure and improving its preservation potential.

Some ants do not necessarily stay in the ground when making their nests, but will get a little bit higher by occupying soft, decaying stumps or trees and other dead wood. One such ichnologically important ant is known popularly (or unpopularly) as the Florida carpenter ant (*Camponotus floridanus*). Homeowners in urban areas may know these ants and their traces a little too well, considering how their wood-boring lifestyle, combined with nest building of termites, can weaken the structural integrity of a house. Carpenter ants, unlike termites, do not consume soft wood as food. Instead, workers tear it apart with their mandibles and remove the pieces to clear tunnels, galleries, and other passageways in a nest; this nest can also extend into the ground (Robinson 2005). Moreover, termites tend to eat wood along its grain, whereas carpenter ants both cut along the grain and across it (Suiter 2009). Their wood-altering mandibular strength can actually be felt by their powerful bite, another reason why these ants do not see much love in the southeastern United States. Pest-control specialists detect these ants by rapping on wood and listening for a clicking sound, which is produced by carpenter ants snapping their jaws in response to intrusive vibrations. One of the more interesting aspects of these ants, though, is how one caste—workers—is split into minor workers and major workers. Minor workers, which are smaller than major workers, start work on the nest soon after a queen has established a colony; the major workers later succeed these. Hence tunnels and chambers constructed by minor workers may be modified or intersected by those of the major workers, and nest structures will show definite progressions in size based on the maturity of a colony. Moreover, this ant prefers to nest in preexisting spaces, sometimes provided by unoccupied termite galleries. The resulting traces are thus composite, representing the work of at least two species (Chapter 4).

Unfortunately, nests of the Florida carpenter ant have not been described in an ichnological way. As a result, ground nests of a related species (*C. socius*) are explained here for comparison. Like many other ant nests, these were examined by first making metal and plaster casts of nest interiors, which revealed structures as deep as 60 cm (24 in), with horizontal chambers connected by vertical shafts (Tschinkel 2005). The size of a nest, as measured by both area covered and volume, depends on the number of workers present. This means that nests with only a small number of minor workers are modest, whereas those with lots of major workers are correspondingly robust. Chamber size and number of lobes coming off chambers increase with the number of workers too (Tschinkel 2005). With enough activity, carpenter ant borings may become mixed in with the traces of wood-boring beetles, carpenter bees, termites, and other insects that thrive in woody substrates. As a result, an ichnologist may have to sort out each distinctive trace to discern who was chewing through the wood and when.

Last, although not native to Georgia, the imported red fire ant (*Solenopsis invicta*), originally from South America, is worth noting as a tracemaker on the Georgia barrier islands. This is because of its extensive nest making and alteration of terrestrial sediments, as well as how its mere presence in a terrestrial ecosystem affects the behaviors of native tracemakers. Fire ant colonies are often easily spotted as large, 10–50 cm (4–20 in) tall mounds of bare soil in the midst of low-lying vegetation. Their identity can be tested readily by sitting on them, which will provoke as many as 250,000 worker ants to instantly flow out of the nest. Fire ants both bite and sting. They first grasp offenders with their powerful mandibles, followed by rapidly and repeatedly injecting venom with their abdomens. Understandably, people dislike these ants. Many other mammals feel the same way, and pointedly avoid areas infested with fire ant colonies, meaning their tracks will not co-occur in the vicinity of extensive nests. The nests themselves, though, are quite interesting ichnologically, and researchers have characterized these through their castings. (And if there ever were ant nests begging to have molten zinc poured down them, these would be it.)

At first glance, fire ant nests seem quite complicated because of the dense arrangement of vertical shafts connecting to short, horizontal chambers, both of which are only 1–2 cm (0.4–0.8 in) or less wide. In actuality, they follow a basic plan seen in other ant nests in which vertical shafts lead to chambers. The difference is that more of these components are packed into a smaller space than in most ant nests (Tschinkel 2003, 2004), which explains how the nest accommodates so many ants close to the surface and ready to attack.

Trace fossils of ant nests are known from the geologic record, with the ichnogenera *Attaichnus* and *Parowanichnus* used for some structures resembling ant nests (Genise et al. 2000; Smith et al. 2009). No doubt more such fossil nests will be found once the search images provided by casts of modern nests are applied to the geologic record. Considering how ants have so vastly affected terrestrial sediments since their evolution in the Mesozoic, the likelihood of paleontologists discovering some fantastic fossil ant nests is high.

Ant Lions

With ants providing such a large amount of the biomass in terrestrial ecosystems of the Georgia coast, something certainly must prey on them. Sure enough, many species of animals (and a few plants; Chapter 4) will eat ants, but two species of insects on the barrier islands in particular, *Myrmeleon crudelis* and *Vella americana*, should be singled out for their depredation of ground-dwelling ants. Popularly known as ant lions, the larval forms of these animals make a living by consuming ants and other small insects unfortunate enough to fall into their ichnologically distinctive traps. Ant lions are larvae of a particular group of neuropterans, which are insects with beautiful net-like wings, such as lacewings and dobsonflies (Capinera 2008). Hence these are excellent examples of holometabolous insects that change from ground dwelling to aerial forms within their lifetimes, which in turn translate into distinctive phases of tracemaking activities. Size differences between *M. crudelis* and *V. americana* help to distinguish them, but they can also be identified on the basis of their traces.

The smaller of the two species, *M. crudelis*, makes conical, concave pit traps by burrowing in downwardly oriented spirals, ejecting loose sand with its head as it burrows and deepens the pit. Ant lions are discriminating in their use of sediments, tossing out larger-sized particles and keeping finer-grained sand, thus altering the size range of sediments in the uppermost part of a soil (Lucas 1982). Interestingly, these insects burrow by pulling with their back legs and abdomens, pointing their rear ends in the direction of movement: in other words, they burrow butt first. Ant lions burrow in a wide circle, roughly defining the outer edge of the pit. They then move inward in a spiral, flicking out sediment along the way until the pit deepens. Once pits are made, they are funnel-like in shape and further denoted by a raised rim of sediment around the burrow. Burrows are about 2–7 cm (0.8–2.8 in) wide and 2–5 cm (0.8–2.0 in) deep, with a short (< 10 mm [0.4 in] deep and < 2 mm [0.1 in] wide) vertically oriented shaft at the bottom of the pit (Figure 5.5a,b). Ant lions sit inside these burrows with their disproportionately huge jaws at the top, and just wait for gravity-challenged prey to tumble down their way. Once an ant or other small insect walks over the edge of the pit, it will fall down the steep slopes. Ant lion larvae make the pit walls with high enough angles (20–30°) that, when combined with their composition of loose, dry, fine-grained sand, make escape for any small, flightless arthropod nearly hopeless. Sand grains that fall down the slopes of these traps are immediately flicked out of the pit bottom by the ant lion larva, which also lets an ichnologist know whether a burrow is occupied or not. Flicking is done with the head and jaws, which snap upward in reaction to inert objects blocking the burrow entrance, but is also triggered by sand grains dislodged by a prey insect. Sand flung out of a burrow also adds to a small-scale avalanche on the pit wall, hastening the downward slide of a struggling victim. Once an insect reaches the bottom of the pit, it is snatched rapaciously and sucked dry of its juices. Victims' carapaces, drained of their precious bodily fluids, are then discarded unceremoniously outside of the burrow entrance, serving as further sign (or warning, from a prey–insect perspective) of an occupied burrow. The bottom shaft also serves as a pupation chamber, from which the flying adult phase will eventually emerge.

The preservation potential of such traces in the geologic record is probably low, which is seemingly verified by no reports of their trace fossils. Nonetheless, these should be included in search images of possible insect trace fossils, especially when considering the abundance and close spacing of modern examples, which portends well for preservation in the geologic record. Also noteworthy is how ant lions are responsible for sediment advection, in which they purposefully reject larger particles in favor

TERRESTRIAL INVERTEBRATES 197

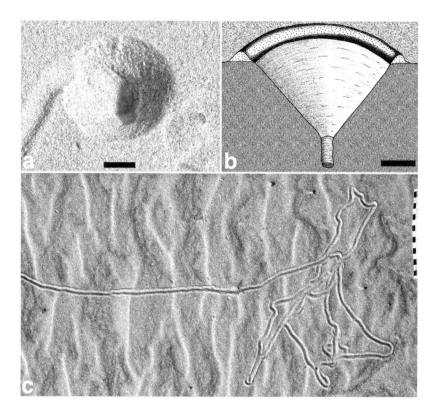

Figure 5.5. Ant lion traces. (a) Surface view of burrow funnel made by *Myrmeleon crudelis*, with attached collapsed burrow: Cumberland Island. Scale = 1 cm (0.4 in). (b) Oblique and vertical section of typical ant lion burrow. Scale = 1 cm (0.4 in). (c) Shallow, collapsed horizontal burrow of another species of ant lion (*Vella americana*), lending to its nickname, doodlebug: Cumberland Island. Scale in centimeters.

of finer-grained sediments (Lucas 1982). As a result, these insects alter the grain-size sorting in the uppermost few centimeters of any area they occupy in great numbers.

Other common ant lion larval traces encountered in sandy substrates are shallow, meandering and looping horizontal burrows or trails visible on dry sand surfaces. These traces look like random doodles made by a child's finger (Figure 5.5c), and in fact are why this larger of the two species of ant lion, *V. americana* (and other species of *Vella*) are nicknamed doodlebugs. Mark Twain (in *The Adventures of Tom Sawyer*) and others have mentioned doodlebugs in their writings, demonstrating the influence of terrestrial invertebrate ichnology on Southern literature, however slight that might be. Burrowing is again done backward relative to the tracemaker's anatomy, as its posterior leads the way when plowing through the sand. Back-dune meadows, the landward sides of coastal dunes, and other open, sandy areas make up the places where the doodlebug actively hunts for prey in the sand; contrast this behavior and tracemaking with that of the passive predation of *M. crudelis*. These ant lion larvae are about 1–1.5 cm (0.4–0.6 in) long; hence they also make proportionately larger traces than *M. crudelis*. Their shallow burrows are also astonishingly long: one I followed on Cumberland Island was more than 10 m (33 ft)! These factors, coupled with their more active lifestyles, mean that an individual doodlebug affects more sediment during its larval life than that of a smaller, sedentary ant lion species. The latter, though, probably have a greater effect on near-surface sediments through their sheer abundance. Regardless, doodlebug traces are also rather shallow and thus easily eroded, implying that such traces might be quite rare in the geologic record. Nonetheless, I would be happy to be proved wrong on this.

Bees

Although ant lions are indeed fascinating tracemakers on the Georgia barrier islands, they are easily overshadowed by the variety of traces made by bees, which along with wasps belong to the evolutionarily linked group Apoidea. Bees and wasps had a common ancestor at one time, probably in the middle of the Mesozoic Era; that common ancestor, in turn, was shared evolutionarily with ants (Grimaldi and Engel 2005). Bees and wasps, though, clearly evolved along different routes in comparison to ants, but some lineages likewise developed eusocial behavior with caste systems—such as workers, drones, and queens—which of course affects the relative complexity of their traces. Like many insect groups, bees are quite diverse and cannot be neatly summarized behaviorally. Furthermore, although we often equate bees with honeybees and hives, bees are represented by tens of thousands of species. Most of these are solitary or primitively eusocial bees, not the fully eusocial bees that have queens, drones, and workers in huge colonies (Michener 2000; MacGavin 2001). Nonetheless, the vast majority of traces made by bees that interest ichnologists are those related to reproduction. Hence the following treatment mostly discusses bee reproductive traces, while also mentioning any other noteworthy trace-inducing activities.

Bees are prolific tracemakers on the Georgia coast, represented by a variety of ground-burrowing, leaf-incising, and wood-boring species. Among these tracemakers, though, perhaps the best known are the halictid bees, commonly known as sweat bees because of their annoying attraction to human perspiration, although they are more properly labeled as miner bees in recognition of their burrowing prowess. On the Georgia barrier islands, the most common halictids are *Agapostemon virescens* and other species of *Agapostemon*, which are small bees, less than 1 cm (0.4 in) long, with beautiful metallic green thoraxes and yellow-and-black-striped abdomens.

If you also see such a bee burrowing, it is most assuredly a female. The purpose of the burrow is for brooding, which involves making chambers for development of offspring (called cells); provisioning cells with food, in most instances a ball of nutrient-rich pollen; lining cells with a waxy substance to decrease water intrusion and fungal growth; capping cells to prevent intruders, some of which get in regardless, and most of whom are wasps (discussed later); and covering the entrance to the burrow. The burrow top has a small, 3–5 mm (0.1–0.2 in) diameter and circular cross section and is denoted by a short pile of sediment (tumulus) excavated from the burrow that may or may not be covering the burrow entrance (Figure 5.6a). The amount of sediment indirectly conveys the length of the burrow, as its diameter stays nearly constant throughout: in other words, more sediment corresponds to a longer burrow. How long? Some can be as deep as 30 cm (12 in), although burrow lengths depend largely on each individual bee and its species, but is also affected by the host sediment and other ecological factors (Eickwort 1981; Michener 2000). Millions of years of natural selection have led to halictid bees today that avoid burrowing into saturated sand, for example. Bee larvae and pupae, like many terrestrial insects, drown quickly once immersed in water, and even slightly moist conditions will cause fungal attacks on their brood cells (Fellendorf et al. 2004). Halictid bees are also incapable of boring into consolidated clay, wood, rock, or sediments with large grain sizes that might be too big for them to move; thus they solely dig into sand. Halictid bee burrows are typically oriented vertically to obliquely, depending on whether these burrows are made in a horizontal surface or an inclined bank, respectively (Eickwort 1981; Michener 2000). Vertical or near-vertical shafts have horizontal and lateral

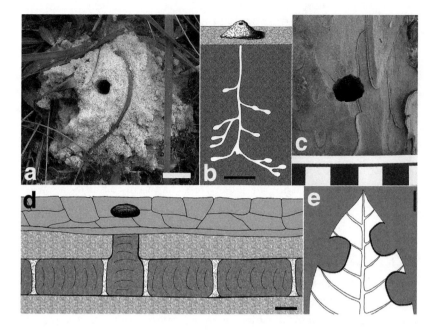

Figure 5.6. Bee traces. (a) Surface view of sediment mound and burrow entrance made by halictid bee (probably *Agapostemon virescens*) that had excavated and exited a burrow: Sapelo Island. (b) Longitudinal view of halictid bee burrow (*Agapostemon* sp.) with single vertical shaft connecting to narrow tunnels and brooding cells, made by multiple mother bees; adapted from Eickwort (1981). Scale = 10 cm (4 in). (c) Surface view of nest entrance (and exit) for eastern carpenter bee (*Xylocopa virginica*), bored into wood: Little St. Simons Island. Scale in centimeters. (d) Carpenter bee (*Xylocopa* sp.) brooding chambers, depicted as a horizontal row just below the wood surface. Scale = 1 cm (0.4 in). (e) Traces of leaf-cutter bee (*Osmia* sp.), made as marginal incisions on leaves. Scale = 1 cm (0.4 in).

tunnels branching off this main axis (Figure 5.6b). In turn, tunnels contain the slightly expanded, ovoid brooding cells, which are the sites of food provisioning and egg laying by the female who, as in other species, seems to be doing all of the work. In some instances, a sequence of multiple cells will occupy each lateral tunnel, increasing its length.

Cells are made water resistant by an application of wax, produced by the Dufour's gland in most bees. Some recent research on this waxy substance suggests that it also acts as a chemical cue for a female bee to find its individual burrow (Hefetz 2008). This is a sensible adaptation, considering that brooding chamber entrances tend to look alike, especially if crowded together by the thousands in a nesting ground. Some halictid bee species make dense nesting grounds, with hundreds of brooding burrows per square meter (11 sq ft), which means their searches for burrows need some sort of cheat sheet. This situation of many nests crowded together makes it tempting to say halictid bees are social insects, and accordingly interpret their traces as evidence of group behavior, but not quite. In actuality, little to no cooperation is taking place. Each bee is digging and provisioning its own brooding cell, and they just happen to be closely spaced to one another in their nearly identical nests because of the dry, flat, predictable, and boring terrain—optimal real estate for halictid bees. (An analogy between halictid bee nests and human suburbs comes to mind.) In some instances, however, multiple mother bees will use the same burrow shaft made by a single bee, then will dig their own lateral tunnels and brooding cells from that shaft.

Trace fossils of halictid bee brooding cells and burrows have been interpreted from the geologic record as the ichnogenera *Celliforma*, *Cellicalichnus*, and others (Genise et al. 2000; Melchor et al. 2002; Martin 2006b; Chapter 10), although distinguishing these trace fossils from similar structures made by wasps (explained later) is sometimes challenging. Nonetheless, continued research on modern halictid bee nests should help to further test whether bees or bee-like tracemakers were responsible for forming what are clearly insect brooding structures.

In contrast to halictids, some ground-nesting bees on the Georgia coast are eusocial and make complex subsurface nests through cooperative behavior. This is exemplified by bumblebees, which consist of various species of *Bombus*. Ground is broken for a new bumblebee nest by a queen, who in the spring awakens from her overwintering slumber with the intention of founding a new colony. Just before the onset of winter, she will have dug a short ground burrow and hibernated in it; such a structure is a hibernaculum, or hibernation chamber (Goulson 2003; Robinson 2005). Once active, she searches for a suitable nesting place, which on the Georgia coast might be an abandoned small mammal burrow in open and sandy upland areas, such as a back-dune meadow. The nest also must be near lots of flowers, for reasons explained later. Nest building starts with her digging a short, vase-like, wax-lined chamber known as (appropriately enough) a honey pot, and filling it with honey. Pollen from flowers in the vicinity is added to the honey pot, and this food is used to sustain her as she collects more pollen for the start of the colony. Once more pollen is gathered and consolidated into balls, she begins making brooding cells—not of sediment but mostly of pollen and wax—on top of these food sources, deposits eggs in the cells, and seals them. (Incidentally, these eggs did not originate from immaculate conception, but were fertilized by a long-dead male bumblebee from the previous summer or fall.) As is typical for holometabolous larvae, eating, and lots of it, quickly follows hatching: bumblebee larvae thus act like human teenagers, but with the main difference that the larvae are free of distracting thoughts of reproductive activities. During nearly all of larval feeding and growth, the larvae do not excrete, retaining waste materials in a blind gut (Alford 1975). Once pupation is imminent, the feces come out all at once and are plastered by the larvae onto the walls of their pupation chambers for reinforcement.

Differentiation of castes within the colony depends on the amount of feeding, as the better-fed larvae develop into queens, and the not-so-well-fed ones become workers, who are also female. Once workers emerge, more pollen is collected, the queen lays more eggs, more brooding cells are made, and the colony builds upon its older foundation. The appearance of new queens then triggers the laying of eggs that develop into the nearly useless males, who exist only briefly for the purpose of perpetuating the species. With the coming of winter, the old colony dies and the new queens will dig hibernacula, starting the cycle anew. A single bumblebee found dead in a burrow, however, is not necessarily a no-longer-overwintering queen. Several species of flies prey on bumblebees by laying an egg into a bumblebee abdomen; the hatched larva somehow causes the bumblebee to burrow and then consumes it from the inside (MacGavin 2001; Moore 2002). This coerced activity ensures a safe place for the fly larvae to pupate once the bumblebee has died. So is the resulting burrow a trace of the adult bee, the fly larvae, or both? Please discuss.

What traces of all of this activity might be preserved in the fossil record? As far as ichnologists know, not many: the construction of various parts of a bumblebee nest from wax, pollen, and feces dooms it to a lower preservation potential in comparison to a nest or brooding chamber made in or of sediment. In this respect, hibernacula burrows are more likely to have been fossilized, but these are small, shallow structures that may not escape the ravages of erosion. One factor leading to a higher probability of finding both types of traces, though, is that overwintering queens often make their hibernacula proximal to the nest, so the discovery of one trace may lead to finding the other. An additional trace left by a bumblebee nest is a high concentration of pollen grains, condensed in bee feces (Teper 2006). Pollen, some of which fossilizes, then could be used to discern feeding habits and biodiversity of plants in habitats near a former nest site.

Carpenter bees, represented on the Georgia coast by the eastern carpenter bee (*Xylocopa virginica*) and southern carpenter bee (*X. micans*), are perhaps the most ichnologically interesting of bees on the islands, aside from all of the other bees. This situation is largely because of their adaptations for boring into dead wood, hence their woodworking nicknames. Their borings are multipurposed, serving as both places of brooding and overwintering. An eastern carpenter bee looks similar in size and overall appearance to a bumblebee, but is slightly smaller, about 2 cm (0.8 in) long, and lacks the yellow banding present on the abdomen of the latter, going instead for basic black (Howell 2006). Eastern and southern carpenter bees live in a wide variety of woody materials, including those parts of houses made of wood, resulting in their status with homeowners as pests. These bees chew into the wood but do not ingest it: much like the way humans deal with chewing tobacco, they spit out the macerated remains as they progress. Chewing produces 1.0 cm (0.4 in) wide, circular to oval holes and horizontal tunnels, which are sometimes straight or "T" shaped (Figure 5.6c,d). The initial tunnel, normally made perpendicular to the wood surface, is about 1.5 cm (0.6 in) long. Tunnels used primarily for brooding cells mostly follow the wood grain and are slightly wider than the entrance tunnel, about 1.2–1.8 cm (0.5–0.7 in), and can be 4–48 cm long (1.6–19 in) long. Tunnels contain as many as 40 cells, each of which is about 3.3 cm (1.3 in) long (Gerling and Hermann 1978). In these tunnels, bees use chewed wood mixed with saliva to construct partitions between the cells, starting with the one farthest down the tunnel and working their way back to the entrance. Eggs and provisions (pollen balls) are deposited in each cell as a bee progresses outward, and each chamber is sealed before she exits the tunnel. Once the larvae hatch, they consume the provided food, grow into bigger larvae, and pupate. The end of pupation normally coincides for all of the brood, and the newly adult bees then gnaw their way through the partitions and climb over one another on their way out of the tunnel and into the real world of pollination. Surprisingly, this may not be the last time they see the inside of these wooden tunnels, as adult bees often return to the same place for overwintering (Gerling and Hermann 1978; Robinson 2005). The resulting traces, if preserved in the geologic record in fossil wood, then may be discernible as having occurred before pupation (partitions intact) or after pupation (partitions broken).

Leaf-cutting bees and the closely related mason bees—all belonging to the family Megachilidae—are best represented in North America by species of *Osmia,* some of which are also in Georgia. These bees are similar to carpenter bees in affecting plant substrates, but the descriptor "leaf-cutting" makes it obvious what they affect most often. Using their mouthparts, these solitary bees make neat, nearly circular to half-moon incisions on the margins of plant leaves (Figure 5.6e), sometimes provoking wound responses from the plant (Chapter 4). Leaf parts, some of which may also be chewed, are then used as building materials for brooding cells. In these, female bees lay an egg, provide some food, seal off a cell, and make another cell. These bees are either lazy or resourceful, depending on one's perspective, stealing pollen balls from other bees and using naturally occurring cavities in plants—including those made by carpenter bees—as templates for cells. Some species have even been documented nesting in dead snail shells (Cane et al. 2007), which would make for an interesting trace fossil if preserved. Mason bees, however, use mud and other sediments to build brooding cells on hard surfaces, such as rock.

Of the traces left by leaf-cutting bees, their cookie-cutter-like incisions on the margins of leaves are actually the most likely candidates for preservation in the geologic record, as long as the leaves fall into an anaerobic sedimentary environment, like a pond or lake. The recent discovery of trace fossils in Eocene and Miocene fossil leaves attributed to megachilid bees gives hope that more such identifications may come from the fossil record (Wappler and Engel 2003; Sarzetti et al. 2008). A key trait of these trace fossils was their slight asymmetry, which is also seen in traces made by modern leaf-cutting bees. These incisions are not perfectly circular, but have distinctive kinks along their paths (Figure 5.6e). Megachilid nests are more fragile than whole leaves and less prone to staying intact once weathered, so I do not hold out much hope for fossil equivalents to show up any time soon. Nonetheless, I also welcome being proved wrong on this prediction.

Wasps

Wasps, which are commonly seen on the Georgia barrier islands, are among the greatest predators of the insect world. More important, in an ichnological sense, some of these wasps make numerous brooding traces in a wide variety of substrates, using sand, mud, paper (chewed from wood pulp), and parts of living plants, such as branches and leaves. Most wasp predation is done insidiously but maternally, in which the mother wasp provides food for her young by laying eggs on or in another arthropod. This victim is usually another insect, but is sometimes a spider, in a form of insect retribution. In some instances, the mother wasp stinging and paralyzing its significantly larger prey precedes egg laying; the prey is then still alive (and thus fresh) for the hungry wasp larvae once they hatch a few days later. After the larvae chow down on their hapless victim, they pupate. Insects that behave this way are called parasitoids, in recognition of the way this behavior resembles parasitism but with one key difference: the host is killed by the larval wasps, whereas a parasite keeps its host alive long enough so that multiple breeding cycles can take place (Godfray 1994; Whitfield 1998). Parasitoid behavior is also expressed in a few other insect orders, such as dipterans (remember the fly that preys on bumblebees?) and coleopterans, but is most developed in wasps. (Parasitoid behavior is one of my favorite examples from the insect world of how horror films with alien antagonists' larvae bursting out of human chests are actually inspired by insects in our backyards.) Parasitoid wasps and other insects can be further categorized as egg, larval, pupal, or adult parasitoids, depending on what stage they attack in their insect prey.

What does all of this have to do with ichnology? Well, parasitoid wasps often have to make a brooding chamber, and in many instances these are burrows with cells large enough to hold both the paralyzed prey and the hatched larvae. Given that, probably the most ichnologically important of wasp tracemakers are collectively called sand wasps or digger wasps, belonging to the family Sphecidae, which burrow to accommodate their predatory lifestyles. Sand wasp burrows can be simple, inclined tunnels that lead to an expanded chamber—similar to those made by halictid bees—or quite complex, with multiple branches off a main shaft leading to many brooding cells (Evans and O'Neill 2007). These cells, however, are not stacked end to end as in a halictid bee burrow. More than one brooding cell attached to a single tunnel can be interpreted in several ways: (1) a single wasp made a single tunnel, from which it also constructed multiple brooding cells; (2) two or more wasps shared the same tunnel and made their own respective brooding cells; or (3) multiple generations of wasps reused the same tunnel over several years. Such traces can thus reflect an organization that might be interpreted as the work of social insects, but is normally the result of a solitary wasp, or more than one solitary wasp, making traces in the same burrow.

On the Georgia coast, two common species of ground-dwelling wasps serve as good ichnological examples for similarly behaving wasps: the cicada killer (*Sphecius speciosus*) and Carolina sand wasp (*Stictia carolina*). Both species prefer to burrow in open areas with well-drained, sandy soils, which means their traces are common in back-dune meadows or clearings adjacent to maritime forests. Cicada killers in particular like to lurk near deciduous trees, which not coincidentally is also where cicadas hang out. Like many nesting insects, they make the majority of their traces seasonally, in early–late summer. This means that brooding burrows are occupied by different stages of growth through the fall (larvae) and over the winter (pupae). Adult wasps then emerge in the late spring or early summer, ready to mate (or, at least look for mates), and females later hunt for food for their prospective offspring. In an example of predator–prey coevolution, cicada killers and many other predatory wasps time their emergence as adults with the arrival of their preferred prey (Hastings 1986; Godfray 1994).

The cicada killer, as might be surmised from its name, is the more formidable of the two species, and although large by wasp standards, it is still smaller than its preferred prey, the dog day (annual) cicada (*Tibicen auletes*), and other cicada species it might pursue. Cicada killers do indeed sting and paralyze their prey to provide food for their brood, but are then faced by aerodynamic challenges in carrying this living larder to their burrows and brooding cells: the cicadas are too bulky for the wasp to just pick up and fly away. Consequently, a typical scenario is like so: the wasp stings the cicada, which is often in midsong, thus permanently halting its mating ritual; the paralyzed cicada falls out of its tree; the wasp flies down to the cicada on the ground; the wasp grasps and drags the cicada to its burrow entrance; the cicada is stuffed down the burrow to the brooding cell, where an egg is laid on the cicada; and the brooding cell is sealed (Evans and O'Neill 2007). These wasps may also get an aerial assist by pulling the cicada to a nearby tree, usually the same one it fell from. Beating its wings furiously, the wasp will use a combination of walking and vertical flying up the tree trunk to gain some height before taking off for the burrow with its payload underneath. Some variation of this action may be repeated as many as 20 times, because the mother wasp lays as many as 20 eggs and makes a separate brooding cell for each one. Only once all of these cells are cached and the eggs are laid does she seal the burrow and fly away.

Hence the traces of a cicada killer include not just the burrow with associated, multiple brooding chambers, but the drag marks of the cicadas associated directly with the minute tracks of the wasp. The main shafts of their burrows are 2–3 cm (about an inch) wide—large enough to accommodate fresh, plump, juicy cicadas—and normally about 15–25 cm (6–10 in) long, but can be as much as 1 m (3.3 ft) long, and nearly 15 cm in vertical depth (O'Neill 2001; Evans and O'Neill 2007; Figure 5.7a). If this sounds like a lot of work on the part of the mother cicada killer, it is: the weight of sediment moved in making a brooding burrow can be as much as 1,000 times that of the wasp (Coelho and Weidman 1999; Coelho and Holliday 2008). The burrow shaft occasionally levels out into a horizontal tunnel before continuing downward again at an incline; this change in slope helps to retard and trap any flowing water that might enter the burrow from above. The burrow also must be inclined, rather than vertical, because digging wasps rake sediment behind them and out of the burrow to keep it open. In contrast, a vertical burrow would cause sediment to fall back into the hole as the wasp dug.

At the top of a burrow, an apron of sediment is on one side of the burrow entrance as spoil piles from the digging. These wasps dig initially with their heads (mouthparts, mostly) and forelegs, and with further progress will push the sediment underneath the body to the hind legs—which have specialized spurs to assist in shoveling—and out of the burrow. This apron also may have a central trench leading into the burrow, formed by the back-and-forth motion of the wasp; it backs out of the burrow with sediment, and then enters it headfirst. This trench is slightly wider than the body width of the tracemaker, about 1–1.5 cm (0.4–0.6 in).

As mentioned previously, a cicada killer burrow normally has multiple brooding chambers, with 3–4 clustered at the end of the burrow, but others may be added through secondary tunnels branching from the main burrow shaft. Brooding chambers can be quite large, as more than one

208 LIFE TRACES OF THE GEORGIA COAST

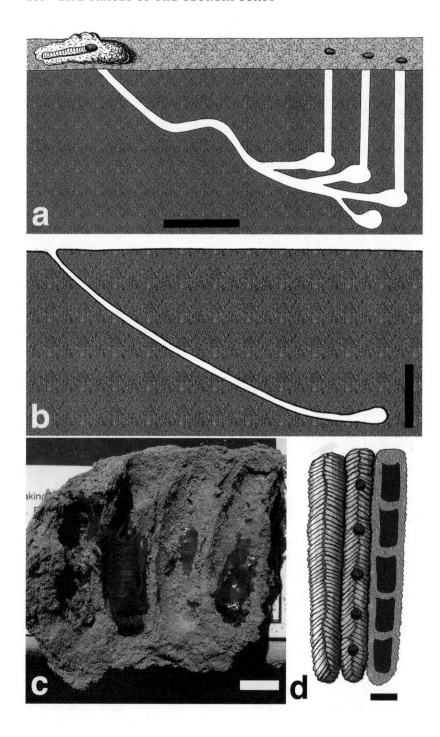

cicada might be placed in them to feed ravenous larvae. In fact, chamber size may correlate with the sex of the hatchlings: female eggs receive two to three cicadas, whereas male eggs only receive one immobilized prey item. Adult females will sometimes share burrows, but they may also fight for exclusive control. Cooperation is more likely if the wasps are related to one another, a form of hymenopteran sisterhood (Pfennig and Reeve 1993). Consequently, in instances where females co-nest, the resultant trace is composite, but with multiple tracemakers of the same species. New adult wasps, fresh out of their cocoons, dig straight up from their brooding chambers, rather than following their mother's previously dug burrow shaft (Lin 1978). As a result, a cicada killer nest that has gone through one life cycle will often hold a large number and variety of subsurface traces representing the works of many individuals, yet is still not considered the work of a social insect.

Of the traces of wasp behavior, the burrows, brooding chambers, and emergence burrows are most likely to make it into the geologic record. The rule of thumb in ichnology for fossilization of subsurface traces is "deeper is better," which means traces further down in a sediment are less prone to erasure through erosion by surface processes, such as wind and rain (Chapter 9). What ichnologists would do well to remember, however, is how individual parts of a solitary wasp burrow system might be wrongly interpreted as the work of different species of insects. Likewise, these collective, compound traces could be incorrectly discerned as a composite trace, made by more than one species of insect.

Carolina sand wasps (*Stictia carolina*) are somewhat similar to cicada killers in their digging of slanted burrows and brooding chambers, with

Figure 5.7. *Facing.* Wasp traces. (a) Longitudinal view of the sediment mound, trough, burrows, and brooding cells of the cicada killer (*Sphecius speciosus*); included are vertical burrows made by adult wasps once they emerge from cells. Scale = 10 cm (4 in). (b) Longitudinal view of the burrow and brooding chamber of the Carolina sand wasp (*Stictia carolina*). Scale = 10 cm (4 in). (c) Nests of the black-and-yellow mud dauber (*Sceliphron caementarium*), broken open in longitudinal section, with dark pieces of pupal cases: Little St. Simons Island. Scale = 1 cm (0.4 in). (d) Nest of the organ-pipe mud dauber (*Trypoxylon politum*), with an exterior view of a newly formed tube (left), a tube in which holes show where pupating wasps emerged (middle), and an interior view with stacked brooding cells (right). Scale = 1 cm (0.4 in), but widths and lengths of tubes can vary considerably.

the main difference being that their burrows are slightly smaller, less than 2 cm (0.8 in) in diameter. This size difference is at least partially a function of their smaller prey, tabanid flies, also known as horse flies. As might be figured from their common name, tabanids pester horses, so Carolina sand wasps are also sometimes called horse guard wasps (O'Neill 2001; Evans and O'Neill 2007). Once stung, paralyzed horse flies are flown by sand wasps to their burrows and placed in brooding chambers, where eggs are deposited on each body. The burrows and brooding chambers of *Stictia* species are similar to those of other solitary wasps, constructed as single obliquely oriented tunnels, 30–110 cm (12–43 in) long, and 15–30 cm (6–12 in) deep, ending with a 2 by 3.5 cm (0.8 by 1.4 in) brooding cell (Evans and O'Neill 2007; Figure 5.7b).

Eusocial wasp tracemakers in coastal Georgia terrestrial environments are also ichnologically noteworthy, and two groups in particular: paper wasps and mud daubers. Paper wasps, such as *Mischocyttarus mexicanus*, construct brooding cells using masticated wood, which a solitary mother wasp chews and then pastes together into regularly sized and directly adjacent brooding cells. This collection of 3–30 brooding cells is suspended by a thick paper cord from an overhanging surface and looks much like a papier-mâché chandelier. In maritime forests, these wasps attach their nests to the undersides of cabbage palm (*Sabal palmetto*) leaves (Hermann and Chao 1984), although most people in the southeastern United States are probably accustomed to seeing these structures on the window frames or eaves of buildings.

Similar paper-like constructions of nests are made by colonies of the bald-faced hornet (*Dolichovespula maculata*), which actually is a yellow jacket and not a true hornet (Robinson 2005). These aerial nests, once encountered, often inspire people to suddenly take up running as a physical activity. Nonetheless, if you decide to stay and study these nests (preferably abandoned ones), you would see them as 25 cm (10 in) wide, football-shaped structures of considerable complexity. In contrast, the eastern yellow jacket (*Vespula maculifrons*) constructs ground nests, using a combination of paper and soil displacement to form nests that can hold more than 14,000 brooding cells (MacDonald and Matthews 1981). Their nests are denoted by a 1.5 cm (0.6 in) wide spherical hole at the surface; sediment displaced about a centimeter (0.4 in) outside of the hole; nests as many as 30 cm (12 in) in diameter; and numerous yellow jackets flying in and out of the hole. (This is yet another practical example of when ichnology, and observation skills in general, are handy to have.) The southern yellow jacket (*V. squamosa*) makes either ground or aerial nests, but given the choice of usurping a nest already made by a colony of eastern yellow jackets, they will take that. In one study, as many as 15 percent of all *V. squamosa* nests began as nests of *V. maculifrons* or other wasps (MacDonald and Matthews 1984). These nests can become rather large, containing as many as 11,000 brooding cells. Like social bees, a queen is responsible for the colony and is the only one to overwinter: this means that any given active colony is annual, and mostly constructed in the summer. Yellow jackets, though, will make nests on top of abandoned ones, constructing multiannual and compound traces that become huge after just a few iterations.

Although their paper construction translates to a low preservation potential, the sheer massiveness of such wasp nests may aid in their fossilization. Thus far, though, only one trace fossil (*Masrichnus*), coming from the Eocene–Oligocene of Egypt, has been speculated to be a possible underground paper wasp nest, similar to those made by modern species of *Vespula* (Genise and Cladera 2004).

Mud daubers are solitary wasps, and three widespread species living in Georgia and the barrier islands are the black-and-yellow dauber (*Sceliphron caementarium*), the organ pipe mud dauber (*Trypoxylon politum*), and blue dauber (*Chalybion californicum*). The first two wasps make nests and, as can be discerned from their name, they use wet mud as a medium for building these. The blue mud daubers, much like the southern yellow jackets, simply depend on the work of the other two, and reuse nests of those species, which they provision with spiders (Landes et al. 1987). Female wasps of either the black-and-yellow or organ pipe dauber species collect mud by visiting the edges of puddles or other available areas with moist, muddy substrates. They then grasp it with their mandibles, transfer it to their anterior legs, and fly off with their sediment load, leaving barely noticeable excavation marks and leg impressions. Once at the nest site, which is typically a hard, vertical surface (such as a tree trunk), they attach and mold the mud into cylindrical brooding cells, each separated by thick walls. Nests of both species may initially seem similar, but they differ in their organization and overall architecture.

Black-and-yellow mud daubers construct ovoid brooding chambers, all lateral to one another and separated by mud walls; one female may make as many as 25 cells (O'Neill 2001). These cells, like those of many other wasps, are stuffed with paralyzed spiders, and eggs are laid on these before the female seals each chamber. Exteriors of these nests may simply look like masses of dried mud stuck to the interiors of hollow trees, porch eaves, or other hard, relatively flat and elevated surfaces. Nonetheless, a closer look may reveal subtle sculpting along the side of each cell, and broken cells will have smooth, oval interiors (Figure 5.7c). Their overall outward appearance may also seem like pieces of pottery made by coiled clay, and can be admired similarly for their craftsmanship.

Nests of the organ pipe dauber, in contrast, superficially resemble the pipes projecting from organs played in places of worship, although these are rather corrugated on their exteriors. These tubular structures are either by themselves or clustered adjacent to one another (Figure 5.7d). The single most important difference between these nests and those of yellow-and-black daubers, however, is that the organ pipe dauber stacks its brooding cells on top of one another (Brockmann 2004). Hence each individual tube contains multiple brood cells, whereas those of the yellow-and-black dauber only have one cell per cylinder. Organ pipe daubers also provision each brood cell with paralyzed spiders (they prefer orb weavers), with one egg per chamber, and chambers are partitioned from one another: no sharing is allowed. In cross section, this series of tubes will show partitions perpendicular to their long axes. Visible lines mark boundaries between each tube, causing adjacent tubes to double the wall thickness compared to the outermost ones. In the making of each type of nest, mud balls brought to the site are pushed against the previous one, left–right–left, making an en-echelon pattern on the tube exterior. In contrast, the interior is smooth walled. In some instances, differently colored (compositionally distinct) mud in separate but adjacent tubes will indicate separate sources and construction times for each tube, which also implies that different female wasp tracemakers were responsible. Holes made by emerging wasps also may dot the side of a tube, showing how many wasps successfully made it through pupation.

Because of their sedimentary medium, mud dauber nests may be preserved in the geologic record; indeed, one trace fossil attributed to such wasps is *Chubutolithes* from the Eocene–Miocene of Argentina (Genise and Cladera 2004). Such trace fossils would have required special conditions for burial, however, because they are often positioned above the ground and attached to objects that may not become incorporated into sediments.

Two-for-one traces caused by wasp brooding in plants, as well as reactions by the plants to these invaders of their tissues, are galls, discussed previously (Chapter 4). Gall-forming wasps are nearly all from the family Cynipidae, and many of these form galls in oak branches. Galls get started when females lay eggs into the plant tissue. The eggs then hatch into hungry larvae that start consuming their surroundings, and the plant responds by expanding and toughening tissues around the infestation (McGavin 2001). On the Georgia coast, these traces consist of visible swellings on the distal branches of live oaks (*Q. virginiana*), and are typically made by species of *Andricus*. Some gall-forming wasps, however, attack oak leaves as well as branches; one gall wasp (*Disholcaspis cinerosa*) actually alternates generations in its reproduction between leaves and branches (Morgan and Frankie 1982). Some gall-forming wasps are adapted to specific plants and make distinctive enough galls that species can be identified by just their traces.

In summary, wasp traces that are most likely to make it into the geologic record are deeper subterranean burrows and brooding chambers, followed by mud nests, and then galls; paper nests are only expected to be preserved in the fossil record under exceptional conditions (Genise and Cladera 2004).Other rarely preserved traces are wasp pupal cases or cocoons, which also require special conditions for making it past the fossilization barrier. Very simply, these structures must be either lined or filled with fine-grained sediment to make molds of them, which then must be followed by rapid cementation. Not surprisingly, fossil cocoons credited to wasps are relatively rare and are especially uncommon in older rocks, although excellent Late Cretaceous examples have been found in Montana (USA) and Patagonia, Argentina (Martin and Varricchio 2011; Genise et al. 2007).

Beetles

Beetles are the most diverse of all insect groups, and their ichnology reflects this diversity. Because beetles are so extraordinarily common in terrestrial environments of the Georgia barrier islands, coverage of their traces here must be brief and highlight only the most ichnologically significant tracemakers. Tracemaking standouts among beetles of the Georgia

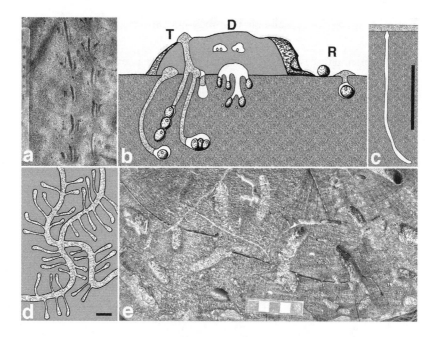

Figure 5.8. Beetle traces. (a) Typical large beetle trackway in sandy soil on the edge of a maritime forest: St. Catherines Island. Scale in millimeters. (b) Traces associated with three behavioral categories of dung beetles: tunneler (T), dweller (D), and roller (R); eggs are depicted as tiny ovals on or near dung balls. Based on a figure by Bertone et al. (2005). (c) Longitudinal view of a tiger beetle (*Cicindela* sp.) larval burrow. Note the enlargement near the top of the burrow to accommodate the larva as it hangs out, waiting for victims. Based on descriptions and figures by Choate (2009) and Platt et al. (2010). Scale = 25 cm. (d) Idealized schematic of borings in wood caused by the Southern pine beetle (*Dendroctonus frontalis*), in which S-shaped tunnels made by adult beetles serve as brooding chambers, and branches are where the brood radiate from tunnels. Note how borings made by offspring widen further away from the adult tunnels. Scale = 1 cm. (e) Frass (chewed wood) used as backfill material in beetle borings, in a cross section of a former live oak (*Quercus virginiana*) trunk: Sapelo Island. Scale in centimeters.

barrier islands include dung beetles (Scarabaeidae), tiger beetles (Carabidae), rove beetles (Staphylinidae), and wood- and bone-boring beetles (Scolytidae and Dermestidae, respectively). Other beetles on the Georgia barrier islands certainly make trackways, burrows, brooding chambers, and other traces, but examples of the preceding groups will provide a healthy overview of noteworthy beetle contributions to making traces.

Among the most commonly encountered beetle traces on the Georgia coast are their trackways, which can be spotted on nearly any clean, sandy surface in dunes and back-dune meadows, but are also common in maritime forests (Figure 5.8a). Most of these trackways are made in the early morning, just before, during, and after dawn. Moreover, their abundance is seasonally dependent, as they are more prevalent during the spring and summer. The probable reasons why beetles might be out for a walk near a beach or other open areas are twofold: they are gathering moisture from morning dew or they are hunting for food, some of which may be live prey. Beetle trackways are identifiable as repeating patterns of six impressions, corresponding to the number of walking legs. For adult beetles, two pairs of these impressions are more or less equal in size, with one pair (in front) close to the midline of the trackway, whereas the last pair registers two parallel lines, either dashed or continuous. Such imprints are caused by the backward-pointing part of a beetle's rear legs contacting the sediment surface, which the beetle sometimes drags along behind the other walking legs. Surprisingly, some larval beetles also can make trackways, but these normally only have pinpoint impressions made by their small legs, and may display a central drag mark from their relatively long abdomens.

No one has yet attempted to distinguish beetle species on the basis of trackways, let alone on the Georgia coast, but this seemingly impossible feat could be done through a combination of field observations and experimentation. Simply watch identified species make tracks, carefully measure and otherwise describe the trackways, and look for where trackways may overlap in sizes and traits. Painstaking and exacting work, for sure, but such distinctions could assist in insect surveys of beaches where endangered or otherwise rare species of beetles might be located (Knisley and Schultz 1997).

Many burrowing beetles dig during their larval stages, although adults are also responsible for some of this bioturbation. Some of their more distinctive traces are sinuous burrows containing half-moon, crescentic lines or meniscae. Meniscae, mentioned previously (Chapter 4), are formed by active backfilling, in which the tracemaker packs sediment behind it as it progresses. For example, some beetle larvae turn around completely in their burrows and compress sediment behind them with their flattish heads against the back of the burrow, thus forming a meniscus (Counts and Hasiotis 2009). The resultant burrows are similar to those made by cicada nymphs, but normally have smaller widths.

Dung beetles are among the most ecologically important animals in any terrestrial ecosystem, especially if those ecosystems contain large mammals capable of producing sizeable, steaming piles of feces. Fortunately for human visitors who like to keep their shoes tidy, the Georgia barrier islands are rich in dung beetles: nearly 20 species have been identified from Blackbeard Islands (adjacent to Sapelo) and Cumberland Island alone (Fincher 1975, 1979; Fincher and Woodruff 1979). On Cumberland, these beetles are needed because of copious amounts of solid waste produced by invasive species on the island, such as feral hogs (*Sus scrofa*) and horses (*Equus caballus*). Dung beetles have been around much longer than the Georgia barrier islands, though, and dung beetle trace fossils co-occur with dinosaur coprolites in Cretaceous Period rocks of Montana (Chin and Gill 1996; Chin 2007) or otherwise are preserved in ancient deposits (Laza 2006; Chapter 9).

The primary results of dung beetle behavior constitute a sort of recycling that everyone should appreciate. Indeed, whenever discussing their ecology in my classes, I always praise these insects and tell my students how they should begin every day by offering thanks to dung beetles. Dung beetles, which belong to the evolutionarily related group Scarabaeidae (scarab beetles), are also referred to as tumblebugs in Georgia because adults are often seen rolling a neatly spherical ball of dung across a ground surface. And why would they be working so hard at moving feces? As with other industrious insects, these beetles are just trying to feed their young: the dung ball is a provision for a brooding chamber. Different dung beetle species, however, do this in different ways, and are divided into three groupings based on behavior (Figure 5.8b): (1) tunnelers (also known as paracoprids), represented by species of *Onthophagus* and *Phanaeus*; (2) dwellers (endocoprids), which include species of *Aphodius*; and (3) rollers (telecoprids), best exemplified by species of *Canthon* (Bertone et al. 2005).

Tunnelers, as is apt for their descriptor, will do the following: dig deep, inclined, and J-shaped burrows into the sediment directly below a heap of feces; construct an enlarged terminal brood chamber from the main burrow shaft; stuff the chamber with a dung ball; lay eggs on the dung balls; and backfill the burrow. The latter is done to protect the nest from anything that might like to steal the dung ball (after all, it is filled with precious nutrients) or prey on the eggs and larvae. Hence their traces are compound, formed by distinctive phases of behavior, yet connected with one another and made by the same individual tracemaker. Dwellers actually feed and breed within a fecal pile, making brood chambers in the feces or directly below it: With so much dung to eat, why move? Rollers differ from tunnelers and dwellers by tearing off chunks of dung, then shaping these into balls. However difficult it might be, imagine grabbing a piece of cookie dough and then rolling it into a ball; this is what beetles do with their soft foodstuff. Roller beetles will then push the dung balls to another area, holding them with their hind feet and pushing backward on their front two pairs of feet, sometimes riding on top of the ball and looking much like little circus performers. The trackways caused by this activity, although I have never seen them, would be delightful to encounter. These would show beetle tracks mostly made by the front two pairs of feet, but pushing backward, accompanied by an erratic impression made by the dung-ball-rolling print in the middle of the trackway. Once roller beetles get their hard-earned prizes to where they are needed, they are buried in shallow burrows.

Dung beetle traces are thus quite varied with accordingly different preservation potentials: obviously, the burrows and deeply buried chambers of tunnelers will have the best likelihood of making it into the geologic record. Nonetheless, vertebrate coprolites containing traces of coprophagy (feces eating) have been documented from the Triassic and Cretaceous (Wahl and Martin 1998; Chin and Gill 1996; Chapter 9). A subtler sign of the intertwined relationship between dung beetles and large herbivores in some ancient terrestrial ecosystems is evidence that, when any sizeable feces producer went extinct, its beetle cleanup crew went with it. For example, extinctions of North American and Australian megafauna at about 11,000 and 40,000 years ago, respectively, coincided with the extinction of a few dung beetle species, which must have specialized in consuming certain species' feces (Martin 2005; Johnson 2009a). The dependency of these insects' reproductive cycles on the end products of vertebrate digestion is still observable today in many dung beetle species that are choosy about what types of feces their kids should eat.

Just to get away from manure for now, let us instead consider the colorfully striped tiger beetles as tracemakers on the Georgia barrier islands. Tiger beetles, consisting of various species of *Cicindela,* are active predators of other insects as adults and leave many trackways in dunes and back-dune meadows. Their larvae, however, are ichnologically more important than the adults. Although tiger beetles are so named because of tiger-like patterns on the adults, their larvae are also voracious ambush predators. Amazingly, larvae of some species of *Cicindela* make open, vertical burrows more than a meter (3.3 ft) deep, although the burrows of species along the Florida and Georgia coasts are more likely to be about 10–75 cm (4–30 in) long (Pearson et al. 2006; Choate 2009; Figure 5.8c). Burrow diameters are not as impressive as their depths, and are only slightly greater than tracemaker widths, less than 1 cm (0.4 in) wide and with circular outlines. If a larva stays in the same burrow, it will gradually expand it in accordance with growth, widening it after completing each molt. These burrows function similarly to those of ant lions and some spiders, where the larva sits near the top and waits to nab insects, spiders, or even small crustaceans that walk by the burrow. One key trait of a tiger beetle larval burrow, though, is a slight expansion just below the burrow top, a space that accommodates hook-like projections near the end of the abdomen. These hooks hold the larva in place to prevent falling down the burrow; meanwhile, it places its head and thorax at the burrow entrance, its jaws ready to grab prey (Pearson and Vogler 2001). Thus if a complete fossil tiger beetle burrow were found, minimal larval length could be calculated by measuring the distance from the burrow top to the distal end of this bulge. Profiles of burrows will be mostly straight but may also curve slightly along the way, especially toward their bottoms where they become more oblique and J shaped.

Some tiger beetles, along with a few species of rove beetles, are among the few insects adapted to living in marginal-marine environments, such as salt marshes and beaches. The beach-dwelling larvae dig burrows as low as the upper intertidal zone, but typically live in foredunes just above the high-tide rack line. If burrows are inundated by a high tide or waves, larvae block the burrow top with sand and wait for the water to subside before resuming their predatory ways. Larvae also overwinter (hibernate) in their burrows, and after two years pupate in chambers at the base of the burrow (Pearson and Volger 2001). In the spring of their third year, they emerge as fully mobile adults and resume hunting prey, but this time covering much more ground and hence making tracks instead of burrows. As a result, many beetle trackways observed in dune and back-dune areas may first be regarded as those of tiger beetles. These can be distinguished from tracks of the larger and heavier scarab beetles, which accordingly leave bigger tracks and trackways. Because of their narrow ecological range, disturbances such as human development or habitual trampling of shoreline environments often cause tiger beetle populations to decline or become locally extinct. Fortunately, much of the Georgia barrier island coast is relatively undeveloped, so species still left include *C. dorsalis* and *C. hirticollis*, which are normally on beaches. In contrast, *C. striga, C. marginata*, and *C. trifasciata* are more common in salt marshes, although they also are seen on beaches (Choate 2003, 2009).

The simple vertical burrows made by tiger beetle larvae closely resemble the trace fossil *Skolithos*; indeed, some specimens of *Skolithos* found in formerly terrestrial Triassic deposits have been interpreted as the possible work of tiger beetles or similar insects (Netto 2004). Nonetheless, more detailed studies of modern tiger beetle burrow forms and tracks are probably needed before trace fossils by these or similarly behaving beetles can be diagnosed from the geologic record.

Tiger beetle larval burrows are often accompanied in upper intertidal areas of the Georgia barrier islands by the smaller burrows of rove beetles. Rove beetles are represented worldwide as a family (Staphylinidae) by more than 50,000 species, and accordingly inhabit a wide range of ecosystems (Frank and Thomas 2009). Rove beetles native to the Georgia coast are about half the size of tiger beetles, so their burrows and trackways are also half the diameter of their coleopteran compatriots. These burrows are normally vertical to near-vertical shafts, with some Y shaped, having two entrances at their tops (Ratcliffe and Fagerstrom 1980). In upper intertidal areas, rove beetles mainly function as scavengers of dead and decaying material deposited at the high-tide mark.

Freshwater beetle tracemakers include burrowing water beetles (*Hydrocanthus* spp.), which are in Lake Whitney on Cumberland Island (Frick et al. 2002), and likely live in other freshwater environments of the Georgia barrier islands. Burrowing water beetles are fully aquatic, and even pupate in the water; larvae and adults burrow into the bottoms of temporary or permanent water bodies. Unfortunately, almost no information is available about their burrows, but these must be as minute as the beetles, which range from 1–5 mm long and only a few millimeters wide. Water beetles of all sizes, however, are known to make swimming trackways along sediment bottoms, although these have not yet been documented in freshwater environments on the Georgia barrier islands. Some such trackways, belonging to the ichnogenus *Warvichnium*, are preserved in the fossil record, including in lake deposits from the Permian of Germany and the Pleistocene of Lithuania (Seilacher 2007; Uchman et al. 2009).

Wood-boring beetles, also known as bark beetles, make traces that are easily distinguished from those of other insects that chew their way through tough plant tissues. The most common wood-boring beetle in the southeastern United States is the Southern pine beetle (*Dendroctonus frontalis*), which despite its native status is considered one of the greatest enemies of pine trees in the region (Meeker et al. 1995). Pine beetle borings start with female beetles making a single tunnel in wood, in which they lay their eggs; this tunnel then serves as the brooding chamber. The larvae are born to chew wood, radiating away from the central tunnel upon hatching. These tunnels stay along the same plane within the wood, and are typically just under the bark surface. The resulting pattern is a bushy one, with a central tunnel interconnecting with many branches that become larger distally, a direct result of the larvae growing bigger as they eat more wood (Figure 5.8d). Tunnels later made by adults can be "S" shaped too (Meeker et al. 1995). Similar to carpenter bees, beetle tunnels are often backfilled with chewed woody material (called frass), easily seen in exposed tunnels (Figure 5.8e). Scolytid beetle borings on Georgia barrier islands are, more often than not, found on the south side of a tree trunk, facing the direction of maximum sunlight. This positioning warms up the beetles to give them the energy they need to eat more wood. As can be imagined, these traces have excellent preservation potential in the geologic record, as all one has to do is look at petrified trees for similar-looking trace fossils. Indeed, wood-boring beetle trace fossils of the ichnogenus *Paleoscolytus* have been interpreted in Mesozoic and Cenozoic fossil trees, giving some insights on the evolution of this behavior in beetles (Ash and Creber 2000; Labandeira et al. 2001).

Although some people might be concerned about wood-boring beetles for their home-wrecking abilities, a far worse one on the Georgia coast is an invasive species, the red bay ambrosia beetle (*Xyleborus glabratus*). This beetle is contributing to the decline of an important plant species, red bay (*Persea borbonia*). After these beetles bore into a red bay tree, they introduce one of their food sources, a fungus (*Raffaelea lauricola*), which then infects the tree and contributes to its death (Fraedrich et al. 2008; Zomlefer et al. 2008). This is an example of a farming trace, as the beetle makes the tunnels with the intention of growing and later harvesting the fungus in these. As of this writing, red bays were in big trouble on all of the Georgia barrier islands because of this foreign threat.

Other wood-boring beetles specifically attack tree roots, such as the live-oak stump borer (*Archodontes melanopus*), mentioned previously (Chapter 4). This beetle lays eggs in the soil near the base of a live oak; upon hatching, the rather large, 3.5 cm (1.4 in) long larvae bore into the roots, provoking wound reactions in the tree (galls) or sprouts from a stump (Solomon 1995). As a result, ichnologists studying root traces in the geologic record (Chapter 4) should also check for similar close associations between these traces and those formed by insects that bored into roots.

Dermestid beetles, also known as carrion beetles, have also received some attention from ichnologists and paleontologists because of their effects on vertebrate material. Dermestids are scavenging beetles that strip flesh from bones and often leave distinctive borings from such activity (Robinson 2005). Dermestid beetle adults, once they encounter a delectable vertebrate carcass, will lay eggs on it; once these hatch, the beetle larvae start eating soft tissues and bone alike. The traces they leave on bone are quite varied, and include grooves (mandible marks), surface furrows, tunnels, pits, and frass composed of chewed bone (Martin and West 1995; Paik 2000; West and Hasiotis 2009). Once they ready themselves for pupation, these beetles bore into a bone to make a flask-shaped pupation chamber, where they stay until adulthood. Dermestid beetles have been proposed as the tracemakers for pockmarks observed on Late Jurassic and Cretaceous dinosaur bones (Rogers 1992; Hasiotis 2004; Roberts et al. 2007; Britt et al. 2008), although some trace fossils in dinosaur bones are also attributed to bone-eating termites (Dangerfield et al. 2005). In a fortuitous two-for-one deal, ichnologists even interpreted dermestid beetle borings in a Jurassic theropod dinosaur coprolite filled with bone, a great example of a composite trace fossil (Chin and Bishop 2004). In studying modern bones, though, care should be taken in distinguishing invertebrate traces from tooth marks of predators, scavengers, and rodents who may be just getting a little more calcium in their diets (Chapter 8).

Cicadas

Cicadas, such as the periodical cicada (*Magicicada septendecim*) and dusk-calling cicada (*Tibicen auletes*), are marvelous burrowers in terrestrial environments of the Georgia barrier islands. Cicada nymphs hatch from eggs

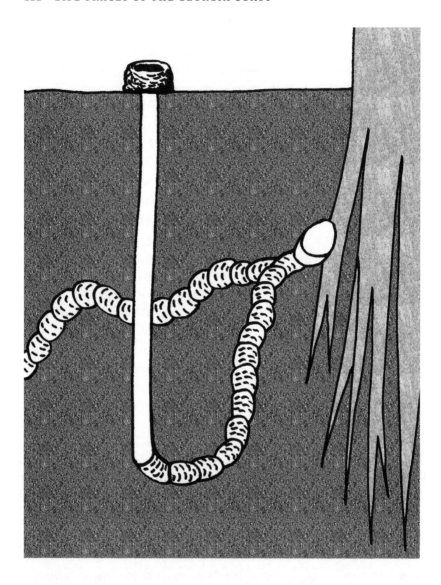

Figure 5.9. Cicada nymph traces: meniscate (backfill) structures caused by burrowing, feeding chamber associated directly with roots, and vertical emergence burrow ending with a short turret at the ground surface. Based on illustrations by Krause et al. (2008) and Smith and Hasiotis (2008). Not to scale, but emergence burrows are normally 50–100 cm long.

laid on trees and start burrowing as soon as they fall to the ground from their former nurseries. As mentioned earlier, periodical cicada nymphs spend most of their lives (13–17 years) underground feeding on sap from deciduous tree roots, hence they are among the most important tracemakers in maritime forest soils. This bioturbation is typically limited to the uppermost 50 cm or less of a given soil profile, although rambunctious nymphs of some cicada species can burrow more than a meter (O'Geen et al. 2002; Smith and Hasiotis 2008). Numerous burrows over generations of these long-lived cicada nymphs result in many overlapping burrows, lending to rather complex patterns in sediments (Figure 5.9).

Cicada nymphs make burrows similar to those of some beetle larvae, forming many horizontal and oblique burrows marked by definite meniscae. The way nymphs make these burrows, however, differs considerably from those attributed to beetle larvae. As they move forward in a burrow by digging with their rear legs, they flip around 180° and pack the burrow with their forelegs, thus making a meniscus (Smith and Hasiotis 2008). As a result of this acrobatic movement, the burrow width is slightly wider than that of the cicada nymph, allowing enough room for it to turn around. This also imparts a pinch-and-swell outline to a burrow along its length, with the wider parts indicating where the cicada flipped (Smith and Hasiotis 2008). Consequently, cicada burrows are typically greater than 1 cm (0.4 in) in diameter, whereas most beetle larvae burrows are narrower.

Changes in behavior during cicada development also cause radically different orientations in burrows. For example, normal day-to-day burrowing around tree roots imparts seemingly random horizontally and obliquely oriented burrows. Once it is nearly time to mature into an adult, though, cicadas dig vertical burrows straight up to the surface. As they bid adieu to their underground world, they leave many open, vertical shafts in their wake (Luken and Kalisz 1989; Smith and Hasiotis 2008). Short, sediment-made turrets may also cap these emergence burrows. Emergence burrows form en masse as a brood emerges; again, these burrows are about 1–1.5 cm (0.4–0.6 in) wide. Nymphs then climb the nearest tree—probably the same one from which they were most recently sucking sap—molt, and leave behind brownish-golden husks. These cast-off skins are often seen clinging onto tree trunks in late spring or early summer on the Georgia barrier islands. Annual cicada species, such as the dusk-calling cicada, make emergence burrows every year, which then overlap with differing broods of periodical cicadas.

Cicada trace fossils, once rarely recognized, are now becoming more commonly interpreted. For example, the ichnogenus *Naktodemasis* is directly comparable to the backfilled burrows of cicada nymphs (Smith and Hasiotis 2008; Smith et al. 2008). Moreover, fossilized feeding chambers attributed to cicada nymphs in Cenozoic rocks of Argentina were given the ichnogenus name *Feoichnus* (Krause et al. 2008). An exciting aspect of this trace fossil was its co-occurrence with root trace fossils, showing a paleoecological relationship similar to that seen in modern sap-sucking cicadas. In a few previous studies, backfilled burrows in terrestrial deposits identified as possible cicada nymph burrows have been assigned the ichnogenus *Taenidium*, seen in Pleistocene deposits of Sapelo Island (Gregory et al. 2004).

Mole Crickets

Orthopterans, which include grasshoppers, locusts, crickets, and related insects, are prolific tracemakers through their herbivory, although such traces may have low preservation potential or be difficult to perceive. Nonetheless, probably the most ichnologically distinctive orthopteran traces on the Georgia coast are burrows made by mole crickets, consisting of three species of *Scapteriscus*: the southern mole cricket (*S. borellii*), tawny mole cricket (*S. vicinus*), and short-winged mole cricket (*S. abbreviatus*). Mole crickets are relatively recent denizens of the Georgia coast, having been accidentally introduced from South America to the United States through the port of Brunswick, Georgia, at the end of the nineteenth century (Walker and Nickle 1981; Capinera and Leppla 2008). Despite their nonnative status, mole crickets have adapted admirably to the terrestrial ecosystems of the southeastern United States, which is at least partially attributable to their impressive burrowing abilities. In fact, their bioturbating skills inflicted much harm on the recreational facilities of at least one of Georgia barrier islands (Jekyll), where they began to alter the grassy habitat of a golf course to the point of uselessness for its human patrons. Mole crickets earned their nicknames not only because of their burrowing prowess, but also for their distinctive forelegs, which are spade shaped and have finger-like projections (dactyls) that look very much like the digging implements of moles.

TERRESTRIAL INVERTEBRATES 225

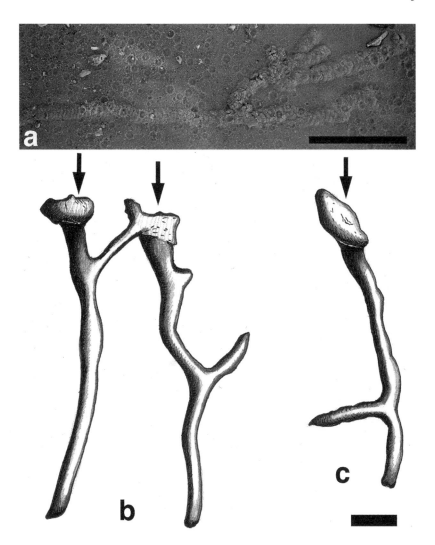

Figure 5.10. Mole cricket traces, made by species of *Scapteriscus*. (a) Exterior view of a mole cricket tunnel with branching at its distal end, originating from a dune but ending in the upper intertidal zone: Cumberland Island. Scale = 10 cm (4 in). (b) Tawny mole cricket burrow, horizontal view, with a Y shape near the top from two entrances (arrows). (c) Southern mole cricket burrow, with a Y shape near the end of the burrow, arrow pointing to entrance. Both (b) and (c) based on fiberglass casts figured by Brandenburg et al. (2002). Scale = 5 cm (2 in).

Why do mole crickets burrow? The answer depends on the species and their life habits. Southern mole crickets, for example, are carnivorous, so they are seeking prey, whereas tawny mole crickets eat plant roots. But perhaps the most surprising aspect of mole cricket burrows is that males use them as resonating chambers, from which they project their songs to attract comely female crickets (Ulagaraj 1976; Nickerson et al. 1979). Under ideal conditions, their songs register more than 90 decibels and can be heard from 1.5 km away! Burrows are mainly constructed as open, shallow, horizontal tunnels, often evident at the ground surface as upraised ridges. Better defined ones in moist, fine-grained sand or firm mud show a chevron-like pattern on the exterior, and interiors are nearly circular in cross section but often contain a backfill structure (Figure 5.10a,b). Tawny cricket burrows are about 8–10 mm (0.3–0.4 in) wide, 50–70 cm (20–27.5 in) long, and almost always Y shaped because of dual openings that meet to form a single tunnel, although they can have other tunnels branching from an entrance (Brandenburg et al. 2002; Figure 5.10c). Tunnels of the southern mole cricket, however, are wider, 1–1.5 cm (0.4–0.6 in). These burrows only have one entrance and also branch into a Y shape, but further down the tunnel (Figure 5.10c), normally within about 10 cm (4 in). Southern mole cricket burrows are also considerably shorter than those of tawny crickets, but deeper because they are hunting other infauna. Tunnels may also be slightly wider (flared) near surface entrances, a trait that enhances their resonance: an example of where ichnology and acoustics overlap. If mating is successful, female mole crickets make semispherical, 3–4 cm (1.2–1.6 in) wide egg chambers, constructed laterally to main tunnels and 5–40 cm (2–16 in) below the surface (Capinera and Leppla 2008). If an egg chamber is encountered, it would have been made in late spring (April–May), and the eggs often hatch within a month. Like all orthopterans, mole crickets are hemimetabolous; hence eggs hatch into nymphs that look like miniature adults (and yes, they start burrowing right away).

The habitat range of mole crickets is quite wide in terrestrial ecosystems, although they prefer moist substrates. Their burrows even occur in the upper parts of beaches, overlapping with burrows made by ghost crabs and other marginal-marine invertebrates. In terms of their trace fossil record, paleontologists have also suggested that some trace fossils, such as certain forms of the ichnogenus *Spongeliomorpha*, may be linked with mole crickets (Metz 1990; Melchor et al. 2006). These, however, are difficult to distinguish from those made by beetles that also enjoy moist, muddy sediments. Nonetheless, the likelihood of mole cricket burrows in the geologic record is probably quite good; mole cricket body fossils have been reported from Early Cretaceous rocks of France (Perrichot et al. 2002), pointing to a minimum age for when we might see their burrows.

Caddis Flies

Caddis flies (Trichoptera) are holometabolous insects common in freshwater environments throughout Georgia, and are exclusively aquatic throughout all of their development. This means their traces can be used as definitive ecological indicators, although bottom-dwelling caddis fly larvae represent their most important tracemaking stage. These larvae tempt a large number of freshwater predators as potentially small and tasty morsels, so they must conceal themselves in unappetizing guises, which is accomplished by binding coarse-grained sand or plant debris around their burrows. These disguised burrows do not fool ichnologists, though, because the selective use of building materials by the tracemakers produces anomalous objects in an environment containing more homogenous, finer-grained sediments. Caddis fly larvae, like ants and ant lions, are hence responsible for sediment advection, and can be used to explain odd, localized occurrences of coarse-grained sand in otherwise fine-grained sediments of freshwater environments.

Although modern examples of caddis fly traces have not been studied on the Georgia coast, trace fossils attributed to caddis fly larvae are interpreted from Permian deposits, as well as Mesozoic and Cenozoic rocks. For example, the ichnogenus *Terrindusia* is attributed to trichopteran tracemakers (Bromley et al. 2007), and more such trace fossils should be found once search images improve for their modern traces.

Termites

Termites are no doubt the most famous consumptive lovers of wood, easily rivaling or surpassing their ant, bee, and beetle companions. Termites (Isoptera) are eusocial insects distantly related to cockroaches: think about how a few species of roaches are also called wood roaches because of that feeding preference. Like other eusocial insects, termites are polymorphic, with different body plans linked to castes within a colony, consisting of primary reproductives (which includes a queen), secondary reproductives, workers, and soldiers. They differ from eusocial hymenopterans, though, in being hemimetabolous (McGavin 2001). Primary reproductives have a winged stage (alates); once a year they fly about in huge numbers (swarming), with most looking for love in all the wrong places. Nonetheless, a few males and females mate and start a colony with a few workers, which are followed by secondary reproductives and soldiers. Worker termites in the colony form tunnels in wood by chewing through it with their mandibles, and only they do this. Hence any termite traces identified in either modern or fossil examples can be considered as the handiwork of workers.

Termites are much-loathed tracemakers in Georgia terrestrial environments because of their taste for woody substrates, which unfortunately compose the framework of many people's homes. Nine species of termites have been detected so far in Georgia (Scheffrahn et al. 2001), a high number, but expected for its warm, temperate climate. Of those nine, six species have been identified from the Georgia coast and on a few barrier islands: *Kalotermes approximatus, Cryptotermes brevis,* the Formosan termite (*Coptotermes formosanus*), *Calcaritermes nearcticus, Incisitermes minor,* and the southern drywood termite (*Incisitermes snyderi*). An indeterminate number of species of *Reticulitermes* (at least three, perhaps more) are also found throughout Georgia, but the eastern subterranean termite (*R. flavipes*) best represents this genus. Some termite species are nonnative to Georgia, coming from the western United States, Asia, or countries south of the United States, and were considered invasive species once they started eating people out of house and home.

Ichnologically and ecologically, Georgia termites can be divided into two categories: subterranean, in which they make nests underground and eat rotting wood; and drywood, where nests are made in dead wood above ground, including timbers. Examples of subterranean termites include *C. formosanus* (which, as one might gather, is from Southeast Asia), whereas drywood termites are represented by *C. brevis* and *I. minor,* both invasive. The only two native drywood termite species in Georgia are *K. approximatus* and *I. snyderi* (Scheffrahn et al. 2001). Incidentally, the Formosan termite is often labeled as the most destructive termite in the world, and hence it deserves a special sort of wrath for its ichnological acumen. Georgia termites, like other termite species, are best suited for year-round or seasonally humid environments conducive to the formation of molds and subsequent decay of wood. Moreover, subterranean termites adjust the depths of their nests on the basis of outside temperatures: too cold or warm conditions above the surface will cause a colony to shift downward for protection. As a result, ichnologists can often interpret overall climate and other paleoecological conditions based on the presence of termite trace fossils in a given assemblage, along with dung beetle or hymenopteran traces (Genise et al. 2000; Chapter 10).

Termite nests in wood are readily distinguishable from those of other wood-boring insects because of their depth and ubiquity. Most insect borings stay relatively close to the outer surface of the wood, whether in a living tree (near the bark or outermost cambium) or a piece of dead wood. In contrast, termite borings run well into the interior of the wood, generally following the wood grain. These borings consist of galleries and cylindrical, intersecting tunnels, each of which is open in places, but also containing frass backfilled with wood fibers. Tunnels connect with broader, open galleries used by the queen for egg laying and raising of young by the workers (Abe et al. 2000). Perhaps surprisingly, termites cannot digest the cellulose in wood on their own, so they employ symbiotic bacteria in their tiny guts to do this for them, using fermentation. By-products of this digestion are huge volumes of methane, making up as much as 5 percent of the global methane budget (Engel et al. 2009). Methane (CH_4) is a greenhouse gas, with excellent insulation properties in the lower atmosphere; hence the collective behavior of termites and their symbiotic bacteria has an influence on global climate. Now that's a trace!

As mentioned earlier, subterranean termite nests in the southeastern United States are best represented by the native species *R. flavipes*. Alates of this species each year start flying and seeking sex in January, although this can continue through the spring. Females will descend to the ground surface, shed their wings—making their ground-dwelling existence permanent—and look for a potential nest site near delicious rotting wood, while still being pursued by persistent (and also wingless) males. Once a female–male pair has decided to make more termites, they make a royal chamber (no, I am not making that up) in the soil. In the nascent colony, hatched eggs become workers, and the workers in later instars can morph into soldiers or alate nymphs; the latter become secondary reproductives. Nevertheless, this cycle may take 5–10 years; hence a mature termite colony, if preserved in the geologic record, may represent many generations of primary and secondary reproductives (Abe et al. 2000; McGavin 2001).

Tunnels and galleries of *R. flavipes* are heavily influenced by preexisting conditions in the sediment and other surroundings, and often follow previously made tunnels, spaces formed by plant roots, burrows of other insects or earthworms, and crevices of any kind (Pitts-Singer and Forschler 2000; Lee et al. 2008). In other words, termites, despite their reputation for drilling through whatever wood might be in their way, actually follow a path of least resistance. Locating a nest within and adjacent to many woody barriers, as opposed to open, sandy areas, is also important for defending a termite nest against ants. For example, experiments indicate the woodland ant (*Aphaenogaster rudis*) can quickly invade and decimate a colony of *R. flavipes* if unimpeded by physical barriers (Buczkowski and Bennett 2008). Thus wood, however decayed, can prevent or slow down the progress of ants intent on taking over a termite colony. Termite species in the southeastern United States, though, mostly restrict their activity to alteration of shallow subsurface sediments and wood interiors, and none makes the large, multimeter-high termite nests so famous in the landscapes of Caribbean islands, South America, southern Africa, and Australia. These nests, case hardened by termite saliva, are among the most spectacular and oldest functional insect traces in the world. In fact, radiocarbon dates calculated for some nests in Africa indicate their use by generations of termite colonies for more than 4,000 years (Moore and Picker 1991).

On a more ecological note, termites and other wood-boring insects, such as ants, bees, and beetles also help us to better understand how a dead tree is actually teeming with life, filled with a great variety and vast number of wood-boring insects. These trees are thus important microhabitats within terrestrial ecosystems, providing places of increased biodiversity, ecological niches, and nutrient cycling. Woodpeckers and other insectivores are especially thankful for trees filled with tracemaking insect larvae, pupae, and adults, and will leave their own identifiable predation traces in decaying wood that may contain the objects of their interest, an example of traces leading to more traces (Chapter 8). Also, most wood can float in the sea, which is why ichnologists should learn to identify insect borings, thus helping to separate these traces from those made much later by marine invertebrates (Chapter 6). Consequently, an ichnologist examining either modern or fossilized wood should look at the crosscutting relationships of the traces, as well as the sizes and forms of the traces: the older traces made by insects should be intersected by the younger traces of marine borers. Moreover, insect traces should riddle the entire thickness of the wood and be filled with frass or other traces pointing toward an insect origin. On the other hand, marine borings should be only on the outside of the wood and lack frass, and may sometimes contain bodily remains of their tracemakers.

Termite lineages probably evolved from their cockroach-like ancestors in direct response to the expansion of woody habitats and substrates during the Mesozoic Era (Engel et al. 2009). As a result, isopterans have likely been around for more than 100 million years. Although their body fossils are relatively rare in Mesozoic deposits, fossil nests have been identified from the Triassic, Jurassic, and Cretaceous (Hasiotis and Dubiel 1995; Bordy et al. 2004; Hasiotis 2004). Nonetheless, some of these identifications are controversial, and require more study (Genise et al. 2004; Bromley et al. 2007). Future finds of possible fossil termite nests likely will be enhanced by further and careful examination of modern examples, some of which are offered by species on the Georgia barrier islands, whether native or invasive.

Lots of Insect Traces, Little Study

Despite what may seem like an extensive coverage of the daunting number and variety of insect traces of Georgia coast environments, I will surmise that the not-so-gentle reader, especially the more entomologically inclined, will no doubt think of the many insect traces of the Georgia coast that were not mentioned in this all too brief synopsis, and triumphantly exclaim, "But you forgot about [fill in the blank]!" These critics are right, of course, and their objections are duly noted and supported. Nonetheless, I will also point out that some insect tracemakers were not actually forgotten, just omitted. Additionally, I encourage anyone entomologically and ichnologically inclined to go out and do such research and documentation for themselves, rather than point fingers of blame. After all, the islands are there, along with their insect traces. Most importantly, though, I hope the preceding coverage has imparted a realization of the abundant and diverse types of insect traces in freshwater and terrestrial ecosystems of the Georgia barrier islands and similar environments. Another aspiration of this ichnological summary is to cultivate an appreciation for how much these insects can teach us all about the behaviors of the earth's most successful group of animals, leaving their marks at nearly all scales of observation.

NO LEGS, NO PROBLEM: ANNELIDS AND GASTROPODS

Annelids and gastropods comprise an unusual pairing for zoologists because these two groups are not directly related to one another, and accordingly are quite varied in their respective habitats, with little ecological overlap between them. Nonetheless, as an ichnologist, I am lumping them together here because as terrestrial tracemakers they are united by their lack of an anatomical trait shared by every other invertebrate tracemaker discussed thus far: legs. Hence representative tracemakers are discussed in terms of the various traces that legless invertebrates can make in freshwater and terrestrial environments on the Georgia barrier islands. In other words, a lack of legs does not mean a lack of traces.

Oligochaetes (Earthworms)

Most people are well acquainted with annelids. Now, if you are shaking your head no to that statement, then you are admitting to having never seen an earthworm, which even in urban areas make themselves apparent by the thousands on paved surfaces after a particularly soaking rain. Annelids are segmented worms with complex internal anatomies, and are composed of three major groups: hirudineans, which are blood-sucking leeches and their kin; polychaetes, or so-called bristle worms because of their prominent parapodia ("false legs"); and oligochaetes, which include earthworms (Moore and Overhill 2001). From these three groups, polychaetes are represented abundantly in marine environments (Chapter 6), whereas the other two are more typical of terrestrial and freshwater environments. Oligochaetes are diverse, represented by about 5,000 species worldwide, although more than a third of all oligochaete species in North America are exotic, a result of unintentional or purposeful introductions (Hendrix 1995). A small number of earthworms are native to terrestrial and freshwater environments of the Georgia coast, but these are most abundantly represented by species of *Diplocardia*, particularly *D. mississippiensis*, the most common native earthworm in southeastern United States, and *Lumbricus rubellus*.

Earthworms are fascinating animals for their tracemaking abilities. In fact, Charles Darwin was inspired enough by their ichnological dimensions to write a book about them, which was also his last (published in 1881): *The Formation of Vegetable Mould through the Action of Worms with*

Observations on their Habits. In fact, Darwin actually conducted neoichnological experiments on earthworms. In these tests, he placed flat, circular "wormstones" in his yard. Then, over the course of years, he measured the progressive depth of burial of these stones, which was the result of earthworms overturning the underlying soil (Pemberton and Frey 1990). I should point out that this sort of patient, long-term, observation-based experimentation involved the use of entire, living organisms in an outdoor setting, required no external funding for expensive laboratory equipment, and resulted in only one publication. All of which neatly illustrates why Darwin would have never been awarded tenure under current academic standards in biology.

Anyway, as part of his studies of worms, Darwin also shouted at them—one can only speculate what—and played tones from a tin whistle, bassoon, and piano for them. Again, were these Victorian-era classics, or raucous sea ditties inspired by many years of sailing on the *Beagle*? Although it might seem strange that Darwin included auditory stimuli as part of his study of worms, he was simply being thorough by trying to test all aspects of worm behavior and how it affected their burrowing. (Surprisingly, he was onto something about the effects of sound on worm burrowing, as explained later.) His pioneering work on worms and their neoichnology was underappreciated for a long time, and even Darwin, in a moment of self-deprecation, regarded it as "a subject I have perhaps treated in foolish detail." Nonetheless, Darwin, ever the gradualist, recognized early on the important large-scale and long-term ecological effects of these small-scale organisms and their daily activities, in which they impart an ichnological overprint on terrestrial landscapes (Chapter 3).

Indeed, the amount of bioturbation of most terrestrial soils by oligochaetes is considerable, and essential for the recycling and renewal of nutrients in a terrestrial soil. Oligochaete traces include burrows, fecal castings, aestivation chambers, and surface trails. Earthworms are deposit feeders, ingesting sediment through one end and excreting waste as fecal castings through the other. These animals consume sediment as they burrow, and burrowing is accomplished by displacing sediment—rather than active digging—in which they use peristaltic contractions and extensions that travel along their lengths, transmitted from forward to rearward segments (Lee 1985; Edwards and Bohlen 1996). So if you watch a worm move along

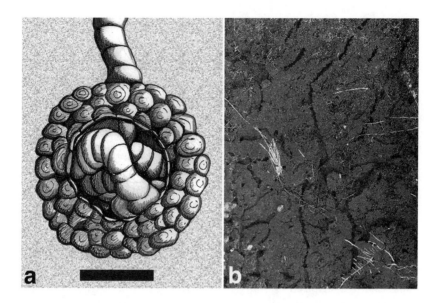

Figure 5.11. Earthworm traces. (a) Estivation chamber connected to a vertical burrow, showing the pelletal exterior and fecal string within the chamber; composite rendition, based on traits of modern and Pleistocene specimens figured by Verde et al. (2007). Scale = 1 cm (0.4 in). (b) Earthworm burrows made just below the ground surface, showing mostly simple horizontal tunnels, but with some intersecting and others turning downward (vertical); tunnel widths are 5–12 mm (0.2–0.5 in): Cumberland Island. Photo by Carol Ruckdeschel.

a surface, you are also witnessing its burrowing motion. Earthworms also grip the sides of their burrows with short bristles (setae). Moreover, anyone who has handled a live earthworm can attest to its sliminess, caused by mucus secreted onto the outside of its body; this then lubricates the burrow walls and aids in its movement through the soil. As a result, earthworm burrows have thin (1–3 mm thick) linings and clay minerals aligned parallel to these. Linings provide sufficient means for distinguishing oligochaete burrows from those of insects, as insects do not produce mucus. Individual earthworm burrows, however, are probably difficult to pick out of any given bioturbated soil horizon. Nonetheless, the mixed sediment observed in the top 50 cm (20 in) of any soil profile can be attributed confidently to earthworm burrowing, and may form the background pattern that is then intersected by insect and mammal burrows.

Earthworm aestivation chambers are remarkable traces, in that they reveal adaptive behaviors of earthworms that evolved from what were originally water-dwelling ancestors. Aestivation is a form of stasis undertaken by some animals in the event of a prolonged environmental stress, such as drought or winter cold. With earthworms, one such stress is a drought that dries out the uppermost layers of soil; this triggers a protective response in a worm, which is to retreat from the surface and make an aestivation chamber. Chambers are semispherical structures reinforced by fecal castings, and may be connected to or otherwise closely associated with a 1 m (3.3 ft) deep vertical burrow leading to the chamber (Figure 5.11a; Lee 1985; Verde et al. 2007). Once inside this damp microhabitat, a worm will coil into a ball and become quiescent until moist conditions outside the chamber give it a reason to come out and resume its normal lifestyle of vigorous dirt eating. Even during a period of severe heat and high evaporation, the sediment surrounding the chamber may desiccate, but a well-sealed chamber will continue to sustain an aestivating worm. Using our ichnological common sense, we can hypothesize these traces, being deeper than most traces made by earthworms, should have good preservation potential. This supposition is borne out by a few trace fossils interpreted as earthworm aestivation chambers (Verde et al. 2007). Some fish and amphibians likewise share aestivation as a behavior, and are known to make it through several years of drought by staying in their burrows (Chapter 7).

Because of the ubiquity of earthworms in terrestrial environments of the Georgia coast, most people who have lived there can attest that earthworms cannot tolerate saturated sediments, and will quickly drown if left immersed. This is why earthworms will emerge from their underground habitats and wriggle out after heavy rains, a response greeted with delight by their avian predators. Yet earthworms also require damp sediment for proper respiration and production of mucus; as a result, earthworms staying on sun-baked sidewalks constitute a deleterious strategy. On the Georgia barrier islands, though, the most commonly observed earthworm surface traces are their trails. These traces are found on the edges of puddles following rainstorms that exceeded the infiltration capacity of the soil. Earthworm trails are about 0.5–1 cm (0.2–0.4 in) wide, meandering to straight, shallow grooves that often cross one another (Figure 5.11b). As a result, budding neoichnologists should pay careful attention to crossover points, and not jump to the conclusion that the trails are branched. If similar-looking traces were actually formed by branching, these are more likely the result of multiple probes by a single tracemaker from a central point. Earthworms will do this, in which their forward halves expand outward in one direction while the rest of the body stays in place; they then retract this portion and continue moving in another direction.

Interestingly, earthworms are rather sensitive to underground vibrations, which relates to their ichnology. Avid fishers in the southeastern United States, who bait hooks with these wrigglers, take advantage of this adaptive trait in earthworms by placing a wooden stake in the ground, then rasping a steel slab on the stake top. This produces a low-frequency grunting sound that, when transmitted through the soil, causes earthworms to flee toward the surface, where the happy fisher grabs fistfuls of worms. Why do they do this (the earthworms, that is)? Behavioral biologists have interpreted it as protective behavior in reaction to an earthworm's most dreaded predator—not the American robin (*Turdus migratorius*), but the eastern mole, *Scalopus aquaticus* (Mitra et al. 2009). As moles burrow through sediments, they impart similar low-frequency vibrations that, over thousands of generations of earthworms, resulted in selection favoring those worms that react with alarm to such subterranean stimuli. This insight is not new, though: in 1881, the ever-observant Charles Darwin noted, "It is often said that if the ground is beaten or otherwise made to tremble, worms will believe that they are pursued by a mole and leave their burrows." Even long before Darwin, earthworm predators on the surface had figured this out, as terrestrial turtles and seagulls have been observed stamping their feet, causing earthworms to emerge from their burrows and providing squirming, slimy treats (Kaufmann 1986; Mitra et al. 2009). As fascinating as this behavior might sound, the resulting escape burrows would be even more interesting to interpret from the fossil record, especially if accompanied by nearby mole burrows or vertebrate tracks directly associated with such burrows.

As mentioned before, earthworm trace fossils have been interpreted from the fossil record. These include their burrows, which are allied with the ichnogenus *Taenidium* (Netto 2004), fecal castings, and aestivation chambers, the latter given the ichnogenus *Castrichnus* (Verde et al. 2007). Various simple surface trails also have been attributed to terrestrial oligochaetes, especially those preserved in floodplain deposits (Uchman et al. 2004), where earthworms might have left their subterranean habitats after high surface runoff events, such as rainstorms or floods. Such traces, however, are difficult to separate from those made by similarly sized nematodes, gastropods, and even some insect larvae (Uchman et al. 2009). Consequently, trace fossils should always be interpreted with several potential tracemakers in mind as alternative hypotheses. Single answers run the risk of facile labeling ("worm trail"), a scientifically inadequate label for what are often complicated traces made by a huge variety of animals, and as a result of a many interacting ecological factors.

Gastropods (Snails)

Gastropods, or snails, are common tracemakers in terrestrial and freshwater environments of the Georgia barrier islands, although these have not been studied at all for their ichnological significance. When people normally talk about gastropod tracemakers in the context of the islands, their attention is focused on the numerous and gorgeous examples they see in marginal-marine environments, such as intertidal sand flats and salt marshes (Chapter 6). Understandably, their smaller, drabber, and relatively less abundant land-dwelling cousins are often ignored in favor of those who live on and in beaches. Nonetheless, I have often watched snails make lots of trails on the sands of dunes and back-dune meadows, and figured they need some loving attention. After all, their trails overlap with traces of insects, ghost crabs, toads, lizards, mice, raccoons, and many other animals in these coastal environments. Gastropods are also reported from some freshwater environments on Georgia barrier islands, such as Lake Whitney on Cumberland Island (Frick et al. 2002), although these are few, and again are poorly studied. Among these freshwater gastropods are mimic pond snails, such as the American ribbed fluke snail (*Pseudosuccinea columella*), as well as species of *Physella* and *Planorbella*.

If you see a snail on land, and if it shows no signs of obvious distress with this habitat choice after your watching it for a few hours, it is probably a pulmonate gastropod. The adjective *pulmonate* (similar to "pulmonary") reveals their anatomical adaptation for getting around on land without a water source, namely lungs. These animals, like ghost crabs, descended from fully aquatic ancestors, but are now adapted to breathing air; moreover, their shells maintain internal aquatic moisture levels by retarding water loss. Most snails living in freshwater environments, though, are also pulmonates. Terrestrial slugs, which are also gastropods, have taken this

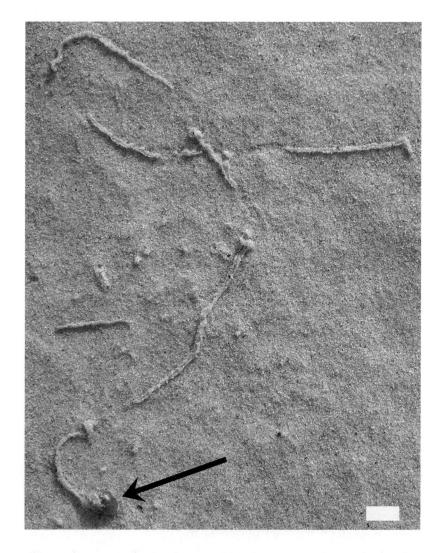

Figure 5.12. Pulmonate gastropod trail in back-dune meadow, with the tracemaker caught in the act (*arrow*): Cumberland Island. Scale = 1 cm (0.4 in).

evolution a step further (if a gastropod can indeed take a step) and even managed to forgo shells. Like most gastropods, pulmonates lay down a smooth highway for moving across rough surfaces, composed of a thin, only a few millimeters-wide line of shiny mucus, often identified readily by sharp-eyed homeowners and gardeners in more urban settings. These same search images can be applied in back-dune meadows and dunes of the Georgia coast, particularly in the early morning (Figure 5.12), just after gastropods have been out on walkabouts. These trails are mostly seen as straight to gently curving gossamer ribbons crisscrossing sandy surfaces, but also as spots where minute amounts of sand are clumped together, caused by a combination of sand sticking to mucus and changes in gastropod movement. Gastropods move through a wave-like series of extensions and contractions from the front to back, which pull the body forward; this will be explained further in relation to marine gastropods (Chapter 6). Pulmonate gastropod traces are often subtle, but can be noticed after watching snails make these trails.

Although the ephemeral nature of pulmonate gastropod trails in most terrestrial environments means their preservation potential is practically nil, a fascinating behavioral trick that might be gleaned from their trace fossils is how individual snails can track one another by following trails. With this purposeful movement, they detect physical characteristics of another gastropod trail by feeling the ridges with their foot, as well as by smelling chemical cues in the mucus trail (Barker 2001). One might never have imagined that gastropods have their own form of ichnology, but there it is, waiting for us to admire and study, and to look for its equivalent in the fossil record.

Terrestrial and freshwater gastropod trace fossils, such as their locomotion trails, are rarely interpreted from the geologic record other than ichnologists stating, rather unhelpfully, "possible gastropod trail." One understandable challenge in identifying such trace fossils is their overall resemblance to trails made by worms of all sorts, such as oligochaetes and nematodes. Nonetheless, careful descriptions of modern examples can probably help to better sort these out so that the snails are separated from other legless tracemakers.

MANY LEGS MEAN MANY TRACKS (AND MAYBE BURROWS): MYRIAPODS

Myriapods ("many legs") include millipedes (Diplopoda) and centipedes (Chilopoda). Of these, millipedes are common arthropod tracemakers in terrestrial environments of the Georgia coast, particularly in maritime forests, back-dune meadows, and dunes. Millipedes are especially important parts of maritime forests, helping with decomposition of forest litter

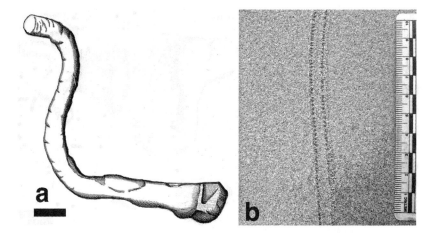

Figure 5.13. Millipede traces. (a) Burrow of a western U.S. species of millipede (*Orthoporus ornatus*); based on a plaster cast figured by Hembree (2009). Scale = 1 cm (0.4 in). (b) Trackway in a back-dune meadow: Wassaw Island. Scale in centimeters.

by eating much of it; in fact, they even have been called model detritivores (Crawford 1992). Of the two groups of myriapods, millipedes are the more important ichnologically, owing to their ability to make trackways and burrows. Yes, most millipedes burrow, a seemingly unlikely behavior considering their lack of obvious digging tools in their anatomy. Nonetheless, expectations aside, burrows they make, as explained later. Moreover, the sheer number of millipede legs compared to other terrestrial arthropods implies that their trackways should be unmistakable. How many legs do millipedes possess? This of course depends on the millipede species, but all millipedes have two pairs of legs per body segment behind their heads; so making an approximation of leg number is a matter of counting body segments, then multiplying by four. This number requires some subtracting, though, as the segment immediately behind the head lacks legs, and several segments after that only have one pair each. With all of that anatomical arithmetic in mind, most millipede species own hundreds of legs, although some shorter ones have fewer than a hundred.

If you have only seen millipedes walking about and in moist, organic debris of maritime forests, think about how they might avoid predators and desiccation. Does the word *burrow* come to mind? Indeed, this is how most millipedes cope with stressful surface conditions, and they construct surprisingly varied burrows. These burrows are made not by digging, but by the millipedes, powered by their many legs, ramming their heads into sediment, or otherwise pushing apart preexisting cracks or other open spaces (Hembree 2009). In other words, they have legs, and they know how to use them. From an evolutionary perspective, however, burrowing is not surprising, as myriapods were among the first colonizers of terrestrial environments, which we know through their trace fossils, discussed later.

Sadly enough, burrows made by modern millipedes on the Georgia barrier islands remain unstudied. However, a western U.S. species of millipede (*Orthoporus ornatus*) may provide some clues about the appearances of such burrows (Hembree 2009). These structures are vertically to obliquely oriented and simple tubes, but often with compacted walls caused by the millipede head butting the burrow walls as it pushes itself through the sediment. Burrows have a variety of geometries, consisting of U-shaped, J-shaped, and otherwise discrete forms. These burrows are mostly open, but can have backfilled portions toward their bottoms. Where burrows have only one opening, these are normally J shaped and have an enlarged chamber (Figure 5.13a). These chambers either accommodate the millipede so it can turn around and move back up the burrow, or supply sufficient space to invite another millipede to join it (if you know what I mean). Sometimes plugs of backfilled sediment block entry to these chambers, providing for some privacy in cases of successful millipede courtship.

Millipede trackways, like those of beetles and their six-legged kin, are commonly seen on clean, sandy surfaces in maritime forests, back-dune meadows, and coastal dunes (Figure 5.13b). Although these traces may have lower preservation potential compared to millipede burrows, they nevertheless should not be ignored in any survey of invertebrate trackways. Look for narrow trackways—only a few millimeters to a centimeter wide—that, at first glance, may look very much like trails left by legless animals. Look closer at these, and you will see fine, closely spaced pinpoints making up the outside of the trackway, producing a stitching pattern. Look even closer, and the two pairs of legs associated with each body segment make themselves apparent as miniscule breaks in between two impressions on each side of the trackway. Millipedes move along a surface in a wave-like motion, with each pair of legs pushing both up and back, from the front to the rear. With increasing speed, spaces between track sets become more obvious, and the length of the animal is more easily discerned as a gap between a full set of legs and no legs at the front of the animal. Millipede resting traces also may be connected to trackways, showing where a millipede coiled tightly, thus rendering a beautiful spiral in the sand outlining its body.

Trace fossils connected to myriapod tracemakers consist of both burrows and trackways, and go back well into the Paleozoic Era. For example, trace fossils from Late Ordovician terrestrial rocks in Pennsylvania were interpreted as myriapod burrows (Retallack and Feakes 1987; Retallack 2001), as were traces in Devonian rocks of the United Kingdom, assigned to the ichnogenus *Beaconites* (Morrisey and Braddy 2004). Myriapod trackways have also been proposed on the basis of numerous closely spaced impressions within trackways, and given the ichnogenus name *Diplichnites* (Morrisey and Braddy 2004). Some of the makers of these trackways were also frighteningly massive when compared to modern species. Based on body proportions calculated from trackway widths, some myriapods may have been as long as 1–2 m (3.3–6.6 ft)! The most likely makers of these trackways were arthropleurids, an extinct group of myriapods that were among the largest known land-dwelling invertebrates (Briggs et al. 1979; Pearson 1992; Martino and Greb 2007). Although no modern myriapods approach such sizes on the Georgia coast (which comes as a relief to some people), their traces can still give insights about how these animals interact with their terrestrial ecosystems.

SUMMARY OF FRESHWATER AND TERRESTRIAL INVERTEBRATE TRACES

This chapter may have given an overall impression that vast amounts of information are available about freshwater and terrestrial invertebrate traces and tracemakers on the Georgia coast. This is only partially true, though, as historical emphasis in ichnological research has been placed on invertebrate tracemakers of the marginal-marine environments (Chapters 2, 6). As a result, I encourage future emphasis on the study of freshwater and terrestrial invertebrate tracemakers of the Georgia barrier islands. Crayfish, arachnids, insects, earthworms, gastropods, and myriapods are all important tracemakers in nonmarine environments that deserve more attention than received thus far. Such investigations should be done in the spirit of discovery and learning more about these animals, who often live not just on the Georgia barrier islands but in our backyards, lending more to a sense of everyday ichnology (Chapter 11). On a broader scale, geologists and paleontologists can use their modern traces as models for similar ones in the geologic record, helping to figure out whether an ancient sediment was on land or not. In other words, wherever these tracemakers leave their traces, they provide us with answers to the basic paleoenvironmental question, "The sea, or not the sea?"

6

MARGINAL-MARINE INVERTEBRATES

THE DEATH SPIRAL THAT ONLY SPIRALED

No matter how many times I visit the beaches of Sapelo Island, its sediments always hold surprising traces. On this particular morning in June 2008, my wife, Ruth, found the wonder-inducing trace of the day and directed me to it, just before she left to show some of our companions a fresh sea turtle trackway further south on the beach. This was also well before my ichnologic partner in crime solving, Andy Rindsberg, of the University of West Alabama, joined me. What could be more interesting to an ichnologist than an expectant sea turtle's trackway and her probable nest structure, made mere hours before? Something much more primeval beckoned us, a tug that pulled me back into the Paleozoic Era instead of the mere Mesozoic. The latter was when sea turtles evolved from their land-dwelling ancestors, but the Paleozoic connected with a time before any four-legged animal walked on land. The beguiling trace of the day was made by a large adult limulid (*Limulus polyphemus*), popularly known as a horseshoe crab, that most archaic of marginal-marine arthropods available for us ordinary, Cenozoic-bound humans to view.

Figure 6.1. The horseshoe crab that almost made a taphichnion on a beach of Sapelo Island. (a) Trackway of a large adult limulid (*Limulus polyphemus*) which was trying to return to the sea, but turned the opposite direction and eventually stopped, making a looping pattern. (b) Close-up of same limulid, with markings made by its telson in a last effort to move itself. Scale (knife) = 5.8 cm (2.3 in) long.

Throughout their ancestry and for more than 400 million years, limulids have looked on the world with compound eyes. They shared Paleozoic seas with their long-lost brethren, the trilobites, as well as nautiloids, bivalves, gastropods, brachiopods, polychaete worms, sharks, and a host of other active bottom-dwelling and free-swimming animals. Limulids, however, not only saw these animals and their seascapes during the Paleozoic, but also witnessed primordial landscapes and the evolution of life on land. Also embodied in their modern descendants is a behavioral legacy that hints at how other marginal-marine invertebrates and their relatives may have colonized terrestrial ecosystems. For example, each year from April through June along the east coast of the United States, female limulids will crawl onto beaches by the thousands to lay their eggs in the beach sands. These females are closely followed and vastly outnumbered by male limulids, who seek to fertilize their eggs (Brockmann 1990; Shuster et al. 2003). Two unique aspects of this mating ritual for marginal-marine invertebrates are where these limulids mate (on the beaches) and how they fertilize eggs (externally). The lack of these behavioral traits in other marginal-marine invertebrates is considered the result of evolutionary distance. Time and extinctions have thrown out all other species that used to have the same behaviors, and we are lucky indeed to witness today such an ancient mode of reproduction through modern limulids.

The overall trace of the adult limulid on the Sapelo beach that morning was of what paleontologists sometimes nickname a death spiral, and the limulid's body was still at the end of the trace. This one had probably not come ashore as a slave to its genes to reproduce, though, but was washed up by a high tide that turned only a few hours before Ruth spotted its immobile carapace near the shoreline. Its tracks changed in character as it moved from saturated sand to drier sand up the beach. Drag marks caused by its pointed feet and the edge of its broad head shield (prosoma) pulled cohesive sand with each movement and recorded subtle alterations in its progress. The trackway consisted of patterns of eight impressions—four on the right and four on the left, reflecting its eight walking legs—bounded by lines drawn with the outside edges of its prosoma. The limulid had completed a nearly oval path punctuated by notably sharp turns. Its trackway originated in the high-tide zone and ended in its present location well above the low tide (Figure 6.1a).

From the perspective of the limulid arriving from the ocean, it floated, swam, and then walked onto the beach from the left, went up the beach slope for nearly a meter, turned slightly to the right to rest briefly in a small depression that retained some sea water in a shallow pool, then left the depression once this water drained into the underlying beach sands. This decision, made out of necessity—its external gills needed wetting—was where the animal's desperation began. It turned abruptly to the right and made a tight but open loop, one that reflected both its body size and inflexible exoskeleton. (Think of a hard-cased piece of luggage turning on an airport carousel as an analogy for this restriction of movement.) It then turned even more sharply to the left, this time in the direction of the sea, and paused briefly in the drying sand. Its pointy tail (telson) helped it to shove off from the left, and as a consequence of this movement, its broad head shield pushed up a plate of sand on its right side. It then lurched forward and slightly to the left. Nevertheless, this choice came too late, as the slow-moving and nearly exhausted limulid did not quite catch the lowering tideline. It made one more turn, another tight loop to the right, before walking back up the beach and away from the life-sustaining sea, ending where Ruth found its body. Signs of distress were recorded in the sand as it attempted to change direction one last time: its telson had moved back and forth like a windshield wiper, the dying limulid equivalent of a snow angel made by children in more northern climates (Figure 6.1b).

"Death spiral" is an expression applied to various aspects of human life. Most graphically, it occurs when an airplane stalls and begins a spin downward from which it cannot recover, normally resulting in the death of the pilot and passengers. More metaphorically, it can also be an economic condition, such as when banks face increasingly dire losses that reach a point of no return (unless aided by the government, of course). In sports, this term is also used in figure skating, when one partner of a skating duo spins tightly while holding onto a nearly horizontal partner, who then spins centripetally in a much wider circle. The physical and technical danger of this move is the inspiration of its nickname in this instance, but because of the fleetingly similar traces this movement leaves on icy surfaces, this death spiral most closely resembled what was before me in the warm sands of Sapelo.

Paleontologists had seen the most famous of such traces in the Solnhofen Limestone of Bavaria, Germany, a shallow lagoonal deposit formed during the Late Jurassic. These trace fossils, consisting of a series of tracks similar to the modern ones before me, were not renowned just for the traces themselves, but for what lay at the end of each trackway: a fossilized limulid. The firm, muddy sediments of the lagoon bottom had preserved both the final steps of the tracemaker and the tracemaker itself, a rare instance where ichnologists can link these two with absolute certainty and without fear of undue skepticism from their nonichnologically inclined peers. Furthermore, the trackways normally preserve the same final attempts toward continued life: a looping pattern that closed in on its starting point, a sort of spiraling that represented the transition from life to death.

Such traces are termed taphichnia, traces made just before the tracemaker was buried, and in some lucky instances (for paleontologists, that is), the tracemaker body is still at the end of its trace (Frey et al. 1987a). Traces made just before these limulids began to die are called fugichnia (escape traces), which implies that the tracemaker was trying to escape death or some other unpleasant stimulus that prompted movement toward better conditions. Even more finely defined than taphichnia and fugichnia are equilibrichnia, or equilibrium traces. These are traces made by an animal that is adjusting itself away from a less comfortable (but not necessarily life-threatening) setting, such as the disturbances of sediment left by a bivalve as it moves its burrow upward after a layer of sand is dumped on it, or the splatter pattern imparted on a carpet by a dog shaking itself vigorously to dislodge water (or worse) from its fur. (The latter effect is more easily observed with lighter-colored carpets and darker sediments shaken from the fur.) In short, a fossil limulid at the end of its trackway can be back-tracked to show the following events in a sort of paleontological version of the movie *Memento:* death, dying, desperation, and mere discomfort, all briefly interrupted in places where the tracemaker stopped to rest (these are known as cubichnia) before making its next decision, one of which was its last.

Only a few other fossil animals have been found in trace fossils of their making. One was the burrowing beaver (*Palaeocastor* spp.) of the Miocene Epoch in the midcontinental United States. *Palaeocastor* sometimes formed its own subterranean version of a death spiral, a huge, several-meter-long helical burrow that was nicknamed the Devil's Corkscrew when first encountered by awed farmers (Martin and Bennett 1977). Scratch marks on the walls of the burrows hinted at its tracemaker, but only later, when beaver bones were found at the bottom of the spiral, did paleontologists make the connection between the trace and tracemaker. Figuratively speaking, the vertical dimension of the trace fossil was more akin to the aeronautic death spiral in its geometry. Similarly, and reaching back into the Mesozoic about 95 million years, the small Cretaceous ornithopod dinosaur *Oryctodromeus cubicularis* was found in a semihelical and burrow-like structure that matched its body size (Varricchio et al. 2007). Two half-grown offspring accompanied the dinosaur, indicating that the burrow was likely used as a den for protecting its young. The skeletons of both *Palaeocastor* and *Oryctodromeus* also held clues they were burrowers; in their shoulder girdles were the attachment sites for powerful muscles needed for digging. Of course, careful examination of the substrate preserving the trace, the anatomy of the tracemaker, and the behavior of the tracemaker could eventually lead to the identity of the likely constructor of any such trace fossil, but debate is cut short when a dead body is found within the trace fossil itself. Nonetheless, paleontologists still allow for the possibility of a secondary occupier, or squatter, that might have inserted itself into the home of the actual tracemaker.

Hence, this limulid and its traces on the Sapelo beach represented an important connection to the geologic past and could serve as a model for comparison to similar occurrences in the fossil record. Accordingly, I eagerly commenced studying the overall pattern and nuances of its trackway and resting traces, walking carefully around the periphery of the trackway so as not to disturb any aspect of its exquisite details. Naturally, like any good teacher and scientist, I wanted to share this find with others, and I soon waved over my colleague Andy, who had been distracted by other traces further up the beach slope. Once he arrived, he was just as taken as I was by the implications and opportunities presented by this modern trace and tracemaker, and we both went to work observing, noting, and photographing the traces and limulid. For nearly an hour, we only interrupted one another to quietly share observations and discussions of certain features we noted of the traces and tracemaker.

Yet in midst of all of this clinical documentation, an irksome feeling crept into what was an otherwise dispassionate analysis. Perhaps the most basic of questions asked daily in science, one that distinguishes real scientists from mere pretenders, is this one: "How could I be wrong?" The sequence of events spelled out before us seemed too simplistic and final.

Somehow, I was mistaken. The realization of possible error can be an uncomfortable one to a scientist and is the intellectual equivalent of an equilibrichnion leading to a fugichnion, then a taphichnion, which is then followed by the death of an idea. Indeed, science often goes against human nature, which in our evolutionary history selected for self-preservation and preservation of others, an instinct that is easily extended to creative thoughts and concepts. Nonetheless, the ability to indulge in healthy skepticism in ichnology was also selected in our evolution. For example, if people kept misidentifying tracks of game animals tens of thousands of years ago, they and their families would have starved (Chapter 1). In those instances, refusing to let an incorrect idea die, and stubbornly clinging to it in the mistaken belief that somehow it could be right, would have resulted in the same sequence of adjustment, mild discomfort, extreme discomfort, desperate flailing about, dying, and death. My less dire situation, with its increasing sense of uncertainty, had certainly progressed to the point of wanting to adjust it, and I began to express this to Andy.

"Are you about done?" I asked him.

"No, I still have a few more notes to take, then some more photos."

"Oh, OK," I said, although it was not OK.

I had figured out what was wrong, the alternative hypothesis, so to speak: the limulid was not dead. Thomas Huxley, a contemporary of Charles Darwin and one of the best evolutionary biologists of the late nineteenth century, once said, "The great tragedy of science: the slaying of a beautiful hypothesis by an ugly fact." The death of our beautiful hypothesis, however, was also going to result in the death of this magnificent invertebrate. In an admission of unscientific reasoning, I have no idea how I knew the limulid was still alive, but my hunch began to push, then shove.

"Andy," I finally said. "I'd like to put it back in the water, to see if it's still alive." In retrospect, this simple statement seemed to be where animal rights conjoined laboratory experimentation. And in this instance, it was the right thing to say.

Andy, in the middle of writing what was no doubt a brilliant observation in his field notebook, sighed. "All right, just a minute." It wasn't a minute, but several, infinitesimally small in the enormous time span of limulid evolutionary history, like a grain of sand in the beach beneath our feet. Nevertheless, these minutes might have been the difference between dying and death for this single limulid, and I waited anxiously, but while also trying to respect my longtime friend's curiosity, a conflict of feelings that caused me to shift back and forth, unconsciously making interesting equilibrichnia in the sand beneath me.

"OK, I'm done," he finally said. With that, I reached down to pick up the limulid with both hands, one on each side of its prosoma, and turned it upside down. As soon as I flipped it onto its back and in the air, its walking legs moved in response to the release of gravity, almost as if in relief. Andy seemed as astonished as I was not. I smiled, said, "Still alive," walked with our object of study to the water's edge, and set it down on the wet sand. While still in observant scientist mode, though, I set my digital camera to record video footage of the limulid making tracks as it walked into the ocean. Only, it did not. Still exhausted by its landward ordeal, it remained motionless in the surf, probably taking up some water into its parched gills. So I ceased video recording, lifted it a second time, and placed it in water deep enough to fully immerse it and where buoyancy was its friend. It kicked its legs and swam/walked out of view into the murky water, and in self-parody, I began singing "Born Free," altering the lyrics slightly to accommodate the zoological differences between limulids and lions.

The last bit of science to do then was to document its resting trace, or cubichnion, which had revealed it to be a not-quite taphichnion. Andy and I photographed the place the limulid had occupied until just a few minutes ago, and recorded the marks where its telson swept against the sandy surface, its walking legs had churned underneath its body, its gills had dried, and the edge of its prosoma had scraped. Thus the death spiral was turned into a denied death spiral, accompanied by our footprints leading up to and away from its trackway. Only a hypothesis of uncertain beauty had died, but our knowledge had increased with the life span of the limulid.

MARGINAL-MARINE TRACEMAKERS, STARTING WITH SPONGES, BORING AND OTHERWISE

The preceding tale was of a marginal-marine tracemaker that often interacts with tidal environments, and its traces are nearly always found where the animal was actually making the trace. In contrast, traces made by sponges, or more properly poriferans (from Porifera, "pore bearing"), are nearly always found out of place, with the substrate well away from wherever a sponge might have made its traces. Moreover, these seemingly inert colonial animals, once thought of as plants, might not be suspected

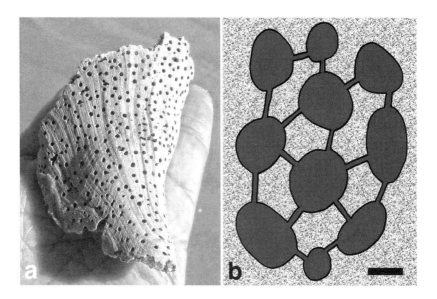

Figure 6.2. Clionaid sponge traces. (a) Borings in whelk-shell fragment: St. Catherines Island. (b) Schematic view of typical clionaid–sponge boring network. Scale = 1 mm (0.04 in).

of behaving, let alone making traces. Even more improbable is the concept of soft-bodied animals leaving traces in hard substrates, such as molluscan shells or rocks. Yet an entire group of sponges (Clionaidae), which is mainly represented by species of *Cliona* (most commonly, *C. celata*), is well known for the bioeroding abilities of its members (Cobb 1969; Rützler 1975; Bromley 1992). These sponges leave distinctive traces that can be readily identified in both modern and fossil shells, corals, or other rock-like substrates on the Georgia coast. Pick up a blackened, beaten, and broken shell on a Georgia beach, and it will very likely bear the densely spaced, interconnected pockmarks of clionaid sponges (Figure 6.2a). These sponges do not just choose dead hard substrates, either. Live sea turtles, such as the loggerhead (*Caretta caretta*; Chapter 9), include clionaid sponges among the many marine invertebrates that attach to their shells and hang on for the ride, as well as live bivalves (especially oysters) and gastropods. In this sense, clionaid traces indicate parasitic behavior, as they can potentially weaken their hosts' shells. Clionaids, if found on living organisms, are categorized ecologically as epibionts, whereas they are epilithic if on non-

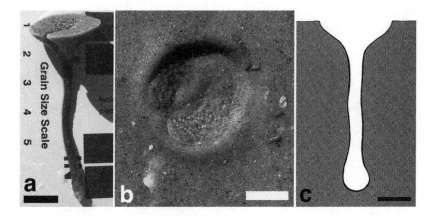

Figure 6.3. The unusual-looking sea pansy (*Renilla reniformis*) and its traces. (a) Sea pansy in lateral view. (b) Top view, with its stalk embedded below in the sand: Sapelo Island. (c) Longitudinal section of an idealized sea pansy burrow, minus the tracemaker. Keep in mind that the stalk can expand to more than twice the depicted diameter, potentially increasing the burrow width. Scale = 1 cm (0.4 in) in all parts.

living rocky material. As organisms that prefer to attach to stable surfaces, clionaids and other sponges are common in Gray's Reef and places on the Georgia coast wherever hard substrates might be exposed (Freeman et al. 2007). Elsewhere, these sponges can be spotty in their distribution, and are most often confined to individual molluscan shells in shallow subtidal areas.

How do these sponges bore? Very simply, they use a combination of collective action and acid. After the colony attaches to a hard substrate, it secretes a tiny amount of an organic acid in specific spots on that surface, which eventually cause chips to spall off. These pieces are then flushed away from the attachment surface through small pores in the sponge (Cobb 1969). All sponges feed by filtering water through the colony; so some of the openings are incurrent (ostia, which pump in water), whereas others are excurrent (oscula, which pump out water). With enough chipping and flushing, the sponge creates numerous holes that allow the colony to spread and secure itself more easily on the hard substrate. These hemispherical holes are then connected with one another like beads on a string, but later these strings join with one another to form three-dimensional networks

(Figure 6.2b). Hence what starts out as mere porosity later becomes permeability, and as bioerosion affects a greater amount of surface area, it causes a positive feedback that accelerates shell breakdown.

Clionaid sponge traces vary considerably according to the species of sponge, the nature of the substrate, and the length of time given to the sponge to erode the surface. Nonetheless, they can be identified as less than 1–3 mm (0.03–0.1 in) wide, circular to semicircular apertures that lead into slightly wider and interconnected subspherical chambers, forming the previously mentioned meshwork. Sponges are not averse to boring into previously bored surfaces either, so often these borings will constitute compound traces made by multiple generations of clionaids, or will merge with other borings to make composite, multispecies traces. Trace fossils that match the preceding morphological description and were clearly made in hard substrates belong to the ichnogenus *Entobia* (Bromley 1970, 1992; Bromley and D'Alessandro 1984). *Entobia* and a suite of other trace fossils provide important evidence for the presence of hardground communities, or organisms ecologically adapted for living on and in hard substrates (Chapter 10).

THE ANTHOZOANS AMONG US

Probably the most alien-looking tracemakers I have encountered at low tide on the Georgia sand flats are the anthozoans *Renilla reniformis*, known more commonly as sea pansies. These tracemakers are purple, fleshy, discoidal objects with an outline like a Valentine's heart, so on a sand flat, they stick out like a sore thumb (and have about the same diameter as one). Once curiosity overcame fear of the unknown, I picked up several and was surprised to see that each also had a short stalk attached to its bottom surface, which was used to anchor it in the sand (Figure 6.3a). Part of their odd appearance is imparted by their lack of a mouth, eyes, or other sensory organs we normally associate with animals. Nonetheless, the lack of these attributes is understandable, because the sea pansy is not a single organism, but a colony of polyps.

What is a polyp? This is one of the growth stages of cnidarians, or so-called stinging animals (and not to be confused with hymenopterans, the stinging animals of terrestrial environments; Chapter 5). Cnidarians include true jellyfish (scyphozoans), hydrozoans, and anemones and corals (anthozoans), all of which have stinging cells called nematocysts; these are why some jellyfish strike fear in beachcombers, snorkelers, and divers alike. Cnidarians also have two stages in their reproduction, a free-swimming medusa stage and a sessile polyp stage. Although many cnidarians are colonial, some hydrozoans and anthozoans are solitary. In a sea pansy, the 1–5 cm (0.4–2 in) thick stalk (peduncle) attached to the bottom surface of the colony is actually a huge polyp. This polyp, which is thicker on both ends and thin in the middle, extends down into the sand to secure the main part of the colony (the heart-shaped pad), which withdraws slightly into the surface. Once the tide is up and the colony is covered by water again, the small feeding polyps project above its surface to gather food through suspension feeding (Ruppert and Fox 1988). Traces made by these colonial animals are short, thin, vertical shafts, less than 10 cm (4 in) long and connected to shallow, bilobed depressions at their tops (Figure 6.3b,c). Moreover, the vertical shaft should have a slight protuberance at its base, representing the distal swelling of the stalk. Realistically, the top depression of such a trace is easily eroded with each incoming tide, so it should not be expected in fossil examples. Nonetheless, if present, this part would be a key identifying trait of its tracemaker.

More notable burrowers among Georgia anthozoans are several species of anemones, among them the sea onion anemone (*Paranthus rapiformis*). The sea onion anemone superficially resembles a small, pickled onion to some people. Others, however, argue it looks more like a head of garlic. The specific name (*P. rapiformis*) does not help resolve this disagreement either, as *rapiformis* means "turnip-like." (The controversy over its vegetative look-alikes, no doubt, rages on.) Anyway, it is a member of Actinaria, an evolutionarily related group of anemones, also known as soft corals in contrast to their calcified kin. More important for ichnological purposes, some species in this group are well known for their burrowing.

The sea onion anemone, which is a solitary animal, has a basal disc used either to attach to a hard surface (usually a bivalve or gastropod shell) or to anchor in sand. In the absence of solid substrates for attachment, it burrows, covering most of its body and leaving its tentacles exposed at the sandy surface. In order to burrow, it uses its malleable basal disc (called a physa) to poke into the sediment, expand, anchor, poke deeper, expand, and anchor. This is a peristaltic movement, lubricated by mucus, which it repeats as necessary until it is safely ensconced in the sand. Through this activity, it forms a mucus-lined vertical burrow 2–3 cm (0.8–1.2 in) wide at the top and 7–8 cm (2.8–3.1 in) long, although it can be as long as 30–35 cm (12–13.8 in) (Frey 1970b; Ruppert and Fox 1988). In instances where it burrows deeply, the animal is quite elongated and looks like a carrot, thus adding fuel to the root-vegetable affinity argument. The distal end of the burrow might be slightly expanded, caused by the anchoring of the physa; otherwise the burrow is cylindrical. Unfortunately, extraction of the tracemaker causes the burrow to collapse, so this structure cannot be examined readily. Furthermore, this easy breakdown of its burrows suggests their low preservation potential, although the deeper parts might stay both intact and identifiable. Even more ephemeral, though, are bouncing marks these anemones leave on sandy substrates. Yes, that's right: when washed out of its burrow or otherwise detached, the anemone contracts into its ball-like form and glides with the currents, rolling and skipping along the bottom until it becomes still enough so it can burrow.

Probably the best known of burrowing anthozoans on the Georgia coast, though, is the North American tube anemone (*Ceriantheopsis americanus*). This anthozoan is actually not a true anemone, as it belongs to a group called Ceriantharia, whereas anemones are all within Actinaria. Taxonomy aside, the "tube" part of its common name is accurate, seeing that this species makes 1.5–4 cm (0.6–1.6 in) wide and 40–50 cm (16–20 in) long, mucus-lined vertical tubes in intertidal and subtidal sands (Figure 6.4), similar to those of the sea onion anemone (Frey 1970b; Ruppert and Fox 1988). This species, though, is also nicknamed the sloppy gut anemone (no, I am not making that up), its colorful appellation owing to its elastic, entrail-like body and the thick mucous lining of its burrow. This more or less permanent burrow serves as both dwelling and feeding trace. Once in its burrow, the animal extends its tentacles above the substrate for food gathering. This tracemaker is also remarkable for its rapid reactions within its burrow, moving upward if touched by a subsurface molluscan from underneath, or retracting downward if manhandled at the sedimentary surface. Similarly, sediment dumped on top of the burrow by a storm or other disturbance will prompt the tracemaker to adjust its burrow upward, whereas erosion will cause it to move downward. Either action leaves a web of concentric lines of contrasting sediment called spreite. Compared to the sea onion anemone, the tube anemone has a robust burrow that resists erosion, and thus it has excellent fossilization potential. Even if the tube vanishes over time, disruption of the surrounding bedding caused by

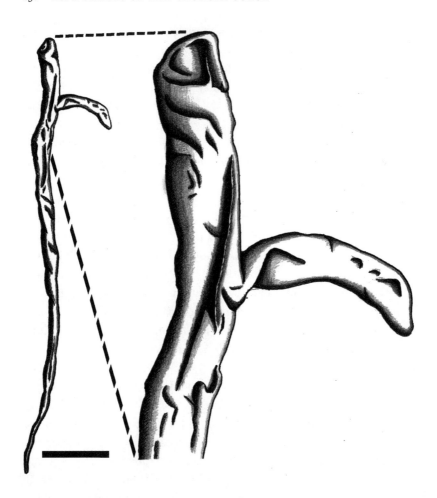

Figure 6.4. Burrow of the North American tube anemone (*Ceriantheopsis americanus*), with overall form (*left*) and close-up of top portion (*right*). Scale (left) = 5 cm (2.5 in). Based on a figure by Frey (1970b).

the tracemaker's up-and-down movements would be easily discernible. Trace fossils that resemble these structures have been identified from the geologic record and are termed *Conichnus*; accordingly, these are often interpreted as the work of burrowing anthozoans (Chapter 10).

Although not nearly as common as in subtropical and tropical areas, anthozoans are important tracemakers in Georgia subtidal environments because of their vertical burrows and alterations of bedding. Additionally, because anthozoans are typically marine organisms, their traces can be included with other paleontological evidence that may favor the interpretation of marine conditions in otherwise nonfossiliferous rocks (Chapter 10).

ANNELIDS AND OTHER MARINE WORMS: PLENTY OF POLYCHAETES

Whenever I have engaged in conversations concerning annelid tracemakers in marginal-marine environments, polychaetes are the segmented worms mentioned most often. Other annelids, such as oligochaetes (earthworms) and hirudineans (leeches), are much more common in terrestrial or freshwater environments (Chapter 5), whereas polychaetes have adapted and diversified admirably to aquatic environments with marine salinities over the past 500 million years or so. Polychaetes make most of their traces through burrowing, and these animals are passive carnivores, scavengers, or deposit feeders, or are active predators, seeking out live food items by burrowing in the sand or swimming just above the surface. As a result, their burrows can indicate a wide range of behaviors. Moreover, in Georgia coastal environments, polychaete burrows are exceedingly abundant. For example, one survey of an emergent, nonvegetated mud bank of a Sapelo Island tidal creek had more than 1,000 burrows/m^2, most of which were credited to the polychaetes *Heteromastus filiformis* and *Nereis succinea* (Basan and Frey 1977).

Of the polychaetes on the Georgia coast, though, tube worms (*Diopatra cuprea*) are probably the most readily observed burrowers in intertidal and subtidal sandy sediments. These worms adorn their vertical burrows by adhering bivalve shells, plant debris, and sediment to parchment-like burrow linings (Figure 6.5a). They particularly like using shells of the small, thin-shelled dwarf surf clam (*Mulinia lateralis*) and coquina clam (*Donax variabilis*). In life position, these burrows are vertically oriented and range from 15–50 cm (6–20 in) long and at least 2 cm (0.8 in) wide, expanded because of their molluscan accouterments (Howard and Dörjes 1972; Howard and Frey 1973; Morris and Rollins 1977). The added shelly parts are concentrated in the top 10–15 cm (4–6 in) of the burrow, which fortunately makes it more visible to people searching for their burrows. This upward attachment of debris, however, is not done as a favor to ichnologists, but is a behavioral adaptation the worm uses to reinforce its burrow in the unstable and shifting substrate of lower intertidal and subtidal areas (Meyers 1970; Howard and Dörjes 1972). Nonetheless, even the longest and strongest of *Diopatra*

260 LIFE TRACES OF THE GEORGIA COAST

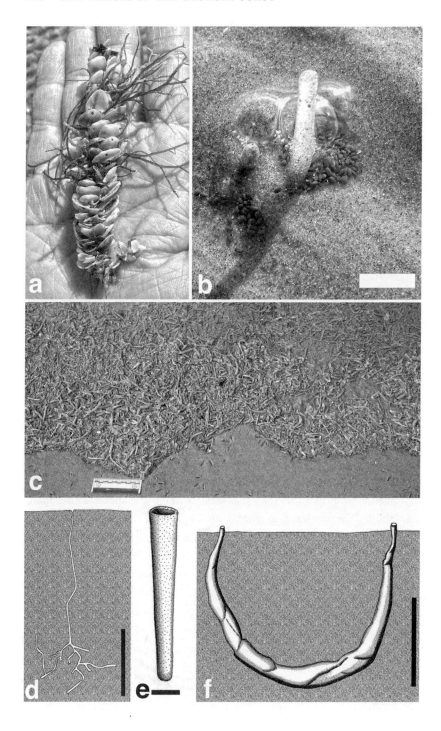

burrows will get washed out of the sand by strong tidal currents or waves, and these collect on sand-flat surfaces, prone, fragmented, and removed from their original environmental context. One of the great advantages of most trace fossils over body fossils for interpreting paleoenvironments is that most trace fossils are in situ (Chapters 1, 10), but these transported burrows are rare examples of how some can be moved, deflated, and reoriented. Hence ichnologists should take special note of the orientations of linear shell debris concentrations in the geologic record they think may be polychaete worm tubes. Vertical arrangements are more likely to be in place, whereas horizontal ones more likely underwent a journey, however brief. Have such shell-laden burrows been found as trace fossils? Yes, and it is appropriately named *Diopatrichnus* in honor of the tube worms who make such similar traces (Kern 1978; de Gilbert 1996).

Another polychaete tracemaker whose abundant traces will be seen much more often than its tracemaker is the parchment worm (*Onuphis microcephala*). Parchment worms are so called because of their paper-like burrow linings, which are mucoidal sheaths with sandy exteriors that look much like drinking straws (Figure 6.5b). Like those of tube worms, these burrows are arranged vertically, but are only about 3–6 mm (0.1–0.2 in) wide and 15–30 cm (6–12 in) long (Howard and Dörjes 1972; Howard and Frey 1973). Similar to *Diopatra* burrows, parchment worm burrows can be eroded from their original positions in intertidal sand flats and scattered elsewhere. In some instances, I have seen masses of these lifeless, collapsed

Figure 6.5. *Facing.* Polychaete worm traces. (a) Tube worm (*Diopatra cuprea*) burrow, dislocated by erosion, but still intact because of its bivalve shell reinforcement, consisting of algae and dwarf surf clams (*Mulinia lateralis*): Sapelo Island. (b) Parchment worm (*Onuphis microcephala*) burrow, in place and projecting above the surface of an intertidal sand flat, along with an accumulation of fecal pellets below: St Simons Island. Scale = 1 cm (0.4 in). (c) Transported mass of parchment worm burrows accumulated on intertidal surface: Wassaw Island. Scale = 10 cm (4 in). (d) Capitellid thread worm (*Heteromastus filiformis*) burrow; note downward branching, a diagnostic trait. Based on figure by Howard and Frey (1975). Scale = 10 cm. (e) External view of trumpet worm (*Pectinaria gouldii*) burrow, preserved as a sandy, cemented tube. Based on figure by Howard and Frey (1975). Scale = 1 cm (0.4 in). (f) External view of the mucous-lined and U-shaped burrow of the other parchment worm (*Chaetopterus variopedatus*), placed in its original position within a sandy bottom. Based on a figure by Frey (1970b). Scale = 10 cm (4 in).

tubes in sand-flat ripple troughs at low tide (Figure 6.5c). The fossilization potential of such reworked burrow deposits is probably low, but possibly could be recognized by anomalous, thin sandy layers within relatively muddy beds. If left in place, however, the slight difference in grain size of the burrow lining versus the surrounding host sand—a result of polychaete selectivity—would be discernible as a thin, vertical tube in the rock record. Such a trace is normally assigned the ichnogenus *Skolithos*, although this name may have been overused and abused by geologists when naming other vertical burrows from the fossil record (Chapters 4, 5, 10).

Both tube worms and parchment worms are passive carnivores, in that they build a burrow and stay in it, waiting for prey to wander by, comparable to burrowing spiders and ant lions (Chapter 5). Other polychaetes, though, are a little more active in their seeking of food. For example, the common clam worm (*Nereis succinea*) swims freely above the surface, eating any debris that it may encounter. Nonetheless, it also burrows, and in a wide range of marginal-marine ecosystems, although its burrows have not been described in any detail.

One of the more common and best studied of burrowing polychaetes of the Georgia coast is the capitellid thread worm (*Heteromastus filiformis*). This worm, which prefers to live in the muddy environments of salt marshes or beach–marsh transitions, is known as a head-down deposit feeder. When feeding in this manner, it orients itself vertically, ingests sediment at its anterior end, and dumps piles of sediment-filled fecal pellets at the surface. As a result of this conveyor-like activity, thread worms, especially when abundant, are responsible for sediment advection, much like the action of ant lions or earthworms in terrestrial environments (Chapter 5). Thread worms make narrow (< 1–2 mm), vertical, mucus-lined burrows that can extend 20–30 cm (8–12 in) down into the sediment (Hertweck et al. 2007). Upper parts of the burrow may spiral, whereas distal ends may branch, then branch again as the worm probes into sediment, seeking organics. This worm is well adapted for sticking its head into nasty, sulfur-rich, anoxic sediments. So how does it breathe? However improbable it may sound, it uses its rear end. When not producing fecal pellets, its posterior portion absorbs oxygen from the sediment surface through gill-like appendages (Schäefer 1972; Bromley 1996). The overall form of its burrow is like an upside-down tree, with a central vertical shaft at the top and numerous branches projecting from the base of the shaft (Figure 6.5d).

In the fossil record, trace fossils with forms similar to thread worm burrows would be the ichnogenera *Skolithos* or *Gyrolithes* for the upper part and *Chondrites* for the lower part (Gingras et al. 2004; Hertweck et al. 2007). If preserved, their copious fecal pellets would also constitute trace fossils, but these might be difficult to distinguish from other invertebrate-made pellets, whether these are feces or pseudofeces. Sedimentologically speaking, though, these worms are important for changing the character of muddy sediments in marginal-marine ecosystems along the Georgia coast, giving a glimpse of the effects of similarly behaving polychaetes in the geologic past.

The shimmy worm (*Nephtys bucera*) is a vigorous carnivorous polychaete, burrowing into intertidal and subtidal sands in search of invertebrate prey, such as palp worms (*Scolelepis squamata*), coquina clams (*Donax variabilis*), and mole crabs (*Emerita talpoida*), among others (Ruppert and Fox 1988). As its common name implies, it undulates in both its swimming and burrowing. When burrowing, it pushes its head into the sand, anchors with its mouth, then pulls the rest of the body into the sediment (Clark and Clark 1960; Clark 1962). Such worms would leave visible wakes of disturbance that cut across laminae and other bedding; otherwise their burrows might not be easily described, because soft, saturated sand should simply collapse behind them.

Miscellaneous polychaete tracemakers in subtidal environments on the Georgia coast include a variety of other burrowing worms, including palp worms (*Scolelepis squamata*); bloodworms (*Glycerin dibranchiata*); trumpet worms (*Pectinaria gouldii*); and another so-called parchment worm (*Chaetopterus variopedatus*). Palp worms are suspension feeders that make 1 mm wide and 15 cm (6 in) long, vertical burrows from which they stick out their heads and use appendages called palps (hence their name) to gather suspended organics from the water. Although relatively small compared to other polychaete burrows, numbers matter, as these worm burrows can be densely populated and palp worms tolerate living mere millimeters away from one another (Ruppert and Fox 1988). Bloodworms are different as tracemakers, in that they do not use entire-body peristalsis to burrow, but only their pharynx. This part of the mouth apparatus rapidly intrudes the sediment, where it expands and anchors, then pulls back into the body, which causes the rest of the worm body to move forward into the sediment (Ruppert and Fox 1988). Bloodworms, which earned their common name the hard way (they bleed profusely when baited on hooks by fishermen), make 5 mm (0.2 in) wide, long, sinuous vertical to oblique burrows that often branch into broad Y or U shapes. These burrows are lined with mucus, which is used to stick sand to the side of the burrow. Likewise, trumpet worms also use mucus (a popular substance amongst polychaetes), around which they cement sand, reinforcing their burrows. Another common name for these polychaetes is the ice-cream cone worm, owing to its conically shaped burrow linings (Figure 6.5e). These linings, which somewhat resemble caddis fly larval cases (Chapter 5), are rather fragile, composed of a thin layer of sand grains, but the cone increases in both grain size and overall dimensions to as long as 6 cm (2.4 in) as the worm grows (Howard and Frey 1975). The other parchment worms (*Chaetopterus variopedatus*), not to be confused with *Onuphis microcephala*, make mucus-lined, paper-like burrows, which are U shaped, 2–3 cm (about 1 in) wide, and as long as 60 cm (24 in). The shape of the U varies, though, depending on the spacing of the two openings, which can be 15 or 30 cm (6 or 12 in) apart (Figure 6.5f; Frey 1970b; Bromley 1996).

Nonpolychaete, nonsegmented worms in shallow marginal-marine environments of the Georgia coast, such as nemerteans (ribbon worms) and nematodes (roundworms), are also responsible for a large number of traces, most of which are burrows. For example, the silky ribbon worm (*Cerebratulus lacteus*) burrows while seeking out prey in subtidal sandy substrates (Frey 1970b; Gingras et al. 2008a). This worm can be quite long, about 1.2 m (4 ft), the largest ribbon worm known in the eastern United States. As an active infaunal predator, it executes sneak attacks while burrowing, attaching its proboscis to its prey and then enveloping it. This ribbon worm hunts after razor clams (*Tagelus plebeius* and *Ensis directus*), so its burrows—which show up as disruptions of sand laminae—may be connected directly to those made by burrowing bivalves. The silky ribbon worm is similar to the shimmy worm in that it uses it proboscis to initially anchor in the sand, then burrows by using peristaltic movement (expansion and contraction) along its body. Its burrows are typically shallow, and horizontal to slightly oblique (Frey 1970b).

Polychaetes and other worms in marginal-marine environments of the Georgia coast are numerous and varied, and produce a correspondingly huge number and diversity of traces, most of which are burrows. Understandably then, a cliché in ichnology is that inexperienced geologists, especially those unenlightened by the life-altering qualities of ichnological knowledge, will label any nondescript, cylindrical structure in a sedimentary rock as a worm burrow. This glibness may actually be correct in that general sense, but encourages further questioning. For example, is it a burrow made by a polychaete, nemertean, nematode, priapulid, or hemichordate? (Yes, there are many different worms.) If it is a polychaete burrow, does it have a definite lining or not, associated sand, bivalve shells, or other debris? Is it oriented vertically, obliquely, or horizontally? Does the burrow have an overall geometry other than a simple tube, such as a broad U shape with multiple openings, or a branching pattern? Such levels of inquiry take what was an easy answer—worm burrow—and transform it into an intellectual exercise, helping geologists to realize that naming something is not necessarily the end of learning about it.

GASTROPODS AND CEPHALOPODS, INCLUDING THE LION OF THE TIDAL FLAT

Predation is a fact of life on the Georgia coast, although it often takes place in unseen and silent ways, directly belying the snarling flurry of motion and dramatic swelling of adrenaline-pumping music we may expect to accompany life-and-death struggles. With this concept in mind, a quieter sort of predation happens year round below the surfaces of shallow intertidal and subtidal sands of the Georgia barrier islands, yet we innocently view its traces as a happy opportunity for personal adornment. Look closely on most Georgia beaches at low tide and you will see many bivalve and gastropod shells with perfectly round, small, < 1–2 mm diameter holes, most of which go cleanly through the shells (Figure 6.6a). Look even closer and you will notice these holes are beveled, like short megaphones, with their wide ends on the outside of the shells. These beautifully sculpted openings offer an easy means for the passage of strings or other cords, lending themselves well to the making of necklaces that show off an impressive biodiversity of mollusks.

Thus it may be somewhat disconcerting to the creators and wearers of such a piece of jewelry to find out that this string of beautiful shells is actually a necklace of death. The holes are traces of predation, made by naticid and muricid gastropods, in particular the common moon snail, *Neverita duplicata* (formerly known as *Polinices duplicatus*). These gastropods are smooth shelled, light colored, glossy coated, and roundish—all features

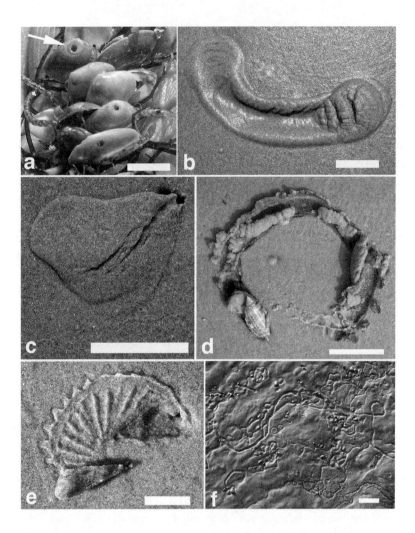

Figure 6.6. Intertidal gastropods and their traces. (a) The deadly calling card of the common moon snail (*Neverita duplicata*), evident as beveled drillholes (arrow)in these small bivalves (*Mulinia lateralis* and *Donax variabilis*); a parchment worm (*Diopatra cuprea*) then used their shells to reinforce its burrow: Sapelo Island. Scale bar = 1 cm (0.4 in). (b) The shallow horizontal burrow of a moon snail on the prowl, in which the gastropod is directly underneath the bulge to the right: Sapelo Island. Scale bar = 5 cm (2 in). (c) The subsurface presence of a barely buried knobbed whelk (*Busycon carica*), indicated by its subtle trapdoor trace: St. Catherines Island. Scale bar = 5 cm (2 in). (d) A gorgeous horizontal burrow and trail made by a gorgeous gastropod, the lettered olive (*Oliva sayana*): St. Catherines Island. Scale bar = 5 cm (2 in). (e) The trace of a common Atlantic auger (*Terebra dislocata*), imitating a breakdancer while seeking moisture: Sapelo Island. Scale bar = 1 cm (0.4 in). (f) Mud snails (*Ilyanassa obsoleta*), doing what they do best, making numerous meandering, looping, and intersecting trails on a mudflat surface: Cumberland Island. Scale bar = 5 cm (2 in).

that lead to lunar comparisons—and they are the top predators below the sand-flat surfaces. Indeed, Robert Frey (Chapter 3) often referred to them as the lions of the tidal flat. Their smooth and semispherical forms are not just aesthetically pleasing to us, but functional, allowing for a mode of hunting that combines stealth with tenacity. While burrowing just underneath sandy surfaces in intertidal and subtidal areas, they seek out stationary or otherwise more slowly moving prey (Figure 6.6b). When you are a mobile predator, snail or not, relativism is your friend: you do not have to be fast, only faster than what you want to kill. The targeted subjects of moon snails are typically bivalves, but can include other gastropods, including its species. In other words, no social taboos against cannibalism exist in the subsurface world of the sand flats.

A moon snail on the prowl for live fodder depends on its huge, muscular foot for implementing an attack. Potential victims are probably identified through chemical detection of mucoidal trails in the sediment, documented in some predatory marine gastropods (Shaheen et al. 2005). The moon snail moves by alternate inflation and deflation of its mucus-lubricated foot, which results in an anchoring and pulling motion of the gastropod toward its intended target (Bromley 1996). This expansion is accomplished by sucking in water through its foot: pick up one of these snails and you will see it weep as its foot contracts and expels the water, a common behavioral response in marine gastropods. Traces of *N. duplicata* simply hunting, and not necessarily killing, are evident as shallow, horizontal, curving to meandering burrows. Once a moon snail is next to its prey, though, its foot surrounds the bivalve or gastropod, effectively trapping it. Immobilization is followed by the secretion of organic acids in a selected spot, which weakens the formerly protective shell of the bivalve or gastropod. Soon after this, the moon snail radula (a hard part near its mouth) is deployed as a circular drill, rotating in a tight spiral that cleanly rasps into the partially dissolved place on the shell, forming a wider opening. Because the radula is applied in a series of arcs and at an angle, the opening inside the shell is narrower, causing the boring to have a conical shape. Imagine rotating a knife at a 45° angle with respect to the top surface of a cake to make a similarly shaped trace. The drilling location can vary, but if applied to a bivalve is normally close to its umbo, where its adductor muscles hold the shell tightly together. Once the drilling has breached the inside of the shell, the adductor muscles relax, further breaking down the clam's defenses, and the moon snail proboscis slides through the drill hole. The moon snail then feeds on the bivalve or gastropod while it is still alive.

Although the preceding predatory behavior sounds ruthlessly efficient, some individuals of *N. duplicata*, like all animals, make mistakes. For example, they may attack dead shells, or chose a prey item too large for effective handling, so it escapes. Furthermore, one of the few defenses a bivalve might have against these attacks is shell thickness. A thicker shell means more time and energy must be expended by the moon snail, which increases the chance that the prey will wrestle out of the grasp of its captor. In such instances, this attempted act of predation leaves its mark as well, an incomplete drill hole that failed to penetrate the shell. As a result, multiple perilous struggles and escapes may be recorded by incomplete drill holes. For example, an attacked bivalve may get away from a moon snail, only to be recaptured, subjected to a resumption of drilling in a different spot, and then escape again (Kelley et al. 2003).

These drill holes not only tell us about behavior, but about predator size. Studies of naticid and muricid gastropods show a positive correlation between drill hole diameter and driller body size (Kelley et al. 2003). Moreover, successful predation is apparently predicated by size similarities between naticids and their chosen prey, giving another dimension to the aphorism "pick on someone your own size." Too small of a prey requires a disproportionate effort relative to the amount of derived nutrition. On the other hand, too large of a prey increases the risk of energy expended on a meal that, once attacked, simply waves goodbye (in a molluscan sort of way) and leaves. Hence moon snails make Goldilocks-like decisions, looking for prey that is just right. In turn, this behavior is a result of an evolutionary history in which individuals who made bad choices were eliminated from the gene pool.

Timing of predation with regard to tidal cycles, however, is not revealed by drill holes. Unlike many marine invertebrates that must remain immobile and buried in moist sands during low tides to avoid desiccation, naticids continue their hunting by burrowing, and hence can encounter prey during any time of the day or night. As a result, naticids are prolific tracemakers, leaving extensive burrows and trails in their wakes, and numerous borings in shells that later reveal themselves on beaches, ready for use as necklaces or earrings. Nonetheless, their traces serve as reminders of the daily dance between predators and prey that happens below us on beaches of Georgia and elsewhere, much of which is invisible, yet whose outcomes are still detectable.

Interestingly, in slightly deeper waters offshore, smaller but similar drill holes are made by some species of octopi. A hungry octopus will ensnare a gastropod or bivalve with its eight sucker-laden arms, and then, like a moon snail, drills into the shell with a rotary motion using its radula. This cephalopod drill hole differs from those of naticids, however, in its oval outline and its straight, up-and-down walls, making a cylinder in negative space, rather than a bevel (Bromley 1993). Once the drill hole breaches the shell, the octopus injects a poison into the interior that also acts as a digestive enzyme, turning the bivalve from a semisolid meal into a liquid one that is then simply sucked out through the hole. With bivalves, an octopus may also pull on both valves in an attempt to wear down the adductor muscles. If this strategy works, drilling becomes unnecessary for feeding, and the traces of octopus predation are subtler. The primary clues of such death dealings are shell middens that accumulate near an octopus lair, a refuse pile that indicates both the proximity and voracity of an octopus (Walker 1990). Such middens, however, would be difficult to detect along a Georgia shoreline because of the constant movement of sand and mud through tidal currents and waves that break them up, much as a trash pile is blown apart by a stiff wind. These shell accumulations are more likely to be seen while snorkeling over a reef in subtropical areas just south of Georgia, such as in Florida or the Bahamas. When you find these middens, also look for a cubbyhole just above the heap of shells, which may hold a well-camouflaged but wary octopus staring back at you, perhaps wondering how you knew it was there. Shells in the pile may also have octopus drill holes in them, leaving no doubt about the identity of the tracemaker.

A group related to naticid gastropods, the murex snails, or muricids (Muricidae), are also well known for their predation on other molluscans. Given their common names, the thick-lipped oyster drill (*Eupleura caudata*) and the Atlantic oyster drill (*Urosalpinx cinerea*), not much imagination is needed to figure out what they prey on, how they do it, and what sorts of traces they make. The Atlantic oyster drill goes after juvenile oysters in particular, drilling neat holes into their thin shells by using a combination of alternating physical and chemical attack similar to that used by moon snails: radula scraping and acid (Manzi 1970; Harding et al. 2007). These predators are not obsessed just with young oysters, and will go after other bivalves or barnacles if these are available in a marsh. Also like moon snails, both species are cannibals, although only female snails have been observed eating their own kind (Manzi 1970). Amazingly, some shell chips scraped from the victim can be swallowed by an oyster drill, pass through its gut, and show up in its feces (Carriker 1977). Hence, a close examination of suspected gastropod feces (modern or fossil) may not only reveal whether the tracemaker was preying on other molluscans, but that it was using a drilling technique to acquire food.

Knobbed whelks (*Busycon carica*) are also active gastropod tracemakers, whether through predation marks they leave on bivalves, plowing through sand to form surface trails, or burrowing. With regard to their predatory proclivities, knobbed whelks attack bivalves by chipping their shells. In doing this, the whelk uses the outermost edge of its shell (its lip) to push apart both halves of its bivalve prey, cracking the shell edges and exposing the bivalve's soft insides for the whelk's proboscis. Traces of whelk predation are subtle; they are normally visible only as mirror-image chipping of shell edges, but they can also show up as breaks on a whelk's lip (Carriker 1951; Dietl 2004). Their burrows, however, are more easily spotted, showing up as large-scale disruptions of otherwise smooth sand-flat surfaces at low tide. Their resting traces likewise stick out as trapezoidal flaps of cohesive sand, looking very much like trapdoors (Figure 6.6c). Once this flap is carefully excavated, a whelk is revealed, quickly pulling in its own personal trapdoor (operculum) to cover its soft parts from you, its potential predator. This behavioral reaction, probably selected for millions of years because of hungry shorebirds, is also not so far-fetched in response to other large upright bipeds: humans. One day at low tide on Cabretta Beach of Sapelo Island, I was surprised to encounter several people from the Hog Hammock community (Chapter 2) out gathering whelks for dinner. They had been using ichnological techniques (happily unaware they were called that), looking for the right-sized and right-shaped outlines in the sand flat that denoted the subsurface presence of whelks. Their whelk-filled buckets attested to their prowess, and I wondered how many generations of African Americans on Sapelo had honed and passed on these practical skills.

Among the more commonly encountered gastropod trails and shallow burrows on the Georgia coast are those of what is probably its most beautifully shelled gastropod, the lettered olive (*Oliva sayana*). The lettered olive has a long, smooth, glossy shell with a short spire and large whorl making up most of it, and is decorated with gorgeous mottled patterning. Its smoothness, like that of *N. duplicata,* aids considerably in its ease of burrowing. In contrast to the more ovoid moon snails, though, its long shell is streamlined for traveling long distances through the sand in its hunt for small bivalves (such as *Donax variabilis*), mole crabs (*Emerita talpoida*), or other infaunal treats (Ruppert and Fox 1988). Once a lettered olive finds an intended prey, its foot wraps around the animal until it suffocates, then it enjoys a freshly killed meal. When moving on or just under a sand-flat surface, a lettered olive extrudes its soft, mucous-laden foot to completely cover its shell, which also decreases the abrading effects of the surrounding sand. Then, by alternately anchoring and contracting its foot, it pulls the shell forward with its spire pointing behind it.

Lettered olive traces vary in accordance with whether they are surface trails, or shallow burrows in emergent or submerged sand (Figure 6.6d). If trapped by a low tide, their surface trails will nearly always segue into shallow burrows, in which their foot first penetrates the sand surface and is followed by the shell. In some cases, their burrows may also begin as circular to oval resting traces, from which they plow through the cohesive sand just over them. If they breach the surface with the tops of their shells, this variation may confuse ichnologists, who then argue whether the resultant trace is a burrow or a trail. (The answer is both, but that might be too accommodating for some.) Lettered-olive burrows made on emergent sand flats are about 1.5–3.0 cm (0.6–1.2 in) wide and can be as much as a meter (3.3 ft) long. They are curved (arcuate) to straight and have raised ridges with corrugated exteriors, which sometimes form false meniscae that have nothing to do with backfilling but are caused by shear stresses exerted by the moving animal against the wall of the burrow. If the sand is cohesive enough, chunks of this burrow wall drop onto the outside of the burrow, further revealing the trail beneath. Any rotational motion by the gastropod tracemaker as it turns also widens the burrow. These attractively textured burrows and trails are easy to find and identify in the lower intertidal zones of beaches. In contrast, traces made by the same tracemakers in submerged sands are vaguely defined and just as easily confused with those made by other gastropods, such as moon snails. This is a result of saturated sand simply collapsing around a lettered olive as it plows through the sediment, hence leaving a slightly raised and smoothly outlined burrow in its wake. Inexplicably, these gastropods also sometimes emerge from their protective burrows and start cruising across the sand surface, leaving a ribbon-like trail of mucus-bound sand behind them. Where populations are high enough, intersecting burrows and trails made by different lettered olives will also make false branching. Consequently, ichnologists who find similar-looking trace fossils are cautioned to scrutinize any initially identified meniscae and branching that might be associated with suspected gastropod shallow burrows and trails.

The common Atlantic auger (*Terebra dislocata*), also a carnivore, is a small, high-spired gastropod that makes simple trails and shallow burrows along sand surfaces at low tide, but also forms some odd little traces worth mentioning for their distinctiveness. These traces start with a depression definable as an isosceles triangle, reflecting the overall shell profile of the tracemaker. From there, the trace gets really interesting, making a swath narrower than the length of the triangle and composed of alternating fine ridges and valleys, bordered by a smoothly curved groove with sprocket-like teeth on its outside (Figure 6.6e). Where the tracemaker is still present in such a trace, it will have turned almost 180° from its original position, or will be otherwise well on its way back to its original resting trace. Think of the break-dancing move that involves spinning in place on one side, and that is what the auger is doing, in its own legless way. The detailed features of this trace are formed through the progressive stopping and starting of the shell (making the alternating ridges and valleys), the dragging of the lip (the smoothly curved groove), and the lowermost end of the shell (the sprocket teeth), all being driven by the snail's muscular foot.

The innocently named baby's ear (*Sinum perspectivum*) was given its moniker because of its flattish, smooth, and coiled shell, which to some people resembles a baby's ear (albeit a dismembered one). Yet this naticid gastropod, like its distant cousin *N. duplicata*, is also a vicious predator. It has a flat, streamlined shell that slides efficiently through the sand, where it preys on small bivalves, such as the dwarf surf clam (*Mulinia lateralis*) or green jackknife clam (*Solen viridis*). Like moon snails, it uses its muscular foot, generously lubricated by mucus, to push and pull its way through the sand. Where it differs from these snails, though, is in its lack of an operculum and a much smaller shell, which means its foot is permanently extracted while burrowing or enveloping prey (Ruppert and Fox 1988). Burrows of this species, as far as I know, have not been described or otherwise distinguished from those of similarly sized gastropods, but the flatter profile of the shell should result in a horizontally compressed burrow profile relative to those of moon snails.

Salt marsh gastropods, in their own way, constitute keystone species in Georgia salt marshes because of their important contributions to trophic relations and nutrient cycling in these ecosystems, manifested by their traces. For example, billions of marsh periwinkles (*Littoraria irrorata*), mentioned previously (Chapter 3; Figure 3.1), graze constantly on algae and fungi growing on the stalks and leaves of smooth cordgrass. This action eventually tears down these plants, unless the gastropods are culled by predation. Fortunately for the cordgrass, marsh periwinkles have crab predators that will not leave them alone, such as the Atlantic mud crab (*Panopeus herbstii*) and the formidable, shell-crushing blue crab (*Callinectes sapidus*). In fact, periwinkle movement up the stalks is timed with increasing tides, which bring in the swimming crabs; thus their upward migration is more linked to predator avoidance than fear of immersion. Interestingly, periwinkles attacked by crabs that manage to survive are normally larger individuals and farther up in the high marsh, away from open water. This situation means that shell size and distance from the sea can lend advantages to the snails (Kneib 1997). Hence healed scars carried on periwinkle shells, which are both traces of predation by the crab and of successful escape by the snail, can reflect relative ecological positions of the periwinkles in a marsh. Other than these, marsh periwinkles do not leave distinctive traces with good fossilization potential, other than their broader contribution to the productivity of a Georgia salt marsh, a trace that is simultaneously overt and subtle (Chapter 3). A common associate of the marsh periwinkle is the coffee-bean snail (*Melampus bidentatus*), which grazes on both living and dead cordgrass while encouraging further fungal growth, thus benefiting itself and marsh periwinkles (Graça et al. 2000).

Marsh periwinkles and coffee-bean snails, however, stay mostly clean by sticking to the above-ground portions of smooth cordgrass. So if you want to get really down and dirty with your gastropod ichnology, then you simply must watch the awesome tracemaking abilities of the two most abundant gastropods directly affecting salt marshes and other muddy intertidal surfaces. These gastropods are the appropriately named mud snail (*Ilyanassa obsoleta*) and common mud snail (*Nassarius vibex*). Together, these species are responsible for most of the surface trails and shallow burrows observed in marshes and marsh ecotones. Thus they warrant special attention for their ichnological contributions.

Mud snails, despite their lack of limbs and the sophisticated sensory organs possessed by vertebrates, manage to track one another by following their own species' surface trails, a behavior noted earlier in some pulmonate snails (Chapter 5). The mucus on these trails imparts a microtopography that is felt and recognized by each snail, as well as a scent that normally attracts others of their species. This alteration of their surroundings causes aggregations of large numbers in small areas (Bretz and Dimock 1983), and a concomitant increased density of their meandering, looping trails (Figure 6.6f). Their trails are primarily feeding traces, in which they slide along muddy surfaces with abundant growths of diatoms and ingest these (along with sediment) as deposit feeders. Their digestive systems then sort out nutritious organics from inorganic sediment.

Mud snail trails also might reflect both the health of individuals and social interactions (and reactions) based on health. For example, parasitic flatworms (trematodes) can infect mud snails, then badly afflicted snails will err off course and head for higher ground, well above the intertidal zone (Curtis 1987 2004). Hence trails identifiable to this species found outside of their normal range could represent unhealthy specimens that have been body snatched by internal invaders, similar to how bumblebees are coerced into burrowing for the benefit of their parasitoid hosts (Chapter 5). Not coincidentally, these higher areas are also places where these snails can be more easily eaten by decapods—such as amphipods or ghost crabs—or birds, which then unwittingly ingest the trematodes or otherwise come in contact with the trematodes' free-swimming larvae, shed in a mucus-lined snail trail. In other words, these snails act as intermediate hosts before the flatworms reach the promised land of a decapod or avian gut. Remarkably, mud snails can sense through biochemical cues that certain individuals are diseased, and subsequently avoid them (Ruppert and Fox 1988). Healthy snails, in their shunning of sick ones, then leave trails that move away from those of their trematode-bearing brethren. So can this combination of ichnology and disease ecology be applied to the fossil record, where gregarious gastropod trails can be quite abundant on bedding planes? Probably yes, but defining trails in which snails picked out infected individuals, then subsequently made avoidance trails, would require meticulous study and

Plate 1. A collection of tracemakers in a salt marsh, with smooth cordgrass (*Spartina alterniflora*) in the background and eastern oysters (*Crassostrea virginica*) and mud snails (*Ilyanassa obsoleta*) in the foreground: Sapelo Island.

Plate 2. Important plants of a typical maritime forest on a Georgia barrier island, consisting of live oaks (*Quercus virginiana, background*), with Spanish moss (*Tillandsia usneoides*) hanging from it (*right*) and saw palmetto (*Serenoa repens, foreground*) in the understory: Jekyll Island.

Plate 3. *Facing.* Sea oats (*Uniola paniculata*), with its roots holding down the main part of a coastal dune: Sapelo Island.

Plate 4. Ant lion larva (*Myrmeleon crudelis*), extracted from its burrow in a dune: Cumberland Island. Scale = 1 cm (0.4 in).

Plate 5. Scarab beetle (adult), species unknown, walking in a back-dune meadow: Sapelo Island.

Plate 7. *Facing.* Female Carolina sand wasp (*Stictia carolina*) exiting its brooding burrow in a dune: St. Catherines Island. Scale bar = 1 cm (0.4 in).

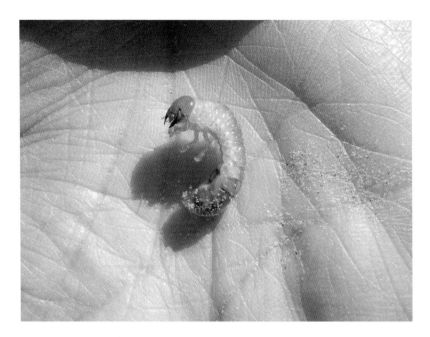

Plate 6. Scarab beetle (larva), species unknown, extracted from its burrow in a dune: Little St. Simons Island.

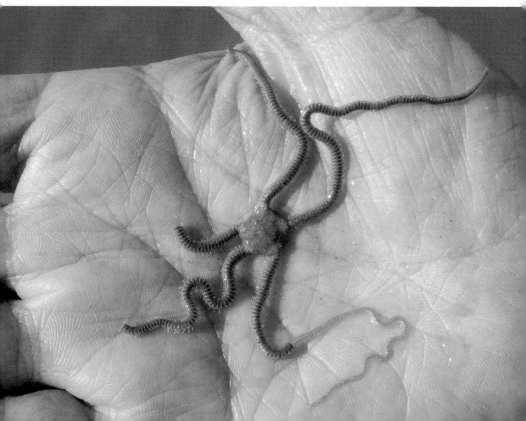

Plate 8. *Facing top.* Bloodworm (*Glycera dibranchiata*), exposed through breakage of its agglutinated burrow and washed up into the intertidal zone of the beach: Sapelo Island. Scale in centimeters.

Plate 9. *Facing bottom.* Brittle star (*Hemipholis elongata*), with three broken arms but regeneration on one arm, exposed in an intertidal runnel: Sapelo Island.

Plate 10. *Above.* Giant Atlantic cockle (*Dinocardium robustum*) on an intertidal sand flat at low tide: Sapelo Island.

Plate 11. Ghost crab (*Ocypode quadrata*) resting and rehydrating in the lower intertidal zone of a beach: Sapelo Island. Scale in centimeters.

Plate 12. Carolina ghost shrimp (*Callichirus major, top*) and Georgia ghost shrimp (*Biffarius biformis, bottom*), unhappily exhumed from their burrows in the lower intertidal zone of a beach: St. Catherines Island. Scale = 1 cm (0.4 in).

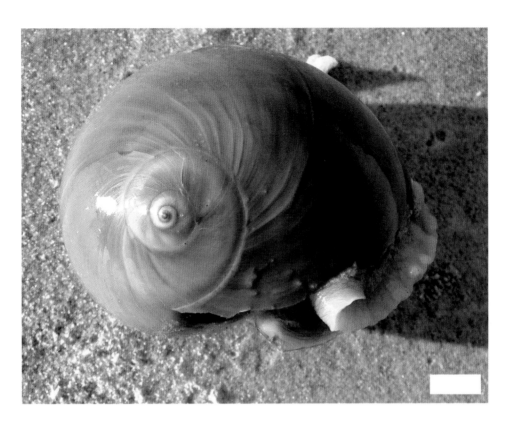

Plate 13. *Facing top.* Young (1- to 2-year-old) horseshoe crab (*Limulus polyphemus*) crawling in the lower intertidal zone of a beach: Sapelo Island.

Plate 14. *Facing bottom.* Profile view of a beach mole crab (*Albunea paretii*) taken out of its burrow in a sand flat at low tide: Sapelo Island.

Plate 15. *Above.* Common moon snail (*Neverita duplicata*) on the intertidal zone of a beach: Sapelo Island. Scale = 1 cm (0.4 in).

Plate 16. Mud fiddler crabs (*Uca pugnax*), composed of males (big claws) and females (small claws) near or partially in their burrows in a salt marsh at low tide: Sapelo Island.

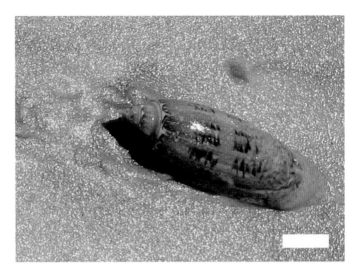

Plate 17. *Top.* Lettered olive shell (*Oliva sayana*) making a trail on the intertidal zone of a beach: Sapelo Island. Sapelo Island. Scale = 1 cm (0.4 in).

Plate 18. *Bottom.* Marsh periwinkles (*Littoraria irrorata*) doing what they do best, grazing on stalks of smooth cordgrass (*Spartina alterniflora*) in a salt marsh: Sapelo Island. Scale = 1 cm (0.4 in).

Plate 19. Male sand fiddler (*Uca pugilator*), defiantly wielding its mighty claw while on a supratidal sand flat: Cumberland Island. Scale = 1 cm (0.4 in).

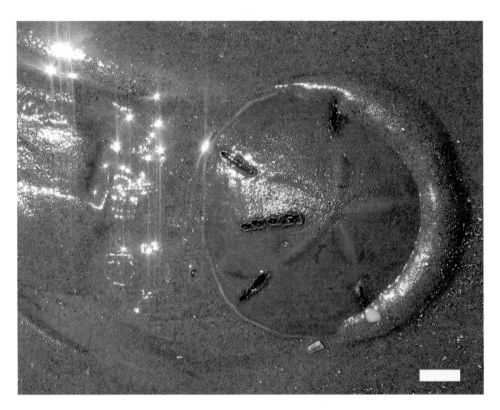

Plate 20. Keyhole sand dollar (*Mellita isometra*) making a shallow burrow in the lower intertidal zone of a beach: Sapelo Island. Scale = 1 cm (0.4 in).

Plate 21. *Below.* Sea cucumber (*Thyone* sp.) exhumed from its burrow (*left*) next to one still in its burrow, with one end protruding (*right*) in lower intertidal zone of a runnel: Sapelo Island. Scale = 1 cm (0.4 in).

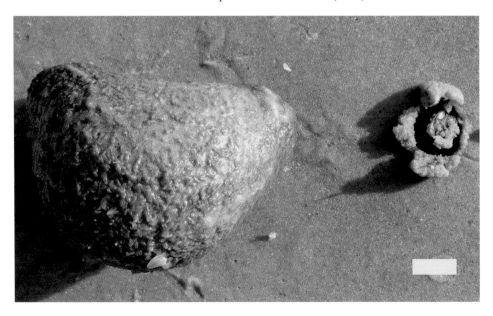

Plate 22. Sea pansy (*Renilla reniformis*) out of its burrow, which was in the lower intertidal zone of a beach: Sapelo Island.

Plate 23. Lined sea star (*Luidia clathrata*), barely immersed in intertidal zone of beach: Sapelo Island. Scale = 1 cm (0.4 in).

Plate 24. Juvenile alligators (*Alligator mississippiensis*) at den entrance, on island in freshwater pond: Sapelo Island.

Plate 25. *Below.* Brown anole (*Anolis sagrei*) in open grassy area: Jekyll Island.

Plate 26. Diamondback terrapins (*Malaclemys terrapin*), here in an aquarium, but normally living in salt marshes: Jekyll Island.

Plate 27. *Below.* Gopher tortoise (*Gopherus polyphemus*), in captivity but near a burrow of its own making: Jekyll Island.

Plate 28. *Left.* Diamondback rattlesnake (*Crotalus adamanteus*), slithering through a maritime forest: Ossabaw Island.

Plate 29. *Below.* Yellow rat snake (*Elaphe obsoleta quadrivittata*), making a trail on inland dune sands: Cumberland Island.

Plate 30. Nine-banded armadillo (*Dasypus novemcinctus*) in open grassy area: Cumberland Island.

Plate 31. Black vulture (*Coragyps atratus*), scavenging on dead fish in intertidal zone of beach: Jekyll Island.

Plate 32. *Facing.* Sandhill crane (*Grus canadensis*) walking along the edge of a maritime forest: St. Catherines Island, Georgia.

Plate 33. Royal terns (*Sterna maxima*) in the intertidal zone: Jekyll Island.

Plate 34. *Below.* Sanderling (*Calidris alba*) probing for delectable infaunal invertebrates, making both tracks and beak marks: Jekyll Island.

the application of spatial analysis techniques not currently used in ichnology (Chapter 11).

The common mud snail, also called the eastern nassa, is often confused with mud snails, but is slightly different morphologically, behaviorally, and ichnologically. These snails are most often shallow burrowers, lurking just below the sedimentary surface, and can emerge en masse from a muddy sand or sandy mud when feeding. Common mud snails have even been seen somersaulting, however improbable that might sound. This sudden upward movement is done in response to touch or fluids exuded by predators, such as other molluscans or sea stars (Gore 1966; Stenzler and Atema 1977). Rolling is accomplished by extension of its foot behind its shell, which it then uses as a handle to flip itself over as many as 5–10 times in succession. Sadly, traces of this behavior have not been documented. Indeed, these would be nearly impossible to discern from the fossil record if an ichnologist was unaware of this behavior in modern gastropods: the words *acrobatic* and *gastropod* are rarely used in the same sentence in either biology and paleontology classes. Shells of the common mud snail also take on second lives after they die, as tiny hermit crabs living in salt marshes quickly pick these up.

Marginal-marine gastropod trace fossils, especially naticid and muricid drill holes in the shells of other molluscans, have been well studied, and provide a robust database for the analyzing the evolution of gastropod predation since the Mesozoic Era. In fact, these trace fossils, assigned to the ichnogenus *Oichnus*, are so easy to identify and measure, that paleontologists have used K–12 schoolchildren to perform this task as citizen science (Hansen et al. 2003). On the other hand, other gastropod trace fossils, such as their burrows and trails, are a bit trickier to separate from those made by similar-sized polychaetes, bivalves, and arthropods, and require careful examination for anatomical clues that can help to narrow down the tracemaker. For example, the presence of leg or parapodia impressions in a shallow burrow or trail would definitely exclude gastropod tracemakers. Nonetheless, gastropods can make surprisingly complex trails and burrows that have all of the appearance of involving appendages. Likewise, invertebrates with legs and parapodia can also leave obscure clues of their anatomy in a trail or burrow, especially if sediment holding the trace is fluid or otherwise does not hold sufficient detail. Either circumstance can lend to an incorrect identification of the tracemaker; ichnologists need to be initially cautious before saying "snail" or "not a snail."

BURROWING, BIODEPOSITING, AND
BIOERODING BIVALVES, BIG AND SMALL

An impressive suite of bivalves is responsible for much of the bioturbation of intertidal and subtidal sediments along the Georgia coast. Most of these bivalves are infaunal suspension feeders, in which they burrow themselves to a near-vertical position, anchor with a muscular foot in the sand or mud, and send two siphons to the surface. In suspension feeding, one siphon pulls in water with suspended organic matter, whereas the other ejects wastes (feces); hence these are called incurrent and excurrent siphons, respectively (Dame 1996). Because of this life habit, one of the more common bivalve traces produced by suspension feeding is a Y-shaped burrow. Bivalve burrowing is normally done through protrusion, in which a bivalve foot, like that of gastropods, is used to penetrate the sediment, anchor, and pull the rest of the body (including the shell) behind it (Trueman et al. 1966; Bromley 1996). A burrow then typically collapses behind a bivalve as it passes through the sand. Consequently, many bivalve burrows are preserved as vague disruptions of sand that are normally noticed as linear paths breaking through differently colored sand layers. Bivalve burrows made in more homogeneous materials with little sediment contrast, such as a thick, organics-rich mud, are more difficult to detect unless the siphon traces are seen at the mud surface.

The most common suspension-feeding bivalves in intertidal sandy areas are small, 1–2 cm (0.4–0.8 in) long coquina clams (*Donax variabilis*) and dwarf surf clams (*Mulinia lateralis*), both of which are also known for their rapid burrowing. Larger bivalves of Georgia coastal environments are Atlantic cockles (*Dinocardium robustum*), often found by both humans and seagulls on exposed sand flats (Chapter 1), and southern quahogs (*Mercenaria mercenaria*), which are common in more muddy environments adjacent to beach–salt marsh transitions. Of course, many other bivalve species are productive tracemakers, but for the sake of brevity only a few will be described here.

Take the disk dosinia (*Dosinia discus*), for example. It is a thin, glossy, smooth-shelled bivalve that is also adept at burrowing. How fast can it burrow? Its rate of movement through sediment has actually been recorded as 29 cm/h (11.5 in./h), albeit under laboratory conditions (Gingras et al. 2008b). This speed outpaced other bivalve species native to the same area of Georgia, yet is probably not going to inspire "Disk Dosinia Racing Nights" at Georgia coast pubs. Unlike suspension-feeding bivalves, *D. discus* is more of an infaunal deposit feeder, acquiring food by ingesting sediment during its burrowing. Like many bivalves, though, its burrows are only describable as horizontal, oblique, and vertical disturbances through sediment, and are not readily distinguished from those made by suspension-feeding bivalves. Additionally, its shell often bears the drill holes of its main predator, the moon snail *Neverita duplicata*, traces that make its valves all the more attractive for jewelry-hungry beachcombers.

Razor clams, a general category of bivalves based on their similar shapes or relatedness, include a number of species, such as the stout razor clam (*Tagelus plebeius*), the purplish tagelus (*Tagelus divisus*), the Atlantic jackknife (*Ensis directus*), and the green jackknife (*Solen viridis*). These bivalves are all thin, much longer than wide, and rectangular in outline, although slightly curved along their lengths. Because of their knife-like shapes (reflected in some of their common names), they easily slice into soft sediment and are fast burrowers. Indeed, if exhumed, do not blink, as they are seen only briefly before quickly burying themselves. Their vertically oriented burrows can be quite deep, limited only by the lengths of their siphons, and are flattish in profile. However, burrows are not left open and will readily fold in behind a bivalve as it slides through the sediment.

The incongruous ark (*Anadara brasiliana*) is an odd-looking bivalve compared to most others discussed here, having a fuzzy, purple, organic exterior, a layer called the periostracum. Its roughly corrugated shell is similar to that of the Atlantic cockle, but is noticeably smaller. This species makes some of the most interesting traces I have seen produced by a bivalve on Georgia sand flats, for which there are a few trace fossil analogues. If stranded by a low tide on a sand flat, this ark does not burrow, but will use its foot to pull its shell along the surface so that it becomes ever so slightly buried, protecting it from surface predators (birds, mostly) and desiccation. This movement, though, is rotational and causes a 3–4 cm (1.2–1.6 in) wide circular depression, outlined by a thin rim of sediment and a raised bump in the middle, looking much like the impression of a tiny doughnut (Figure 6.7a). The length of the bivalve roughly defines the radius of the circle, so a 1–1.5 cm long ark will make a 2–3 cm wide depression. I have also seen two such depressions directly next to one another, made by the same bivalve that moved over to one side and tried to bury itself one more time. The trace fossil equivalent would be preserved as a filling of the depres-

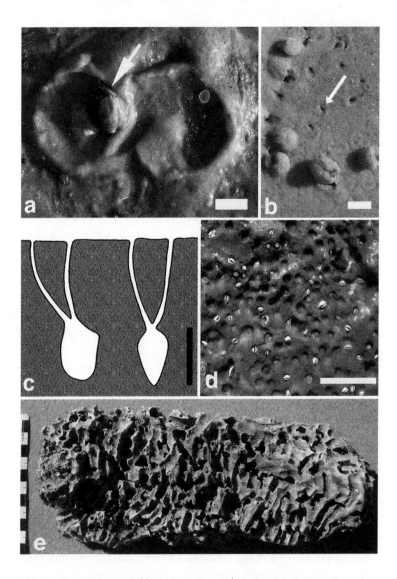

Figure 6.7. Bivalve traces. (a) Incongruous ark (*Anadara brasiliana*; indicated by arrow), having just made two linked doughnut-shaped depressions on beach sand: Sapelo Island. Scale = 1 cm (0.4 in). (b) Dwarf surf clams (*Mulinia lateralis*), some shallowly buried, stranded by a high tide on a sandy beach and capped by sand, with others more deeply buried and only apparent from paired siphon holes (arrow): Jekyll Island. Scale = 1 cm (0.4 in). (c) Longitudinal sections of dwarf surf clam (*Mulinia lateralis*) burrows with paired siphons, minus the clams and showing two possible views (compare to Figure 6.7b). Scale = 1 cm (0.4 in). (d) False angelwings (*Petricola pholadiformis*), bioeroding bivalves caught in their traces in the firmground of a relict marsh: St Catherines Island. Scale = 5 cm (2 in). (e) More bioeroding bivalves, but this time the wedge piddock (*Martesia cuneiformis*), with its handiwork, carved in driftwood: Ossabaw Island. Scale in centimeters.

sion, and hence would be a plug-like and positive-relief circular structure with a dimple in its center (Chapter 10). Incongruous ark shells, especially those of recently dead animals that still have their periostracum, also often bear predation traces from whelks, in which the periostracum is worn off a small area on the edges of either valve and is accompanied by chipping marks. This is where the whelk put its lip onto the ark shell and broke the edges of its valves. Once the ark succumbed to this technique and slightly opened, the whelk blocked the opening with its own shell, inserted its foot, and started chowing down on ark innards. Like many victims of gastropod predation, though, some arks escape these attacks and their shells grow around the scar.

Some of the most abundant bivalves on most Georgia beaches are coquina clams (*Donax variabilis*), which are also incredible escape artists, owing to their streamlined shells and burrowing skills. Under normal conditions, coquina clams make short, vertically oriented Y-shaped burrows, with the lower part of the Y made by the foot and shell, and the upper parts by the siphons. These small, less than 1 cm (0.4 in) long suspension feeders live in the roughest of all marginal-marine environments—sandy intertidal zone of beaches—where the constant back-and-forth motion of the water changes their living space with each incoming wave. To cope with such high-energy conditions, this clam, if exhumed by a sudden swash, opens its valves slightly, then sticks out and vibrates its foot to loosen sand grains underneath it. In short, it makes its own localized quicksand pit, into which it gratefully sinks out of view from potential predators. This feat is most impressive when done collectively, in which hundreds of exposed bivalves will simultaneously vanish into the sand immediately after exposure by a hefty wave. In some situations, though, these clams find themselves stranded near the high-tide mark on the berm of steeply inclined beach, where their attempts at burrowing are all for naught. Once water has drained from the sand, they lack the size and strength to do much more than cover their top halves. Consequently, this part of the beach will be dotted with minute, sand-capped shells of these and other small bivalves, sticking out of an otherwise smooth sandy surface.

Perhaps the two most interesting aspects of these bivalves are how they are attuned to the sounds of crashing waves and move accordingly in response to these (Ellers 1995a), and their ability to migrate with both the tides and seasons (Ellers 1995b; Ruppert and Fox 1988). Wait, clams can hear? Yes, but not in the sense we do; instead, coquina clams react to the low-frequency vibrations imparted by crashing waves by leaping into the water and surfing their way up or down the beach in harmony with the tides. Louder is better, too, as higher-decibel waves correlate with more clams jumping out of the sand and into the water (Ellers 1995a). With regard to seasonal migrations, coquina clams will shift up from the lower intertidal zone—where they live during the fall—onto the higher parts of a beach in the winter. This means coquina clam traces, once identified, should take into account the time of year when made, allowing for finer distinctions in their interpretation.

Another common small burrowing bivalve is the dwarf surf clam (*Mulinia lateralis*), which lives mostly in shallow, muddy environments of Georgia estuaries but can be found in the intertidal zones of sandy beaches too (Figure 6.7b). Dwarf surf clams, like many bivalves, are suspension feeders, staying buried in the sediment while sending lengthy siphons to the sediment surface to suck in organics from the water. Their burrows are Y shaped and finely defined (Figure 6.7c), formed by a pair of thin siphons and a small shell (Howard and Dörjes 1972). Nonetheless, their small sizes are outweighed ichnologically by their numbers, as tens of thousands of individuals have been known to occupy a square meter (10.8 ft^2) of sediment (Rhoads and Germano 1982); multiply these numbers by two for their siphon traces in the same area. With such high densities, they can completely homogenize the uppermost 3–5 cm (1.2–2 in) of the sediment, and end up excluding most other infaunal organisms. Dwarf surf clams are a favorite food item of adult horseshoe crabs, which have even shown a preference for digging out and eating larger individuals when given a choice (Botton 1984). Hence traces of both animals may be directly associated. As mentioned earlier, their dead shells are also common components strengthening the burrows of the tube worm *Diopatra cuprea* (Meyers 1970).

More broadly, burrowing bivalves are often responsible for forming either escape or equilibrium burrows. For example, if a load of sand or mud is abruptly dumped on top of suspension-feeding bivalves by a storm, bivalves accordingly adjust their burrows upward, with an urgency based on survival instincts and depth of burial. Bivalves also burrow downward if sediment is scoured above them, a behavior readily demonstrated to anyone who has dug up a razor clam or coquina clam, set it on a sediment surface, and watched it disappear. Such traces are preserved as linear and vertical disturbances of differently colored or textured laminae that travel from a definite starting point, whether moving up or down, and forming chevron-like patterns. In the fossil record, the interpreted behavior and its tracemaker are often confirmed by the presence of a fossil bivalve at the end of the trace, a fugichnion that changed into a taphichnion.

The ribbed mussel (*Geukensia demissa*), mentioned earlier for its role in the deposition of mud in salt marshes (Chapter 2), is a sedentary bivalve that attaches itself with fine threads (byssae) to firm mud. It also associates with clumps of smooth cordgrass in the low salt marsh. Because it is stationary, one may wonder how it can produce traces worthy of any note. Yet it, along with another sessile and epifaunal bivalve, the common oyster (*Crassostrea virginica*), is largely responsible for creating an entire landscape composed of traces, namely feces. Normally, clay- and silt-sized particles would stay in suspension indefinitely, especially in the Georgia coast, with its constant movement of strong tidal currents. In a sort of positive feedback, ribbed mussels and oysters pull the clay particles out of suspension, then digest the organic goodies, and produce neat, mucus-bound little packets that are the hydrodynamic equivalent of sand (Smith and Frey 1985; Figure 3.5b). As a result, mud gets deposited in salt marshes as long as these two species of bivalves are living there.

Bioeroding bivalves on the Georgia coast include firmground and hardground dwellers, such as the Atlantic mud piddock (*Barnea truncata*), also known as the fallen angelwing; the angelwing (*Cyrtopleura costata*), which evidently never fell; the false angelwing (*Petricola pholadiformis*), which may or may not have fallen, depending on whom you believe; and the wedge piddock (*Martesia cuneiformis*). If you run your finger along the shell of any of these bivalves, you might be reminded of a file used for grinding against a hard surface, and you would be right. These bivalves and their kin—pholad and teredinid clams, respectively—are drillers, which rotate the roughly corrugated ribbing of their shells against a hard surface to wear it down. Of course, because their shells are made of mere aragonite ($CaCO_3$), they have their limits and will not attempt to bore into anything harder than their shells. One of the more helpful aspects of such bioeroding bivalves is that they often leave their shells in their borings, an extremely useful clue for linking these tracemakers with their traces (Figure 6.7d).

Angelwings are relatively rare on the Georgia coast in comparison to fallen and false angelwings; hence borings of the latter two are the most commonly encountered in firmgrounds, such as relict marshes (Morris and Rollins 1977). These borings are similarly oriented, vertical to steeply oblique, and shaped like cylinders with bulbous ends. Nonetheless, they differ in size and depth. On average, fallen angelwings borings are deeper and slightly wider than those of false angelwings. For example, false angelwings, which normally bore to a maximum depth of about 11–13 cm, can dominate the uppermost 20 cm or so of a given firmground. In contrast, fallen angelwings can penetrate to about 30 cm depth, thus taking over the firmground underworld, in a *Paradise Lost* sort of way (Morris and Rollins 1977). In both borings, the distal end expands slightly, which is where the bivalve stops its downward progress and calls this place home.

The hooked mussel (*Brachidontes recurvus*) is another bivalve that commonly co-occurs with these boring bivalves, but is an attacher, not a driller. Hence its traces will be nearly invisible, consisting only of the affected area on the surface where it attached itself. Nonetheless, these bivalves, along with their bioeroding cohorts, will aggregate on firmground surfaces in great numbers, hence their co-occurrence with borings can help to identify firmgrounds or hardgrounds, which are geologically significant horizons (Chapter 10).

Perhaps surprisingly, some pholad bivalves on the Georgia coast also drill bones of both modern and fossil terrestrial vertebrates (Howard and Frey 1975). This means that the bones of long-extinct terrestrial mammals, such as mammoths or ground sloths (Chapter 2), can be unearthed by currents on the seafloor and promptly bored by modern marine bivalves, causing both environmental and temporal mixing. Such bones also reveal attached oysters and other sedentary marine invertebrates that were looking to settle on a hard surface, yet did not mind the antiquity or provenance of these bones. For example, I recall seeing a Pleistocene ground sloth skeleton, recovered from the Altamaha River near the Georgia coast, whose bones were covered with oyster shells: an odd juxtaposition of modern and ancient, land and sea. Interestingly, other bivalves, sponges, and additional bioeroding organisms had not yet bored these bones, implying that they were only recently exposed and made available for exploitation by marine invertebrates.

On the Georgia coast, the most common wood-boring species is the wedge piddock (*Martesia cuneiformis*), which makes relatively small-diameter and shallow borings in driftwood (Figure 6.7e). In the geologic record, former woodgrounds associated with marine environments contain the distinctive traces and sometimes shells of similar wood-boring bivalves and a few other invertebrates. The ichnogenus applied to fossil bivalve borings is *Teredolites,* named after a modern species of wood-boring bivalve, *Teredo navilis* (Bromley et al. 1984; Gingras et al. 2004). Similarly, the ichnogenus *Gastrochaenolites* occurs in former marine firmgrounds and hardgrounds and is attributed to bioeroding bivalves. Although both trace fossils are behaviorally and morphologically similar, each was made originally in quite different substrates. Consequently, they are examples of substrate-dependent ichnogenera, where the original paleoenvironmental context of their tracemaking is essential for their proper diagnosis (Chapter 10).

Marginal-marine bivalves of the Georgia coast produce a large variety of traces because of their wide diversity of sizes, shapes, and behaviors, and this diversity stems from their long evolutionary history over more than 500 million years. As a result, bivalves are adapted to a range of substrates and environments, with their traces occurring in the energetic, shifting sands of the intertidal zone to static, solid firmgrounds and hardgrounds. Trace fossils attributed to bivalves are accordingly quite varied too, consisting of probable suspension-feeding burrows (*Polykladichnus*), deposit-feeding burrows (*Protovirgularia, Hillichnus*), resting traces (*Lockeia*), and borings in rock (*Gastrochaeonolites*) or woodgrounds (*Teredolites*). Moreover, through biodeposition of feces, a few bivalves, such as ribbed mussels and oysters, are responsible for much of the mud in salt marshes and other coastal environments. Through a combination of knowledge about bivalve anatomy, behavior, and substrate preferences, their traces can be interpreted in surprising detail, aiding in clearer interpretations of their trace fossils (Ekdale and Bromley 2001; Bromley et al. 2003; Chapter 10). Study of their traces and trace fossils can thus belie any preconceptions of these animals as simple, passive filter feeders, and instead reconstruct them as pursuing mobile lives filled with excitement.

LIMULIDS AND THEIR MANY TRACES

The horseshoe crab *Limulus polyphemus,* which provided an opening for this chapter via one individual and its agonized trackway, is an excellent example of a single species of marginal-marine tracemaker where its traces change considerably throughout its growth. Small, meandering trails and shallow burrows made early in a limulid's life later change to larger and distinctive trackways and resting traces. Hence they are similar to some

284 LIFE TRACES OF THE GEORGIA COAST

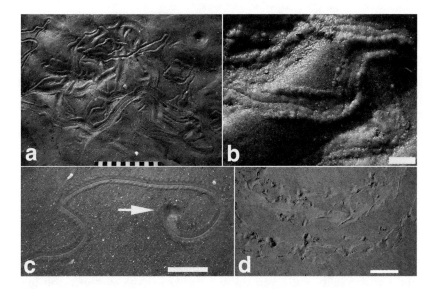

Figure 6.8. Limulid traces, juveniles and adults. (a) Small, meandering trails of juvenile limulids, showing central grooves from telson, inner pathways, and levees. Scale in centimeters. (b) Beaded parts of limulid trail, the result of their little legs pushing back wet, cohesive sand. Scale bar = 1 cm (0.4 in). (c) Resting trace of a juvenile limulid (arrow) attached to a trackway, buried under the sand where the limulid is staying hydrated. Scale bar = 5 cm (2 in). (d) Trackway of a full-sized adult, leaving little doubt of the identity of its tracemaker. Scale bar = 5 cm (2 in). All traces from Sapelo Island.

insects in the way growth stages of the same individual animal, through a combination of anatomical and behavioral differences, can result in radically different traces.

Interestingly, the trails and shallow burrows of juvenile limulids have been compared to trace fossils made by Paleozoic trilobites. This is not such a far-fetched notion, especially if you take a look at a young limulid and see the startling resemblance between these and their long-extinct marine cousins. In fact, marine biologists often refer to freshly hatched limulids as trilobites, a nod to this morphological similarity (Botton et al. 2003b). Functionally speaking, the anatomy and size of these little invertebrates, which are < 5–10 mm (0.2–0.4 in) long (including telsons), do not allow them to make many traces other than shallow burrows and trails (which often merge with one another). Because limulids are the only marine invertebrates to come up onto land (the beach, that is) to lay and fertilize eggs, juvenile limulids and their traces are common in intertidal sand flats, especially runnels. At low tide, these juveniles are on the move, plowing along the sand and mud surfaces with their tiny prosomas. Rippled crests become hills and ripple troughs are valleys from a small limulid's perspective; hence most of the their trails are concentrated in the troughs, with occasional pioneers crossing ripple crests. Juvenile limulid trails are slightly wider than their tracemakers, with slight ridges of sand on outside borders and a central, shallow groove (Figure 6.8a); the ridges are made by the outer edges of the prosoma and the groove by the telson (Martin and Rindsberg 2007). Once they encounter more saturated sand, the plowing action of the prosoma pushes the sand on top of it, causing trail making to switch over to burrow making. Regardless of whether these traces are trails or shallow burrows, they meander, loop, and cross over one enough to make patterns that would have been the envy of Jackson Pollack.

In my opinion, much can be learned about trilobite burrowing behavior by simply kneeling down on a Georgia sand flat and watching juvenile limulids move on and in the sand. Under a thin film of water, they glide along a sedimentary surface, easily making ribbon-like trails with slight furrows and subtle levees, until they encounter an obstacle, such as a shell fragment, piece of wood, or *Diopatra* burrow. Such barriers cause them to stop, back up slightly, and smoothly perform right-angle turns. However, if they walk out of the water and onto more cohesive sands, their trails change, developing deeper furrows and higher levees. Levees also may become beaded, formed by small pads of sediment pushed back by the rear legs (Figure 6.8b). If each of these traces were seen separately in the fossil record, the tracemaker might be difficult to identify as the same species, let alone the same growth stage, unless a paleontologist had spent some time watching these modern analogues in action.

Once these juveniles become a little bit bigger—say, 2–5 cm (0.8–2 in) long—their shallow burrows and trails become more recognizable as trackways: linear to meandering, ribbon-like paths dotted by leg impressions on the side and bisected by a shallow groove, the latter caused by a pointy telson (Figure 6.8c). These locomotion traces become especially more interesting as one follows them on an exposed sand flat because these smaller limulids, like smaller ghost crabs, cannot stay out of the water for long. Limulids have proportionally large gills on the lower parts of their bodies; hence they dehydrate quickly if stranded on a sand flat between tides, especially during a Georgia summer. What do they do in such instances? They hunker down into saturated sand to rehydrate, although they do not stay on the surface to do this: instead, they burrow. Once their bodies are slightly under the sand surface and in wet sand, they move their legs rapidly below them, which liquefies the sand—making it easier for deeper burial—and circulates water over the gills. Once the limulid's legs stop moving and the pore water drains, the surrounding sand collapses onto and around the animal. The resultant trace of this rehydrating behavior is a circular to ovoid depression at the end of a trackway, below which lies a juvenile limulid (Figure 6.8c).

Adult limulid trackways, as mentioned at the start of this chapter, are rather impressive traces once found. Their trackways are commonly encountered on the upper intertidal zones of beaches, such as the berm or even into the dunes, reflecting where limulids are often inadvertently placed by high tides or waves. They then try to find their way back to the sea, done partially by sight but also by moving downslope. Owing to the unpredictable topography of dunes, a wayward limulid becoming disoriented in these environments is completely understandable. Adult limulid trackways formed out of the water will have two sets (right and left) of four tracks from the walking legs, a central telson drag mark, as well as lateral and anterior drag marks caused by the sides and front of the prosoma, respectively (Figure 6.8d). Trackway widths vary according to the size of the limulid, which can be quite large: I have seen some more than 30 cm (12 in) wide, and 40 cm (16 in) wide trackways are not unheard of. Indeed, Georgia limulids are the largest of their species or any other limulids in the world. This circumstance is probably related to the broadness of the Georgia Bight (Chapter 2), which provides plenty of food and hiding places for limulids to grow bigger with age. How old? Some zoologists have estimated they may live as long as 25–30 years, which is a while for an invertebrate (Shuster et al. 2003), so the real number is probably significantly less, more like 15 years.

Adult limulid trackways are especially abundant during mating season (May–June) as females crowd into tidal channels and crawl up on beaches in large numbers, and the males follow them. Males will crowd around one female, competing to fertilize her externally laid eggs, which sometimes results in a male even latching onto her back and catching a ride until she picks a spot for egg laying. (Female limulids are distinguishable by a notch

Figure 6.9. Mating traces of adult limulids, in which a male piggybacked onto a female: Sapelo Island. The female was in the depression before the male entered the scene from the lower left. The male latched onto the back of the female, with each leaving telson impressions next to one another (arrows). They then moved together, spiraling and looping clockwise out of the depression, making superimposed trackways. Photograph by Stephen Henderson.

on their opisthosoma that allows the front part of the male prosoma to fit: a trailer hitch, of sorts.) This coupling means that ichnologists will see double: two superimposed, slightly out of phase trackways are made, one slightly wider (the female) and the smaller one on the interior (the male), with two telson drag marks (Figure 6.9). The female trackway will almost assuredly have prosoma impressions as well, seeing that her male burden caused her to temporarily take on some added weight. After each partner has spilled its gametes into the sand, their tracks will then diverge, perhaps never to meet again.

In short, a single species of limulid is responsible for a notable variety of traces made from birth to mating to death, all of which may be applicable to similar traces made by their close relatives and long-lost marine arthropod cousins, especially trilobites. Furthermore, some trace fossils have been identified as products of limulid activity, such as trackways (*Koupichnium*) and resting traces (*Arborichnus, Limulichnus*). Later on, the importance of traces made by younger limulids will be connected further to trace fossils that are often attributed to polychaete worms or other nonarthropod tracemakers (Chapter 10).

SMALL CRUSTACEANS, CONSTANTLY MIXING, SOME BORING

Beneath the feet of beachcombers and romanticists who walk along a Georgia beach is a seething, churning world of moving sand, with each grain shifting slightly, all of it put into motion by millions of tiny crustaceans. These crustaceans—amphipods and isopods—affect sediment by pushing apart individual sand grains and grazing on algae on and in between grains. The overall effect of this activity on beach sand is to transform its crisply defined bedding to a fuzzy, stirred version, looking much like an out-of-focus photograph. This sediment mixing is called cryptobioturbation by ichnologists because of its barely observable quality (Chapter 3). Yet such a poorly observable effect speaks of minute changes carried out on a large scale, resulting in the unseen alteration of great volumes of sediments.

Amphipods, also known as beach fleas for their small sizes and hopping ability, are extremely abundant on Georgia beaches. In one study on St. Catherines Island, they composed 12–22% of all animals in the intertidal to shallow subtidal sands (Prezant et al. 2002). Given such high numbers and cumulative burrowing effects, can individual amphipod burrows also be seen? Yes, and fortunately several detailed studies on amphipod ecology and their burrowing, including those of species in Georgia, provide the means for identifying their small and somewhat obscure traces (Croker 1968; Howard and Elders 1970; Dörjes 1972; Gingras et al. 2008a).

Georgia amphipods include many species, but only the five most abundant burrowers will be covered here: *Acanthohaustorius millsi, Parahaustorius longimerus, Lepidactylus dytiscus, Neohaustorius schmitzi*, and an unspecified species of *Haustorius*. Howard and Elders (1970) related how these amphipods species varied along a beach profile: *P. longimerus* and *A. millsi* dominate the lower beach; *L. dytiscus* takes over the middle beach; and *N. schmitzi* and *Haustorius* are the amphipod rulers of the high beach. Apparently, some vertical partitioning is at play too, as *N. schmitzi* tends to live in the uppermost 2–5 cm (0.8–2 in) of beach sands, whereas *Haustorius* burrows more deeply, going to about 6–6.5 cm (2.4–2.6 in). This vertical difference may not seem like much to us, but is a significant distance for amphipods, some of which are only about 5 mm (0.2 in) long. Hence through such horizontal and vertical separation, many species can occupy specialized niches in the uppermost 10 cm (4 in) of beach sand. As a result, knowledge of preferred ecological zones of amphipods, depth of burrowing, burrow forms, and physical sedimentary structures—such as bedding and ripple marks—that form in those same zones can all help to identify the makers of these petite burrows. One caveat to keep in mind, though, is how some of these species overlap one another, both horizontally and vertically. Thus there is no magic bullet for identifying a specific part of a beach in the geologic record on the basis of a single presumed amphipod burrow.

What do their burrows look like, and how are they made? In accordance with their small body sizes, amphipod burrows are only a few millimeters wide, and most are backfilled, denoted by meniscae composed of differently colored sand. As they burrow, amphipods use some of their little legs to hold open the burrow (they do not cheat by using mucus to support the walls), and other legs to pull sand grains to their mouths. These grains are then gleaned for whatever algae or other organics might be on them, and discarded grains are packed behind them as the amphipods progress through the sand. From there, individual burrowing patterns of different species vary. For example, the upper-beach amphipod *N. schmitzi* makes mostly short, 2–5 cm (0.8–2 in) vertical burrows that often connect with the beach surface, concentrating its efforts in the uppermost layers of sand. In contrast, *Haustorius* burrows curve downward to about the same depths, but turn to become more horizontal (Howard and Elders 1970). The middle-beach amphipod *L. dytiscus* tends to burrow like the upper-beach *N. schmitzi*, forming mostly vertical burrows, although slightly deeper, to as much as 6.5 cm (2.6 in). The lower-beach amphipods *P. longimerus* and *A. millsi* dig even deeper, which makes sense as an adaptation for avoiding the erosive effects of high-energy waves. *P. longimerus* constructs 10 cm (4 in) long and mostly vertical burrows with slightly curved deviations along the way, looking nearly spiraled. *A. millsi* digs to similar depths, but forms straight burrows, most of which are angled 45° relative to the surface. How fast do amphipods dig? *Haustorius* and *Acanthohaustorius* have been clocked at burrowing rates of 5 cm/h and 1 cm/h, respectively (Gingras et al. 2008a)—impressive speeds for such small animals in sand. Interestingly, *N. schmitzi* is often found in the water above sand surfaces, and occurs throughout shallow sands of the upper to lower beach (Croker 1968). This suggests that it might be using the surf to waft it over wider ranges, similar to how coquina clams depend on waves for transportation.

As mentioned earlier, but perhaps needs reemphasis, amphipods are abundant and prolific burrowers. This means they can homogenize sediments to the point where all but the last few individual burrows are erased by burrowing. This situation also implies that the only clues to their identity in a suspected ancient beach deposit may be those last-formed backfilled burrows in a trace-fossil assemblage. Also, because amphipods make up so much of the biomass in most Georgia beach sands, they are a ready source of food for shorebirds, in some instances providing about 10 percent of their calories (Grant 1981). Hence amphipod burrows also may be closely associated with trackways and beak marks of shorebirds (Chapter 8).

Another group of diminutive crustaceans that live in Georgia beach sands are marine isopods, sometimes called sea lice. (Between mentions of beach fleas and sea lice, I would not blame anyone for feeling a little itchy about now.) As crustaceans, marine isopods are in the same taxonomic category as terrestrial pill bugs, known popularly as roly-polies because they roll into little balls for protection. Marine, sand-dwelling isopods in Georgia include *Ancinus depressus, Chiridotea caeca, Cyathura polita, Exosphaeroma diminutum*, and *Sphaeroma quadridentatum*, among many others (Menzies and Frankenberg 1966; Ruppert and Fox 1988); most of these species are about 1–2.5 cm (0.4–1 in) long. Beach isopods, like amphipods, feed on organic material in between sand grains, but take full advantage of their small sizes by swimming through water-filled spaces between grains in the uppermost few centimeters of a wet, saturated beach (Gingras et al. 2008b). This means that traces of their movements are barely recorded as grain-by-grain shifts of sand, and thus exemplify cryptobioturbation.

More likely candidates for preservation of marine-isopod trace fossils are borings left in wood by the isopods *Sphaeroma destructor, Limnoria tripunctata,* and *L. lignorum. S. destructor* derives its species name for eating into and disintegrating the wood of sailing ships, docks, and other wood exposed to ocean water, in which it is joined by *L. tripunctata* and *L. lignorum* (also called gribbles) as partners in crime. The cumulative effects of this isopod-inflicted bioerosion cause a thinning of dock pilings within the tidal range, looking as if marine beavers chewed around them. As far as I know, though, trace-fossil equivalents of these borings have not yet been interpreted, although some modern examples of *L. lignorum* borings have been described on the west coast of the United States (Gingras et al. 2004). These borings were 2–3 mm (0.1 in) wide, about 10 cm (4 in) long, and mostly cylindrical, but a few were also U shaped. If preserved as trace fossils, the cylindrical ones would be classified as either *Trypanites* or *Teredolites*, and the U-shaped ones as *Caulostrepsis* (Chapter 10).

DECAPODS DIGGING DEEPLY OR DEALING DUROPHAGOUS DEATH

Marginal-marine decapods and their tracemaking on the Georgia coast could have constituted not one, but two books on their own. Indeed, hundreds of peer-reviewed articles have been written about these tracemakers since the 1960s, and the importance of their burrows as ecological and paleoecological indicators is well established. Decapods, first introduced when we considered the ichnological importance of freshwater crayfish (Chapter 5), broadly include crabs and shrimp. For example, fiddler crabs (*Uca pugnax, U. pugilator, U. minax*) and squareback crabs (*Sesarma reticulatum, Armases cinereum*) probably number in the billions in Georgia salt marshes and other marginal-marine environments, and their various traces easily outnumber their living populations. The bountiful traces of a related species, the ghost crab *Ocypode quadrata*, cannot help but be seen by anyone glancing at the upper parts of Georgia beaches and coastal dunes. Likewise, thousands of burrows made by species of callianassid shrimp (*Biffarius biformis* and *Callichirus major*) are clearly visible during any given low tide, dotting Georgia beaches all along the coast, even on developed islands like Tybee, St. Simons, and Jekyll. At low tide, the trackways and shell drag marks of hermit crabs are almost impossible to miss on both sandy and muddy surfaces. Less visible, but nonetheless present, are rapidly burrowing mole crabs (*Albunea paretii, Emerita talpoida,* and others), and many other small, burrowing decapods on sand flats and in muddy salt marshes. So it is with much discipline and restraint that I will focus only on the salient points about the most prolific decapod tracemakers, as well as ones most likely to leave lasting impressions, as represented by analogous trace fossils.

Fiddler Crabs and Mud Crabs

Fiddler crabs, consisting of three species of *Uca* (*U. pugnax, U. pugilator, U. minax*), belong to a related group of decapods (Ocypodidae, or ocypodids) that all evolved from a common marine ancestor, but diversified into specific marginal-marine niches. Fiddler crabs are specifically associated with salt marshes and transitional environments around salt marshes (Chapter 5), where any one-time visitor will be assured of seeing thousands, some waving their claws at those passing by. Perhaps the most interesting evolutionary aspect of fiddler crabs is how they demonstrate habitat partitioning, an evolutionary pathway that was apparently driven by sediment grain size and other ecological factors (Coward et al. 1970; Robertson and Newell 1982). The mud fiddler (*U. pugnax*) prefers to live in muddy substrates, such as in the low to high marsh, whereas the sand fiddler (*U. pugilator*) is associated with sandy substrates, like beach–marsh transition areas or storm-washover fans. The red-jointed fiddler (*U. minax*), like *U. pugnax*, also lives in muddy substrates, but normally burrows into the banks of tidal channels. This is a fiddler crab that prefers to live on the edge (of ecological parameters, that is): it is tolerant of freshwater, lives in mud with low oxygen content—corresponding to high amounts of organics—and is often in the highest parts of a salt marsh (Frey 1970b; Whiting and Moshiri 1974; Basan and Frey 1977). All three species of fiddler are ichnologically united by their digging open burrows and forming balls of processed sediment from their scraping of intertidal surfaces.

Without a doubt, fiddler crabs are fantastic burrowers. Burrow densities can be > 200 burrows/m^2, which also gives a sense of their salt marsh populations. Not all burrows are occupied at the same time by individual fiddler crabs, however, but most have an occupant or two. Burrow cross sections are circular and maintain a constant diameter—only slightly wider than the fiddler crab carapace—until they reach the bottom of the burrow. Here burrows may expand slightly to allow turnaround space for the crab, or (more romantically) a place for two crabs to mate in privacy, away from the prying eyes of ichnologists. Similar to some other burrowing decapods, they roll sediment into little balls with their claws and place these outside of the burrow entrance. Some crabs even construct 5–10 cm tall turrets rising above the marsh surface (discussed previously [Chapter 5, Figure 5.2b] and revisited soon).

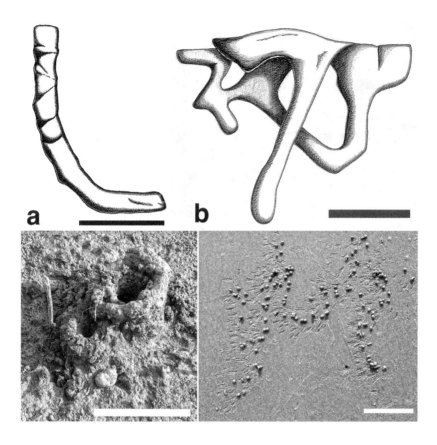

Figure 6.10. Fiddler crab traces. (a) Sand fiddler (*Uca pugilator*) burrow. Based on fiberglass cast figured by Frey and Howard (1969). Scale = 5 cm (2 in). (b) Mud fiddler (*Uca pugnax*) burrow. Based on fiberglass cast figured by Basan and Frey (1977). Scale = 5 cm (2 in). (c) Pellet-lined turrets of a mud fiddler crab burrow; these are probably joined below the surface to form a U shape: Sapelo Island. Scale = 5 cm (2 in). (d) Sand fiddler scrape marks, accompanied by balls of processed sand: Sapelo Island. Scale = 10 cm (4 in).

Burrows of sand fiddlers (*U. pugilator*) and mud fiddlers (*U. pugnax*) are easily identifiable in the field by which substrate they choose, but these burrows also have slightly differing forms (Aspey 1978; Frey et al. 1984; Basan and Frey 1977). Sand fiddlers burrow sideways and in a manner similar to that of ghost crabs, which is not surprising, considering their close evolutionary relationship. Sand fiddler burrows are 1–2 cm (0.4–0.8 in) wide and are obliquely oriented relative to the marsh or storm-washover fan surfaces. Overall shapes of their burrows are J or L shaped, nonbranched, and 15–75 cm (6–30 in) deep (Figure 6.10a). Some burrows also have slightly bulbous enlargements at their bottoms, which often contain water, even at low tide (Aspey 1978). Sand fiddler burrows are most common in sandier, high-marsh environments dominated by halophytes such as *Juncus*, *Salicornia*, and *Distichlis*, and hence their burrows may be closely associated with root traces (Chapter 4). Furthermore, wherever sand and mud fiddler territories overlap, mud fiddlers will dominate, with burrow densities reaching nearly seven times those of the sand fiddlers (Teal 1958; Aspey 1978).

Mud fiddler burrows are only slightly different from sand fiddlers in size and form, with the same range of diameters, but in general are more complex (Figure 6.10b). These burrows begin as simple J- or L-shaped structures, but can be more ambitious in their replicating the alphabet by also becoming U shaped (Basan and Frey 1977). The lowermost part of a mud fiddler burrow, which can reach depths of 60 cm (24 in), is more curved, making the bend of a J or U, whereas the same part of a sand fiddler burrow stays more horizontal, creating the sharp bend of an L (Frey and Basan 1978). Most complex burrows, however, are the works of more than one fiddler, in which older burrows are reoccupied and modified, or intersected by other burrows, causing a misleading false branching. This ichnological insight is affirmed by observations of mud fiddlers that constantly fight over and usurp one another's burrows. (The species name *Uca pugnax*, as in "pugnacious," is purposeful and is closely mimicked by the name of *U. pugilator*.) For example, whenever I approach a marsh at low tide, these crabs scramble for refuge in the nearest burrow, which is not necessarily their own and sometimes contains another crab, who then becomes annoyed by such intrusions.

Mud fiddler burrows are further distinguished from those of sand fiddlers by occasional pellet-lined turrets, which can extend as much as 10 cm (4 in) above the marsh surface (Figure 6.10c). These turrets (or chimneys) look superficially similar to crayfish towers (Figure 5.2), but are much smaller and more cylindrical. What purpose do these serve? Some biologists have proposed that they help to prevent flooding of the burrow at high tide (Aspey 1978), but others have also emphasized how they could exclude intruding crabs. For example, in one species of fiddler crab in eastern Asia (*U. arcuata*), females were more than twice as likely to build these burrows, presumably to keep out unwanted male crabs (Wada and Murata 2000). The heavy marsh crab (*Sesarma reticulatum*) also makes turrets around its burrow entrances, but its burrows are noticeably wider, 2.5–4 cm (1–1.6 in) and less common (Frey 1970b; Basan and Frey 1977).

The red-jointed fiddler (*U. minax*) is the largest fiddler crab species in Georgia, but is not as common as the other two because of its specialized habitat, which is in the upper, elevated parts of marshes that receive more freshwater. Like other fiddlers, it is an accomplished burrower. Its burrows, however, differ from those of other fiddlers in their width, depth, verticality, straightness, and simplicity. Red-jointed fiddler burrows are 2.5–3 cm (1–1.2 in) wide, 30–65 cm (12–26 in) deep, and do not resemble any letters in the English alphabet in their overall forms, although the Greek alphabet may provide a better analogue. Nonetheless, the upper part of the burrow is an inclined (20–30°) tunnel, 2–5 cm (0.8–2 in) long, that connects to a straight to slightly sinuous, vertical shaft that makes up most of the burrow (Basan and Frey 1977). Red-jointed fiddler burrows are ordinarily nonbranched too, although they may change in diameter along the length of the burrow; wider parts correspond to chambers that were occupied by a crab or two.

Like most crabs, fiddlers walk sideways, leaving characteristic V-shaped, eight-legged trackways, although these are much smaller (about 1–1.5 cm, or 0.4–0.6 in wide) than trackways of mature ghost crabs. Fiddlers are also not prone to taking long walks along beaches or dunes, as these crabs stay fairly close to their burrows. Additional fiddler traces are their scrape marks and feeding pellets formed by grazing on algae, which grows on sand and mud surfaces (Frey et al. 1984; Figure 6.10d). In such instances, these fiddlers are deposit feeding, although they are also well known as detritivores. Normally these traces are just outside burrows, a coincidence that is not, as crabs are conserving their energy by building homes near their food source.

The so-called mud crabs of the Georgia coast, whose low salt marsh habitats are revealed by this sedimentary grouping, include the squareback marsh (or wharf) crabs (*Armases cinereum*, also known as *Sesarma cinereum*); the white-clawed mud crab (*Eurytium limosum*); the heavy (or purple) marsh crabs (*Sesarma reticulatum*); and the Atlantic mud crab (*Panopeus herbstii*). Although these crabs overlap ecologically, their burrows can be set apart on the basis of form and size.

Of these, the squareback marsh crab is commonly seen in terrestrial ecosystems, wandering over from adjacent marshes (Chapter 5). This behavior is prompted by its omnivory, because it consumes high-marsh plants (*Iva frutescens*) as well as spiders, aphids, and terrestrial arthropods that live on these plants (Ho and Pennings 2008). Squareback marsh crabs, unlike most crabs living in salt marshes, are weak burrowers, which explains why they are seen more often than their close relative, the heavy marsh crab *S. reticulatum* (Seiple and Salmon 1982). Burrows of white-clawed mud crabs are about 3–5 cm (1.2–2 in) wide, and consist of two or more slightly inclined tunnels about 50–60 cm (20–24 in) long. These tunnels, however, are connected to 20–30 cm (8–12 in) long vertical shafts, making them look like a squashed "Y" (Basan and Frey 1977). Other burrowing crab species often modify white-clawed crab burrows, taking advantage of these larger (and free) burrows. Purple marsh crabs, given sufficient cover provided by smooth cordgrass, excavate as many as 30 burrows/m^2 in the middle of a marsh or along creek banks, although this number may decrease abruptly in unvegetated parts of a salt marsh (Teal 1958; Koretsky et al. 2002). These crab burrows normally have more than one entrance, which connect to a short, nearly horizontal tunnel. This tunnel follows into the main part of the burrow, a 2–5 cm (0.8–2 in) wide and 15–30 cm (6–12 in) long vertical shaft, although these can be as deep as 60 cm (24 in) (Basan and Frey 1977; Seiple and Salmon 1982).

The Atlantic mud crab is also known to share burrows with purple marsh crabs, both species showing some mutual tolerance (Allen and Curran 1974; Koretsky et al. 2002). Despite its small size, the Atlantic mud crab is a voracious predator, going after ribbed mussels, marsh periwinkles, and even small hermit crabs inhabiting periwinkle shells. It is such an important predator of periwinkles that some ecologists have proposed that it indirectly affects the growth and production of smooth cordgrass in a marsh. After all, if left unchecked, periwinkles will overgraze cordgrass, decreasing the amount of marsh vegetation available for other organisms (Silliman et al. 2004). So although one may not ordinarily think of it this way, these crabs are ecologically analogous to wolves in Yellowstone ecosystems, which keep elk populations down and thus help prevent overgrazing of forests there. In accordance with its smaller size, the Atlantic mud crab digs relatively short, U-shaped burrows that are only about 1–2 cm (0.4–0.8 in) wide and 5–10 cm (2–4 in) deep. Nevertheless, these burrows

MARGINAL-MARINE INVERTEBRATES 297

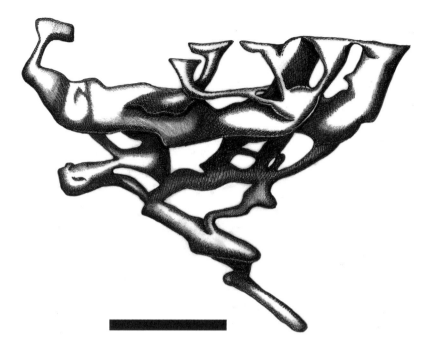

Figure 6.11. Atlantic mud crab (*Panopeus herbstii*) burrow system, originally attributed to the snapping shrimp (*Alpheus heterochaelis*), a frequent roommate in its burrow. Based on a fiberglass cast figured by Basan and Frey (1977). Scale = 10 cm (4 in).

can connect with one another to make complex systems more than 70 cm (28 in) long (Basan and Frey 1977: Figure 6.11).

A common cohort of mud crabs is the snapping shrimp (*Alpheus heterochaelis*) of coastal Georgia, once thought of as a burrower, but alas is not. This shrimp is so called because of its sonic defense technique, in which it uses a large cheliped to produce a burst of sound ("snap") that knocks out or kills small invertebrates and fish that get too close to it. (No one knows if the science fiction writer Frank Herbert, author of the Dune trilogy, was aware of this animal and its awesome auditory abilities.) This deadly sound is made by opening and closing shut its claw rapidly, displacing water and generating a bubble that collapses just in front of the claw. Humans standing near a salt marsh can easily hear the resulting pop, which they should then associate with death. Yet despite owning a large, powerful claw, killing other animals, and often hanging out in burrows, snapping shrimp do not themselves burrow. Instead, they cohabit burrows dug by the Atlantic mud crab (*Panopeus herbstii*). In one study, snapping shrimp occupied about 11 percent of more than 1,000 crab burrows in salt marshes of the eastern United States (Silliman et al. 2003). Apparently, living together works out well for both species—the crabs do not prey on the shrimp, and the shrimp keep away burrow intruders—so it qualifies as mutualism. Nevertheless, no one has documented how many accidental discharges by snapping shrimp have resulted in injuries or deaths of their homemakers. Oddly enough, burrows made by other species under the same genus *Alpheus* are well documented, and burrows mistakenly attributed to the Georgia snapping shrimp were described in loving detail by ichnologists in the 1970s (Basan and Frey 1977). All of this implies that further research is needed on the ichnology of snapping shrimp in general.

So now you know that fiddler and mud crabs are exceedingly common in marginal-marine environments—especially salt marshes—and that these crabs make an extraordinary number and variety of traces. Nonetheless, few trace fossils have been connected specifically to these crabs and their ancestors, although some were recently identified in Pleistocene deposits on St. Catherines Island (Chapter 10). If found in the fossil record, though, many of these trace fossils are easily classifiable as the ichnogenera *Psilonichnus*, *Thalassinoides*, and *Arenicolites* (Chapter 10).

Ghost Crabs

If forced to list my top ten tracemakers of the Georgia barrier islands, ghost crabs (*Ocypode quadrata*) would always be in the top five; just please do not make me choose the other four just yet. Ghost crabs are examples of so-called transitional animals, which creationists insist do not exist, living quite happily between the land and sea and with adaptations for both realms. Yet there they are, and represented by eight closely related species living in temperate to tropical coastal environments throughout the world. These somewhat sizeable crabs, adults of which are about 10–15 cm (4–6 in) wide, have anatomical traits that clearly represent a lineage in the middle of an evolutionary shift from one habitat (marine) to another (marginal-marine and terrestrial). Among these traits are weakly developed lungs and atrophied gills, which show that they are not suited for a life entirely on land, nor would they live long if completely submerged. They also have tall, vertically oriented legs with points on their ends (dactyli), which are well adapted for running on land, but not so good for swimming. Because they live amid two worlds, though, ghost crab traces are excellent ecotone indicators (Chapter 3).

So why are they called "ghost crabs," rather than "really widespread tracemaking ecotone-indicator crabs" (other than decreasing the number of syllables)? Their common name is likely related to several factors. For one, they are well camouflaged, as their carapace colors and patterns match those of beach sands so well that a ghost crab sitting still in a dune or beach sand will be almost invisible. Secondly, if they are spotted, they flit about at incredible speed, seeming to float across the beach surface. Thirdly, and most importantly from an ichnological standpoint, their traces are seen far more often than the tracemakers themselves, and they are most lively on the surface at night. As a result, many ghost crab traces represent nighttime activities, which makes their traces all the more valuable for interpreting and documenting their unseen behavior.

Ghost crab burrows are nearly everywhere one looks along almost every Georgia beach, although they are less common on overdeveloped and overpopulated beaches. Their open burrows have circular outlines and range considerably in diameter, from < 1 to 8 cm (< 0.4–3.1 in), depending on the age of the crab that made the burrow. Interestingly, small-diameter burrows tend to cluster near the position of the low tide, whereas large-diameter burrows are more common in foredunes and back-dune meadows (Frey and Mayou 1971; Hill and Hunter 1973; Duncan 1986). This distribution of burrow sizes is also related crab ages, but is tied in more specifically to physiology, as explained later.

Aspiring ichnologists will also note that some closely spaced pairs of ghost crab burrow openings have the same diameter, provoking an investigation of their subsurface geometry. When in pursuit of such ichnological knowledge, I heartily recommend not sticking fingers or other personal appendages down such holes. Rather, use an inert remote sensing tool, such as an errant piece of cordgrass, which may then go in one hole and poke out the other, showing their direct association. These paired holes, more often than not, are slanted toward one another at about 30–50°, and join about 10–20 cm (4–8 in) below the surface, connecting to a near-vertical shaft that continues downward for a meter (3.3 ft) or more (Figure 6.12a) The end of the burrow is slightly enlarged (bulbous), which allows room for the crab to turn around in its burrow, or if it's lucky, to share the space with a

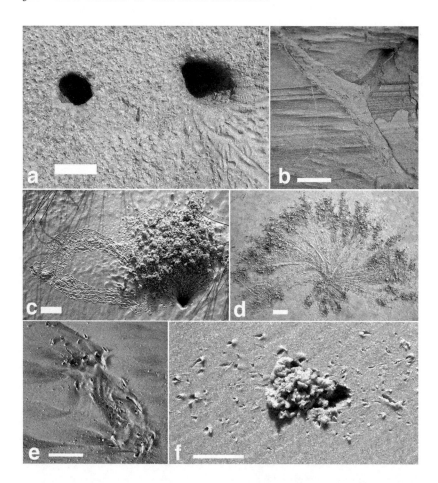

Figure 6.12. Ghost crab (*Ocypode quadrata*) traces. (a) Paired holes associated with the top of a Y-shaped burrow, with tracks connected to the active entrance (*right*): Sapelo Island. (b) Longitudinal section of a Y-shaped burrow in a dune, partially filled with sand: Sapelo Island. (c) Sand ball pile outside a ghost crab burrow and associated with its fresh tracks: Jekyll Island. (d) Radiating pattern of sand around a burrow (about 1 m [3.3 ft] wide), caused by a ghost crab flinging clumps of sand excavated from its burrow: Sapelo Island. (e) The pause that refreshes: a resting trace in which a ghost crab rehydrated its gills: Sapelo Island. (f) Excavations and trackway made by a ghost crab in search of coquina clams (*Donax variabilis*) and dwarf surf clams (*Mulinia lateralis*): Jekyll Island. All bar scales = 5 cm (2 in).

mate, however temporary that might be. The overall vertical profile of the burrow is thus like a large, upright Y (Figure 6.12b). If unpaired, a single burrow will have a high-angle, straight to slightly curved shaft that again leads to a bulbous terminal chamber, becoming more J shaped. Similarly, an incompletely formed Y-shaped burrow, in which the vertical shaft portion was not formed, may have more of a broad U profile. (However, no M, C, or A shapes have been thus far discerned in investigations of these burrows.) Like many burrowing animals, ghost crabs usurp other ghost crab burrows, either as a more permanent domicile or, if caught in the open and startled by a large, upright biped, as the nearest safe hole for protection. This results in what must be some considerably unpleasant exchanges if the original burrow maker is home.

A common theme of subaerial burrows in general is that they often serve an important physiological function for the burrow occupant, almost like adding an extra organ (Chapters 5, 7, 8). Ghost crab burrows are no exception to this generalization, and in fact these animals would quickly die out locally or become extinct as a species if they were not allowed to burrow. Because these crabs are adapted for living in marine-salinity water or on land, they run the risk of either drowning or desiccating, respectively. Burrows neatly solve both problems. First of all, when a high tide causes the water level in the burrow to rise, a ghost crab still may not want to leave its burrow: after all, the outside world has lots of predators. Instead, it simply moves up the burrow, plugs the entrance, and sits in an air pocket for the duration of the high tide. During low tides, the crab can retreat to greater depths in the burrow, where a humid, even-temperature microhabitat is maintained, even during daytime hours of a hot Georgia summer. Deeper burrows may even reach the water table of a low tide. Of course, a deep burrow also decreases the risk of predation, especially at night, when yellow-crowned night herons (*Nyctanassa violacea*), raccoons (*Procyon lotor*), and other vertebrates come out to hunt for delectable decapods (Chapter 7). Furthermore, burrows are used during the winter for alleviating the energy-sapping qualities of colder weather: even coastal Georgia has nighttime low temperatures approaching or below 0°C (32°F) from December through February. In such instances, ghost crabs block their burrow entrances with sand to maintain a more hospitable climate than outside.

Size also matters when it comes to ghost crab burrows, as the larger adult crabs have more strength, energy, and endurance to dig burrows that fit their body sizes. They also are more capable of sustained forays outside of their burrows without the need to rehydrate. Hence they can afford to make their burrows farther away from the shoreline in relatively drier backdune sands. In contrast, smaller, juvenile ghost crabs dehydrate quickly if outside their burrows for long, which also leads to their tiring, slowing down, and becoming more vulnerable to predation. As a result, small crabs dig burrows much closer to the shoreline to save them a long trip to the water; hence their smaller-diameter burrows are more often on the lower part of the beach (Hill and Hunter 1973; Dörjes and Hertweck 1975). Like many burrowing animals, ghost crabs make burrows appropriate for their body diameter, so burrow cross-sectional areas are positively (and tightly) correlated with body mass: the larger the area, the bigger the crab. Consequently, measurements of burrow diameters and relative positions along a shoreline can be used to estimate population structure of this species in a given area and how it might relate to ecosystem gradients.

Ghost crab burrows are also easily noticed from a distance along Georgia beaches by closely associated pyramidal sediment piles, linear patterns of moist sand, and scrape marks. Sediment piles, which are sometimes composed of stacked, loose balls of sand, are carried out of a burrow, one ball at a time (Figure 6.12c). As a fun exercise in actualism, I have lain down on my belly near an active burrow, stayed still, and watched ghost crabs perform this progressive excavation. First they start below the surface, rotating their claws (chelipeds) to rake and sculpt sand out of the burrow. When they exit their burrows, they are walking on only 6–7 of their eight walking legs. The other legs are carrying a clump of sand under their carapace, as if it were a loaf of bread, football, load of laundry, or various other cultural metaphors. What happens next depends on the crab and its mood. It either drops the sand as a ball just outside the burrow entrance, or it walks a short distance, and flings the sand away from its body with a quick, flicking motion. The former action produces sand-ball piles, the latter a radiating pattern of scattered sand, in which the burrow is in the center of the pattern (Figure 6.12d).

Regardless of whether the sand is concentrated in piles or dispersed away from the burrow, each type of trace holds some behavioral significance for the crabs. The sand piles serve as visual cues for a crab to find its burrow among all of the other ones along a beach, the crab equivalent of adding an orange ball to a radio antenna on a car. From a crab's perspective, these pyramids offer easy-to-spot landmarks along an otherwise monotonous beach landscape, and experiments done on crab recognition of markers bear out their effectiveness (Linsenmaier 1967). On the other hand, sand tossed onto a broad area outside of a burrow probably represents territorial marking, in which a male crab spreads its scent, warning off competitors or attracting mates. Similar to their fiddler crab relatives, ghost crabs are enthusiastic in both fighting and flattering.

Directly connected to burrows are ghost crab trackways. During normal walking, crabs leave four-by-four depressions in their trackways, forming a sort of V pattern in which the open part of the V points in the direction of movement (Figure 1.6b). Like most crabs, ghost crabs prefer to walk laterally, but can also move obliquely forward or backward. Regardless of their direction of movement, they are perched on the distal points of their legs, which make marks that look much like commas. The tails of the commas are also oriented in the direction of movement, and sediment mounds are on their exteriors. Owing to the ability of ghost crabs to rapidly change speed and direction, their overall trackway patterns vary greatly, particularly in trackway width. Accordingly, trackway widths—unlike their burrows—should never be used as accurate indicators of crab size, although end members of any given size range can be discerned from the average spacing between impressions on either side of a trackway. Trackway spacing can be measured two ways: (1) pace, which is the distance between alternating sets of legs, measured from the same point; and (2) stride, which is the distance between the same set of legs, but again, measured from the same point. (This basic tracking technique will be revisited in detail when discussing vertebrate tracks; Chapters 7–9.) When running, ghost crabs use leg 3 on one side as a means of support while simultaneously thrusting forward legs 2 and 4 on the opposite side (Burrows and Hoyle 1973). With faster running, legs 2 and 3 alternate on either side, meaning only two pairs of tracks may show up in such trackways.

Just for fun, pick any one of thousands of ghost crab trackways and follow it along a beach and through the dunes, and a picture of crab decision making will emerge, replete with a spectrum from torpid boredom to sheer panic. Pacing will decrease where a crab slowed down, and these tracks may even end with a shallow depression that outlines the body of the tracemaker where it stopped and lay down. Increases in spacing show where crabs accelerated to top speeds, which can be astonishing in various species of *Ocypode*; 1.5–2 m/s (5–6.5 ft/s) (Hafemann and Hubbard 2005). As a form of equalizing and accounting for differences in size versus speed, zoologists have compared animal body lengths of animals with how many of these are covered per second. Using this measure, ghost crabs cover about 100 times their body lengths per second, and thus are the third-fastest land animals in the world, only bested by two species of tiger beetles (Chapter 5). So if moon snails are the lions of the Georgia tidal flats, then surely ghost crabs are the cheetahs.

Pay attention to other traces along a ghost crab trackway, and additional oddities will manifest. For example, a central, discontinuous drag mark may show up, seemingly intermittent in its use. This is made by the larger of the two claws (superior cheliped), and may represent acoustical signaling. Yes, that's right: ghost crabs talk to one another, using their claws to send vibrations into the sand, which are transmitted to other crabs (Horch 1975; Clayton 2001). Ghost crabs are also well known as both scavengers and predators (Wolcott 1978, 1988), but in the former role, they feel no reason to refrain from stealing, as revealed by their traces. In one instance, I tracked a ghost crab to its burrow because I noticed a large drag mark on one side of the trackway, and became curious what might have made it. The drag mark came to a stop at its burrow, then resumed, moving away from the burrow but associated with the trackway from a different ghost crab. These tracks ended at another burrow, and the source of the drag mark was revealed: a partially eaten dead fish. Both crabs were scavenging, but one neighbor unwittingly fed the other, a scenario that probably is more common than we know, but easily documented from their traces.

Probably the most behaviorally complex of ghost crab traces, though, are their so-called resting traces. Other ichnologists, coastal biologists, and geologists had noticed these distinctive traces on the Georgia coast and elsewhere, but they had not been described or otherwise explored further. I became interested in them through the most seemingly innocuous (but important) of scientific questions, posed by my wife Ruth one June morning on a Sapelo Island beach. Her question—"What is this?"—led to a review of these traces and how neoichnology can be used to predict previously unobserved behaviors. Enough with the preamble, you say, What do these traces look like? Very simply, these are shallow, less than 1 cm (0.4 in) deep impressions that define, in beautiful detail, the outlines of a ghost crab carapace, its eight walking legs, and both chelipeds (Figure 6.12e). In these traces, the walking legs were folded slightly under the body, and the chelipeds were simply placed on the sand surface. Knowing the chelipeds are on the anterior part of the body, these traces show exactly which way the crab was facing when it sat down on the moist sand surface. The central parts of such traces also have vague, fuzzy outlines to the proximal parts of the legs, almost as if some sort of bubbling happened there. The oddest aspect of these resting traces, though, was their possession of tracks only leading away from them, not to them. How did the crabs get there, then? Did they walk in one way, and then step exactly in those same tracks on their way out? Were they out surfing and got rafted in by waves? Or did they fall in from above, executing perfect landings from incredible leaps, or having been accidentally dropped by predatory birds?

The hypothesis goes like this, and is helped considerably by the additional knowledge that a full moon was out the previous night, corresponding with a spring tide. This higher-than-normal tide on a summer night meant that some ghost crabs had to abandon previously dug burrows—located in the berm—and take refuge higher up the beach, in the dunes. Meanwhile, waves crashed onto their meticulously crafted burrows and sand piles, erasing their homes and markers, respectively. While in the dunes, they made the first of two types of resting traces, which produced rougher outlines of their bodies. In these, their upper surfaces merged visually with the sand and thus neatly deterred predation while outside of their burrows. But because of their location in the dunes, they were also farther away from their normally saturated sediments, so they grew thirsty. Thus, as soon as the high tide receded, they rushed to the surf zone, sat down, and allowed gentle waves to wash over their bodies, enabling the pause that refreshes and rehydrates, making the second of two types of resting traces. The tracks leading to this resting trace were washed away by the waves as they tumbled around the crab. The less distinct outlines around the center of the trace are where slight gaps between the legs and the carapace allowed water to enter the gills; hence these traces were made by the exchange of water and air into and out of the crab. Once the crab's gills were properly wetted, it lifted itself off the surface and walked away, ready to dig more burrows, warn more rivals, find more mates, eat more food, and otherwise conduct crab business. As a result, to just say such traces simply indicate resting is a grossly oversimplified diagnosis of what is actually a wide range of simultaneous behaviors (Martin 2006a).

Of course, hypotheses must be testable, and this one was no exception. About one month later, on the same beach early in the morning, I saw a ghost crab run from the dunes down to the surf zone, crouch down, and become immobile. Amazingly, I was able to walk up to the crab and take close-up pictures as it made the same traces seen the month before. As many amateur and professional nature photographers on the Georgia coast know, ghost crabs are notoriously skittish, running or disappearing in a burrow with the slightest glance or twitch. So the key to their intimate portraiture is keeping in mind that an exhausted, dehydrated ghost crab will have no choice but to pose for a picture. Moreover, the crab will also leave a gorgeous outline of itself, the trace an added bonus.

Ghost crab traces also can reveal something about their dual ecological roles as both scavengers and predators. The previously mentioned pilfering of a dead animal, indicated by a drag mark and tracks, is certainly an example of scavenging revealed by traces. Sometimes the choice of where to dig a burrow can also indicate food choices. For instance, I have seen crabs set up shop by excavating a burrow directly under a fish, limulid, or other relatively large, stinking carcass. As a result, such burrows and nearby sand piles may also have vulture tracks on top of them (Chapters 8, 9). On the other hand, ghost crabs also leave traces divulging their predatory leanings. During some low tides in the early morning, you might see roughly chiseled and shallow depressions in the wet, freshly exposed intertidal sand, accompanied by finely broken bivalves (such as coquina clams, *Donax variabilis*). Directly associated with these pits are ghost crab trackways (Figure 6.12f). Nighttime, a flashlight, and a full moon help to directly observe how these are made. The crabs walk to the newly emergent sand, dig a few centimeters into the surface with their claws, and pull up scrumptious little clams, which they break and put up to their mouths for slurping soft parts. Ghost crabs also make surface marks and pellets through deposit feeding, in which they scrape up sand with their claws, eat algae in that sand, and leave behind feeding pellets (Robertson and Pfeiffer 1982).

As might be imagined, given the winning combinations of sheer abundance, deeply dug shafts, and occurrence in environments with rapid burial rates, ghost crab burrows have excellent fossilization potential. Indeed, trace fossils similar in size and form to these burrows are assigned to the ichnogenus *Psilonichnus,* and are in rocks from the Jurassic to the Pleistocene, although not all of these are attributed specifically to ghost crabs (Chapter 10). Nonetheless, because modern ghost crabs occupy such narrow ecotones, their trace fossils are valuable as general indicators of shoreline proximity. Furthermore, closer looks at the frequency and distribution of differently sized fossil burrows also may point to relative positions within that narrow zone, such as the lower part of the beach (closer to the intertidal zone = smaller burrows) or dunes and back-dune meadows (farther from the intertidal zone = larger burrows). Unfortunately, no ocypodid trace fossils other than burrows—such as trackways, sand piles, or resting traces—have yet been documented, but their abundance in modern environments bodes well for at least a few of these having made it into the fossil record.

Ghost Shrimp

Further downslope from the ghost crab traces and in the intertidal zone are the traces of callianassid shrimp. These burrowers, otherwise known as ghost shrimp and belonging to two species—the Carolina ghost shrimp (*Callichirus major*) and the Georgia ghost shrimp (*Biffarius biformis*)—are the most impressive of subsurface burrowers on the Georgia coast. Indeed, their burrows are iconic for many geologists, matching the sizes and forms of trace fossils originally made in shallow-marine environments worldwide (Goldring et al. 2007; Chapter 10). Callianassid burrows are remarkably abundant on intertidal sand flats, sometimes numbering more than 400/m^2 and are identified as broad, low-lying cones of sand, 5–10 cm (2–4 in) wide, with a central, millimeters-wide hole, looking much like miniature shield volcanoes (Figure 6.13a,b). Watch them closely, and occupied burrows will also erupt, except instead of lava, out will pour water filled with small, cylindrical, and dark brown fecal pellets that could easily be mistaken for chocolate sprinkles. (Please do not substitute with these in any cupcake recipes, though). Such activity shows a callianassid shrimp is occupying the burrow, and the shrimp is just below its top, flushing out its home. The small cone at the surface, formed by this irrigation of fine quartz sand and fecal pellets, is just the start of a long and complex burrow system. This narrow opening descends downward as an open vertical tube for about 15–20 cm (6–8 in), until it opens into a 2–3 cm (about an inch) wide open shaft, where the shrimp is located (Figure 6.13c). Think of the narrow tube as a drainage pipe leading from the actual home (burrow) of the ghost shrimp,

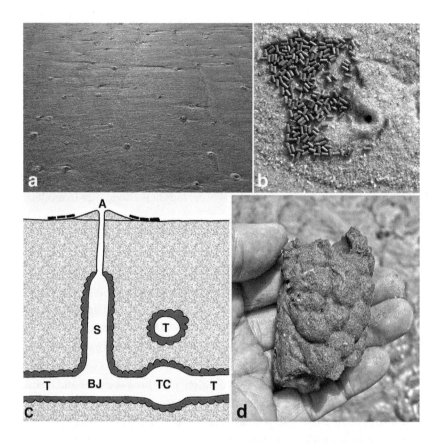

Figure 6.13. Ghost shrimp traces. (a) Intertidal beach surface, dotted by many ghost shrimp burrow mounds: Sapelo Island. (b) The surface expression of an active burrow, accompanied by mucus-packaged fecal pellets: St. Catherines Island. Pellets are about 3–5 mm (0.1–0.2 in) long. (c) Longitudinal and cross-sectional views of typical ghost shrimp burrow systems, showing top sand pile with fecal pellets, constricted burrow aperture (A), main burrow shaft (S), expanded burrow junction (BJ) leading into tunnels (T), turnaround chamber (TC), and knobby (pelletal) exterior on shafts and tunnels. (d) Close-up of the pellet-reinforced wall of a ghost shrimp burrow: Sapelo Island.

which it makes by shooting a thin stream of water upward from the top of the burrow proper. This part of the burrow is much like the neck of a typical wine bottle, such as that used for chardonnay or pinot noir.

A callianassid shrimp burrow is a masterpiece of construction, composed of smooth, mud-lined walls on the interior, and robust, rounded balls of sand and mud pasted onto the exterior by shrimp mucus (Figure 6.13d). These balls act like bricks in reinforcing the burrow so that it resists erosion by waves and tides; under normal conditions, burrows remain stable while sand shifts all around them. As can be imagined, these walls also keep out most predaceous infaunal invertebrates. Furthermore, these marvelous burrows are not just simple vertical pipes, but reach down to the subsurface, joining with horizontal tunnels that branch into complicated three-dimensional networks, a jungle gym of burrows that spread throughout cubic meters of sand and mud. How far down are these networks? Astonishingly deep: at least 2–3 m (6.6–10 ft) in most, with depths as much as 5–6 m (16–20 ft) recorded from modern burrows and their trace fossil equivalents. In such massive complex burrow systems, ghost shrimp can do whatever they wish in their secret societies, far below our surface world. However, small glimpses of these catacombs are revealed by beach erosion when previously made, perhaps decades-old burrows—evident as knobby-walled and stubby tubes—poke out of the intertidal sands. In these, horizontal branches and burrow junctions are sometimes expressed, giving a sample of burrow complexity. Similar to crayfish burrows (Chapter 5), burrow junctions are expanded relative to tunnel widths, reflecting additional space needed for a stiff-bodied shrimp to change direction and go down another tunnel.

How do ghost shrimp make these burrows? Mostly through the power of their digging claws, which are the first two appendages on the shrimp. They scrape, sculpt, and compact sediment as they move through it: no backfilling structures or other types of meniscae are formed through active filling by the tracemaker. If older burrows are filled by sand or mud, these are likely abandoned burrows that were passively filled from above. Callianassid shrimp are also sediment processors, ingesting sand and mud for organics as they excavate.

Considering the extensive amount of subsurface real estate covered by callianassid burrow systems, few should be surprised to discover these shrimp share their homes with a few other animals. Probably the best known of such squatters is a small crab, *Pinnixa cristata*, which has an elongated body that better allows its movement in tunnels and shafts (Bishop and Bishop 1992). Interestingly, though, these crabs prefer to stay at the top of the burrow systems near the thinner chimney, which may be related to availability of light, and hence algal food (Manning and Felder 1989).

A related species, *P. chaetopterana*, does not just limit itself to callianassid burrows, as it also is a frequent cohabitant of polychaete burrows (McDermott 2005).

As discussed before, the linkage between these modern callianassid burrows and their trace fossil analogues was made quite a while ago, and on the Georgia coast (Hoyt et al. 1964; Hoyt and Hails 1967). The thick knobby walls and depth of burial of the original traces made these burrows predisposed for both fossilization and easy visual identification. Their trace fossils, which include the ichnogenera *Ophiomorpha* and *Thalassinoides*, have been found in Pleistocene sediments on the Georgia coast and in sand ridges of the Georgia coastal plain. Surprisingly, fecal pellets attributed to callianassid tracemakers (or animals similar to these) have been interpreted from the fossil record, and were given the ichnogenus name *Favreina* (Kennedy et al. 1969). Because these animals only live in shallow subtidal and intertidal environments, their burrows are sure signs of a former coastline (Chapter 10). Moreover, trace fossils attributed to callianassid shrimp or closely related decapods have been interpreted in rocks from over the past 250 million years or so, from the Permian Period to the Pleistocene Epoch (Chapter 10).

Hermit Crabs

Hermit crabs are among the most productive tracemakers on the surfaces of intertidal and subtidal environments on the Georgia barrier islands, yet these decapods deserve better recognition and understanding of their tracemaking. First of all, one small misnomer needs to be revealed about hermit crabs: biologists do not classify them as true crabs (Brachyura). Brachyurans are decapods that include all of the ocypodids (fiddler and ghost crabs), fully marine crabs (nearly all of those included on seafood menus), and most terrestrial crabs. On the other hand, hermit crabs are members of the group Anomura, decapods that share a common ancestor with brachyurans.

Hermit crabs leave clearly visible trackways on emerged and submerged sand flats, as well as more subtle patterns on formerly occupied gastropod shells, the latter caused by repeated dragging across silicate sands (Figure 6.14). Their use of gastropod shells and relative mobility means that the continual recycling and redistribution of these shells in marginal-marine environments constitute a trace too. Long-dead gastropod shells, even fos-

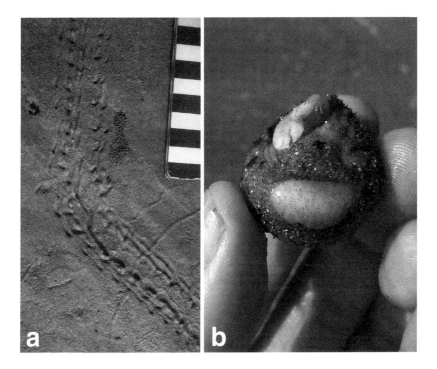

Figure 6.14. Hermit crab traces. (a) Trackway in an intertidal sand flat. Note the central drag mark made by its borrowed gastropod shell: Sapelo Island. Scale in centimeters. (b) The worn spot on a gastropod shell denoting the former presence of a hermit crab in that shell, which is still there: St. Catherines Island.

sil ones exhumed along the Georgia coast, thus take on many lives under the guidance of their secondary occupiers. Appropriately then, paleontologists who study hermit crabs are both taphonomists and ichnologists, easily spanning both realms (Frey 1987; Walker 1989, 1992).

Most of the hermit crabs on the Georgia coast belong to the genus *Pagarus*, such as the long-wristed hermit (*P. longicarpus*), hairy hermit (*P. annulipes*), and flat-clawed hermit (*P. pollicaris*), although the striped hermit (*Clibanarius vittatus*) is also common. The long-wristed hermit is normally seen near salt marshes, as it prefers marsh periwinkle shells, handily provided for these hermits once the periwinkles die. Other hermit crab species will likewise gravitate toward (and are adapted to) habitats in which gastropod shells are plentiful and suitable for their body sizes and shapes.

Hermit crabs normally acquire gastropod shells without having to kill, and much like vultures, become attracted to sick and dying gastropods, then wait for them to unloose their mortal coils. Once a shell is empty, though, all forms of passivity cease, and hermit crabs fiercely compete for these, especially shells that fit the given body size of the crab. Too big of a shell means too much energy will be needed to lug it about, even when aided by underwater buoyancy. Alternatively, too small of a shell causes body parts to stick out of the shell, daring predators to snatch them from their ill-fitting homes. Hermit crabs, like all animals, also grow during their lifetimes, creating a constant need to discard a too-small shell for one that fits better. But just like used clothes, a hand-me-down shell soon becomes another hermit crab's treasure, and wherever shells and hermit crabs coexist, a dead shell soon becomes mobile and a tool that affects the crab's tracemaking.

Sadly, only one fossil trackway attributed to a hermit crab, assigned the ichnogenus *Coenobichnus*, has been verified from the geologic record (Walker at al. 2003). Hermit crab wear marks on fossil gastropod shells, however, are numerous, with examples dating back to the start of the Mesozoic (Chapter 10). Furthermore, other trace fossils have revealed that at least a few arthropods were acting much like hermit crabs by carrying shells on their backs more than 500 million years ago (Hagadorn and Seilacher 2009). Perhaps the most spectacular evidence of hermit crab behavior from the fossil record, however, comes from conjoined body fossils, in which an Early Cretaceous fossilized hermit crab was found inside an ammonite shell (Fraaije 2003), but one can hope their trace fossils—which should be abundant—will show up in more profusion as soon as geologists know what to look for.

Various Burrowing Crustaceans: Mud Shrimp, Mole Crabs, Mantis Shrimp, and More

A veritable potpourri of other burrowing crustaceans accompany the most well known of crustacean tracemakers on the Georgia coast. Because the quantity and diversity of these tracemakers are so high, we will only look at the most ichnologically significant ones, which include mole crabs, mantis shrimp, and mud shrimp. Other than their shared crustacean heritage, what they have in common is their burrowing abilities.

Mole crabs, such as the beach mole crab (*Albunea paretii*) common mole crab (*Emerita talpoida*), and square-eyed mole crab (*Lepidopa websteri*),

Figure 6.15. Mole crab traces. (a) Frontal view of a mole crab (*Lepidopa websteri*), burrowing with its rear legs and abdomen down: Sapelo Island. (b) Trackway of the same mole crab connecting an abandoned burrow to an active one, which is occupied by the crab. Scale = 5 cm (2 in).

are aptly labeled because of their marvelous burrowing abilities. For many years, I had read about these tracemakers, seen photographs of them, and lived vicariously through whispered tales of mole crab antics from my ichnological colleagues. Tragically, I had never seen a live one, though, let alone its burrowing. So it was a most fortuitous day in the summer of 2004 when I finally experienced the thrilling sight of a common mole crab ensconced in a newly dug burrow at low tide on a Sapelo Island sand flat. Curious about its fabled burrowing acumen, and wanting to be impressed (with a name like "mole crab," it had better be good), I put it to the test by digging it out and setting it down on the sand surface. It immediately positioned itself so that its abdomen pointed down, and its rear legs began cycling toward the front of its body (Figure 6.15a). The already-wet sand around it suddenly became more liquefied, and within seconds the mole crab slid into its self-made quicksand and neatly disappeared. Ever the scientist, I repeated the experiment by digging out the crab and placing it on sandy surface again, and it quickly buried itself once more. (Yes, then I stopped right there, not wanting to exhaust my test subject just to get a statistically significant sample.) As is typical in ichnology, once the search image for their burrows was established, I started seeing mole crabs and their burrows throughout the sand flat the rest of that day. In one instance, I even found where a crab abandoned one burrow to make another, leaving a tiny, rarely seen trackway between its two refuges (Figure 6.15b).

The most common mantis shrimp on the Georgia coast is *Squilla empusa*, a crustacean that is well known for its impressive burrow systems.

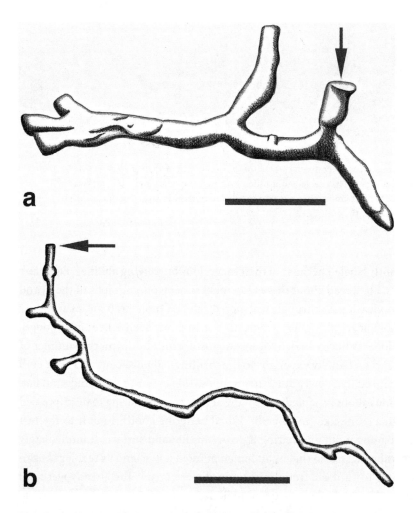

Figure 6.16. Burrows of mantis shrimp (*Squilla empusa*) and mud shrimp (*Upogebia affinis*). (a) Mantis shrimp burrow system, with arrow indicating burrow entrance. Scale = 15 cm (6 in). (b) Mud shrimp burrow, again with arrow indicating burrow entrance. Scale = 50 cm (20 in). Both illustrations based on fiberglass casts figured by Frey and Howard (1969).

Mantis shrimps are not true shrimp, and in fact are not even decapods. They also are not related to praying mantises, but are nominally aligned with these insects because of their scary-looking, saw-toothed chelipeds. This resemblance is not a coincidence, but a result of convergent evolution, where natural selection culminated in appendages well suited for grasping and holding prey securely in place while eating it alive. Nonetheless, because mantis shrimp are also burrowers, their chelipeds aid in burrowing. Their burrows are often in the same places as those of mud shrimp, dug into the banks of muddy tidal creeks in the low marsh, but differ considerably in overall size and form. Mantis shrimp burrows are 2–4 cm (0.8–1.6 in) wide and can be as much as a meter (3.3 ft) long, are mostly horizontal, but have branching, obliquely oriented shafts attached to the main tunnel (Figure 6.16a). These branches add broad U-shaped parts to the overall burrow system. Burrows function as homes for mantis shrimp, as well as places for ambush predation and escaping predation.

The coastal mud shrimp (*Upogebia affinis*) is a common excavator and occupier of burrows in tidal channels and creeks of low salt marshes on the Georgia coast. Upogebiid shrimp are well known worldwide for their burrows; for example, I have seen many modern burrows made by a species of upogebiid shrimp in lagoons of the Bahamas. These distinctive traces, composed of two intersecting U-shaped burrows fortified by thick, pelleted walls, are matched by fossil examples in nearby Pleistocene outcrops. Because modern upogebiid shrimp are restricted to such environments, their fossil burrows provide paleoecological snapshots of Bahamian lagoons of the recent past (Curran and Martin 2003). Similarly, the mud shrimp of Georgia only burrows in muddy marsh environments, so trace fossils matching the forms of modern mud shrimp burrows in older Holocene or Pleistocene sediments would define the former location of a low marsh (Chapter 9). Often the burrow entrances of *U. affinis* accompany the open holes made by other burrowing species on exposed tidal-creek banks seen at low tide. Mud shrimp burrows, however, reveal how very different these are from those of their neighbors. Casts reveal their burrows as 1–2 m (3.3–6.6 ft) long, but thin, at only 1–2 cm (0.4–0.8 cm) wide, and with small protuberances and branches added along the main burrow axis (Figure 6.16b; Frey and Howard 1969). Their burrows, which are often dug into the sides of tidal-creek banks, are mostly horizontal, but can descend nearly a meter over the course of the tunnel.

Durophagous Decapods: Shell Crushers and Peelers

A few species of decapods are tracemakers of a different kind than all of the preceding, in that they leave large, easily noticeable fractures or scars on the edges on molluscan shells. These traces are evidence of durophagy (*duros*, "hard," and *phagos*, "eating"), which is the predatory consumption of hard-bodied animals, whether these animals are shelled or otherwise possess mineralized tissues protecting soft parts. The previously described actions of naticid and muricid gastropods, as well as those of octopi, are excellent examples of durophagy, albeit surgically precise compared to feats of shell or bone crushing performed by other predators. In this respect, we might normally think of sharks, hyenas, alligators, or other fearsome animals with enormous bite forces, cracking through hard parts with apparent ease. On the other hand, we probably neglect thinking of crabs. Yet these invertebrates, representing a fine lineage that has been around for the last 150 million years or so, include some formidable shell crushers.

Durophagous decapods that live in the shallow waters of the Georgia coast include blue crabs (*Callinectes sapidus*) and stone crabs (*Menippe mercenaria*). These crabs are quite adept at shell destruction, using their powerful claws (chelipeds) to break down protective barriers of potential food animals. Gastropods are frequent targets of crab predation, and the hunters become the hunted when crabs attack predatory gastropods, effectively defining another strand in the food web of Georgia subtidal environments. Whelks in particular are favorite morsels for crabs; these gastropods, as mentioned earlier, leave their own marks on the edges of bivalve shells from predation, although these may be easy to miss. Crab traces, on the other hand, have no subtlety whatsoever. Their traces differ from those of whelks like clues left by a burglar who uses a pry bar to open a door, versus one who rips off half of the door. A predatory crab that encounters a savory whelk will first use both claws to latch onto the lip of the gastropod's outermost whorl. One claw acts as an anchor, holding the prey close to its body, while the other claw pulls so that the whelk's shell begins to rip. This tearing of the shell leaves a ragged edge that, if continued, exposes the innards of the gastropod along a broad front, allowing for further attack, consumption, and (of course) death.

Nonetheless, just as with naticid gastropods, not all crab predation is successful, as some crabs choose a larger or stronger gastropod. In such instances, a firmly grasped lip will fracture and peel away, and the gastropod escapes and lives to see another day, while the crab is left holding a useless piece of shell. The recuperative powers of these assaulted gastropods then leave their own traces of healing (Walker and Brett 2002). The mantle of the wounded whelk secretes new shell material outward from the damaged lip

MARGINAL-MARINE INVERTEBRATES 317

Figure 6.17. Crab predation traces (arrows) recorded as scars in the shell of a knobbed whelk (*Busycon carica*), from which new shell material grew: Cumberland Island. This was a very lucky whelk, having escaped at least two attacks.

after these sublethal encounters, showing a slightly thinner shell projecting from an irregular edge (Figure 6.17). Some doubly lucky whelks will bear two scars and signs of repair, each representing a successful escape, and subsequent recovery from two crab attacks, which may have been separated by months or years.

These traces made by crabs remind us how a single gastropod shell may tell a history of life, death, and life after its death. For example, I have seen shells that may bear one or two healed scars from crab attacks, but also have a telltale drill hole of a naticid gastropod or an octopus near its apex. In these examples, the gastropod's luck ran out, but its misfortune also provided its predator with sustenance and a home later for multiple hermit crabs. If a living hermit crab is not currently occupying the shell, its former presence is easily interpreted from a smooth, polished area near the base of the shell. On the Georgia coast, rubbing against quartz and other silica-bearing minerals in the beach sands causes this shiny spot, in which the softer, calcium carbonate shell is worn down by repeated dragging of the shell by the hermit crab. A closer look at the shell may even reveal two overlapping spots, indicating at least two successive hermit crab occupants. Bryozoan colonies, barnacles, juvenile oysters, or other encrusting animals may have partially covered one of the wear marks, only to be worn down again in the same place. In other words, every dead bivalve or gastropod shell on a Georgia beach becomes not just a common name or species name identifying its original owner, but a rich story embellished by ichnological evidence.

ECHINODERMS: DANCING WITH THE (SEA) STARS AND SAND DOLLARS

Echinoderms are exclusively marine invertebrates characterized by their fivefold (pentameral) symmetry, calcareous endoskeletons, and spiny skin (*echinos,* "spiny," and *dermos,* "skin"). Echinoderms include so-called sea lilies (crinoids), sand dollars and sea urchins (echinoids), sea cucumbers (holothurians), sea stars (asteroideans), and brittle stars (ophiuroids), as well as a host of extinct groups (cystoids, blastoids, and edrioasteroids) that enjoyed their heyday during the Paleozoic Era. Many echinoderms are terrific tracemakers, particularly those that burrow, and the long geologic ranges of some echinoderm groups mean their trace fossils are common and easily identifiable. Unfortunately, the Georgia coast has only a small number of echinoderm species compared to more tropical areas, yet some burrowers are worth noting for their distinctive traces and effects on sediments.

Most echinoderms are stenohaline organisms, living within a narrow range of salinity, and they function and proliferate best under full marine salinity (35 ppt). Because environments associated with the Georgia coast are more marginal marine, with salinities dipping to well below 35 ppt, the number and variety of echinoderms—and accordingly their traces—are somewhat limited, although a few species are adapted to brackish water (Turner and Meyer 1980). Nonetheless, several of the most commonly encountered echinoderm tracemakers on the Georgia coast are the margined (or royal) sea star (*Astropecten articulatus*); the lined sea star (*Luidia clathrata*); the blood brittle star (*Hemipholis elongata*) and burrowing brittle star (*Amphipholis gracillima*); the keyhole sand dollar (*Mellita isometra*); the mud heart urchin (*Moira atropos*); and a few species of sea cucumbers (*Thyone briareus*, *Synapta inhaerens*).

Sea stars are fierce carnivores in sandy subtidal environments, only surpassed by predatory gastropods for their voraciousness. For example, gut contents from the margined sea star have revealed the remains of more than 90 invertebrate species, most of which were infaunal molluscans (Wells et al. 1961). Unlike some other sea stars, though, species of *Astropecten* do not have suckers on their podia. This means they cannot pull apart bivalves, and instead must swallow an entire animal to digest it. Because their prey must be small enough to fit through their mouths, they tend to eat mostly juvenile infaunal bivalves and gastropods. In contrast, banded sea stars attack other echinoderms, such as smaller individuals of *Astropecten* or brittle stars, although they likewise lack suckers and have to consume whole animals. Both banded and margined sea stars make shallow burrows while looking for food, coordinating the movement of hundreds of tube feet on their ventral surfaces for moving sand grains out of their way, as well as pushing and pulling their arms. A study of margined sea star hunting methods suggests they probably use chemoreception to detect prey, and once they find a molluscan mother lode, they concentrate their foraging activities, thus making more traces in a smaller area (Beddington and Mclintock 1993). Interestingly, their seeking of prey seems to follow a diurnal pattern, so a concentration of their surface trails and shallow burrows probably indicates feeding that happened during the

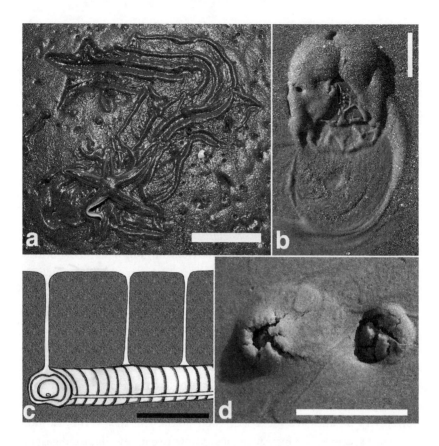

Figure 6.18. Echinoderm traces. (a) Lined sea star (*Luidia clathrata*) resting trace (*upper left*) and trail, ending with the sea star itself (*lower left*): Sapelo Island. Scale = 5 cm. (b) Trail and upraised sediment caused by shallowly burrowed keyhole sand dollar (*Mellita isometra*): Sapelo Island. Scale = 5 cm (2 in). (c) Longitudinal and cross-sectional views of traces made by the deep-burrowing mud heart urchin (*Moira atropos*), including meniscate (backfilled) burrow, its associated drain, and vertical shafts. Based on figure by Bromley and Asgaard (1975) of similar traces made by the heart urchin *Echinocardium cordatum*. Scale = 10 cm (4 in). (d) Top view of a sea cucumber (probably *Thyone briareus*) in its burrow, with either end forming the upward limbs of a U: Sapelo Island. Scale = 5 cm (2 in).

early morning or twilight. Nonetheless, the resulting predation traces of sea stars are difficult to categorize, especially if viewed in a vertical section of sediment, which might be identified only as a general mixing of sedimentary layers.

In contrast, sea star surface trails and resting traces are spectacular. These exquisite traces consist of shallow, star-shaped depressions connected directly to multiple-grooved trails, with each trail corresponding to a sea star arm (Figure 6.18a). Such traces are most often made by sea stars seeking pools of standing water, done in response to a low tide and their subsequent stranding on emergent sand flat and mudflat surfaces. For the resting traces, the sea stars simply wiggled their tube feet into the saturated sand below them, which allows for a shallow burial. If tried after much of the water has drained from the sand, though, a sea star may not get far, and then has to move laterally. Hence both the resting traces and trails may be classified as equilibrichnia, but in degrees related to environmental distress: stationary traces are first attempts to alleviate dehydration, whereas trails represent more desperate measures. To see such traces, I was once lucky enough to find areas on Sapelo Island where hundreds of banded sea stars had washed up onto the lower intertidal zone, affording an abundance of stunning star imprints and trails. (Yes, I tried to throw them back into the water, but there is only so much one ichnologist can do.) Trails also represent varied movements of the tube feet coordinated with the arms, in which the sea stars orient their limbs so that they lead with one or two, and the others fold together behind it as it glides along a sandy surface. These traces are best seen at low tide along a Georgia beach in areas with some mud content, which helps to preserve the finer nuances of their complicated movements.

Ophiuroids, including the aptly named burrowing brittle star, include some excellent burrowers. This may come as a surprise when you examine its not so robust anatomy, consisting of a central disk (its main body) and five spindly arms. Nonetheless, burrowing brittle stars bury themselves just below the sediment surfaces. They do this not by digging, but rather by sliding into saturated sands by pushing their thin arms into the sediment, then pulling the rest of their bodies deeper into the substrate. In this movement, the brittle star stabilizes itself with two arms below, and stretches three arms toward the sediment surface so they stick out, similar to an "I surrender" pose, but using an extra arm. Why does it do this? One explanation is that the three arms gather suspended matter in the water; its tube feet then relay this material to the mouth on the central disk (Howard and Frey 1973). While a brittle star moves downward, its burrow collapses behind it, but once at rest, the overall fivefold form of the tracemaker is evident. Ophiuroid traces are normally seen as subtle disturbances of sedimentary layers, although they will show up well in radiographs, especially if directly associated with a brittle star (Howard 1968). Once well protected by the sand, burrowing brittle stars also deposit feed, ingesting whatever organic material might be in the surrounding sediment (Gielazyn et al. 1999). With this behavior, they may move more laterally, making their deposit-feeding traces more difficult to identify than suspension-feeding ones.

Other brittle stars, such as *Hemipholis elongata, Ophiophragmus filograneus,* and *Microphiopholis gracillima,* are likewise burrowers and deposit feeders in Georgia subtidal sediments. In one study of *O. filograneus* and its tolerance of salinity differences, its burrowing activity was used to reflect how well it was responding to more or less salt in its water (Turner and Meyer 1980). In another study, the burrowing speed of *H. elongata* was measured, although "speed" is probably not the best term for describing its moving 0.05–0.08 cm/h (0.02–0.03 in./h) through sediment (Gingras et al. 2008b). In other words, to say this is a snail's pace is an insult to gastropods. Just as in terrestrial environments, though, open and abandoned burrows of other animals make inviting places for squatters (secondary occupiers), and some brittle stars succumb to such temptations. For example, the blood brittle star can be found in the burrow of the predaceous polychaete worm *Diopatra cuprea,* although it has enough instinct to hang out in the lower part of the burrow (Ruppert and Fox 1988).

Echinoids are echinoderms prized by many beachcombers, as these include sand dollars and urchins. Normally, though, sand dollars are found as inert (i.e., dead) specimens, bleached white and lacking their tube feet. Similarly, sea urchins are most often represented by denuded, hollowed, and broken pieces. So do you want to see live echinoids? Then ichnology will be your friend, as the keyhole sand dollar (*Mellita isometra*) is easily discovered in intertidal sands by recognizing its shallow burrows and attached trails. While keeping in mind the fivefold symmetry of echinoderms in general, look for a symmetrical expression of five oblong holes, radiating from a central point, on an otherwise smooth sand flat surface at low tide. These holes are sometimes surrounded by a roundish outline of upraised sand, about 10–12 cm (4–5 in) wide; in some examples, a smooth trail of about the same width will connect directly with this bump (Figure 6.18b). Carefully scoop from underneath these traces, and in this handful of sand you will find a beautiful, purple, and living keyhole sand dollar. Turn it over, take a look at its flat surface, and watch carefully: hundreds of short hairs on that surface will start moving in waves, looking much like closely cropped grass moved by a breeze. These are the tube feet, which, despite their brevity, help a sand dollar to burrow. Now set it flat side down on a moist sandy area, and watch it burrow. The tube feet underneath start levering in unison, moving the body either straightforward or in a slight curve, with one edge pushing into the sand. Minutes later, depending on the vigor of the sand dollar and the dampness of the sediment, its body will be mostly concealed by a thin layer of sand. Come back in an hour or so, and the sand dollar will be completely buried, and sand will have collapsed around the holes in the central part of its body, hence revealing why it has the common-name descriptor "keyhole." This splendid (and perhaps unexpected) burrowing aptitude of these echinoids is partially related to their tube feet, but also because of their streamlined, blade-like profiles, in which their bodies slice effortlessly into wet sands.

A few sand dollars, like some gastropods, also bear scars of former predation. These survivors have body outlines that look malformed compared to their normal circularity, but tissues aggravated by an attack heal over so that the originally ragged, broken edges become smooth again. Most such assaults are attributed to bony fish, which sometimes take a bite out of a sand dollar that is enough to satisfy their appetites without killing their prey (Nebelsick 1999). These echinoids, like plant leaves preserved with wound responses after insect attacks (Chapter 5), are fine examples of two-for-one traces, especially if found in the fossil record, where they provide valuable information about predator–prey relations (Walker and Brett 2002).

The mud heart urchin (*Moira atropos*) is also a burrower, but lives in more offshore environments. In all of my visits to the Georgia coast, I have never seen a live one, but their burrowing habits are well documented (Howard 1969; Hertweck 1972; Bromley and Asgaard 1975; Ruppert and Fox 1988). How does such an ungainly, oblong, and inflexible animal manage to burrow (and burrow well), but without the benefit of legs? It accomplishes this feat through a combination of mucus and moveable spines. However improbable it might seem, some of these spines provide the mucus, secreting a slimy envelope that enables an urchin to glide easily through the mud or sand; its body is then pushed forward through leverage by other spines. Aided by this mucoidal lining, the echinoid's organic and calcareous skeleton neatly avoids abrasion from any silica-rich sand mixed into the sediment. The mud heart urchin also uses mucus to reinforce a thin, periscope-like vertical shaft that extends about 15 cm (6 in) above and connects to the sediment surface (Figure 6.18c). This shaft serves a dual purpose: bringing in more oxygen-rich water from above into the burrow (thus acting more like a snorkel than a periscope), and acting as a conduit for tasty organic material that might be on the sediment surface. To get this material, the urchin has a large number (as many as 62) of long and flexible tube feet that stretch up to the surface and pull this food down to the mouth (Bromley and Asgaard 1975). These tube feet also made the shaft in the first place, and they poke more such shafts through the sediment as the urchin moves forward. Other tube feet on the urchin help to gather food from the sediment in front of the burrow, move sediment behind the urchin, and drain excess water from the burrow below it. Mud heart urchins can burrow quickly, considering they are only using a combination of tube feet and mucous to move through sediment; in one study, they were clocked at about 2 cm/h (0.8 in./h) (Gingras et al. 2008a).

The progressive, subsurface movement of the urchin imparts a complicated and compound trace, consisting of a main tunnel bearing a backfilled (meniscate) fill; a drain structure on the bottom of the tunnel; and a series of vertical shafts extending above it. In cross section, these burrows may even show the heart shape of the tracemaker (a Valentine's heart, that is), cinching its identity. In Georgia, mud heart urchins are responsible for more than half of the bioturbation of offshore sediments 7.5–15 cm (3–6 in) below the sediment surface, meaning these animals are important sediment mixers in such environments (Howard 1969). The upper parts of their burrows, though, are normally eroded and reworked by currents and waves. This means the vertical shafts may not be preserved, although the main, backfilled tunnels should be present.

Sea cucumbers, or holothurians, are echinoderms that are probably not well known for their tracemaking abilities, yet two species on the Georgia coast, *Synapta inhaerens* and *Thyone briareus*, live in burrows in shallow subtidal and intertidal sand flats there. Both sea cucumbers form U-shaped burrows by first pushing their heads into the sand, using by a series of peristaltic contractions and expansions to work their way down, and then move up to the sediment surface again. What is different about these movements compared to most other burrowing invertebrates is how these animals extend the bottom half of their long bodies downward, which is then followed by the top half. Through this series of actions, they eventually bend downward in the middle of the body so that the head and anus are near the surface, thus making a U shape with a bulge in the center and thinner widths at each end of the U. The burrow form then reflects this shape, and if viewed from one end, the cross-sectional diameter of a sea cucumber alternates between triangular and circular (Howard 1968; Bromley 1996). Once buried, the sea cucumber mouth pokes out one end of the U-shaped burrow, where it projects a beautiful, arbor-like net to pull in water for suspension feeding; meanwhile, the other end pumps out processed sediment as feces. If emergent, both ends will poke out of the sediment as a closely spaced pair (Figure 6.18d). Once made, these burrows can become as mobile as the sea cucumber. For example, if sediment is dumped on top of a feeding sea cucumber, whether from a big wave or during a storm, the animal pushes its body upward to maintain contact with the sediment surface, which also causes its burrow to move up. The structure formed underneath the final position of the burrow—spreite, mentioned earlier with cnidarian burrows—consists of a succession of concentric lines, with each line representing the former outer boundary of the burrow. Such traces thus provide a snapshot of sea cucumber behavior in reaction to a rapid change in their original environments.

On the basis of the preceding modern analogues, many trace fossils are attributed to echinoderms. Some of these are strikingly obvious, such as sea star resting traces with gorgeous fivefold symmetries impressed on bedding-plane surfaces (*Asteriacites*), seemingly proclaiming their stelleroid origin with pride. Subtler traces would be vertical burrows made by ophiuroids, which might be distinguished wherever the disturbances made by three arms up and two arms down are preserved. Ophiuroids, however, are also credited for making resting traces similar to those of their stelleroid cousins, with some such trace fossils dating back more than 400 million years ago to the Ordovician Period (Mikuláš 1990) or a mere 300 million years ago to the Carboniferous Period (Mángano et al. 1999). Even the walking traces of ophiuroids, showing where they moved themselves with their arms along a seafloor surface, have been interpreted from Cretaceous rocks (Bell 2004). On the other hand, the U-shaped burrows made by holothurians are potentially confused with those made by polychaete or hemichordate worms, as explained soon. As a result, care must be taken when attributing such trace fossils to sea cucumbers, although burrow width might be a good way to distinguish these. (Let's just say that few worms can get as fat as a sea cucumber.) Regardless of their tracemakers, though, most U-shaped fossil burrows belong to the ichnogenera *Arenicolites* or *Diplocraterion*, with the latter possessing spreiten, and thus indicating upward or downward adjustments of the burrow. Wide, nondescript backfilled burrows, which may or may not have a siphonal trace associated with them, are also associated with heart urchins. Some of these ichnogenera include *Scolicia* and *Bichordites*, among others (Plaziat and Mahmoudi 1988; Donovan et al. 2005; Seilacher 2007). In short, echinoderm trace fossils are well represented in the fossil record, and traces made by modern echinoderms of the Georgia coast provide ready models for comparison to these.

HEMICHORDATES: OUR WORM-LIKE RELATIVES

After a long, hot day in the field, a neoichnologist might be walking along a Georgia sand flat at low tide and notice a large pile of wet, sandy sediment that, through some sort of delirium, takes on the subliminal appearance of soft-serve ice cream (Figure 6.19a,b). This trace is from an acorn worm, most likely the golden acorn worm (*Balanoglossus aurantiactus*). Acorn worms are hemichordates, a group that shares a common ancestor with vertebrates, so as worms, they are actually more closely related to humans than polychaetes.

Acorn worms are deposit feeders, and make U-shaped burrows that they use to more easily process large amounts of sediment. These burrows are about 5–8 mm (0.2–0.3 in) wide, supported by a thin layer of mucus, and are as much as 40 cm (16 in) deep (Figure 6.19c), but are probably deeper. One end of the U-shaped burrow at the surface has a funnel shape, near where the worm's mouth is located. In a sensible arrangement, its anus is positioned at the other end of the burrow, and is the source of the sediment piles, as these are fecal castings. (The ice-cream analogy probably just became less appealing with that newly added knowledge, especially if such piles are adorned with the chocolate sprinkles of callianassid shrimp feces.) The funnel at the mouth end forms because the acorn worm pushes its head into the surrounding sand to swallow it, prompting a partial collapse in

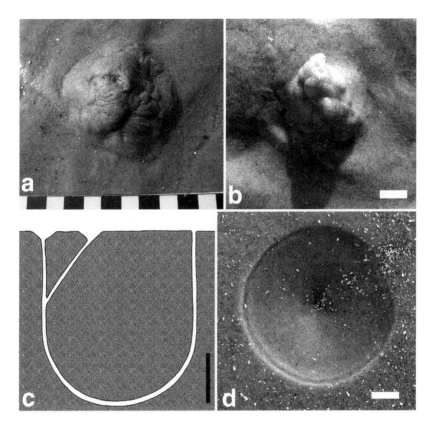

Figure 6.19. Acorn worm traces. (a) Fecal casting by golden acorn worm (*Balanoglossus aurantiactus*) on the sand surface: Sapelo Island. Scale in centimeters. (b) Fecal casting of the much smaller helical acorn worm (*Saccoglossus kowalevskii*): St. Simons Island. Scale = 1 cm (0.4 in). (c) Longitudinal view of a golden acorn worm burrow, showing overall U shape, multiple entrances, and funnel-like openings. Scale = 10 cm (4 in). (d) Funnel-like opening of a golden acorn worm burrow. St. Simons Island. Scale = 1 cm (0.4 in).

the area above the feeding (Figure 6.19d). This depression helps the worm gather more food; organic material then rains in from above and moves downslope to the waiting worm. An upward branch from the main burrow shaft is also made near the funnel, which helps with circulating water into the burrow. Golden acorn worm burrows hence have two entrances at the mouth end, which adds a small Y shape as an accent to the U shape of the overall burrow (Duncan 1987; Bromley 1996). At the other end of the burrow, fecal cast mounds are normally about 4–6 cm (1.5–2.4 in) wide and tall, but normally wider than tall; individual coils are about 5–10 mm (0.2–0.4 in) wide. Interestingly, these worms can become so thin, they can double back in their own burrows. But why would they do this? It turns out that acorn worms are always on the move, and abandon their burrows once an area is mined for its organics. Before leaving a burrow, however, acorn worms reverse position, place their anal ends at the former mouth ends, and leave a fecal deposit at what used to be the burrow entrance: the hemichordate equivalent of fouling the nest.

Similar burrows and fecal castings are made by another species of hemichordate, the helical acorn worm (*Saccoglossus kowalevskii*). The burrows of helical acorn worms, however, are noticeably smaller than those of golden acorn worms (Figure 6.19b). For example, fecal piles are only about 2 cm (0.8 in) wide, flatter, and composed of 1–2 mm wide coils (Ruppert and Fox 1988). As a result, a wandering beachcomber might be able to distinguish burrows between the two species on the basis of size alone. Furthermore, in my experience, the fecal castings of golden acorn worms are far more abundant on the Georgia coast. This perception, though, may be a result of visual bias, considering how their prominent profiles are seen more easily on an expansive sand flat.

In overall form, hemichordate burrows are comparable to a readily recognized trace fossil, *Arenicolites* (Chapter 9), although many other invertebrates, such as polychaete worms and sea cucumbers, commonly make U-shaped burrows too. However, a common problem with U-shaped burrows in the fossil record is that only one arm of the U might be seen in a given vertical exposure of rock, meaning that it can be mistaken as a simple vertical burrow. Furthermore, the funnel or Y-shaped part of the burrow associated with the mouth end of the tracemaker may be absent, having been eroded by surface currents. As a result, careful descriptions of any trace fossil of a U-shaped burrow should be compared to those of their modern equivalents and placed within a proper ecological context. More recently, the ichnogenus *Schaubcylindrichnus* has been linked to acorn worm tracemakers, based on a number of overlapping criteria, including (in one specimen) a sediment mound directly above a funnel-like structure at the top of a U-shaped burrow (Nara 2006). Such studies once again demonstrate the value of neoichnology in providing clues that can help narrow down potential tracemakers of enigmatic trace fossils.

MARGINAL-MARINE INVERTEBRATE TRACEMAKERS: THE ICHNOLOGICAL ICONS OF THE GEORGIA COAST

As mentioned earlier, the majority of neoichnological research done on the Georgia coast from the 1960s through the 1980s focused on marginal-marine invertebrates and their traces. Unfortunately, much of this work ceased after the death of Robert Frey, the retirement of James Howard, and the departure of George Pemberton from Georgia (Chapter 2). In recent years, this lapsed interest has resurged, with works I have done by myself or with colleagues (Gregory et al. 2004, 2006; Martin 2006a; Martin and Rindsberg 2007, 2011), as well as studies performed by former students of George Pemberton (Gingras et al. 2008a, 2008b).

Although marginal-marine invertebrate traces are not the be-all and end-all of neoichnology of the Georgia coast, for many people, these are the most important traces to study. This situation is partly owed to the sheer force of academic legacies, which plays more of a part in research choices than many people may realize, including the scientists doing the research. In such situations, researchers may find comfort in going back to a familiar area and scrutinizing further what has already been studied, hoping to find something new. Nonetheless, the popularity of Georgia marginal-marine traces also relates to the easy comparison of many of these traces to trace fossils in the geologic record. Some of the most iconic and well known of marine-related invertebrate trace fossils, such as *Ophiomorpha, Psilonichnus, Thalassinoides, Asteriacites, Arenicolites, Entobia, Oichnus,* and *Teredolites,* have direct counterparts on the Georgia coast (Chapter 10), making actualism more facile and transparent. Thus it is no wonder why ichnologists, geologists, and paleontologists keep coming back to Georgia. Of course, the extensive, beautiful, and undeveloped beaches of its islands may also play some small part in their enthusiasm.

7

TERRESTRIAL VERTEBRATES, PART I

Fish, Amphibians, Reptiles

THE ALLIGATORS THAT WENT TO THE BEACH

One July morning, on what promised to be another hot day, our merry group of ichnologists walked along Cabretta Beach on Sapelo Island at low tide and stared at the sand, a typical activity for those with our interests. After all, we were keenly interested in the burrows and tracks left by animals in the extensive sand flats of this beach; moreover, we could do it privately, as our footprints were the only human traces added to the assemblage. Thus we were understandably distracted enough to not notice that one member of our group—Jon Garbisch, the UGAMI education coordinator—had disappeared. This fact became apparent as soon as he popped up in front of us, his rematerialization accompanied by a grin—a sort of Cheshire cat in reverse. With this, he then announced happily, "Wait till you see what's around the corner!" No matter how much we pleaded, he would not divulge what could possibly interest all of us so much, especially with so many other distracting traces nearby. Our curiosities thus properly provoked, we quickened our pace at the northern end of the beach and turned sharply left toward a small tidal channel dividing the beach from a salt marsh. The warming air, softened by an occasional sea breeze, seemed to wane slightly as our minds turned inward and our collective vision outward.

After tracking a few times, an awareness of and respect for the ground is cultivated, one uncommon to most people, particularly those who live surrounded by pavement every day. One of the reasons for this heightened consciousness is the negative consequence of not looking ahead (and below): stepping on traces, which overprints or erases valuable information held by each vestige. Accordingly, as we slowed our approach to the tidal creek, our attentiveness grew in inverse proportion, making this particular trace all the more vivid and unforgettable.

There on the still-damp sandy bank was the perfect outline of an adult alligator (*Alligator mississippiensis*). It was a magnificent trace. Well over 2 m (6.6 ft) long, its maker had walked—not crawled—out of the creek and onto the bank, then lowered itself to soak up some of the late morning sun coming from the southeast. All four of its feet had pressed deeply into the sand, whereas the tail and lower part of the head made shallower outlines. Small, circular pockmarks followed the outline of its head, and we did not figure out their meaning until we started thinking ecologically: these were imprints left by drops of water that fell off the still-wet lower jaw of the alligator. A close examination of the indentations left by its torso and head revealed scale impressions, making the animal's appearance more startlingly reptilian and real in our imaginations. A rack line of smooth cordgrass and other debris further helped to outline the left side of its body. Because this part of the trace was upslope, we surmised that the tide was still up, yet shallow, when the alligator pulled itself up on the bank. This interpretation was supported by the lack of clear tracks and tail-drag marks leading up to the resting trace, although these were present going away from it. Those marks showed the rest of the story, where the large reptile pushed itself straight up from its prone position and turned to its right toward the tidal creek. Its thick tail left a wide arc and its feet pushed the moist sand to the left side, forming prominent pressure-release structures on the outer edges of its tracks. These structures, sharp ridges on an otherwise mostly flat sandy surface, simultaneously spoke of the alligator's abrupt right turn and pointed to the next set of tracks, faintly visible in the murky water of the creek. We presumed that buoyancy took over from there, and the alligator no longer walked but sculled away from the site of its resting trace.

We spent nearly an hour measuring, sketching, describing, and photographing the scene, most of it done in silence. We then discussed the lay of the land, the direction of the prevailing wind, the arc of the sun, and the tidal cycle that morning. Once all of this information had been gathered and synthesized, we conjured the preceding scenario, which made sense to all of us. We hypothesized that the alligator had been resting there immediately before we reached the end of the beach, perhaps as little as a half hour before we turned the corner. Jon, of course, had been there before us, so perhaps he had just missed seeing the alligator's armored bulk slide quietly into the tidal channel. From a tracking perspective, then, the alligator getting up and leaving was nearly coincident with our arrival. Had we disturbed the surrounding environment in a way that caused it to interrupt its solar therapy? If we had been more attentive to our surroundings and advanced more quietly, would we have caught its parting glance as it swam up the creek? We did not know for sure, but we all agreed that the elation of finding its traces and the knowledge conveyed by these far exceeded any short-lived glimpse of the actual animal in the water, an exceedingly dull and lifeless story that would have consisted of a single line: "We saw an alligator today." At the same time, the alligator's recently made outline also reminded us who could be the top predator of the Georgia coast at any given moment. Indeed, photographs taken of me squatting next to the resting trace show my feet either barely touching the water or completely out of it, despite my wearing water-hardy sandals that day (Figure 7.1a).

Any time I have taken students to the Georgia coast, I likewise look for alligator tracks and other traces as a way of conveying a similar feeling of awe, respect, curiosity, or dread of these animals that were seemingly sent straight out of central casting from the Mesozoic. Some of these ichnological lessons become amusing, such as one morning on a later field trip with students to Sapelo when I spotted faint alligator tracks accompanied by tail drag marks coming up and onto a beach. I knew my students had been swimming in the surf zone of another nearby beach the previous night, although they did not know I knew. Upon seeing the tracks and tail drag marks, I stopped to direct everyone's attention to them. These traces were vague, a function of the hard-packed intertidal sand, which acted much like nearly set cement. So it took some pointing at details and encouragement to get everyone to the same level of discernment, where they saw what I was seeing. Once they all agreed that these were alligator traces, I gleefully

334 LIFE TRACES OF THE GEORGIA COAST

Figure 7.1. Alligator traces. (a) Resting trace, in which an alligator basked on a tidal-creek bank, and apprehensive author for scale: Sapelo Island. (b) Alligator trackway in coastal dunes, parallel to the shore: Sapelo Island. (c) Alligator trail connecting freshwater pond and salt marsh: Little St. Simons Island. (d) Alligator trails in salt marsh at low tide: Ossabaw Island. (e) Alligator den entrance, about 80 cm (32 in) wide: St. Catherines Island.

expounded on how alligators, despite their reputations for being freshwater animals, often go out to sea and stay in the shallow water just offshore, sometimes finding prey right there. In this respect, they behave more like their crocodilian cousins in Australia, *Crocodylus porous*, also known affectionately as salties for their comfort in salt-infused waters. I mentioned this fact to the students, but also related how alligators sometimes pick out unusual food items from this area, recounting a story by a naturalist who, at a nearby island (Wassaw) had seen alligators in the surf chomping on adult limulids. "Who knows what else they might find out there!" I said cheerfully. The expressions of my students were entertaining to watch, but also betrayed a budding awareness. As the morning went on, they seemed to scrutinize the beach sand more carefully than before, looking for more clues about what had preceded them in that place that morning.

Alligator tracks also can reveal a romantic side that people may have never known about them, that they sometimes go for long walks along the beach. For example, my wife Ruth and I were surprised one morning, again on Sapelo, to find the trackway of a small adult alligator that walked out of a freshwater slough in a maritime forest, only to turn to its left and wander through the dunes. We followed it for nearly a kilometer (0.6 mi) before it turned to its left again and made its way to another freshwater slough. Its trackway was distinctive and easy to see because of its freshness. It had a relatively larger rear foot with four toes that registered just behind or on top of the tracks left by a smaller front foot, which had five toes, and tracks on either side were divided by a sinuous tail drag mark (Figure 7.1b). Its tail was not the only thing dragging, as claws from the rear feet left parallel arcuate grooves between each step. In such instances, alligators are doing a high walk, in which they lift themselves well above the ground and impart a diagonal walking pattern to their trackway. Their thick and heavy tails, however, so useful for powerful back-and-forth movement while swimming, stay on the ground as they walk. As a result, a snake-like sine wave is left in the sand or mud, which in places shows where the keeled tail flopped over to one side or another and left linear drag marks of scales. Sometimes I see such tail drag marks well before their accompanying tracks, which is exactly how I noticed the alligator trackway coming out of the surf, soon followed by my students learning the benefits of ichnologically informed swimming.

Besides provoking primitive predator–prey reactions in humans, alligator traces also serve an essential role in maintaining ecosystems along the Georgia coast. For example, hikers on the less inhabited Georgia barrier islands, such as Wassaw or Little St. Simons, often encounter small freshwater ponds while wandering through maritime forests, which frequently contain alligators. These ponds are seasonal, filling with water during times of increased rain on the Georgia coast (basically hurricane season, from May through October) and becoming dry depressions during the winter and spring months. Alligators are mostly responsible for making these ponds, initially forming them during the start of the rainy season by swirling the saturated mud and sand with their muscular tails (Chapter 3). With enough such movements, numbers, and generations of activity, these low areas in the landscape become places where freshwater accumulates each spring. Accordingly, these wetlands attract their typical plant assemblage (cattails, reeds) and fauna (aquatic insects, crayfish, frogs, salamanders, fish, otters), and thus serve as centers of freshwater biodiversity in the middle of maritime forests. Moreover, alligators walk from pond to pond, and this habitual movement wears down trails through the maritime forest. Once these trails are low enough to be inundated by freshwater, they then serve as connectors between wetlands, better allowing boy–girl fish stories to develop and other such biotic exchanges. Adult alligators also do much of their fishing in salt marshes, and nightly commutes between freshwater ponds and salt marshes sometimes connect these habitats, while also forming distinctive, alligator-wide pathways through salt marshes (Figure 7.1c,d).

Freshwater ponds made by alligators also help to lure in their prey, resulting in slightly fewer fawns, raccoons, or domestic dogs in such areas. Indeed, development of some Georgia barrier islands, such as Tybee, Jekyll, and St. Simons, has filled in many of these alligator-made wetlands, resulting in noticeable population jumps in deer and raccoons, as well as fewer missing dogs. Of course, any overpopulation of deer also leads to overbrowsing of maritime forest vegetation, which, when coupled with decreases in wetland biodiversity, imparts noticeable losses to plant communities. Alligator traces thus have an ecological impact that goes far beyond their academic appeal, and are more than just curios warranting the attention of ichnologists and herpetologists.

Alligator ichnology not only informs us about their role as keystone species on the Georgia coast, but also tells us about their reproduction and adaptations to wintertime conditions. In addition to making large nest mounds, mother alligators dig large den structures on the flanks of freshwater ponds, which they then use for raising their young. The half-moon-shaped entrances to these dens are usually at or below the waterline—depending on the time of year and amount of rainfall—and they connect with tunnels that trend upward into enlarged chambers (Figure 7.1e). These chambers are big enough for an adult alligator to turn around, yet high enough above the seasonal waterline to avoid drowning. Chambers are also sufficiently large so that adult alligators and their offspring have a safe place out of the view of predators that only pick on smaller, less ferocious alligators. During Georgia winters, these dens are also good places to hibernate, maintaining equitable temperatures and humidity while conditions outside become more challenging for alligator physiology.

Alligator dens, when left high and dry, are ichnologically interesting structures to investigate. (No, I do not crawl into them, nor do I send unsuspecting students into them, either.) The area just in front of the entrance is a slight depression excavated by the mother alligator, which becomes filled with water and thus gives baby alligators an easier way to swim in and out of the den. If alarmed by an approaching ichnologist or similarly strange figures, the babies, which may number more than a dozen, begin emitting high-pitched grunts and quickly swim together to this ponded area, where they may vanish below the waterline as they enter the den. The mother may have been out in the open at first, but usually vanishes into the den at the first sign of trouble, leading the way for her brood. From den entrances, tunnels go straight, and then have elbow-like, nearly right-angle turns to either the right or left. Tunnel segments are typically the length of a large adult alligator before turning; the limited lateral flexibility of an alligator body means that they need much room before changing directions.

Citing personal-safety reasons, I have never seen the terminal chambers of these dens. My concern, however, is not so much about the alligators, but the many commensal animals that enjoy the benefits afforded by living in an alligator burrow. Indeed, an alligator den I studied on St. Catherines Island reinforced this point for me one day. While approaching a den entrance above the waterline of a seasonal pond, a meter (3.3 ft) long subadult alligator and I startled each other. As it ran for the den entrance, I stood still and waited for my heart rate to slow down sufficiently and otherwise regain some scientific composure. I then walked up to the entrance and peered into the darkness of the tunnel. Seeing nothing, I extended my camera forward, trusting that the flash would illuminate the darkness of the tunnel and reveal the location of the alligator. It worked, with results even better than intended. When I later reviewed my photos, I was pleased to see the image of the alligator, about a meter in, looking out at me, its eyes reflecting the flash as blue-red points. The unsettling revelation conveyed by my remote-sensing technique, though, was the snake only about 20 cm (8 in) from the den entrance, its previously unnoticed face and eyes also reflected by the flash. This lesson of commensal animals, besides reaffirming my wisdom of not crawling into alligator burrows, hence became indelible, an unexpected addition to my understanding of ecological dimensions of vertebrate ichnology.

Thus with these important lessons imparted by alligators and their traces, we will now enter the world of freshwater and terrestrial vertebrate traces on the Georgia barrier islands, starting with fish, amphibians, and reptiles, and later going on to birds and mammals (Chapter 8). Much of this ichnological treatment requires tapping into the ancient science of tracking (Chapter 1), but is updated with insights on how ichnology and ecology intertwine easily when discussing the tracks, trails, burrows, feces, and other easily visible traces left by vertebrates in terrestrial environments. Vertebrate tracemakers of the Georgia barrier islands were barely mentioned during the 1960s through the 1980s, other than through a few notable studies (Reineck and Howard 1978; Frey and Pemberton 1986). Hence much of what follows is a new synthesis, based on the cumulative knowledge of trackers and other scientists, as well as the fruits of much time I have spent in the field tracking these animals. Hopefully this review will spur on more such studies, improving insights on the mostly hidden lives of terrestrial vertebrates on the Georgia coast.

HOW TO TRACK A FISH

Fish in freshwater environments of the Georgia barrier islands, which are mostly teleosts (bony fish), are relatively uncommon because of the paucity of standing freshwater and flowing freshwater streams (Chapter 2). Freshwater fish consist of a few native species identical to those on the mainland of coastal Georgia and a few nonnative species. For example, fish in the largest freshwater lake on any of the Georgia barrier islands (Lake Whitney, Cumberland Island) include the following species (Frick et al. 2002): the sheepshead minnow (*Cyprinodon variegatus*); striped killifish (*Fundulus majalis*); sailfin molly (*Poecilia latipinna*); mottled mojarra (*Eucinostomus lefroyi*); warmouth (*Lepomis gulosus*); bluegill (*Lepomis macrochirus*); striped mullet (*Mugil cephalus*); western mosquitofish (*Gambusia affinis*); and tarpon snook (*Centropomus pectinatus*). Of these, the mosquitofish is an invasive species, displacing its close relative the eastern mosquitofish (*Gambusia holbrooki*). As one might discern from its name, it was used for mosquito control, with the intended duty of gobbling up mosquito larvae. The tarpon snook is another nonnative fish brought in, but instead for catching and eating by humans.

Despite this low diversity of freshwater fish—a few of which are also adapted for brackish water—some of these fish can make traces of interest. We may not normally think of fish as tracemakers, as their fluid medium does not change shape much (except for freezing), nor does it hold any alterations caused by fish motion. Most fish also have a lack of useful limbs for walking on land; even sea turtles can do this, although not gracefully. Nonetheless, freshwater-fish traces are observable today and freshwater-fish trace fossils have been interpreted from the geologic record. As a result, we will cover their anatomical features, behaviors that may impart traces (especially swimming), and the likely substrate conditions that would preserve their traces. Believe it or not (and you will believe), fish can be tracked, and an astonishing amount of information, such as the size of the fish, its anatomy, probable species, and its swimming mode can be gleaned from a single fish trace.

Most freshwater bony fish have the same basic body plan: a streamlined longitudinal profile that is widest in its middle, narrow at each end (anterior and posterior), and thin when staring a fish directly in its face. Its fins, however, are the most ichnologically important parts of their anatomy to consider, and are, from rear to front, caudal, anal, pelvic, dorsal, and pectoral (Figure 7.2a). Accordingly, swimming modes are then classified by how much of a fish body is involved in the propulsion versus using its fins. These swimming modes can be divided into (1) body and/or caudal fin, otherwise known as BCF locomotion; and (2) median and/or paired fins, or MPF locomotion. BCF swimming ranges from eel-like, whole-body movement, also called anguilliform, to using mostly the tail and beating from the far end of the caudal fin, or thunniform (Sfakiotakis et al. 1999). MPF locomotion, on the other hand, typically employs the beating of the pectoral or other fins more on the middle of the fish body.

Fish swimming traces are made through an interaction of the ends of a fish's ventral fins, particularly on the caudal and anal fins, but sometimes also the pelvic or pectoral fins. If a swimming fish interacts with a lake or stream bottom, and the substrate is sufficiently fine grained to preserve a visible trace, it will make a beautifully incised sine wave, or paired and offset sine waves. A single sine wave is almost always a trace of the caudal fin, in which the bottom edge of the fin slices through the sediment surface. This cut may be only a few millimeters deep, yet should still be visible. Because all waves have mathematical properties, such as wavelength and amplitude,

340 LIFE TRACES OF THE GEORGIA COAST

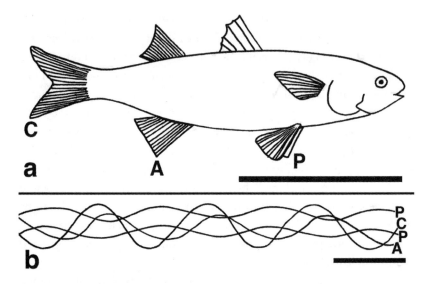

Figure 7.2. Freshwater fish anatomy and how these relate to their trails. (a) Fin anatomy of a striped mullet (*Mugil cephalus*), with caudal (C), anal (A), and pelvic (P) fins indicated. (b) Caudal (C), anal (A), and paired pelvic (P) fin trails that would be made by 23 cm (9 in) long striped mullet swimming along a muddy or sandy bottom. Scale = 10 cm.

these traces are also amenable to quantitative analyses. An observer can note the approximate distance between wave peaks as a way to estimate the length of the fish, and the amplitude for figuring how much the caudal fin was used (Martin et al. 2010). For example, a wavelength of only 5 (2 in) cm and amplitude of 2 cm (0.8 in) in a caudal-fin trail was made by a fish that, if caught, would not warrant any bragging rights, nor taking of photographs of you standing next to your catch. Find a wavelength of more than 50 cm (20 in) and amplitude of 20 cm (8 in), though, and then you may start reaching for your camera and reading some recipes. Naturally, such caudal-fin trails change considerably in form along their course as the fish tracemaker changes direction (sometimes abruptly) or speed. Think of how a tail beating more rapidly within the same horizontal distance will create a series of high-frequency, high-amplitude waveforms, whereas a steady, slow tail beat will correspond to long wavelengths, low frequencies, and low amplitudes. With these basic principles in mind, a fish can be followed and its variations in behavior discerned by looking for such nuances in a caudal fin trail.

Just a moment: What about the paired and offset sine waves? One of these is still a caudal fin trace, and it is the one with the higher amplitude. The lower amplitude trace is typically made by the anal fin, which is positioned in front of the caudal. This means that fish swimming direction can be discerned by looking carefully at these waves. In the direction of forward movement, the peak of the lower-amplitude trace will precede the peak of the higher one by maybe half (or less) of the latter's wavelength. Furthermore, the higher-amplitude trace should also cut across the lower-amplitude one. These two observations make sense when one visualizes the forward motion of the fish: in such movement, the anal fin always precedes the caudal fin, and the caudal fin is always moving in a wider arc than the anal fin. The distance between the caudal fin and anal fin also can be estimated by the amount of offset in the waves, and this proportion then can be used to extrapolate the total length of the fish, all of which involves a bunch of math that will not be explored here and now. If the pelvic fins also become involved in the swimming motion, then these parallel, lower-amplitude waveforms will be inside those of the caudal and anal fin traces as well (Figure 7.2b)

Once all of a fish trail is described, it then becomes possible to figure out which swimming mode was being used, which fins were involved in the propulsion, and a few body parameters, such as the distance between fins. This, in turn, can be used to estimate the size of the fish. Of course, size is important for species identification: after all, a mosquitofish is not going to have its traces confused with those of a striped mullet, unless same-sized juveniles of a striped mullet are acting just like mosquitofish. Species-specific behavior is also important, as some fish will not be persuaded to cruise along a pond or river bottom unless absolutely forced to do so, whereas others are perfectly comfortable with contacting a sedimentary surface.

Along those lines, a few species of freshwater fish from the Georgia coast make prominent traces in addition to trails: semi-bowl-like depressions on the bottoms of pond or slow-moving rivers that serve as nests. Fish make nests? Indeed they do. These structures, which normally are made in water less than 2 m (6.6 ft) deep, are commonly constructed by male warmouths and bluegills, both under the genus *Lepomis* (*L. gulosus* and *L. macrochirus,* respectively). In nest making, male fish of either species will swim close enough to the sediment surface to erode it, but in a continuous loop, which carves out a superficial, circular pit. The width of the pit depends partially on the length of the fish and its degree of body flexion, but normally these are about 30 cm (12 in) wide and less than 10 cm (4 in) deep (Neff et al. 2004; Marcy et al. 2005). These nests are also constructed near submerged objects, such as a logs, stumps, or thick vegetation, and will only be made out in the open if no such cover is available. Depressions are separate from one another, as males then hover over them and wait for a female to swim by, admire their fin art, and hopefully allow fertilization of her eggs, which she then lays in the nest. The males stay above and near the nest afterward and protect the eggs and the resulting baby fish, nicknamed fry. Multiple nests directly next to one another also may have evolved as a type of social behavior to reduce predation of young by larger fish (Gross and MacMillan 1981), although solitary nesting might have its advantages too (Neff et al. 2004).

Many trace fossils attributed to swimming fish have been interpreted from the fossil record, from the Devonian to the Pleistocene, reflecting a wide variety of fish behaviors and anatomies (Martin and Pyenson 2005; Minter and Braddy 2006; Benner et al. 2008; Martin et al. 2010). Most of these trace fossils were formed in marine or marginal-marine environments, but a few notable examples come from freshwater-lake deposits. As of this writing, only one report of fish-nest trace fossils have been interpreted from the fossil record (Fiebel 1987), although these structures should be common in strata from shallow lakes that contained many teleosts. In fact, traces made by fish in freshwater environments of the Georgia barrier islands have yet to be described. This sort of study should be useful, particularly for comparing to some of the more prominent traces left by marginal-marine fish in offshore environments (Chapter 8).

WALKING, HOPPING, AND BURROWING AMPHIBIANS

Amphibians, which include frogs, toads, salamanders, and newts, are mostly unseen on the Georgia barrier islands, although their voices can certainly be heard—especially at night—and their traces are abundant if you know where to look for them. Depending on the amphibian, some dig burrows, shallow and deep, or may make distinctive trackways. Moreover, amphibian traces are closely linked with freshwater habitats, as these animals face a physiological challenge similar to that experienced by most insects: they have a tough time handling salinity. Consequently, their traces are most likely found in maritime forests and near freshwater sloughs or ponds. Occasionally, however, amphibians cross salt marshes, lower parts of coastal dunes, or upper parts of beaches. So do not immediately discount amphibians as tracemakers if their suspected traces are found in the latter two habitats. Nonetheless, skepticism is warranted for any "salamander" trackways interpreted in terrestrial environments where lizards are much better adapted.

Basic Tracking Concepts

Amphibians are the first tetrapod (four-limbed vertebrate) tracemakers discussed in this chapter, and thus they provide an opportunity to introduce some basic tracking concepts. Tracking is, of course, the oldest form of ichnology (Chapter 1), and normally involves describing and interpreting the tracks of terrestrial and freshwater tetrapods. Tetrapods all descended from a specific lineage of lobe-finned fish during the Devonian Period. (Yes, it is true; you and all of your family members are fish, which you may have suspected anyway.) Fossil amphibians from the Devonian are difficult to distinguish from fossil fish because of their anatomical similarities. This evolutionary transition is also marked by the oldest known tetrapod tracks, which likewise are in Devonian rocks (Niedźwiedzki et al. 2010).

Terrestrial and freshwater tetrapod tracemakers may make tracks associated with any of their four walking limbs, but the number of tracks made depends on whether they are bipedal or quadrupedal. All amphibians are obligate quadrupeds, in that they have no choice but to move on four limbs when on land. In contrast, a facultative quadruped is an animal that is normally bipedal, but chooses to move on four limbs, whereas a facultative biped typically moves on four limbs but can go to two if necessary. The front foot in an amphibian or any other tetrapod is also called a manus (plural mani), whereas the rear foot is a pes (plural pedes). Modern amphibians have a maximum of four digits on the manus and five on the pes (Figure 7.3a). The earliest amphibians from the Devonian Period, however, had more than five digits on each foot, a direct result of the number of bones supporting the lobed fins of their fish ancestors. Accordingly, tetrapod tracks from that time have six or more digit impressions (Niedźwiedzki et al. 2010). Digits on any tetrapod are numbered by anatomists in increasing order (I–V) on the basis of proximity to the midline of the body. Thus digit I is the thumb on the manus or the big toe on the pes, although this toe may be quite small or absent on a nonhuman. As shown in amphibian feet, the number of digits on tetrapod feet can be fewer than five. Such reductions in the number of toes are a result of natural selection that weeded out any unnecessary organs or appendages over time. For example, the ancestors of limbless amphibians and reptiles not only had digits on their limbs, they had limbs. Nonetheless, these traits somehow became evolutionary liabilities, never to come back again in descendants of the modern forms.

In order to identify a tetrapod trackmaker, a thorough description of a track must be made, which includes measurements. The most basic measurements of a track include its total width and length, but even such a simple task is fraught with potential error if an aspiring ichnologist wishes to produce consistent results. For one, what part of the track is measured: The plane where it intersects a substrate (the track horizon), or at its bottom (the floor of the track)? Or is it measured across its overall profile, including deformation structures caused by the motion of the animal? The answer is the track floor, or minimum outline, which is more representative of the animal's actual foot width and length than any other aspect of a track (Halfpenny and Bruchac 2002; Figure 7.3b). Nevertheless, even this minimum outline must be measured carefully by looking for where the more horizontal part of the track meets the more vertical part, the latter sometimes called the track wall (Brown 1999). By far the most common error made is measuring where the track intersects a substrate, which results in an overestimation of its size.

Once overall width and length of a manus or pes is recorded, other features, such as the number of digit impressions, lengths and widths of digits, angles between digits, presence or absence of interdigital webbing or footpads, claw impressions, and other anatomical traits can all be noted as well. In sum, these features can help to identify the tracemaker. For example, amphibians do not have claw marks to speak of, but they may have more bulbous ends to their toes, interdigital webbing, and small bumps (tubercles) on their feet. These features collectively distinguish their tracks from those of similarly sized invertebrates, reptiles, or mammals. Amphibian tracks can be confused with those of invertebrates? Yes, they can, as I have seen crayfish trackways I initially thought were made by salamanders, some of which overlap in size and habitats: be cautious.

Pace and stride have already been defined for ghost crab trackways (Chapter 6), and these are also used to describe tetrapod trackways. Nonetheless, other important information can be gleaned from a tetrapod trackway through just a few simple measurements. One is trackway width, also

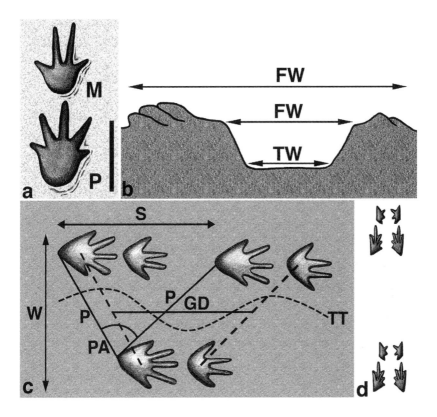

Figure 7.3. Amphibian tracks and trackways, including standard terminology for studying vertebrate tracks. (a) Right-side manus (M) and pes (P) impressions of an eastern newt (*Notophthalmus viridescens*) with 4 and 5 digits, respectively, and associated pressure-release structures. Based on figure by Halfpenny and Bruchac (2002). Scale = 5 mm. (b) Cross section of a track showing its minimum outline and how this should be measured, versus measuring from where it intersects the sediment surface or including its pressure-release structures; TW = true width, FW = false width. (c) Trackway made by a newt or salamander, with pace (P); stride (S); trackway width (W), also known as straddle; pace angulation (PA); and glenoacetabular distance (GD) indicated. Glenoacetabular distance is drawn from midpoints of manus–manus and pes–pes distances (dashed lines). Gait pattern is diagonal walking, and a tail trace (TT) is registered in the middle of the trackway. (d) The other most common gait pattern for frogs and toads, bounding: rear feet are just behind the front feet; all four tracks are closely bunched together; and a noticeable distance separates each set of tracks.

known as straddle, which is measured from the outermost edge of the right and left sides of the trackway; the line bisecting the straddle is, appropriately enough, the midline (Figure 7.3c). Trackway width is partially a function of the animal's body width, but also is affected by pace angulation, which is the angle defined by the midpoints of three successive tracks left by the manus or pes, such as right–left–right. For example, a low pace angulation (about 90°) is characteristic of a sprawling gait, in which the trackmaker stepped well outside of its body width to make a track. This sort of pace angulation is typical of salamander trackways, for example. In contrast, a high pace angulation of nearly 180° is a result of a trackmaker putting one foot almost directly in front of the other, like it walked on a tightrope. In these cases, the straddle can be less than the width of the animal. Another measurement is the glenoacetabular distance, which is the midpoint between the manus–manus and pes–pes distances (Fig 7.3c). Glenoacetabular refers to the shoulder ("gleno") and hip ("acetabular") sockets, thus this measurement is a proxy for torso length in a trackmaker (Schult and Farlow 1992).

All sorts of other measurements of a tetrapod trackways can be taken, but most become superfluous. The most important to remember, however, are pace, stride, straddle, pace angulation, and glenoacetabular distance. From these, a tetrapod's gait—the rhythm of its limb movement—and most body parameters can be hypothesized. Indeed, sometimes an animal can be identified by just its trackway pattern and ecological context, without necessarily looking at the details of its tracks. This sort of macroscopic view is a handy concept in tracking, and is applicable to amphibians, reptiles, birds, mammals, and even invertebrates (Chapters 6, 8).

Gait patterns are important for studying mammal trackways and will be discussed later with this group of tracemakers (Chapter 8), but can be introduced briefly here for interpreting amphibian trackways. In such trackways, variations of only two gaits will be seen: diagonal walking, which is commonly expressed in trackways of salamanders and newts (Figure 7.3c), and bounding, which is normally called hopping when applied to frogs and toads (Figure 7.3d). Diagonal walkers, appropriately enough, make a diagonal pattern with their trackways by moving from one side to another (right–left–right); the body is supported at each phase by the opposite-side manus and pes. In salamanders and newts, these trackways are commonly accompanied by a medial drag mark made by the tail. With bounding, both rear feet land immediately behind the front feet, imparting a pattern of four closely grouped tracks; each group is separated by an intergroup spacing associated with each bound. In this movement, the front legs lift off the ground as the rear legs simultaneously push the animal forward. The animal lands first with the front feet, then followed by the rear. In these patterns, the pes tracks are almost always parallel with one another, but the manus tracks can be either parallel or offset. If offset, this is a half-bound pattern; the former is a full-bound pattern (Halfpenny and Bruchac 2002; Elbroch 2003). Trotting, loping, and galloping are other gaits in tetrapods, but these will be discussed further with mammals (Chapter 8).

Salamanders

Of amphibian trackmakers, salamander trackways are relatively easy to identify, especially if linked to salamander habitats, which are near freshwater environments. Salamanders are diagonal walkers, and in a slow walk, the manus lands in front of the pes on either side, and each manus–pes pair is separated by a small space. In some faster-walking salamanders, however, the pes lands on or exceeds the manus. The torso length of the salamander also affects this pattern, as some salamanders are quite lengthy between their shoulder and pelvic girdles. This means their rear feet never pass their front feet while walking, no matter how hard they try; accordingly, this long-torso trait should be reflected in the glenoacetabular distance of a trackway. As mentioned earlier, a sinuous tail drag mark may be between the tracks, although this part of the trackway could be subtle. In instances where a salamander has poorly developed legs, is in soft mud, or is otherwise not getting itself very far off the ground, it may even leave a drag mark from its belly.

On the Georgia coast, salamander trackmakers include the eastern newt (*Notophthalmus viridescens*) and tiger salamander (*Ambystoma tigrinum*), whose tracks and trackways look similar—four toes on the manus, five on the pes, diagonal walking patterns, and tail drag marks—but are radically different in size. For example, the manus and pes of the eastern newt are only 5 and 6 mm (0.2 and 0.25 in) long, whereas the manus and pes of the tiger salamander are 15 and 20 mm (0.6 and 0.8 in) long, respectively (Halfpenny and Bruchac 2002; Figure 7.3a). Accordingly, their strides differ too: about 2.5 cm (eastern newt) versus 7.5 cm (tiger salamander). Interestingly, the tiger salamander is also apparently restricted to the mainland part of coastal Georgia by the salinity barrier of the back-barrier salt marshes. This geographic isolation, of course, encourages the question how the eastern newt or other amphibians made it to the islands instead (Gibbons and Coker 1978). Nevertheless, if the tiger salamander or its traces were ever found on the barrier islands, these would constitute a nice find, and would provoke some great arguments in herpetological circles.

Another salamander tracemaker is the appropriately named mole salamander (*Ambystoma talpoideum*), which belongs to the same genus as the tiger salamander. This small salamander, which is normally 10–11 cm (4–4.3 in) long, burrows for the purposes of maintaining its moist skin and overwintering, and breeds in seasonal ponds with the onset of wet conditions in the late spring–early summer. What is most interesting about its life cycle, though, is that once its aquatic larval (tadpole) stage is finished, the adult salamander leaves the water and lives most of its remaining life on land, only coming back to the water to reproduce. A few populations, however, have been documented as reproducing while still in their aquatic larval state (Jensen et al. 2008). Mole salamanders live in burrows in maritime forest floors, under rotting logs or other decaying forest debris, where they look for abundant arthropods as food.

Another burrowing salamander, but one different morphologically from the mole salamander, is the two-toed amphiuma, or congo eel (*Amphiuma means*). Congo eels are ravenous predators, eating a wide variety of hard-shelled invertebrates, such as molluscans and crayfish, as well as snakes and frogs (Whitaker and Ruckdeschel 2009). This salamander has short, poorly developed legs bearing (you guessed it) only two toes on each foot, which would make its tracks easy to identify. More importantly, though, they are quite large by today's standards of amphibians, reaching lengths of more than 1 m (3.3 ft). With its long, smooth body and easy-to-miss limbs, the misnomer "eel" is thus understandable. Unlike adult mole salamanders, preferred habitats of the adults are fully aquatic. Thus one of the reasons why they burrow, or occupy an already-made burrow, is to avoid desiccation, especially during droughts. Once in a burrow, congo eels and other amphibians also may use these for overwintering (Pinder et al. 1992).

When an amphibian hunkers down to survive a cold winter, it may hibernate, but if trying to survive prolonged scarcity of water, it enters estivation. Estivation, which also occurs in some insects (Chapter 5), is a state of dormancy in which a fish, amphibian, reptile, or mammal slows down its metabolism so that it can survive long periods of time marked by low amounts of food or water, like during a drought (Pinder et al. 1992;

Hembree et al. 2005; Glass et al. 2009). The difference between hibernation and estivation is related to temperature, as the former only occurs with decreased temperatures, while estivation can (and does) happen during warm conditions. In either case, a mole salamander or congo eel will burrow to avoid such environmentally induced stresses. Congo eels, like many other amphibians, also secrete a cocoon made of mucus to surround it during estivation, which helps to prevent drying out.

Unfortunately, the burrows of both mole salamanders and congo eels are barely described. Congo eels are known to live in burrows more than a meter deep (Knepton 1954), but the width, orientation, and exact ecological contexts of these burrows have not been documented. On the Georgia barrier islands, they likely live in burrows on pond margins or floodplains that intersect the water table in response to droughts, and are known to occupy old crayfish burrows. Intriguingly, their low tolerance for salt water implies that these amphibians, like crayfish, were left behind on the Georgia barrier islands by rising sea level about 11,000 years ago (Chapter 5). Unfortunately, because these amphibians are difficult to locate on the islands, I have no further personal insights on their ichnology. I am also not inclined to investigate congo eels in their burrows after reading one too many descriptions of their excellent biting abilities.

Toads and Frogs

Anurans (toads and frogs) make unmistakable trackways in terrestrial environments and along the edges of ponds or other freshwater bodies. For example, both the southern toad (*Bufo terrestris*) and southern leopard frog (*Rana sphenocephala*) normally move by springing off their rear legs, landing with their front, then their rear feet. This causes a trackway pattern with paired manus impressions directly in front of and inside of the much larger pes impressions, causing four closely spaced tracks. The intergroup spacing can be as much as 50 cm (20 in) in leopard frogs, but is usually only 15–20 cm (6–8 in) in the southern toad. Tracks of each species are also comparable in both size and morphology. Southern toads have four toes in the manus, which is about 1.5 cm (0.6 in) long, and five in the pes, which is 2.5 cm (1 in) long). The pes also has webbing between the toes and occasionally registers small bumps (tubercles) on the heels of both the manus and pes (Halfpenny and Bruchac 2002). Leopard frog tracks are slightly smaller than toad tracks, but can overlap in size. Hence the main distinguishing trait is ecologically based, explained soon. Another consideration is anatomical, such as the presence or absence of tubercles on the tracks, which leopard frogs lack.

In most anurans, digit IV in the pes (the fourth digit counted from the midline of the animal) is elongate relative to other toes, a distinctive trait that also helps to spot and identify their tracks. As befitting their respective terrestrial and freshwater habitats, southern toads are more likely to make their tracks in maritime forests and back-dune meadows, whereas tracks of leopard frogs will be along the edges of freshwater ponds and sloughs. For example, I have often come across southern toad tracks on the sandy roads of Sapelo Island early in the morning and, once encountered, have delighted in using these for nonmammal tracking exercises with my students (Figure 7.4). By far the most common anurans on the Georgia barrier islands, though, are tree frogs (*Hyla* spp.) and peepers (*Pseudacris* spp.), but these rarely leave their above-ground or watery habitats (respectively) to make definable tracks. Nonetheless, if they did move about on the right substrates (wet mud or dry sand), their tracks would be so minute that they could possibly be confused with those of hopping insects. So an attentive ichnologist should look carefully for groups of four tracks (anuran) or six tracks (insect), as well as anatomical traits in each track that might better reveal the tracemakers.

One species of anuran burrower on the Georgia barrier islands is the eastern spadefoot toad (*Scaphiopus holbrookii*). An adult eastern spadefoot is normally much smaller than a southern toad, at about 7 cm (2.8 in) versus 10 cm (4 in) in body length, although they do overlap in size. Nonetheless, spadefoot tracks should be smaller on average. The eastern spadefoot, however, is most famous as a burrower, with feet specially adapted for this task: take a look at the common name again, and Greek aficionados will also appreciate the roots of its genus name. A black, hard projection on each rear foot (the spade) allows it to burrow efficiently, accomplished by synchronizing digging while rotating its body backward and downward (Summers and Delis 1994), a behavior common in burrowing frogs, too (Emerson 1976). These toads, which burrow in well-drained sandy soils of maritime forests, are typically associated with live oaks and similar-sized trees. The spadefoot toad emerges from its burrows nightly to hunt for forest invertebrates, but especially after rain, as saturated burrows give them a reason to go out to eat. Their burrows otherwise serve as protection from predators and help to keep them moist during daytime hours.

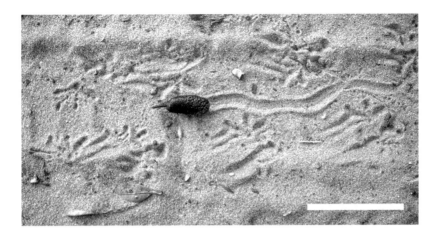

Figure 7.4. Trackway of the southern toad (*Bufo terrestris*), and an unusual one, showing a diagonal walking pattern and drag mark caused by its scat, which was still attached to the toad: St. Catherines Island. Scale = 5 cm (2 in).

Although often described as secretive, these toads can be seen in massive gatherings after fierce rainstorms, lending to their nickname storm toads (no, I am not making that up, and yes, that would make a great name for a band). They are especially gregarious if such storms happen during mating season (March through July), although they are known to mate year round if temperatures remain high enough to encourage such activity (Jensen et al. 2008). Spadefoots are opportunistic in their reproduction, sometimes termed as explosive breeders (which sounds illegal, and on several levels), crowding around temporary bodies of water after a downpour (Greenberg and Tanner 2004). This is where the males make their mating calls to attract females, females are attracted to these calls (or not), mating happens for some lucky toads, and they all return to their respective burrows afterward. Thus abundant tracks will be more likely to form seasonally, and in association with temporarily standing water. Apart from travel caused by the urge to mate, though, they stay close to their burrows, and studies on home ranges indicate movements of less than a meter from a given toad burrow over the course of months (Jansen et al. 2001). Other species of spadefoot toads in drier climates are proficient estivators, staying in their burrows for 7–10 months of a year (Pinder et al. 1992). The eastern species native to the Georgia barrier islands, on the other hand, is not as likely to call upon this hardwired behavior unless during a severe drought.

Spadefoot toad burrows have not been described ichnologically, which may be partially a function of their vague forms. For example, one nineteenth-century researcher termed their burrows "turnip shaped," a descriptor I sincerely hope will never be standardized. Nonetheless, open burrows show elliptical cross sections about 2 cm (0.8 in) at their widest diameters. These toads normally bury themselves only a few centimeters under the ground surface, where the loose sediment then falls into the burrow and effectively hides the toad from its potential enemies, while also retaining moisture. A few researchers, however, have noted they burrow to depths of more than 2 m (6.6 ft: Jansen et al. 1991). Regardless, most spadefoot toad burrows in Georgia maritime forests are only constructed on an ad hoc basis, rather than constituting sturdy, permanent structures. Consequently, the usual shallow, temporary traces would have low preservational potential in the geologic record, whereas burrows with 1–2 m depths are more likely candidates for fossilization.

A last consideration here on amphibian ichnology is the potential overlap between teleost swimming traces and those of amphibians, which becomes feasible once one considers the aquatic, free-swimming larval stage (tadpole) of most amphibians. If tadpoles swam along a sedimentary surface, they also can impart sinusoidal incisions through the beating of their tails. The flexibility of larval amphibian tails, though, means that the resulting sine waves may have high amplitudes and frequencies, and correspondingly short wavelengths. Two other clues that may reveal a larval amphibian tracemaker versus a teleost are size—tadpoles of most Georgia coast amphibians, with only a few exceptions, are much smaller than the teleosts—and the presence of legs: teleosts lack these, just in case you needed reminding. During their development, tadpoles of most salamanders, newts, and anurans go from a limbless to a limbed stage, but at some point they may still retain their tails. Hence the swimming traces of more grown-up tadpoles may also show impressions of these limbs (tracks) on either side of the trail. Remarkably, trace fossils showing just this pattern have been described from Permian rocks in New Mexico, and were appropriately interpreted as larval amphibian swimming traces (Minter and Braddy 2006).

In summary, amphibian tracemakers of the Georgia barrier islands are most likely to leave trackways and burrows, but not much else. Salamander and anuran trackways are readily identifiable and can be distinguished from one another, but amphibian burrows may be difficult to categorize, considering how little neoichnologically oriented work has been done on these. This is an area of vertebrate ichnology that certainly deserves more attention, especially when bearing in mind the relative rarity of amphibian body fossils in most Paleozoic, Mesozoic, and Cenozoic deposits. Amphibian tracks are common in the fossil record, and in fact a world-class fossil track site in Carboniferous Period rocks is in nearby Alabama. With these trace fossils, small salamander-like amphibians (temnospondyls) slogged through wet (but firm) mud in the upper reaches of an estuary, leaving beautifully preserved trackways, tail drag marks, and occasional belly impressions (Buta et al. 2005; Martin and Pyenson 2005). The sheer abundance of these trackways, with nary an amphibian body fossil in sight, speaks to the value of studying modern salamander and newt tracks when trying to interpret the behavior of their long-lost Paleozoic relatives. Other amphibian trace fossils include abundant amphibian estivation burrows from the Permian of Kansas, some with body fossils of their tracemakers still in them (Hembree et al. 2005). In short, such finds bode well for discovering more intersections between the neoichnology and paleoichnology of amphibians.

REPTILIAN TRACEMAKERS AS KEYSTONE SPECIES

Reptilian tracemakers in terrestrial environments of the Georgia barrier islands include one of the most impressive of burrowing vertebrates, the gopher tortoise (*Gopherus polyphemus*), along with the American alligator (*Alligator mississippiensis*), whose ichnology was already featured earlier. Both of these reptiles are keystone species, in that they are essential components of their respective ecosystems, and their respective ecosystems would change quickly with their absence. Moreover, both species alter terrestrial landscapes through their traces, creating habitats used by other tracemaking organisms (Chapter 3). Additional reptilian tracemakers, such as turtles, lizards, and snakes, are also responsible for a wide variety of morphologically and ecologically extensive traces. Most of these are in terrestrial habitats, although a few species occur in freshwater ecosystems too.

The largest behavioral distinction of reptiles as tracemakers compared to amphibians is the need for most to reproduce through enclosed eggs laid on land. This is an evolutionary development that distinguishes amniotes—tetrapods with an enclosed egg or variations on that theme—from amphibians, the latter of which depend on water for their reproduction. As a result, reptile nests can be added to the list of traces made by terrestrial and freshwater tetrapods, and nest structures occur in most major amniotic groups as an evolutionary echo. Eggs, however, are not considered traces because these are extra body parts needed by a developing embryo to live. On the other hand, eggshells can serve as substrates for preserving traces of behavior from hatchlings and egg predators.

Gopher Tortoises

Of course, the ichnological king of the chelonian world is the gopher tortoise (*Gopherus polyphemus*), although biologists are uncertain about whether it is native to the Georgia barrier islands. Nonetheless, this protected species is sometimes transplanted from the mainland Georgia coastal plain —where it is definitely native—to a few barrier islands where they do what they do best, which is burrow, eat, and reproduce. For example, a group of gopher tortoises was recently moved from the proposed site of a gigantic, job-displacing store near Statesboro, Georgia to St. Catherines Island, where they are now free to consume local plants, and make as many burrows and additional gopher tortoises as they please.

Interestingly, this relocation resulted in the gopher tortoises teaching humans about what was needed for their burrows. Well-meaning people associated with the project dug starter burrows for the tortoises, shallow excavations that the tortoises presumably then would have expanded into useable homes. The tortoises, however, ignored the humans' generous efforts and made their own burrows from scratch (literally). Why? The answer came from the preferred orientations of the newly made burrows, many of which had their openings facing the south–southeast. In contrast, human-made excavations were placed randomly. This preferred directionality, shown by the tortoises, favors maximum sunlight striking the burrow entrance first thing in the morning. This warms a tortoise near the entrance and gives it energy for further digging, looking for tasty plants, mating, or brooding. As a result, their human caretakers learned to make initial digs for recently arrived tortoises while also minding the arc of the sun.

Gopher tortoise burrows are incredible traces, made all the more amazing when you see who makes them: a relatively small (25 cm long), shy, drably colored, and slow-moving tortoise. Yet their burrows are among the wonders of vertebrate ichnology and terrestrial ecology. Descriptions

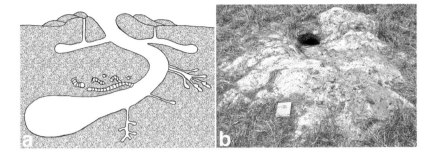

Figure 7.5. Gopher tortoise traces. (a) Longitudinal view of an idealized tortoise burrow showing a twisting tunnel, and including burrows of juvenile tortoises (the two tunnels near the main burrow entrance) and commensal animals. Notice the remaining vertical shafts connected to the juvenile tunnels, indicating how the juvenile tortoise tunnels are reworked from burrows of the Florida mouse (*Peromyscus floridanus*). Not to scale, because that would take up several pages. (b) Entrance of a burrow, showing overlapping aprons of sand excavated by the tortoise and half-moon outline: St. Catherines Island. Scale (notebook) is about 20 cm (8 in) long.

of their burrows mention lengths of more than 14 m and vertical depths of 6 m, which are, simply put, enormous structures that rival insect nests in their effect on terrestrial sediments (Chapter 5). Burrow depth is limited by several factors, but particularly by the depth of available loose sand and the local water table (Means 1982). However, tortoises will tolerate living in partially submerged burrows, which happens often during the winter. Burrows are inclined at relatively steep angles, varying from 20–40°, and although some run straight down, more often they curve to the right or left before terminating in an enlarged chamber (Figure 7.5a; Doonan and Stout 1994). The overall longitudinal profile of a gopher tortoise burrow is analogous to those of solitary wasp burrows and brooding chambers, but of course scaled up by orders of magnitude (Chapter 5).

The torsion of most tortoise burrows is important for several reasons: (1) it slows down potential predators of the adult tortoise or juveniles; (2) decreases the rate of moisture evaporation within the burrow; (3) helps to regulate burrow temperature, which is useful both for overwintering and keeping cool in the summer; and (4) decreases the likelihood of surface fires affecting its occupants. The latter aspect is especially important to think about ecologically, because gopher tortoises' preferred habitats are relatively dry, upland, sandy areas associated with pine forests, particularly those primarily composed of wiregrasses (such as *Aristida stricta*) and longleaf pines (*Pinus palustris*). Such ecosystems are fire adapted, with regeneration of nutrients and plant reproduction encouraged by regular fires. Many animals native to such habitats are also adapted to survive fires, and deep burrowing, or taking refuge in someone else's deep burrow, are certainly effective methods for staying out of harm's way during a ferocious blaze.

Gopher tortoise burrow entrances can be identified by looking for a bare apron of sand, 1–2 m (3.3–6.6 ft) wide and about 20–30 cm (8–12 in) high; then look for a half-moon-shaped hole nearby. The entrance, which is typically about 25–30 (10–12 in) cm wide and 15–20 cm (6–8 in) high, has a flat bottom and a rounded top, reflecting the cross-sectional profile of its main occupant (Figure 7.5b). Burrow widths, however, are close to tortoise lengths, which is a function of the burrow accommodating a tortoise that may decide to make a 180° turn in the main burrow shaft (Doonan and Stout 1994). Recently dug or modified burrows have freshly piled sand outside of them and claw marks associated with tracks, which may be accompanied by shell drag marks in between the tracks. For burrowing, these tortoises have powerful front limbs, flat, flipper-like front feet, and robust claws, all of which are used primarily for burrowing. On the other hand (or foot, rather), the rear feet are employed for pushing their bodies forward while burrowing with the front legs; the hind limbs also push back some of the sediment excavated by the front feet. Their hard shells, well worn from scraping against quartz-rich sands, also tend to locally compact the walls of the burrow, which keeps the burrow open and otherwise prevents the tunnel from collapsing while the tortoise is actively digging.

Male tortoises dig or otherwise use many burrows, more than 30 in some instances, which are within a home range (Guyer and Hermann 1997). Female tortoises use relatively fewer burrows (5–10), but settle into one for nesting. Eggs are normally laid in a small hollow dug less than 15 cm (6 in) deep into the sand apron about 50 cm (20 in) outside the burrow entrance. Clutches consist of 2–10 eggs, which take as long as 100 days to hatch. Like many reptiles, sex determination is governed by nest temperatures, as greater than 30° C produces more females, and less than 30° C results in more males (Burke et al. 1996).

Juveniles then hit the ground digging, and may start their own burrows only a day after hatching. This is a useful adaptation, because mother tortoises, unlike crocodilians, invest no care in their offspring after hatching; hence burrowing prevents juveniles from becoming slow-moving snacks for local predators. Juvenile burrows are as much as 80 cm (32 in) long and sometimes branch from the main burrow shaft of the adult tortoise (Aresco 1999; Doonan and Stout 1994; Epperson and Heise 2003), which, if preserved in the fossil record, would make for distinctive compound trace fossils. Juvenile tortoises, however, sometimes dig along burrows of other species, such as those of small rodents and large insects, thus making composite traces (Chapter 9). Burrows made by older juveniles are identifiable as smaller versions of the adult burrows, generally having widths of less than 15 cm (6 in), whereas hatchling burrows are less than 6 cm (2.4 in) wide. Added to these juvenile tortoise traces may be those of their predators, such as those of the seemingly ubiquitous and always-hungry raccoons (*Procyon lotor*), which seek out hatchlings and excavate them from their burrows. Juveniles and adults alike try to deter predators by placing their burrows in close association with an obstacle, such as a tree trunk, log, or branch.

Because of their extensive and abundant burrows, gopher tortoises are also responsible for creating the homes of many other animals, mentioned previously as commensal species (Chapter 5). As a result, these tortoises are sometimes nicknamed "landlords of the sandhills" and similar appellations. Excepting their lack of rental income, this characterization is not far off the mark: more than 300 species of invertebrates and vertebrates are documented from gopher tortoise burrows (Jackson and Milstrey 1989; Lips 1991). Two notable vertebrates that depend on gopher tortoise burrows for survival are the gopher frog (*Rana capito*) and the eastern indigo snake (*Drymarchon couperi*); the latter has the honor of being the longest snake native to North America, growing to almost 3 m (10 ft). Other vertebrate cohabitants are quite varied, including rattlesnakes, foxes, rabbits (although probably not all together), opossums, skunks, armadillos, rodents, burrowing owls, and lizards, among others. In a survey of more than 1,000 tortoise burrows, a few researchers (Witz et al. 1991) found that reptiles were the most common vertebrate commensal species (53 percent), followed by amphibians (36 percent) and mammals (11 percent). Moreover, some of the smaller vertebrates are important ichnologically, such as the Florida mouse (*Peromyscus floridanus*), which makes its own tunnels branching from the main burrow shaft, and vertical tunnels (chimneys) that connect with the surface (Jones and Franz 1990). In fact, a large number of commensal species add their own burrows to the larger tortoise burrow, resulting in exceedingly complex traces that, if preserved in the geologic record, would take ichnologists a long time to unravel (Figure 7.5a).

Still not convinced of the potential for these structures as complex traces, and their ecological intricacy? Let us talk about commensal invertebrates, then. Endemic to gopher tortoise burrows are flies (*Eutrichota gopheri, Machimus polyphemus*); moths (*Idia gopheri*), including one whose caterpillars scavenge on dead tortoise shells (*Ceratophaga vicinella*); and beetles, most notably a scarab beetle (*Copris gopheri*) that specializes in consuming gopher tortoise dung (Hubbard 1894; Deyrup et al. 2005). Having its own personal waste-management system is a good thing for the tortoise, which sometimes may not want to make the long trip up the burrow to use an outdoor privy. As mentioned before, gopher tortoise burrows also serve as temporary refugia for animals, particularly whenever fires break out. Of course, a tortoise does not have to be home for other animals to use its burrow, meaning that many abandoned burrows become domiciles for secondary occupiers. Because tortoises can live as long as 60 years, the burrowing activities of each individual tortoise over time ensures that they will manufacture thousands of subterranean habitats for other species in a given terrestrial ecosystem.

Although fossil examples of gopher tortoise burrows have not yet been found, newly discovered fossil burrows attributed to reptilian tracemakers are being compared to modern tortoise burrows for added insights. Examples include the recent discovery of probable small dinosaur burrows in Cretaceous rocks of Montana and Australia, which share the gopher tortoise trait of twisting tunnels and expanded terminal chambers (Varricchio et al. 2007; Martin 2009b). The Montana dinosaur burrow even had two differently sized burrows branching off the main burrow shaft, one sized appropriately for a mammal and the others for insects, suggesting that commensal animals also lived in this dinosaur burrow (Varricchio et al. 2007). I took part in that study, and happily confess that gopher tortoises and their burrow commensals were certainly on my mind when we studied the dinosaur burrow at the outcrop: Would this dinosaur burrow hold other directly connected burrows? So it was with much satisfaction when we found the smaller burrows, providing a glimpse of paleoecology underground from 95 million years ago, and directly comparable to our tortoise friends living in the sandy soils of present-day Georgia.

Alligators

Gopher tortoises are surpassed both ichnologically and ecologically on the Georgia barrier islands by alligators, the largest terrestrial reptiles in North America. Alligators form small freshwater ponds; trails between these ponds, which sometimes fill with water and help to connect them; trackways, which include tail drag marks; nest mounds; and dens. Dens, as mentioned earlier, also serve as homes for other animals, with the types of animals determined by whether the dens are underwater, high and dry, or somewhat in between. Of all of these traces, ponds and dens may be quite subtle, especially if the latter have their entrances hidden by water surfaces in ponds. Alligator trails might also be mistaken for those of other terrestrial animals, such as deer or raccoon; indeed, those mammals could easily use these trails too. This leaves alligator trackways as the most obvious traces that can be linked with them. The only possible tracemaker that could be confused with alligators in North America is the smaller American crocodile (*Crocodylus acutus*). This crocodilian, however, is limited to southern Florida, probably because of its intolerance for the colder winters in Georgia. One wayward crocodile, however, was found off the coast of South Carolina in 2008, leading to speculation on where else it landed after departing south Florida: Could it have visited Georgia? Nonetheless, crocodile body fossils are also in Georgia, serving as probable indicators of warmer conditions (interglacial periods) during the Pleistocene.

Alligator trackways are often spotted first where their weighty tails slid along a sandy or muddy surface. Looking much like a snake trail, these traces stand out as narrow, slightly compacted and undulating strips of sand or mud (Figure 7.6a). Because alligator tails are keeled, they often will flop onto each side, leaving linear drag marks caused by scales. Alligators, like all legged reptiles, also have unequal sizes in their manus and pes prints, with the pes considerably larger than the manus. Track lengths and widths for full-sized adult alligators are much bigger than those of any native lizard in North America, and hence unmistakable (Halfpenny and Bruchac 2002). Even hatchling alligator trackways should be larger than those of the most robust lizards on the Georgia barrier islands. With that said, track measurements vary considerably with the size—and appropriately, age—of the alligator. Manus prints of adults have five elongate digits, with four pointing forward; three of these have claw marks (digits I to III). Manus tracks can be 17 cm (6.7 in) long and about half as wide. Pes tracks have only four digits, with claw marks on digits I–III, but also show a prominent, elongate pad posterior to the digits. These tracks can reach nearly 30 cm (12 in) long and 25 cm (10 in) wide (Figure 7.6b) (Whenever I encounter tracks this large, I take a long, slow look at my surroundings, then leave.) Walking strides may exceed 1 m (3.3 ft) in the largest individuals, and trackway width may be about half that. These parameters depend on whether the alligator was doing a so-called high walk (elevating itself above the ground), a more sprawling crawl, or was partially buoyed in the water. When in shallow water, an alligator may do a combination of walking and swimming, in which its feet barely touch the sediment surface; no tail drag mark is recorded in such trackways, either. Along those lines, always keep in mind the ecological context of any suspected alligator tracks.

Freshwater counterparts to traces of failed predation coupled with successful escape, seen in shells of marginal-marine gastropods (Chapter 6), are also recorded in the shells of aquatic turtles living in alligator habitats. Turtles wear their battle scars in the form of healed punctures in their shells, 1–2 cm wide circular pockmarks that match the dental records of alligators. Alligator bite forces have been measured experimentally, though (fortunately) not through biting live animals (Erickson et al. 2003). This empirical evidence is then combined with the diameter and morphology of the tooth marks, along with the spacing and linear arrangements of the teeth. When bite forces and bite marks are linked, better estimates can be made of what constitutes a mere gnawing or a death-dealing chomp. Of course, a living turtle with healed punctures is certainly one that made it past at least one bite.

Freshwater Turtles

Speaking of freshwater turtles, all of these, despite living most of their lives in water, must still obey their original amniotic reproductive habit by laying eggs on land. This requires their digging hole nests, most of which are quite shallow compared to nests of their ocean-faring relatives, such as the loggerhead sea turtle (Chapter 8). Hole nests represent an ancient behavior in aquatic reptiles, and trace fossils of these are documented from more than 200 million years ago in Late Triassic river deposits in Petrified Forest National Park in Arizona (Hasiotis et al. 2004).

Among freshwater turtles on the Georgia coast that make nests are the Florida softshell turtle (*Apalone ferox*); the striped mud turtle (*Kinosternon*

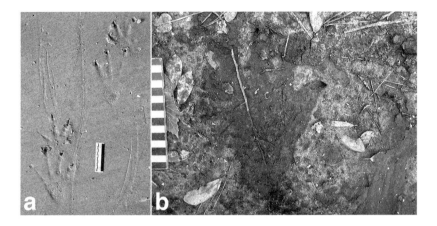

Figure 7.6. More alligator traces. (a) Tail drag mark and tracks on compacted beach sand, in which tracks are visible mostly as claw impressions and drag marks: Sapelo Island. Scale = 10 cm (4 in). (b) Left pes print of a large adult alligator adjacent to partial tail drag mark (*right*), recorded in freshwater mud puddle of a maritime forest: Sapelo Island. Scale in centimeters.

baurii); and the eastern mud turtle (*Kinosternon subrubrum*). These freshwater turtle nests are bowl or vase shaped (looking nothing like a turnip), 8–13 cm (3–5 in) deep, and are often dug in sandy soil well above the local water table (Wilson et al. 1999). This positioning of the nest prevents drowning of the eggs, but nests are located close enough to a water body so hatchlings can quickly escape into their preferred habitat. Clutches for these turtles are small, consisting of only 2–6 eggs.

All three species, when searching for nesting sites (which can take as long as a month), also leave distinctive trackways across back-dune meadows or other sandy areas. Trackways consist of a prominent shell drag mark in the center, with arcuate claw impressions connecting paired tracks (manus–pes) on right and left sides, and the manus impressions just in front of the pes (Figure 7.7). The trackway forms a diagonal pattern with a wide pace angulation compared to most other reptilian trackmakers. Manus impressions have five claws, whereas the pes has four, although fewer may actually register in any given track. To narrow down identification of freshwater turtle tracks to the species level, look for whether the tracks are adjacent to nearby freshwater ponds or sloughs, or more brackish water. For instance, the striped mud turtle spends relatively more time on land than the eastern mud turtle, whereas the latter is more tolerant of high-marsh environments. Nesting for all three species normally happens in June, so look for their trackways then. Any tracks made in preceding months may be those of males looking for mates during the breeding season, which is usually March–May in Georgia, but could be year-round in Florida.

In more northern climates, mud turtles make long estivation burrows in mud banks during droughts or for hibernation during the winter. These burrows can be nearly a meter (3.3 ft) long and not much wider than the body width of the turtle, about 5–10 cm (2–4 in). In Georgia, the eastern mud turtle likewise overwinters in shallow burrows, 2–11 cm (0.8–4.3 in) deep, for an average of about 100 days, which means it spends nearly one third of any given year out of the water (Bennett 1972; Steen et al. 2007). Interestingly, female mud turtles often burrow before and after laying eggs, but especially afterward, and near their egg clutches. Originally, this behavior was interpreted simplistically as the female staying close by to protect the egg clutch from predators. Now, though, this is viewed as a combination of protectiveness and recuperation, as the female is tired from walking for several weeks, digging a nest, and laying eggs. For anyone who has given birth, imagine the comfort of staying in a burrow for a few weeks of rest afterward, hidden away from the surface world.

Lizards

Lizards are common tracemakers on Georgia barrier islands and readily seen in interdune meadows, maritime forests, and other terrestrial environments. Many lizards dig shallow burrows for protection and thermoregulation (explained later), but their most commonly spotted traces are trackways left on dune sands. Lizard trackways are potentially confused with ghost crab trackways, although the presence of a medial tail drag mark and four fewer leg impressions should help an amateur ichnologist to discern these accurately. Fast-moving lizards, turbocharged by morning sunlight during Georgia summers, will sometimes leave only intermittent tail impressions, and footfalls become correspondingly lighter and spaced farther apart. Trackways of fast-moving lizards also show overstepping of the manus impression by the pes on either side, whereas slower movements result in the pes landing behind or on the manus print, as in trackways of salamanders.

Probably the most frequently encountered lizard trackways on the Georgia barrier islands will be those of the six-lined racer (*Cnemidophorus sex-*

Figure 7.7. Trackway of a mud turtle (*Kinosternon* sp.), showing manus (M) and pes (P) impressions, moving from left to right across inland dunes adjacent to freshwater lake (Lake Whitney): Cumberland Island. Scale = 10 cm (4 in).

lineatus), a relatively large lizard, 15–25 cm (6–10 in) long (including its tail), that runs quickly across surfaces throughout back-dune meadows and other sandy areas. Other lizards that make smaller but similar-looking trackways are green anoles (*Anolis carolinensis*), ground skinks (*Scincella lateralis*), various skinks of the genus *Eumeces* (*E. fasciatus, E. laticeps,* and *E. inexpectatus*), and the northern fence lizard (*Sceloporus undulatus*). Lizard tracks are all characterized by five digit impressions on the manus and pes, claw marks, and nearly straight medial drag marks, the last imparted by their stiff tails (Figure 7.8a). Using the fence lizard as an example, its manus and pes are about 2.0 cm (0.8 in) and 2.5 cm (1 in) long, respectively; its running stride is about 8 cm (3 in) and straddle is slightly less than half the stride (Halfpenny and Bruchac 2002). Other lizard trackways can be recognized on the basis of track length, stride, and straddle in proportion to the size of the tracemaker, although some species overlap in their dimensions. For example, fence lizards are 10–19 cm (4–7.5 in) in total length, whereas green anoles are 12–20 cm (4.7–7.8 in), and five-lined skinks (*E. fasciatus*) are 12–22 cm (4.7–8.7 in) (Jensen et al. 2008). In contrast, the broad-headed skink (*E. laticeps*), the largest legged lizard native to the Georgia barriers, can be more than 30 cm (12 in) in total length, hence its tracks and trackways should be noticeable larger than those of all other lizards. Habitat is also important to consider when encountering lizard tracks, as some lizards, such as green anoles and ground skinks, prefer to live in vegetated areas in maritime forests or back-dune meadows.

Locomotion traces of one type of lizard in back-dune meadows, however, are potentially confused with those of snakes. Glass lizards, consisting of various species of *Ophisaurus,* are legless lizards that derived their common name from an effective defense used against predators or people who like to grab them: their tails breaks off when grasped, leaving a twitching (and hence distracting) piece of the lizard. Meanwhile, the rest of the lizard slithers away, free to grow another tail and live its life a little longer. (Think of the interesting traces made by such a predator–prey interaction!) Although these lizards are more common in maritime forests—where they forage in forest debris—their sinuous trails might be seen in sandy areas, and probably look much like those of small snakes.

Widespread lizard traces other than tracks and trails are their burrows, which are slightly inclined (subhorizontal), as much as 20 cm (8 in) deep, and about 2–4 cm (0.8–1.6 in) wide, with half-moon cross sections; the flat part is on the bottom of the burrow entrance (Figure 7.8b). Like gopher tortoise burrows, lizard burrows are often oriented to the south–southeast to help get their occupants going in the morning, the ectotherm equivalent of espresso. As morning sunlight illuminates the front of a burrow, the lizard places its head at the entrance to soak up this heat, raising its body temperature and energizing it for the day. Burrows also work well for keeping warm; for example, six-lined racers, which are famous for loving hot environments, make and temporarily stay in burrows during cool weather, or hibernate in them through the winter (Etheridge et al. 1983). Because six-lined racers prefer loose sand, they sometimes do the equivalent of swimming through it by laterally undulating their bodies. Unlike many burrowing vertebrates, their legs, which are reduced compared to most lizards, are not used for burrowing but instead are held along their sides as they wriggle through the sand. With this motion, the surrounding sand instantly collapses behind the lizard with its forward movement. Such burrowing results in vague disturbances of sedimentary bedding, which if fossilized could only be recognized as a burrow of some sort, but probably could not be linked to a specific tracemaker, let alone a small vertebrate.

Probably the most common burrowing lizard on the Georgia coast is the appropriately named mole skink (*Eumeces egregious*), a 10–15 cm (4–6 in) long lizard that is told apart from other skinks by its red tail and its procliv-

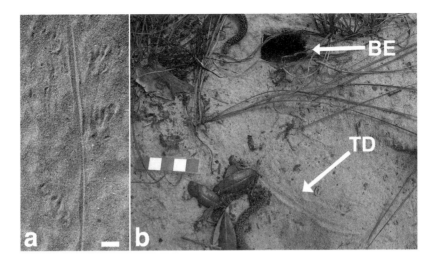

Figure 7.8. Lizard traces. (a) Trackway, probably of a skink (*Eumeces* sp.): Sapelo Island. Scale = 10 cm (4 in). (b) Burrow entrance (BE), probably of a mole skink (*Eumeces egregious*), with tail drag mark (TD) in front. Both traces preserved in sediment mound made by a gopher tortoise: St. Catherines Island. Scale in centimeters.

ity for burrowing, especially if used as a means of escape. In other words, if you see a wriggling flash of red tail before a small lizard disappears under a layer of sand, it was probably a mole skink. Mole skinks are among the many species responsible for making composite traces with gopher tortoises, as their burrows are commonly encountered in the sand aprons outside of tortoise burrows. With these burrows, they also can be ambush predators, sprinting a short distance to snatch any invertebrates strolling by the burrow entrance. Because they have reduced legs and are so adept at burrowing—which they accomplish mostly by rapidly undulating their bodies through loose sand—they almost never leave trackways far from their burrow entrances. As a result, these can be eliminated from consideration for nearly all other lengthy lizard trackways in Georgia coastal environments.

Snakes

Snakes are widespread tracemakers on the Georgia barrier islands, forming many trails through maritime forests, back-dune meadows, and occasionally on dunes; unlike alligators, they have little reason to go to the beach. Among these snakes, the southern black racer (*Coluber constrictor*) is the snake that visitors to the islands are likely to see, as it frequents most terrestrial areas. Trails of this and other snakes are normally discerned as shallow, smooth, sine wave grooves on sandy surfaces (Figure 7.9). Nearly as common as the southern black racer is the eastern garter snake (*Thamnophis sirtalis*), which also ranges from maritime forests to interdune environments, to the shorelines of freshwater ponds; the last of these is a good place to find tasty amphibians. Its trails can be distinguished from the southern black racer by its smaller width and wavelength.

Snakes that tend to live near freshwater ecosystems include the eastern mud snake (*Farancia abacura*) and the peninsular ribbon snake (*Thamnophis sauritus*). The eastern mud snake is especially fond of amphiumas (in a predatory way, that is), and hence stays close to its preferred food source (Jensen et al. 2008). As might be discerned by their common names, mud snakes inhabit wet, muddy environments, such as freshwater sloughs and temporary ponds. Likewise, peninsular ribbon snakes are linked to freshwater habitats, although they are not as comfortable in the water as some other semiaquatic snakes and tend to stay close to shore. Regardless, trails of both species of snakes could exit and enter such water bodies, and can be distinguished from tail drag marks of juvenile alligators by the lack of tracks on either side of the trail.

Of course, the snake that everyone worries about the most, but is least likely to encounter, is the eastern diamondback rattlesnake (*Crotalus adamanteus*), the most massive snake of the Georgia barrier islands. Eyewitnesses have told me about rattlesnakes on the islands that were 2.5 m (8.2 ft) long and 20 cm (8 in) diameter, confirmed by the sizes of their hunting trails, which I have seen in the debris of a many a maritime forest (Chapter 3). As the top predator of the lowermost part of a maritime forest, its use of an auditory warning system (also known as a rattle), and its ability to inject venom, it can do pretty much whatever it wants. If left unmolested by fearful humans, these snakes can enjoy long lives of more than 20 years and grow to legendary sizes. They mostly hunt on the ground, although some have been known to pursue prey up low-lying vegetation. Once they strike, the prey animal may run away, but it quickly succumbs to hematoxin-laden venom and dies, which allows the rattlesnake to eat it without any further disagreement. An interesting series of traces then would be that of a sinuous rattlesnake trail stopping in a coiled form, just in front of a small mammal trackway—say, of a marsh rabbit (Chapter 8)—followed by a high-

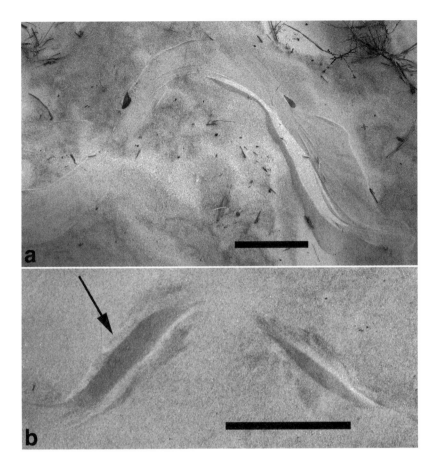

Figure 7.9. Snake trails associated with species and behavior. (a) Trail made by a slow-moving 1.7 m (5.5 ft) long yellow rat snake (*Elaphe obsoleta*, subspecies *E. o. quadrivittata*) registered in dune sand of Cumberland Island. Motion was from right to left, based on cross-cutting of scale drag marks caused by previous movements. (b) Trail made by a fast-moving 1.5 m (5 ft) long coachwhip (*Masticophis flagellum*), also in dune sand of Cumberland Island. Movement was left to right, evidenced by prominent pressure-release structure (*arrow*) where snake pushed against the sand. Scale bar = 10 cm in both photos. Measurements of snakes are of total body length and were made directly by the author and Carol Ruckdeschel.

speed gallop pattern that becomes increasingly shorter, then stops with a resting trace that segues into a taphichnion (Chapter 6). This trackway would be crosscut by the slow, methodical trail of the rattlesnake, which would end at the death trace of the mammal, and show a gradual widening of the snake resting trace caused by the prey item moving down its gullet. Oh well: an ichnologist can dream.

Although not an endangered species, eastern diamondbacks are somewhat uncommon on most Georgia barrier islands, with the exception of Little St. Simons, which has the highest concentration of rattlesnakes of any place in Georgia. So if you see one, and its venomous aspect does not deter any paranoia-driven wishes to inflict harm on it, think of its children. Like many snakes, rattlesnakes give live birth, or are viviparous. Moreover, as burrow commensals of gopher tortoises, a mother rattlesnake may have a litter of newborn snakes in a tortoise burrow. So-called rattlesnake roundups, in which people try to find and kill as many rattlesnakes as possible in a single day, involve pumping smoke or pouring gasoline into gopher tortoise burrows to make rattlesnakes run out of the burrows (Means 2009). (Now that you know how many other species live in gopher tortoise burrows, think of the amount of collateral damage caused by such a practice.) Rattlesnakes often overwinter in burrows of gopher tortoises, armadillos, or other vertebrates.

Many other snake species of the Georgia coast are secondary occupiers of burrows, and are not known to modify these preexisting structures. The closest some snakes of the Georgia coast come to actively making their own burrows is when they push apart forest debris to make shallow tunnels. For example, the northern scarlet snake (*Cemophora coccinea*) does much of its hunting for eggs and small vertebrates under forest litter (Jensen et al. 2008). Hence it could be considered a burrower, but the preservation potential of such traces in vegetative debris is extremely small. Species of egg-bearing (*oviviparous*) Georgia coast snakes are also not known to make hole nests, but instead lay eggs under dead vegetation.

Snake trails may seem like the vertebrate traces that are most easily lumped into one category, defying any further scrutiny from a budding neoichnologist. Please resist these urges of intellectual (or is it ichnological?) laziness. As the gentle reader might have noticed, a few small clues can help to narrow down the species that made a seemingly nondescript trail. Among these hints are (1) habitat, such as the middle of a maritime forest, near a freshwater slough, or in a back-dune meadow; (2) trail widths, from 5–20 cm (2.5–8 in) wide (and if the latter, should you start running?); and (3) wavelength, where longer wavelengths may correspond to longer snakes, depending on variations of their behavior along a trail. With regard to behavior, a slow-moving yellow rat snake (*Elaphe obsoleta*, subspecies *E. o. quadrivittata*) will make a higher-amplitude trail than that of a fast-moving coachwhip (*Masticophis flagellum*), even if they are nearly the same length (Figure 7.9). Although this checklist is not comprehensive, it can aid in identifying probable tracemakers for such trails, especially when using criteria outlined for common snake species of the Georgia coast. Also keep in mind, though, that snakes can change their motion from more of a simple, straightforward sine wave to lateral undulations (Gans 1986), which accordingly make a much wider trail as the body drags laterally. Other than these two basic patterns, no snakes on the Georgia coast are known to sidewind. This is fortunate, as the resultant trails made by such an interesting (but complicated) form of locomotion would be difficult for me to explain.

SUMMARY OF FISH, AMPHIBIAN, AND REPTILE TRACES

The Georgia barrier islands, like many islands worldwide, contain a rich variety of fish, amphibian, and reptilian tracemakers, which can help immensely with providing models for suspected trace fossils. The most ecologically and ichnologically important of these tracemakers are the American alligator and gopher tortoise, but others are significant tracemakers in their formation of burrows, tracks, trails, and nests. Any and all of these tracemakers, however, deserve much more study of their traces, especially in terms of qualifying and quantifying their variations in the context of terrestrial and freshwater substrates.

Perhaps the most intriguing potential application of modern amphibian and reptile traces from the Georgia coast to the fossil record is with regard to interpreting seasonality. As ectotherms living in temperate to subtropical conditions with only occasional freezing during the winter, traces made by these vertebrates of the Georgia islands are made nearly year round. Nonetheless, the islands still reflect the pronounced behavioral differences of winter and summer extremes in temperatures, moisture, and food availability. Indeed, some questions about how amphibian and reptile-like animals adapted to climatic extremes of more high-latitude environments in the geologic past, such as polar dinosaurs (Martin 2009b), can likely be answered through more study of modern traces of their closest living relatives.

8

TERRESTRIAL VERTEBRATES, PART II:

Birds and Mammals

PLAYING 'POSSUM AND DR. BUZZARD

I could not help but notice the dead body of the opossum (*Didelphis virginiana*), a dark lump on an otherwise light brown sandy road, as I traveled through Hog Hammock (Sapelo Island) that last morning of July. Squinting through the dusty windshield of a UGA Marine Institute pickup truck, I slowed the vehicle and unrolled the driver's-side window to get a peek at its corpse. Judging from the shiny flecks of blood around part of the body, the opossum was freshly killed. Its position on the right-center side of the road implied it was walking across the road from the left, a supposition based on the location of a maritime forest (the opossum's probable former home) to the left, or west. So the opossum probably met its demise from a motor vehicle driving north (the same direction I was heading) as it walked from west to east. I noted the time—8:25 AM—and figured that it must have been struck less than an hour beforehand. With this mixture of observations and hypotheses duly noted, I did not pause long. My foot pressed on the accelerator and the truck lurched and rumbled forward in response. Other learning opportunities awaited in the northeastern corner of the island, well into the thickest part of the maritime forest. I wanted to see tracks and other traces there while they remained crisply defined from the previous night.

I had already tracked earlier that morning at the Hog Hammock dump site just south of town, a place frequented by two species of large, scavenging birds—black vultures (*Coragyps atratus*) and turkey vultures (*Cathartes aura*)—as well as raccoons (*Procyon lotor*), armadillos (*Dasypus novemcinctus*), opossums, mice, and rats. So why go to a dump on an island that has long, beautiful, and mostly uninhabited beaches? After a rather odorous experience at the trash heap (and remember, this was late July in south Georgia), the same thought had gone through my mind. Nonetheless, the visit had been well worth the olfactory assault, for I had seen a large number of vulture tracks in a small area, accompanied by tracks and other traces of mammals who had popped in for some human-made food. The vulture tracks in particular intrigued me, as I was entertaining the idea that tracks of the two species might be distinguishable from one another. However, I also knew that accomplishing such a task might require a large sample size of tracks, as well as my watching each species make these tracks. The birds themselves are easy to identify by sight, but as an ichnologist, I wanted much more than to just see a bird, say its name, and move on. I wanted to know how each species walked, landed, and took off; what they ate; and how they related with other individuals within their species and between species. In other words, I wanted to know all of their life secrets. The Hog Hammock dump was the perfect place to make the concentrated observations necessary to answer such questions. Biased sampling? Yes. Unnatural conditions? Sure. Would the tracemakers' behaviors be influenced by a continual human presence? Definitely. Regardless, it was a place to start understanding the basic track forms and behavioral patterns for each vulture species. Afterward, I felt a little closer to meeting that goal, although I was left hungering for answers to the inevitable deluge of questions provoked by any cursory investigation.

I later regretted not stopping and doing a complete forensic analysis of the opossum's untimely death. Nonetheless, I visualized it. First, I would have tried to backtrack the opossum to see where it had been and what it had been doing just before its fatal misstep. How to tell the difference between its tracks and those of abundant and similarly sized raccoons and armadillos living in the same area? For one, opossums have five digits on each foot, with slightly bulbous tips to their toes, most of which are punctuated by claw marks. The most identifiable part of an opossum track, though—the one that screams "'Possum!"—is its backward-pointing first toe (digit I) on the rear foot (pes), which serves as an opposable thumb for grasping branches in treetops. Pes tracks of an adult opossum are 5.5 cm (2.2 in) wide and 5.0 cm (2 in) long, whereas the front-foot (manus) tracks are 5.5 cm (2.2 in) wide and 4.0 cm (1.6 in) long. Although raccoon tracks also have five toes in each foot, these are noticeably larger, and the rear feet leave 10 cm (4 in) long bear-like prints, with all toes pointing forward. Armadillo tracks normally register only three prominent toes in the pes and two in the manus, with marks made by stout digging claws. Opossum trackway patterns are likewise distinguishable from those of raccoons and armadillos. Raccoons usually place their opposite-side front and rear feet next to one another while walking, whereas opossums and armadillos are mostly slow diagonal walkers, putting their same-side rear feet just behind their front feet, with all toe prints mashed together. Last, both armadillos and opossums have heavy tails that drag along the ground, but an armadillo tail more often leaves a thicker impression. Thus when studied in total, traces linked to each animal could be readily separated from one another, and no other mammals on the Georgia barrier islands fit their ichnological profiles.

With opossum-trace search images in mind, I would have followed the drag mark of the opossum body to where it was originally hit by the vehicle, and then tracked it backward from there. Earlier I noted the bare, sandy areas separating the road from the maritime forest just to the west; these places lacked tire tracks and thus afforded clear, unimpeded tracking. Knowing that opossums are largely arboreal, and keeping in mind the large number of fresh dog tracks on the east side of the road—a canine deterrence to opossum progress—this one more likely came from the woods. As a result, I would have glanced to the west to look for where the opossum had entered the road (a method trackers call "cutting for sign"), rather than meticulously studying each individual track. Given enough time, I would have saved the most painstaking work for the maritime forest, looking for compression shapes in the pine needles and live-oak leaves of the forest floor that matched the same opossum trackway pattern. Once found, this trackway may have led to the tree from which the opossum had descended one last time; perhaps the bark recorded newly made claw marks. Last, I would have looked at the tire tracks of the vehicle to see whether the driver had applied the brakes, swerved, or otherwise tried to slow down. In other words, he or she may have seen the opossum, and cared enough to avoid hitting it. Or, maybe in the early morning light, the driver had not noticed it before striking it with one of the front tires or bumper, and instead only heard an odd thump. A slight turn to the right in the tire tracks just past the impact site may have revealed where the driver unconsciously pulled the steering wheel in that direction, a trace made while glancing into the rearview mirror to see what might have caused the aberrant noise.

After imagining all of this, I also wondered whether the opossum had, as its final act, performed a defensive move selected by millions of years of evolution that originally increased its likelihood for survival, only to fail this time. This behavior, which consists of a opossum lying down and acting like it is dead when confronted with a threat, is popularly known as "playing 'possum." Alas, although this lineage of marsupials had survived ice ages, Pleistocene dire wolves, and saber-toothed cats, it was not faring so well against its multiton metallic enemies during the last hundred years or so, even on a sparsely populated barrier island with few paved roads. Playing 'possum was as ineffective a defense against automobiles as that employed by armadillos, which jump up into the air. This aerial feat often places an armadillo at the height of a car's front bumper or grill, which does not work out so well for it, either. Either defense employed against a predator, though, would show up in their respective traces. For example, an opossum would have its trackway end abruptly with a lateral impression of its torso, which in turn would have been followed by its tracks leading away from the "dead" body. Armadillo escape traces, on the other hand, would have started with two closely associated, square-outlined, four-foot patterns of tracks. In one set of tracks, pressure-release structures would indicate where the armadillo sprang upward, whereas the other set would have left landing marks. Although I had not yet seen either pattern indicating these defenses, they were stored in my mind as potential behaviors, in the hope of finding particularly puzzling opossum or armadillo traces.

Still, it was a glorious day to do ichnology, and a healthy amount of it was accomplished in lieu of spending several hours with a dead opossum. The morning consisted of tracking maritime-forest animals, nearly all of which were mammals, but exquisite tracks—complete with scale impressions—and tail drag marks of a large adult alligator (Chapter 7) were thrown into the mix of traces near an ephemeral pond. Prominent muddy depressions spotted near the road were investigated and identified as feral hog (*Sus scrofa*) wallow pits, an assumption confirmed by abundant hog tracks in these pits and scat just outside of them. Two-day-old feral cow (*Bos taurus*) tracks were subtle and obscured by more recent raindrops, but fresh deer tracks made only a few hours before were sharply outlined. Scat of both species was unmistakable. For example, one cow left a huge, fluid patty that would have covered a dinner plate, whereas the much smaller and pinched pellets of deer may have topped off a teacup. The feral cow scat, accompanied by clear tracks, was so fresh that few dung-loving insects had reached it, although I expected that situation to change as the day warmed further. Armadillo and raccoon tracks traveled from maritime forests to sandy roads to mud puddles as they ambled on their seemingly never-ending searches for sustenance or mates.

Reluctantly, I pulled myself away from the maritime forest, and in the afternoon shifted to Cabretta Beach, where the traces of marginal-marine plants, invertebrates, and vertebrates beckoned and, once found, beguiled. Roots and root traces of smooth cordgrass were well exposed in a relict marsh there (Chapters 3, 4), and the sand flats at low tide provided thousands of shorebird tracks and feeding traces. The latter traces were often directly associated with the burrows, trails, and trackways of a wide variety of intertidal invertebrates, such as callianassid shrimp, gastropods, and mole crabs (Chapter 6). If variety is the spice of life, then the traces in front of me that day made for a piquant stew, indeed.

By midafternoon, the July sun beat down on the white sands of Cabretta Beach, and I acknowledged the need for seed time—a respite that allowed for all of the visual stimuli of the day to germinate into more sensible patterns and ideas. So off I went, driving back south along the same sandy road taken in the morning through Hog Hammock. While en route, at about 3:45 PM, I recalled the dead opossum, and wondered how its body had fared since I had seen it seven hours earlier. I narrowed my eyes, looking straight down the road toward Hog Hammock, still about a half kilometer away. A group of black lumps were on the road, shifting positions but concentrated around a small area. Their vulture identities were revealed when one lifted off the ground and was replaced by another from a nearby tree, its broad, black wings displayed. Excitedly, I realized they had found the opossum and had been feeding on it. I found myself becoming reenergized as the heat-induced torpor of the afternoon shrank in the face of the ichnological and taphonomic opportunities in front of me. This was going to be great.

I stopped the pickup truck about a hundred meters from the vultures and turned off the engine. After taking some quick photographs through the windshield, I then stuck my arm out the window and took a few more, the digital camera view screen aiding my aim. Binoculars helped to confirm a minimum of eight black vultures on the ground and a few more in the trees. No turkey vultures were in sight. Approaching the scene would surely cause them to fly away, so before walking toward them, I envisaged how my closing in on their afternoon meal would affect their tracks. First, one by one, they would all turn to face the same direction, preparing for takeoff, with the exception of one or two stubborn malingerers who would try to get in few more bites from the opossum carcass before reluctantly departing. From a standing start, where their feet were side by side, they would hop in half-bound movements (left foot landing slightly in front of the right, or vice versa) while beating their wings. With each hop, their wings would lift them up, hence the offset pairs of tracks would be spaced farther apart further down the trackway. Eventually, and usually within about three or four sets of tracks, each trackway would end with a pair of tracks bearing prominent pressure-release structures that showed a pushing back of the sand, all pointing in the same direction. Even the claw marks would be aligned, and could be used as vectors to resolve the direction of flight at the time of takeoff. In many instances, the trackways would be accompanied by white, liquid trails of their scat, which are common in the takeoff patterns of many birds. Backtracking from these traces of flight should lead me to signs of what they were doing just before the vibrations of the pickup truck reached their feet and warned of my arrival. Thus all of these preintrusion traces would be subtracted from my analysis of the scavenging to give a better picture of how these vultures behaved in the absence of humans.

Sure enough, as I got out of the truck and walked unhurriedly forward, stopping often to watch them, the vultures aligned and took off, one after another. Only one stayed until I was a few meters away, taking advantage of its lone-dining status to nab a few more morsels from the opossum body. Then it too hopped away and flew to a nearby tree. Eight pairs of vulture eyes watched me from above, and I felt their impatient annoyance at my scientifically motivated interruption of their feeding. In a moment of self-

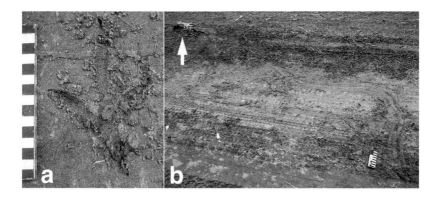

Figure 8.1. Traces of interactions between black vultures (*Coragyps atratus*) and a dead opossum (*Didelphis virginiana*) on a sandy road: Sapelo Island. (a) Left track of one vulture that had stopped in front of the body at one point during its postmortem journey. (b) The meters-long drag mark produced by black vultures fighting over the carcass (*arrow*). Scale = 15 cm (6 in).

awareness, I was amused at the thought of how the whole scene symbolized my mortality. "Not yet," I said to them. With that, I turned to study the postmortem scene in front of me.

It was a mess. The original resting spot of the opossum body could still be perceived, although its outline was alternately obscured and accentuated by repeated compressions imparted by several vehicles that had run over the body. Hundreds of vulture tracks had also trampled the site in a wide radius around the body. Their tracks were about 12–14 cm (4.7–5.5 in) long and had three long, thick, padded digit impressions in front, and a shorter one in the rear that pointed behind the others; all four impressions were punctuated by claw marks (Figure 8.1a). Six or seven vultures had landed a few meters away from the body, walked purposefully toward it, and started feeding. Their numbers were hard to tell, though, because some of their footprints were erased by tire tracks; each approaching car caused them to scatter and return. This first pack of vultures were quickly followed by what I often call the alpha vultures (or bullies), which consisted of two individuals who had arrived a little late to the party, hopped quickly toward the carcass, and tried to chase away the other vultures. A fight ensued, and at least two (probably three) vultures took turns grabbing the opossum body with their beaks and dragging it to another spot before renewed eating. At least three interruptions showed up in the overall pattern of the drag mark of the body, which was nearly 5 m (16 ft) long (Figure 8.1b). Because of this competition, the carcass had been transported to the other side of the road and was now closer to the maritime forest, the opossum's former home. This stop–start pattern reminded me of families arguing at the dinner table, who, after a few feisty exchanges, resume eating in silence for a short time, and then start arguing again. (Not surprisingly, then, I wondered whether any of the vultures were siblings.) At each stopping point, the opossum body had become a little lighter, indicated by a less definite imprint. The greatest amount of sediment disturbance was at the end of the drag mark, and judging from the concentration of overlapping vulture tracks, this was where they ceased most of their squabbling and got down to some earnest eating.

I was astonished at how the opossum body had changed since first viewed in the morning. It was completely eviscerated, consisting only of skin and bones. The vultures were exceedingly thorough, and left almost nothing to waste. Breakdown of the remaining parts of the body would be carried out by a few species of flies, already landing on the skin and presumably laying eggs so their offspring would get a good meal; dermestid (carrion) beetles, which would strip off tiny morsels of flesh from the bones; and aerobic bacteria. I could smell the effects of the latter decomposers, their metabolic processes accelerated by the summertime sun. Although all of the opossum's limbs were held together by ligaments and tendons and still articulated, cars had crushed some of the bones, especially those in the skull, leaving behind a grim two-dimensional death mask. I had also read previously about how scavenging birds often pluck the eyes from a body, but none of those words on a page taught that concept better than the eyeless face in front of me. These vultures were good at what they did, and I left the scene impressed with how quickly they had dispatched the body. This rapidity also provided some insights on the preservation potential of small mammals and similar-sized terrestrial vertebrates in the fossil record, especially if scavengers like these vultures had been present. Given the circumstances, the tracks and other traces of the opossum were much more likely to have preserved, rather than any part of its body. I also mused about how the vultures' ingestion of the opossum temporarily returned its physical essence to the maritime forest from which it had come just that morning.

At the time, I did not know my lack of originality by having an interest in vultures and their behavior. In fact, vultures (or buzzards, if you prefer) have long been important cultural symbols among Native Americans, African Americans, and other people of the southeastern United States. For example, a Cherokee creation myth related how a great vulture once flapped its wings, and the forces imparted by these wing beats created the Appalachian Mountains (Krech 2009): an ichnologically interesting legend indeed. Furthermore, I did not know until just recently that few communities had more keenly preserved vulture awareness than on Sapelo Island, especially in the traditions of the Geechee people in Hog Hammock. For example, famed storyteller Cornelia Bailey, who lives in Hog Hammock and leads cultural tours of the island, has written about the significance of vultures in Geechee culture. She even included the word "buzzard" in the title of her book about growing up on Sapelo (*God, Dr. Buzzard, and the Bolito Man: A Saltwater Geechee Talks about Life*; Bailey and Bledsoe 2000). In other parts of Georgia, Reed-Bingham State Park has an annual Buzzard Festival, including 5-km and 10-km foot races, collectively titled the Road-Kill Run. Similarly, the town of Louisville, Georgia holds a Buzzard Blast every year, which also includes a 5 km (3.1 mi) road race. (The symbolic connection between these races and running from death seems rather obvious.) Just to show the effect this opossum and its vulture admirers had on me personally, I included a photograph of the same opossum carcass in a textbook of mine (Martin 2006d) to illustrate the rapid postmortem changes that could happen to a body before burial. This same photograph (much to the dismay of some students) has been projected in all of its ghastly glory in many of my classroom discussions of taphonomy. I have also used it and photos of the accompanying vulture tracks and opossum-body drag marks in public talks to illustrate the seamless integration between ichnology and taphonomy as sciences. In short, vultures serve an important ecological role as scavengers, they inspire us creatively and recreationally, and they help to educate us.

It was much later, years after that encounter with the opossum and the vultures, and after tracking many vultures on nearly all of the Georgia barrier islands, that I realized how much I had been relearning vulture behaviors well known to many generations of people who had lived on and near the islands. As a result of frequent exposure to these complicated animals, people there had incorporated their knowledge of all things buzzard into an understanding and respect for these birds so disdained by others in modern times. (I test this prejudice occasionally by praising vultures as gorgeous and sophisticated birds, and look for telltale signs of disgust and contempt in a listener.) The lesson learned was that science, although helpful for furthering knowledge, should not divorce itself from other ways of knowing, such as tracking and animal stories, ones that so often feature the birds and mammals of terrestrial environments.

BIRDS IN PARADISE AND THEIR TRACES

Although the gentle reader may be tired of hearing this stated, yet another group of tracemakers on the Georgia coast deserves a book all of its own: birds. This pronouncement is justified because birds (also called avians), including songbirds, corvids (crows, blue jays, and their kin), herons, shorebirds, and raptors, are by far the most prolific, diverse, and behaviorally complex terrestrial vertebrate tracemakers along the Georgia coast. Avians are responsible for tracks, nests, feces, and regurgitants, as well as predation and scavenging traces, in a wide variety of substrates, from intertidal zones to the maritime forests. The Georgia barrier islands are also well known to ornithologists and other bird enthusiasts as places to see a large number of species in a small area, with more than 200 species recorded from the islands over the years. The Georgia coast is thus a place where many life lists are completed by birders looking for that special, rare bird they have never seen anywhere else. Additionally, several of the relatively undeveloped islands, such as Wassaw, St. Catherines, Little St. Simons, and Little Cumberland (Chapter 2), serve as rookeries for some species of shorebirds. For the sake of brevity, however, this overview of avian traces will be split in two, with this part of the chapter limited to birds common to maritime forests, freshwater habitats, back-dune meadows, and coastal dunes. For shorebird traces, look on to the next chapter (Chapter 9).

One way to better understand the anatomies and behaviors of birds in general, which helps to interpret their traces, is to think of them as living dinosaurs. Such a seemingly imaginative exercise is actually not so difficult, because birds, evolutionarily speaking, are dinosaurs. For example, the oldest known bird in the fossil record, *Archaeopteryx lithographica* from the Late Jurassic of Germany, shows a near-perfect blend of anatomical traits of a small theropod dinosaur and a modern bird (Erickson et al. 2009). The hypothesized connection between nonavian dinosaurs and birds went through ups and downs in its support among paleontologists for a long time after the discovery of *Archaeopteryx* in 1861, but became more certain since the 1970s. This acceptance happened as dinosaur paleontologists started looking more carefully at theropod dinosaurs and realized they had many bird-like features. The evolutionary linkage between birds and dinosaurs was especially bolstered since the late 1990s by the discovery of about two dozen species of feathered dinosaurs, most of which come from an Early Cretaceous (130–122 million years old) lake deposit in China (Chang et al. 2009). This number will probably be closer to 30 species by the time you read this sentence, meaning that paleontologists, like birders, like to expand their life lists too. Owing to this evolutionary heritage, researchers increasingly look to birds as models for dinosaur behavior. Accordingly, dinosaur trace fossils—particularly ground nests and tracks—are compared regularly to those of modern birds, and vice versa, although important differences are noted on a case-by-case basis.

Bird Tracks

Although birds are thought of as aerial and arboreal animals, many come down to the ground on a regular basis, and some live there regularly, whether because of flightlessness, ground nesting, foraging habits, or aquatic adaptations. As a result, avian tracks can be found in nearly every place where birds live on the Georgia barrier islands, and these tracks are preserved in a wide variety of substrates. Nevertheless, one cannot find what is not sought, and bird tracks are no exception to this maxim. Furthermore, in my experience with teaching students and nonstudents alike, by far the most common misperception about bird tracks and their trackways is that they all look alike. Alas for such ichnological pessimists, they do not. Avian tracks are not only common but can also be classified on the basis of the arrangement of the digits; sizes of certain toes (especially digit I, explained soon); presence or absence of webbing between the toes; and amount of webbing (Elbroch and Marks 2001). Using these traits as guides, avian tracks fall into four easy-to-distinguish categories: anisodactyl, palmate, totipalmate, and zygodactyl (Figure 8.2). With these track groupings in mind, along with other clues, ecological knowledge, and a little bit of patience, the tracks of nearly 100 species of birds of the Georgia barrier islands can not only be found, but can be reliably distinguished from one another.

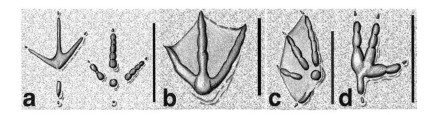

Figure 8.2. Bird track categories, based on overall form, using the terms of Elbroch and Marks (2001) and tracks of bird species native to the Georgia barrier islands. (a) Anisodactyl, both normal (*left*), made by a white ibis (*Eudocimus albus*) and incumbent (*right*), by a wild turkey (*Meleagris gallopavo*). (b) Palmate, made by Canada goose (*Branta canadensis*). (c) Totipalmate, made by double-crested cormorant (*Phalacrocorax auritus*). (d) Zygodactyl, made by great horned owl (*Bubo virginianus*). Scale = 10 cm (4 in) in all figures. All tracks are from right feet.

Anisodactyl tracks consist of three digits (II–IV) pointing forward (anterior) and digit I (also known as the hallux) pointing backward (posterior), so that digits I and III form a nearly straight line (Figure 8.2a). Such a condition is sometimes referred to as a retroflexed hallux, in recognition that, with modification from an originally dinosaurian foot, this digit migrated to the posterior portion of the foot. Such tracks then can be further subdivided on the basis of digit I. If the hallux is reduced, digit III is also raised in its posterior portion so that the metatarsal part of the foot registers; this is also called an incumbent foot. Passerines (often summarized as songbirds), herons, ibises, and raptors make normal anisodactyl tracks (Table 8.1), whereas incumbent tracks are made by a variety of shorebirds (Chapter 8) and ground-dwelling birds, such as bobwhites (*Colinus virginianus*) and wild turkeys (*Meleagris gallopavo*).

In this respect, anisodactyl tracks can reveal much about the life habits of their avian trackmakers. For example, a long hallux is used for grasping branches, in which this digit meets the forwardly pointing three digits to hold tightly onto its perch. You can imitate this behavior by pretending your thumb is the hallux and the following three fingers are others, then try holding onto a tree branch or limb. Bird tracks with this form thus tell an ichnologist that their makers were adapted for maritime forests, back-dune meadows, or similar ecosystems with large trees or shrubs. In contrast, the reduction (or absence) of a hallux would result in a bird falling off a branch as soon as it alights from flight: not a scene one would normally

TABLE 8.1. TRACKS OF COMMON BIRDS IN TERRESTRIAL ENVIRONMENTS OF THE GEORGIA COAST

Common Name	Species	Track Length (cm)
A. Anisodactyl, with prominent hallux impression		
Belted kingfisher	*Ceryle alcyon*	2.8–3.1
Ground dove	*Columbina passerina*	2.8–3.1
House sparrow	*Passer domesticus*	2.8–3.3
Northern cardinal	*Cardinalis cardinalis*	3.2–4.0
Song sparrow	*Melospiza melodia*	3.5–4.1
White-breasted nuthatch	*Sitta carolinensis*	3.8–4.1
Mourning dove	*Zenaida macroura*	4.1–4.7
Northern mockingbird	*Mimus polyglottus*	4.4–5.0
Red-winged blackbird	*Agelaius phoeniceus*	4.3–5.2
American robin	*Turdus migratorius*	4.4–5.5
European starling	*Sturnus vulgaris*	4.5–5.5
Blue jay	*Cyanocitta cristata*	4.4–5.7
Rock dove (pigeon)	*Columba livia*	4.4–5.5
Boat-tailed grackle	*Quiscalus major*	6.0–8.2
Fish crow	*Corvus ossifragus*	7.5–8.5
American crow	*Corvus brachyrhynchus*	7.5–9.0
Green heron	*Butorides virescens*	8.0–9.5
Red-tailed hawk	*Buteo jamaicensis*	9.5–13.5
Cattle egret	*Bubulcus ibis*	10.0–11.5
Snowy egret	*Egretta thula*	10–11.5
Black vulture	*Coragyps atratus*	9.5–14.0
Turkey vulture	*Cathartes aura*	9.5–14.0
White ibis	*Eudocimus albus*	12.0–13.5
Great egret	*Ardea alba*	16.5–18.5
Bald eagle	*Haliaeetus leucocephalus*	16.0–19.0
Great blue heron	*Ardea herodias*	16.5–21.5
B. Anisodactyl incumbent, with reduced or absent hallux impression		
Killdeer	*Charadrius vociferous*	2.5–3.0
Northern bobwhite	*Colinus virginianus*	3.5–4.0
Common snipe	*Gallinago gallinago*	3.5–4.5
Virginia rail	*Rallus limicola*	4.5–5.0
American coot	*Fulica americana*	8.0–9.0
Sandhill crane	*Grus canadensis*	9.5–12.0
Wild turkey	*Meleagris gallopavo*	9.5–12.5
C. Palmate		
Green-winged teal	*Anas crecca*	3.5–4.5
Mallard	*Anas platyrhynchos*	6.0–7.5
Canada goose	*Branta canadensis*	10.0–12.0
D. Zygodactyl		
Eastern screech owl	*Otus asio*	4.0–5.0
Northern flicker	*Colaptes auratus*	4.5–6.0
Pileated woodpecker	*Dryocopus pileatus*	5.5–6.5
Great horned owl	*Bubo virginianus*	8.2–11.5

Note. Environments include areas such as such as freshwater sloughs, maritime forests, back-dune meadows, and dunes. Tracks are organized on the basis of track category and ranked within each category by length. Track lengths from Elbroch and Marks (2001) and my field measurements. Track lengths include hallux impressions for categories A and D, but exclude these for categories B and C. See Table 9.1 for shorebird tracks.

see in nature specials. Try this again with your hands and hold onto a low-lying tree branch, but using only your forwardly pointing four digits and not your thumb. (For personal safety reasons, though, please remember that I said "low-lying" tree branch.) This is not to say, however, that birds with incumbent feet will not be seen in trees, as some species are still good climbers or can fly onto wide branches that serve more as platforms, rather than perches to be grasped. For example, wild turkeys have such feet, yet I have had to crane my neck to look at groups of these birds high up in the branches of live oaks on Cumberland Island. Other researchers have even noted turkeys nesting in trees on Ossabaw Island (Fletcher and Parker, 1994).

Further evolutionary adaptations gleaned from anisodactyl tracks include the presence or absence of webbing, such as a small amount between digits II and IV, termed proximal webbing; these feet are also referred to as semipalmate. The presence of webbing in a bird track is a sure sign that bird is adapted to aquatic conditions, and its degree of adaptation to watery environments can be further elucidated by the amount of webbing. For example, semipalmate tracks with proximal webbing are likely to have been made by wading birds, such as herons and egrets, not swimming birds (explained next). Furthermore, because herons and egrets have four-toed anisodactyl feet with a prominent, retroflexed hallux, they are also excellent at perching, which aids in their nest building and avoiding predators.

Palmate and totipalmate tracks are both characterized by full webbing between the digits, also called distal webbing because it is further out on the foot and near the ends of digits (Figure 8.2b,c). As can be easily guessed, these tracks indicate birds adapted for much of their lives in water. This webbing helps considerably when paddling on the surface or diving, just as snorkelers and scuba divers prefer to use fins when swimming. Webbing, however, results in slower movement on land, hence the pace and stride of a web-footed bird normally is shorter than an equivalently sized bird without webbing. Typical palmate trackmakers include a large number of shorebirds, such as terns and gulls (Chapter 8), but also ducks and geese. The difference between palmate and totipalmate tracks is then denoted by how the hallux relates to the webbing. In palmate tracks, the webbing is only between digits II, III, and IV, with a reduced or absent hallux in the rear of the track. In totipalmate tracks, the hallux is not only connected to digit II by webbing, but points toward the midline of the trackway. These tracks, which on the Georgia coast are only made by brown pelicans (*Pelecanus occidentalis*) and double-crested cormorants (*Phalacrocorax penicillatus*), are nearly unmistakable because of their distinctive shapes and sizes, as discussed later (Chapter 8).

Zygodactyl tracks (Figure 8.2d) are odd compared to most bird tracks, making an X or K pattern as a result of two toes pointing forward (digits II and III) and two pointing backward (digits I and IV). Once zygodactyl tracks are recognized, think "trees," as these footprints are invariably associated with arboreal birds such as owls, flickers, and woodpeckers. More specifically, these tracks indicate feet adapted for vertical movement along a tree trunk, in which the trackmakers grip the surface from two directions as they move up and down. However uncommon such tracks might seem because of their tree-dwelling makers, they are possible, as owls and woodpeckers are known to spend time on the ground searching for or consuming food. Perhaps the most counterintuitive trait of owl tracks, however, is the variability of digit IV, which can point to the outside of the trackway and thus make its left footprint look much like the right footprint of an anisodactyl bird. So if an avian trackway seemingly shows a bird that continually cross-stepped, reevaluate which foot is right and left, and consider an owl as the trackmaker.

Other small but important anatomical details of bird feet may be garnered from their tracks, given the right combination of substrate and behavior. A slow-moving bird stepping into a firm mud or fine, moist sand, for example, may leave exquisitely defined digital pads and scale impressions on those pads. The thickness and number of pads in each digit may help with defining the trackmaker, and are otherwise worth noting if found. These are not, however, necessarily equivalent to the number of bones composing a digit, as we might intuit by looking at digital pads in our hands and feet. Scale or pad impressions are always exciting to find in bird tracks, a skin imprint that reminds us of their dinosaurian ancestry.

Size is also important when identifying avian tracks, as some tracks of related species are geometrically similar but differ greatly in width and length. For example, tracks of both a ground dove (*Columbina passerina*) and great blue heron (*Ardea herodias*) are normal anisodactyl tracks, which have a prominent, retroflexed hallux impression. A great blue heron track, though, is at least three times longer than that of a dove. Other anatomical differences between their tracks are discernible, but size immediately distinguishes them from one another and other species. Likewise, if a downy woodpecker (*Picoides pubescens*) ever landed in fine-grained sand, its zygodactyl tracks would be much smaller than the zygodactyl tracks of a pileated woodpecker (*Dryocopus pileatus*).

Bird Trackways, Including Evidence of Flight

Identifying a bird species from a track or two is perhaps impressive to a novice ichnologist, but learning does not stop with naming. To truly understand unseen bird behavior from their tracks, avian trackway patterns must be studied. "How hard could that be?" one might ask naively. After all, birds, just like humans, only walk bipedally, but some spend a fair amount of time in the air or trees, and in contrast to their human admirers, they have small brains relative to their body sizes. Ah, if only ichnology were so easy, fitting our ideals of simplicity and order, and fulfilling a mechanistic view of animal behavior, all bolstered by a smug sense of superiority. Instead, get ready to be humbled and awed. Bird trackways are actually quite complicated, and often can reveal previously hidden behavioral nuances that normally cannot be witnessed through binoculars or a spotting scope.

Yes, birds walk bipedally, but this is a gross oversimplification of avian locomotion. Try walking on a narrow pathway, like a balance beam or a tightrope, and you will begin to better appreciate how avian anatomies cause trackway patterns far different from those of upright primates. In normal walking, most birds place one foot directly in front of the other, resulting in a tight, linear pattern with a high pace angulation, almost 170° (Figure 8.3a). This pattern is a result of each foot swinging around and from behind in an arc, rather than moving straight forward. This movement neatly keeps the trailing foot from striking the forward foot, which would result in a much shorter trackway. Deviations from the expected trackway pattern are, of course, more interesting. While birds walk, they change their pace along any given path. Birds slow down. Birds speed up. Birds pause. Birds reach over one shoulder with their beaks to scratch their backs. Birds peck at the ground (forward, right, left) while walking. Birds look to their right or left, often abruptly, especially if startled by a rapid movement on the periphery of their vision, or a sharp sound. Birds sometimes move in harmony with other birds of their species, either with a mate or a group, the latter a sort of flock behavior on the ground. In other words, birds will do much more than just walk forward, one foot after another,

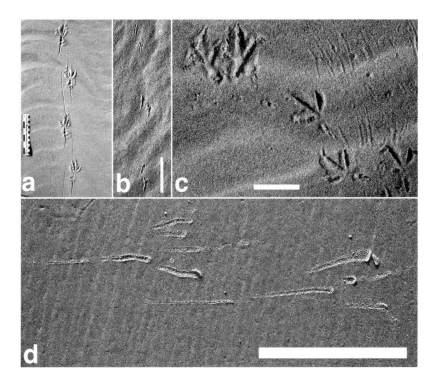

Figure 8.3. Examples of bird gait patterns and flying traces. (a) Normal diagonal walking of boat-tailed grackle (*Quiscalus major*): Sapelo Island. Scale = 10 cm (4 in). (b) Hopping, with offset pairs of tracks, probably of a sparrow (*Spizella* sp.): Little St. Simons Island. Scale = 10 cm (4 in). (c) Paired tracks associated with takeoff of common ground dove (*Columbina passerina*): Sapelo Island. Scale = 5 cm (2 in). (d) Paired tracks of great egret (*Ardea alba*) associated with landing: Jekyll Island. Scale = 10 cm (4 in).

like automatons. They not only behave, they behave complexly, and many of these behaviors are recorded by their tracks. To best detect variations in avian trackways, look for any deviations from an expected straight-line, tightrope sort of left–right, right–left pattern, such as where one footprint seems to lag slightly behind the other, goes off to one side, or drags along the ground a bit longer than the other. These variations usually represent rapidly made decisions, in which a bird may have stopped briefly to peck at a seed or insect, looked around for potential threats, listened to a strange (or alluring) sound, or started to walk more rapidly in preparation for flight.

Avian trackway patterns can be classified as diagonal walking or hopping, but differences in pace and stride help to pick out more sublime behaviors along a given trackway. Moreover, gait patterns often reveal whether their bird tracemakers are more adapted for life in trees or the ground (Elbroch and Marks 2001). Normal, diagonal walking patterns are associated with birds that are comfortable on the ground, whereas bipedal hopping (footprints parallel to and directly next to one another), or skipping (footprints parallel but offset slightly) are more typical of birds that only occasionally visit ground surfaces (Figure 8.3b). Only a few birds show combinations of these gait patterns, in which diagonal walking might be mixed with hopping trackway patterns: examples that come to mind are American robins (*Turdus migratorius*) and the previously mentioned black and turkey vultures. Based on my limited research on avian trackways, birds do not show handedness in their tracks, in which one leg is dominant over the other. Handedness could be discerned either by a consistently longer stride on the right or left side of a trackway, or lateral preferences in decision making. This trait does frequently show up in mammal tracks, though, and is especially noticeable in human trackways.

Once bird tracks and trackways have been properly described, take your analysis a step further by looking more closely at individual tracks, especially those from a larger bird, such as a wild turkey or black vulture. Associated with these tracks will be ridges of sand or mud directly associated with the outside or inside of any given track, or other extra sedimentary features that are clearly connected to the track. These ridges are pressure-release structures, formed as a result of the application and release of pressure from a tracemaker's foot (sensu Brown, 1999). Although vertebrate tracks are often reduced to cartoon outlines or black silhouettes in articles, books, and on the Web—even in peer-reviewed research papers—their oft-omitted pressure-release structures can hold a wealth of information for enriching the interpretation of the tracemaker's behavior. For example, a bird that suddenly stopped and looked to its left will have shifted its weight more to that side, which correspondingly caused an increase in pressure against the outside wall of the left-foot track. This stress pushed sand grains against sand grains, which compressed them closer together and formed layers stacked on top of one another, separated by shear planes. (For those of you from northern climates, think of how shoveling snow causes similar sorts of structures.) Of course, such structures will also depend on the nature of the substrate—sediment moisture, grain size, grain sorting, and so on—but these should be noted in any description of a bird trackway for their potential added information. Certain mammal tracks are especially amenable to analysis of pressure-release structures, but these methods also can be applied to larger birds, such as herons, ibises, wood storks, and sandhill cranes.

The tracking of birds becomes even more challenging than tracking mammals when taking into consideration those birds that combine their earthly trackways with flight. As a result, an avian ichnologist should become familiar with takeoff and landing patterns attached to the ends and starts (respectively) of trackways. Although shorebirds are much easier to study in this respect—lots of sand, open landscapes unencumbered by vegetation, and many trackmakers (Chapter 9)—a few principles are applicable to studying trackways of flighted birds in terrestrial and freshwater environments. First of all, anatomy is the single most important determining factor in any given bird species' takeoff pattern. For example, great blue herons, with their long legs and wide wingspans, prepare for flight from a standing position by simply squatting, flapping their wings a few beats, and straightening their legs. Within seconds, they are off the ground. Their tracks then usually show a normal diagonal walking pattern leading to parallel (side by side) tracks, perhaps some pressure-release structures associated with the up-and-down movement of the bird, and no more tracks (Figure 8.3c). In contrast, black vultures or turkey vultures have much shorter legs and wingspans relative to their bodies; hence they need a hopping start to get airborne. Accordingly, their takeoff trackways may start with the same diagonal walking pattern as a heron, which is then interrupted by a motivation to fly, changing abruptly to a bipedal hopping or skipping pattern. What is important to note in such a transition is the change in stride length between each set of tracks: measure the distance between tracks made by the same foot (right or left) and note if this distance increases or not. If the number gets bigger, then the vulture was probably flapping its wings and becoming slightly more airborne with each hop. Look closely enough, and you might even see feather impressions on a sandy surface, where the bird's vigorous wing beating attempted to imitate a Cherokee legend. Of course, the last pair of tracks also marks where the vulture finally took to the air, but this final set might be preceded by as many as 5–6 sets of bipedal-hopping tracks.

Landing patterns in trackways must be scrutinized similarly, and often can be found by backtracking a bird. With flighted birds that like to forage on the ground, this technique inevitably leads to the start of the trackway, which is where the bird landed. I have had the most success with this method when applying it to corvid and passerine tracks in back-dune meadows, particularly those of fish crows (*Corvus ossifragus*), boat-tailed grackles (*Quiscalus major*), and mourning doves (*Zenaida macroura*), but it can be tried on any flighted bird trackway. How a given bird lands is again dependent on its anatomy. Birds with more cumbersome builds and shorter wings need more bouncing (maybe 2–3 hops) before slowing their forward motion enough so they can bring their feet together and walk normally. On the other hand, birds with longer wings relative to their body size or with long legs—think of songbirds and herons—may almost hover, orchestrating a slow-motion control on their fall to the ground, where their feet touch the ground together or slightly offset, after which they start walking.

Naturally, as any pilot can tell you, landing is a controlled fall, in which gravity is a mortal foe. Nonetheless, with this riskiness and pressure-release structures in mind, landing tracks, if taken in isolation, should never be confused with takeoff tracks (Figure 8.3d). For example, a descending bird almost always approaches the ground at an angle. If you ever see one land straight down, then a nearby hunter probably shot it, in which case its landing trace will either be a tool mark or a taphichnion (Chapter 6). As a bird slows its descent, its feet will reach downward with this angle of approach. If the bird has typical anisodactyl feet with a prominent digit I, the first digits that touch the ground will be the hallux of one foot, quickly followed by the hallux of the other foot. More specifically, the hallux claw on one foot is the first part of the bird anatomy to contact the substrate. As the rest of the bird's weight shifts forward and down, the feet roll onto the front part, causing contact of the metatarsal pads and remaining digits with the substrate, from proximal to distal ends: the claws on digits II to IV are the last parts of the feet to touch the ground. If preserved in a firm mud or fine-grained sand, the resulting tracks are absolutely superb, telling the entire story. Digit I impressions on each foot will have long, sharply defined claw drag marks ending in deep points like inscribed commas; sand will have been pushed forward in the anterior portions of each digit impression (including digit I); and other such structures may be associated with any wobbling, hard landings, or other irregularities associated with the initial placement of the feet, followed by the full weight of the bird.

Any given trackway of a flighted bird may contain both its landing and takeoff tracks, as well as everything that happened in between those two events. Consequently, an ichnologist should test what was just described for landing and takeoff tracks by comparing these to their observations, just to better define these, as well as note species-specific differences in such patterns. Takeoff and landing patterns associated with avian tracks, covered more with shorebirds (Chapter 9), provide many opportunities for such study, given their sheer abundance, diversity, and occurrence on uncluttered sandscapes of the Georgia barrier islands. Bird tracks also help to better understand the behaviors of other flighted vertebrates, such as pterosaurs, the flying reptiles of the Mesozoic Era, and the earliest known birds (Chapter 10). Hopefully this introduction to avian tracks will have convinced the reader that birds not only make tracks, but also tell many interesting stories through their tracks.

Bird Nests

All birds, regardless of whether we divide them into habitat-based categories of land birds or shorebirds, must nest in terrestrial environments. Now, this is where knowledgeable birders will ask, "But what about the marsh wren [*Cistothorus palustris*]?" To this, I would counter with my own trenchant rhetorical question: "Is the marsh wren nest actually in the salt marsh, or above the salt marsh?" The answer is (of course) the latter, as no bird nest would function in its capacity for housing and protecting an egg clutch if immersed twice daily by tides. Nonetheless, marsh wrens are rightfully respected for weaving a nest from stalks and leaves of smooth cordgrass that is then suspended between the tallest of such plants and well above the high tide mark in a salt marsh. Yes, geographically speaking, this bird is nesting in the salt marsh, but all bird nests must stay in contact with air, as eggshells and their membranes function as extra organs for their respiring, encased embryos. Similarly, many shorebirds nest in coastal dunes, although most instinctively construct ground nests and lay their eggs above the mean high tide (Chapter 9).

With those near exceptions out of the way, bird nests are in emergent, terrestrial environments of the Georgia coast, whether in maritime forests, back-dune meadows, or on the banks of freshwater sloughs and ponds.

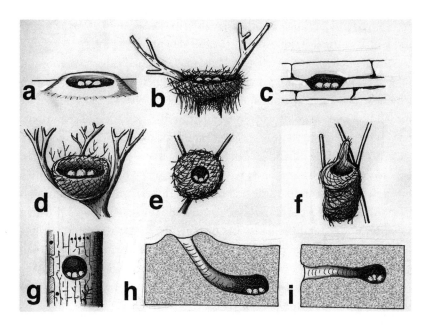

Figure 8.4. Bird nest categories. (a) Scrape, made on ground surface. (b) Platform, which can be in trees or on ground. (c) Crevice, in this instance in rock outcrop. (d) Cup (or saucer), in a bush or tree. (e) Spherical, shaped around a branch. (f) Pendant, hanging from branches. (g) Cavity, bored into a tree. (h–i) Burrows, with one below a ground surface (h) and another as more of a boring into a cliff side (i).

Although nests are traditionally thought of as cleverly arranged, bowl-like weavings of sticks, grasses, and leaves, they are actually quite variable in their construction, placement, and use of substrates. Bird nest architecture, in fact, is one of those instances of avian behavior where the word *ingenious* is applied instead of the more usual insult of *bird brained*. Nevertheless, there are plenty of examples of bird nests that also make us shake our heads ruefully and say, "What were they thinking?"

Avian nests can be named and organized in various ways, but normally fall into categories based on relative vertical position, such as referring to them as ground nests or tree nests; substrates used in making the nests, such as vegetation or sediment; or a mixture of the verticality and substrate modification, such as borings in wood and outcrops of sediment or soft sedimentary rock. Bird nest classifications based on overall form and other architectural traits include scrape, platform, crevice, cup (or saucer), spherical, pendant, cavity, and burrow (Hansell 2000; Elbroch and Marks 2001; Figure 8.4). Most of these nest types are made from loose items found in their environment, although some birds excavate a substrate in order to lay eggs, brood, and raise young. On the Georgia coast, only a few species of birds are known to burrow or bore into sediment or solid substrates.

Scrape nests are the simplest sedimentary structures made by birds related to reproduction, and probably the easiest to link to dinosaurian behavior from the Mesozoic. For example, a few bowl-like structures in the geologic record have been hypothesized as dinosaur nests, which were likely made by scraping the sediment with feet, snouts, or a combination of the two (Varricchio et al., 1997; Chiappe et al. 2004; Meng et al. 2004). Of Georgia coast birds, shorebirds are the most frequent makers of scrape nests (Chapter 9), although some birds common to maritime forests, such as wild turkeys, use their feet to scratch out ground nests in leaf litter before laying their eggs in these slight depressions. Of course, such nests have poor preservation potential in the geologic record, and so far fossil avian nests are relatively rare as trace fossils (Scott et al. 2009). Even the examples of similar nests made by dinosaurs were discerned more from the presence of abundant eggshell material, juvenile bones, and, in a few lucky instances, an adult dinosaur on top of an egg clutch or nest structure (Varricchio et al. 1997).

Far beyond mere scrape nesting, though, is probably the best studied of subsurface birds in North America, the burrowing owl (*Athene cunicularia*). This small owl lives in much of the western United States and Florida, but is not known to live more than one generation in southern Georgia. Apparently, it was either extirpated from the Georgia barrier islands or has not yet made its way north from Florida. A few owls, however, are occasionally brought to the Georgia coast as so-called accidentals, where hurricanes have swept them in from Florida and dropped them onto Sapelo and Jekyll Islands (Parrish et al. 2006; Jen Hilburn, personal communication, 2009). Burrowing is a partial misnomer with reference to these owls, because they normally usurp appropriately sized mammal or gopher tortoise burrows and modify these by digging with their feet. Like gopher tortoises, they prefer well-drained, dry, sandy upland soils, which are best for ensuring the survival of their eggs; each clutch typically consists of 5–10 eggs, but can be as few as two (Martin 1973). Females incubate an egg clutch for about 30 days, which is aided by their staying out of sight (and smell) of predators while in a burrow. Meanwhile, the male is not out carousing with his owl buddies, but is hunting for food to bring back to the burrow each day, which is especially needed after hatchlings demand sustenance. Baby owls start flying on their own after only 40–45 days, which is when their parents teach them how to catch invertebrates and small vertebrates. Burrowing owls are also among the few birds in which tool use has been documented, and the tool they use, however improbable it might sound, is dung. In other words, the trace of one animal is used as a tool by another animal for gathering food (Levey et al. 2004). Burrowing owls love to eat dung beetles, and will grab chunks of nearby mammal excrement and place it as bait in front of their nest holes, where they will lie in wait for any dung beetles that cannot resist the sweet smell of feces (Chapter 5).

Bird nesting takes on a different form on the southernmost islands of the Georgia coast that have 5–10 m (16.5–33 ft) high bluffs, such as Cumberland, St. Catherines, or Sapelo (Chapter 2). Such bluffs formed wherever tidal channels or waves have eroded and exposed outcrops of soft Pleistocene sandstone, providing perfect places for northern rough-winged swallows (*Stelgidopteryx serripennis*) to build their nest cavities. These birds start making these traces by pecking with their beaks, then scratch out initial holes with their feet (Figure 8.5a). They eventually make a horizontal tunnel, which is slightly expanded into a chamber that accommodates the egg clutch, hatchlings, and adults coming in and out of chamber for brooding and raising the young. This chamber also may have a small amount of vegetation added for insulation. Because these holes are excavated in Pleistocene sandstone—albeit poorly cemented—these traces are thus better classified as borings, rather than burrows. Swallow nest cavities are elliptical to circular in outline, about 6–10 cm (2.4–4 in) wide, and can extend 1–2 m (3.3–6.6 ft) into a bank, thus representing a lot of pecking and scratching. Nonetheless, also be aware that northern rough-winged swallows can commandeer bank burrows dug by other birds, such as bank swallows (*Riparia riparia*) and belted kingfishers (*Ceryle alcyon*) (Hill 1988; Parrish et al. 2006). So the preceding information should be tested by direct observations of who does the primary digging, and who moves in later. Furthermore, these structures may be reused and modified by multiple generations of the same species, just like many other vertebrate burrows. The most interesting ichnological aspect of such traces, though, is how they cut across invertebrate

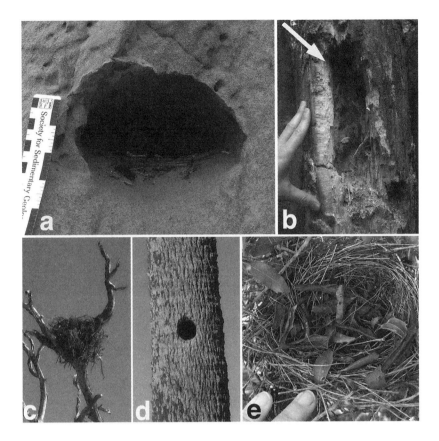

Figure 8.5. Examples of bird-nest structures from the Georgia coast. (a) Nest hole (burrow) of northern rough-winged swallows (*Stelgidopteryx serripennis*), made in Pleistocene sandstone: Cumberland Island. (b) A nearly complete cavity nest (*arrow*) of a pileated woodpecker (*Dryocopus pileatus*) in a not-so-live oak (*Quercus virginiana*), accompanied by many woodpecker feeding traces: Sapelo Island. (c) Platform nest of an osprey (*Pandion haliaetus*) in a maritime forest: St. Catherines Island. (d) Cavity nest of red-bellied woodpecker (*Melanerpes carolinus*): Ossabaw Island. (e) Cup nest of common ground dove, abandoned and now filled with leaves instead of eggs and hatchlings: Little St. Simons Island.

and plant trace fossils within their host sandstones (Chapter 9), thus lending more complexity to the history of tracemakers and time separating their tracemaking. This crosscutting of trace fossils by modern traces is analogous to the effects of boring bivalves on consolidated mud in relict marshes (Chapters 3, 6).

Nest holes in wood may start out as smaller cavities caused by fungal rot or other decay in a dead tree, or as feeding traces made by woodpeckers and other birds that prey on wood-boring insects, such as carpenter ants, bees, beetles, and termites (Chapter 5). In such instances, birds drill into tree trunks with their stout beaks to get at their food source just under the surface. Pileated woodpeckers (*Dryocopus pileatus*) are particularly good at this sort of activity, and their powerful hammering can be heard easily throughout a maritime forest, telling everyone in earshot that deep borings are being made and dinner is served. These woodpeckers, however, also knock on wood to communicate with one another, similar to how ghost crabs use acoustic signaling (Chapter 6). Once a hole is drilled, the woodpecker immediately pokes its long tongue into the freshly made hole and laps up its insect prey. It then may move up or down the trunk of the tree to find another spot to drill, taking full advantage of its zygodactyl feet for these vertical movements. These initial holes are vertically oriented in rows, only a few centimeters across, somewhat rectangular in outline, which narrow inward (Elbroch and Marks 2001). Their edges are defined by grooves, each caused by the beak. Nevertheless, these holes can be quickly modified by the same or different woodpeckers, which return to the same spot and resume drilling for more insects. These starter holes thus can serve as places for easy expansion into much larger, deeper nest cavities, which are characterized by rectangular profiles; vertical orientations; dimensions twice as long as wide, about 10 by 20 cm (4 by 8 in); and depths as much as 50–60 cm (20–24 in) inside the tree (Figure 8.5b). The resulting gourd-shaped cavities then make for excellent hole nests. Their positioning in the largest trees, high above the ground, keeps them away from prying eyes (or sniffing noses) of egg and hatchlings predators. In the warm climate of the Georgia barrier islands, the nest entrance is also more likely to face north, which prevents eggs from cooking from southern sunlight. Male–female pairs of pileated woodpeckers take turns incubating the eggs—usually about four per clutch—over the course of 15–18 days, then raise the hatchlings for 27–30 days before the latter fly off to make their own drill holes and nests (Sibley 2001; Parrish et al. 2006). These nest holes are common in old-growth maritime forests of the Georgia barrier islands, and are difficult to overlook in those ecosystems on Sapelo, St. Catherines, Ossabaw, and Cumberland Islands. These borings, whether used for nests by woodpeckers (or not), also become homes for secondary occupiers, such as small mammals, like mice and flying squirrels. Woodpeckers do not commonly reuse nests, though, meaning that an actively used hole nest is probably a new one (Sibley 2001).

Cavity nests of other woodpeckers are easily distinguished from those of pileated woodpeckers on the basis of size and shape. Woodpecker nest entrances are slightly larger than the diameter of the adult tracemaker; hence smaller holes correspond with smaller birds, such as hairy, downy, red-bellied, and red cockaded woodpeckers. No living woodpecker in the southeastern United States today rivals the size of the pileated, although it was surpassed by the ivory-billed woodpecker (*Campephilus principalis*). This woodpecker, nicknamed the "Lord God Bird" and long assumed extinct, is purportedly still alive in the south-central United States (Fitzpatrick et al. 2005; Hill et al. 2006), but is surely extirpated in Georgia. Furthermore, its nest holes differ noticeably from than those of pileated woodpeckers, having oval cross sections about 11–13 cm (4.3–5 in) wide. (Regardless, some ivory-billed woodpecker nest holes may still be preserved in older trees of the Georgia barrier islands, so these are worth looking for.) As might be expected, a pileated nest cavity is huge in comparison to those of smaller woodpeckers, and is the only overtly rectangular one. Moreover, other woodpecker nest cavities are circular to squarish in outline. For example, the red cockaded woodpecker (*Picoides borealis*) has a nest hole that is almost perfectly circular and about 10 cm (4 in) wide. This endangered species restricts its nesting to mature pine forests, hence extensive logging of pines during previous centuries greatly reduced their numbers. The reason why this woodpecker picks older, living pines is because these are more likely to have been softened by fungal infections, which allow for easier excavation of the wood. What sets apart their nest cavities from those of other woodpeckers are sap wells created by their boring into living trees, marked by extensive flows of sap below the hole. These wound responses are meant not so much to attract insects as deterrents to tree-climbing and egg-eating snakes. In a seeming paradox, the snakes' ventral scales cannot grip onto the sticky sap, so they fall off a trunk once they encounter the sap. On the Georgia coast, red-bellied woodpeckers (*Melanerpes carolinus*) also carve out similarly sized and circularly outlined nest cavities in dead cabbage palms (*Sabal palmetto*; Figure 8.5c).

Nest cavities attributed to woodpecker-like birds have been interpreted from the fossil record in Eocene and Miocene petrified wood (Mikuláš and Zasadil 2004), but are either quite rare or unrecognized. Considering the nearly 40-million-year history of woodpeckers and their ancestors, combined with much petrified wood in the Cenozoic fossil record, and the sizes of modern structures, trace fossils of cavity nests are probably more common than thus far reported.

The most common types of nests made by birds living in terrestrial and freshwater habitats of the Georgia coast are two of those mentioned previously: platform and cup. Platform nests, in which egg clutches are placed on a thick, built-up structure of vegetation (grasses, leaves, sticks), are made by owls, hawks, bald eagles, ospreys, ducks, geese, and other birds that live near or otherwise habituate freshwater sloughs, ponds, and other waterways (Elbroch and Marks 2001; Figure 8.5d). These nests are either on the ground—which is more probable with waterbirds—or a tree branch. When made by larger birds, these nests are hard to miss because of their thickness and width. After all, they must be big and strong enough to accommodate an adult landing on them without knocking out any eggs or nestlings, which would ultimately lead to the extinction of such an inept, clumsy species. Cup nests are the stereotypical bird nests that even urbanites have seen, particularly on branches of deciduous trees during the winter; these have the overall profile of a cup or bowl. Such nests are made of tightly interwoven sticks, leaves, or other materials, including human refuse (paper, string, or plastic), and are sometimes cemented by mud and/or feces. This kind of nest is specially designed by the bird to safely hold the egg clutch and conform to an incubating adult's lower body surface, making for a relatively tight seal. Furthermore, insulating materials, such as moss, or in the Georgia maritime forests, Spanish moss (Chapter 3), may be included in the innermost part of the nest. Many passerines make such nests, and because of how these are shaped for birds of a specific size and egg clutch volume, these often can be linked with a specific species (Figure 8.5e). Also, some passerines, such as the American robin, make a new nest each year (Elbroch and Marks 2001), meaning that an individual tracemaker may be responsible for more than one nest with the same form in a given area.

"Saucer" is sometimes used as a separate category for nests, but such nests are flatter versions of cup nests, hence these are placed in the latter category. Likewise, so-called mud nests, such as those made by barn swallows (*Hirundo rustica*), qualify as a type of cup nest because of their overall morphology. These masons of the avian world mold nests by plastering together pellets of wet mud, in which each pellet represents a beakful of mud from a nearby puddle, flown to the nest site one load at a time. Barn swallows breed and raise their young on the Georgia coast, and have been reported on Jekyll, Cumberland, and other barrier islands (Bellis and Keough 1995). Their handiwork, though, is most closely associated with human structures, especially in the high corners of unoccupied buildings. Of course, an ichnological bonus of any bird that uses mud in its nest construction is the increased likelihood of finding its tracks and beak marks in muddy areas adjacent to wetlands, temporary ponds, or mud puddles.

In short, bird nests constitute terrestrial vertebrate traces that are extremely varied, an ichnological diversity based on how they are made, the materials composing them, and their vertical distribution. Accordingly, preservation potentials for these nests in the fossil record also vary considerably, with burrow, cavity, and ground nests as the most likely candidates for fossilization. Although few fossil bird nests have been reported thus far from the geologic record, hopefully the preceding descriptions will have made their discovery a little more likely.

Bird Feeding Traces

Avian feeding traces that might be observed in terrestrial environments of the Georgia coast include collections of feathers, scratch marks from feet, beak marks, feces, and regurgitants. An accumulation of feathers in a pile, unaccompanied by any other avian body parts, is a sure sign of predation, and can be attributed either to a feral cat or predatory bird, such as a raptor, that systematically plucked a songbird while it ate the digestible goodies (Figure 8.6a). To verify the predator was a bird of prey, look up to see if a tree branch is directly overhead that could have served as a perch, and below for any white, liquid droppings, which are typical of raptors (explained further soon). Scratch marks are common in the floors of maritime forests, where birds have used their clawed feet to kick back sediment or forest debris to reveal infaunal insects or other invertebrates, which are instantly consumed. Perhaps closely associated with these are beak marks, in which the beak pokes into the same substrate to capture an invertebrate attempting to escape. Sizes and shapes of scratch and beak marks naturally vary with the species of birds and their sizes, but scratch marks will be about as wide as the two feet (put together) of the tracemaking bird, which makes them by alternately moving its feet. Likewise, widths and depths of beak marks are slightly less than the maximum width and length of the bird's beak (respectively), depending on the ferocity of the bird.

Other, more long-lasting beak marks that are exceedingly common in maritime forests are those of woodpeckers, sapsuckers, and other birds that enjoy pounding their beaks into woody substrates. For example, a visitor walking through a maritime forest may notice horizontal, evenly spaced rows of 1 cm (0.4 in) wide holes in live oaks, red maples, or red cedars (Figure 8.6b). These traces function as passive traps for insects, in which the wound response of the living tree (Chapter 4) unwittingly conspires with the bird to catch insects that walk across the sticky substance and become mired (Sibley 2001). These rows of holes, also called sap wells, are the traces of sapsuckers, and on the Georgia coast, this is specifically the work of the yellow-bellied sapsucker (*Sphyrapicus varius*). Sapsuckers, by drilling into living trees, are also making traces that can be used more than once, as a sapsucker may return to reopen the wound to cause more sap to flow (Elbroch and Marks 2001).

More subtle feeding traces left on trees by smaller woodpeckers include bark sloughing, in which these birds grab pieces of dead bark with their beaks and pull them off the trunk, revealing yummy wood-boring insects just underneath the surface. One way to test whether you are looking at the traces of bark sloughing, or simply the effects of gravity on decaying bark, is to look for what might have motivated a bird to visit, such as the tunnels, galleries, and frass of wood-boring insects on exposed surfaces (Chapter 5). Thus if you have ever wondered why a dead tree might lose its bark so completely, exposing the outermost cambium and its insect tunnels for all to see, think of how this is a multigenerational trace made by woodpeckers.

Bird feces, the bane of obsessive car owners everywhere, are additional important traces for interpreting avian feeding behavior in terrestrial environments of the Georgia coast. Based on first appearances, bird feces (also called droppings) of different species may not seem distinguishable from one another, consisting of varying combinations of white, liquid urea and finely ground solids. A white cap to any fecal material, caused by uric acid, is one of the distinguishing traits of bird feces, and is absent from those of mammals. Bird feces can be classified easily on the basis of whether they

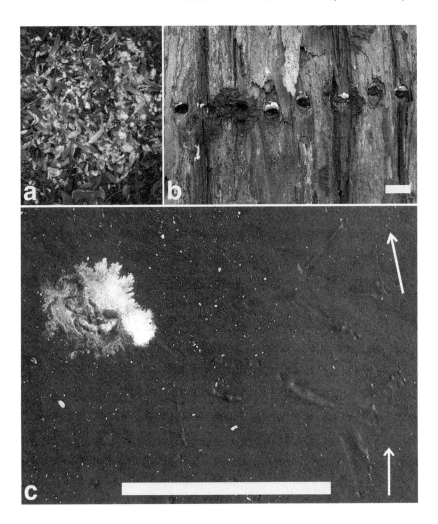

Figure 8.6. Terrestrial bird traces associated with feeding. (a) Pile of songbird feathers, with no other bird parts evident, and just below a live-oak branch, the probable feeding trace of a large predatory bird, such as a hawk: St. Simons Island. Scale (pocket knife, *below*) = 6 cm (2.4 in) long. (b) Drillholes in red cedar (*Juniperus virginiana*) made by yellow-bellied sapsucker (*Sphyrapicus varius*): Little St. Simons Island. Scale = 1 cm (0.4 in). (c) Paired tracks and liquid scat (droppings) of immature great egret (*Ardea alba*) directly associated with its takeoff. Note claw drag marks (*left of arrows*), indicating flight direction of the bird at takeoff toward the top of the photo; contrast with landing tracks of Figure 8.3d: Sapelo Island. Scale = 15 cm (6 in).

are (1) cylindrical and mostly solid; (2) irregularly shaped and semisolid; or (3) liquid (Elbroch and Marks 2001). Herbivorous birds, such as geese, ducks, grebes, and other grazers, produce cylindrical scat. Semisolid scat has some sort of grainy mass holding it together, such as insect parts, seeds, and grains; thus woodpeckers, passerines, and many other types of forest birds make these. Liquid scat comes from carnivores, both predators and scavengers, such as corvids, raptors, herons, and vultures. Because of its fluidity, this scat actually can be sprayed as much as a meter (3.3 ft) behind the bird, sometimes while the bird is feeding (presumably to make more room) and other times just before flying (to lighten the load). Some kayak and canoe enthusiasts also surmise that when herons fly overhead and perform precision bombing with their feces, this is done to get them out of a nesting area. Unfortunately, such anecdotes have not been tested rigorously, nor have many volunteers come forward to participate in such studies. With regard to shedding excess weight for flight, I have seen how nearly every set of takeoff tracks made by herons and egrets is accompanied by a Rorschach-like splatter of white liquid just behind and medial to the tracks (Figure 8.6c). Last, bird droppings are of course useful for discerning diet, but when combined with the ability of most birds to fly, also can give broad information about nearby ecosystems where birds had their meals, versus places where they landed or roosted.

Regurgitants are also known more popularly as cough pellets, based on their shapes and the behavior that produced them ("coughing it up," so to speak). Some predatory birds—most famously owls, such as eastern screech owls (*Otus asio*) and great horned owls (*Bubo virginianus*), but also raptors, herons, and egrets—use regurgitation to get rid of indigestible material in their prey. Owl cough pellets are commonly in maritime forests below their favorite roosts, which can help to define their territory and their diet. Wherever nighttime hunting for small mammals and other prey is good, owls will swoop down, nab their midnight snack, consume the entire animal, digest the useable protein, and then regurgitate any undigested hair and bones into a compact pellet (Sibley 2001). Pellets are normally cylindrical (but rounded at each end); about 2–3 times longer than wide; filled with chitinous insect parts, feathers, fur, small bones, or scales, the latter from reptiles or fish; and they will be much drier (less gooey) than any feces that might come out the other end of a bird. Dried and sterilized owl pellets are excellent sources of small-rodent anatomy lessons, as entire skeletons of mice, voles, or chipmunks can be reconstructed from careful disaggregation of these masses (Hager and Cosentino 2006).

Although feces and regurgitants are polar opposites (alimentarily speaking), they are valuable traces of avian feeding habits, despite having relatively low preservation potential in the geologic record. The liquid sprays of carnivorous birds are probably the least likely to preserve, as these are easily washed away by a single rainstorm. In contrast, pellets filled with bones would probably stand a better chance of becoming trace fossils (Terry 2004), and some owl pellets have been interpreted from the fossil record (Andrews 1990). In between these extremes, the scat of insectivorous or herbivorous birds might be only apparent as unusual concentrations of ground-up insects and vegetation, respectively. Thus it is not surprising to know that avian feces have not yet been interpreted from the geologic record, and neither have some other traces described here, such as drill holes in trees, scratch marks, or feather assemblages. Nevertheless, such a statement is not made with finality, but with the hope of being amended some day.

Dust Baths

Among the more interesting avian traces to encounter in a maritime forest, and normally found on a dry, sandy road, are 15–50 cm (6–20 in) wide, 2–8 cm (0.8–3.1 in) deep, semicircular depressions associated with feather impressions and tracks. These traces are the result of dust baths, which birds use for treating and preventing parasites (such as lice), as well as taking care of their plumage. Birds have uropygial (preen) glands that secrete a type of oil, which is spread more evenly over their feathers by preening. This activity helps a bird in many ways, not least of which is making its feathers water repellent, but also assists with heat conservation, and ensures that feathers look attractive to potential mates (Sibley 2001; Elbroch and Marks 2001). Whenever the oil accumulates too much on any one spot, it needs to be broken down; otherwise it might interfere with flying or maintaining warmth.

Birds form dust-bath traces by alternate scratching of their feet, then hunkering down into loose, dry sediment. They then rapidly move one or the other wing in a way to throw sediment onto the body, all the while fluffing so that the maximum surface area of their feathers is covered. Once done—and in some birds, this seeming ritual could take as much as a half hour—the dirty bird walks a short distance in front of its dust-bath trace, stops, and shakes its feathers. Hence the entirety of a dust-bath trace will ideally include tracks leading up to the depression; scratch marks from the tracemaker's feet; the depression itself; flight-feather impressions from the ventral surface of either wing; feather impressions from the torso (downy feathers, that is); tracks leading away from the depression that show the feet becoming parallel within a short distance; and a fine debris pattern caused by the shaking of loose sediment from the body, which should roughly outline the bird. The parallel tracks also may have pressure-release structures associated with the rapid back-and-forth movement of the bird as it shook. During most moist times of the year, birds accomplish this same goal by using standing water provided by puddles, which also are free of predators they might normally encounter in freshwater ponds. Unfortunately, dust-bath structures probably have an extremely low preservation potential in the geologic record because of their formation in dry sediments. Nonetheless, their soft-sediment equivalents in moist substrates may fare better in the taphonomy sweepstakes. Additionally, sizeable bath structures, made by relatively larger birds, are the most likely ones to be preserved. The most obvious examples of such traces in maritime forests of the Georgia islands would be those of wild turkeys, which make depressions as much as 50 cm (20 in) wide (Elbroch and Marks 2001).

In summary, the preceding section has probably convinced any initially skeptical readers about the variety and intricacy of avian traces in terrestrial and freshwater environments of the Georgia coast, an ichnological richness that rivals or surpasses that of mammals. The wide variety of ecosystems in the relatively small areas of the Georgia barrier islands translates into diverse substrates, foodstuffs, nest-building materials, and other resources that, when combined with diverse bird behaviors, result in a myriad of their traces. Also keep in mind the seasonal nature of many traces, as many migratory birds only stop briefly at the Georgia barrier islands for refueling or reproducing. These themes will be explored more in the next chapter with a close examination of shorebird traces, which present ichnologists with plentiful data but their own unique challenges of interpretation (Chapter 9). Nevertheless, all avian traces, particularly their tracks, are important analogues for interpreting the evolution and behavior of birds from the fossil record of the Cretaceous through the Pleistocene (Chapter 10).

MAMMAL TRACKS, TRAILS, BURROWS, AND MUCH MORE

Reading Mammal Tracks and Other Traces

Because we are mammals and have a long history of preying on other mammals, we can be forgiven for having a natural, keen interest in their traces: this affinity is probably embedded in our genes. Furthermore, terrestrial mammal traces in general are quite varied, including tracks, trails, burrows, wallows, lays, lodges, nests, dens, tooth marks, scraping, scent posts, and feces, to name a few. Similar to bird traces, mammal tracks are nearly ubiquitous in terrestrial environments of the Georgia coast, and with enough patience and skill can be found in nearly every substrate, from soils underfoot to the canopies of the maritime forests. In contrast to reptile and bird traces, however, is the relatively lower biodiversity of mammals on the Georgia barrier islands, which corresponds to a lower ichnodiversity. In other words, although mammal traces are indeed common on the Georgia coast, their variety is somewhat limited. That notwithstanding, each visit I make to the islands yields a new, surprising trace that I did not expect from some of the mammals that live there. Furthermore, large-bodied feral species of mammals, such as horses (*Equus caballus*), hogs (*Sus scrofa*), and cattle (*Bos taurus*), are common on most of the Georgia barrier islands and make enough traces to shout out those made by smaller, native species. Consequently, the study of mammal traces on the Georgia barrier islands requires a bit of looking through rose-colored glasses and imagining a time when these recent invaders were not affecting the terrestrial vertebrate-trace assemblage. Such a view also needs to add the larger, native fauna—ground sloths and mammoths in particular—that were leaving their marks on Pleistocene landscapes, as well as recently extirpated carnivores, such as wolves and bears (Chapter 2).

Naturally, tracks are the default topic of conversation whenever discussing modern mammal traces in a broader sense, so we will start there. Mammal tracks are normally classified on the basis of a digital formula for the manus and pes (Halfpenny and Biesiot, 1986; Elbroch 2003), so just counting toe impressions is an important first step of description (Table 8.2). For example, although hoofed animals make tracks different from those of mammals with softer, padded feet (more on that later), horses and their relatives are the only single-toed mammals (which belong to

TABLE 8.2. TRACKS OF COMMON MAMMALS IN TERRESTRIAL ENVIRONMENTS OF THE GEORGIA COAST

Common Name	Species	Track Length (cm)
A. One digit in manus and pes, hoofed		
Horse (feral)	*Equus caballus*	2.8–3.1
B. Two digits in manus and pes, hoofed		
Hog (feral)	*Sus scrofa*	5.0–5.5
White-tailed deer	*Odocoileus virginianus*	6.0–6.5
Fallow deer (feral)	*Dama dama*	7.0–7.5
Cattle (feral)	*Bos taurus*	9.0–13.0
C. Four digits in manus and pes		
Domestic cat (feral)	*Felis domestica*	3.0–3.5
Gray fox	*Urocyon cinereoargenteus*	3.8–4.3
Red fox (feral)	*Vulpes vulpes*	4.0–4.8
Bobcat	*Lynx rufus*	5.0–5.3
Domestic dog	*Canis familiaris*	3.5–11.0
Coyote	*Canis latrans*	6.0–6.5
Marsh rabbit	*Sylvilagus palustris*	7.5–8.0
Cottontail rabbit	*Sylvilagus floridanus*	8.5–8.8
D. Four digits in manus, five digits in pes		
Mouse (various species)	*Peromyscus* spp.	0.8–1.0
Meadow vole	*Microtus pennsylvanicus*	0.8–1.0
Southern flying squirrel	*Glaucomys volans*	1.0–1.3
Hispid cotton rat	*Sigmodon hispidus*	1.3–1.5
Eastern chipmunk	*Tamias striatus*	1.5–1.8
Marsh rice rat	*Oryzomys palustris*	2.5–3.0
Norway rat (feral)	*Rattus norvegicus*	2.5–3.5
Nine-banded armadillo	*Dasypus novemcinctus*	5.0–5.5
Eastern gray squirrel	*Sciurus carolinensis*	6.5–7.0
Fox squirrel	*Sciurus niger*	6.5–7.0
Round-tailed muskrat	*Neofiber alleni*	Not recorded
Beaver	*Castor canadensis*	12.5–14.0
E. Five digits in manus and pes		
Southern short-tailed shrew	*Blarina carolinensis*	0.5–1.0
Mink	*Mustela vison*	4.5–4.8
Striped skunk	*Mephitis mephitis*	4.5–4.8
Opossum	*Didelphis virginiana*	5.0–5.5
River otter	*Lutra canadensis*	9.0–10.0
Raccoon	*Procyon lotor*	9.0–10.5

Note. Items are organized on the basis of track category and ranked within each category by pes length. Recently extirpated species, such as the Cumberland Island pocket gopher (*Geomys cumberlandius*), eastern wolf (*Canis lupus*), black bear (*Ursos americana*), and Florida panther (*Felis concolor*) are not included. Track lengths are from Elbroch (2003) and my field measurements, but do not include impressions by claw marks or dewclaws, and are for adult animals.

Perissodactyla), whereas two-toed mammals (Artiodactyla) include ungulates, such as white-tailed deer (*Odocoileus virginianus*), feral hogs, and cattle. For these perissodactyls and artiodactyls then, the digital formulae are simply 1–1 and 2–2 for toe counts in the manus and pes tracks, respectively. No mammals north of Central America make three-toed tracks (yes, tapir tracks would be an unusual find on the Georgia coast), so other tracks are described with either 4–4, 4–5, or 5–5 digital formulae. Trackmakers with a 4–4 formula include felids (cats), canids (dogs), and lagomorphs (rabbits), whereas those with a 4–5 formula are rodents of all sizes and shapes, but also include armadillos. Shrews, mink, river otters, raccoons, and opossums represent mammal trackmakers with five digits on each foot.

Perhaps some backing up is needed before talking more about other anatomical traits of mammal tracks. For example, how to tell the manus impression from that of the pes? Size is part of the answer: rodents, raccoons, and bears have rear feet nearly twice as long as their front feet, which help to support their prominent hindquarters. In contrast, hoofed mammals, felids, and canids have slightly larger front feet compared to their rear feet; this circumstance may be related to more musculature, large skulls, antlers, and other accouterments that require allocating more surface area for the front feet. Thus knowledge of relative size differences in manus–pes pairs in mammals, when used in combination with digital formulae, should assist in making preliminary identifications of the manus and pes in a mammal trackway.

Gait patterns are also related to figuring out the manus–pes placement, and an understanding of mammal gaits and their variations is absolutely essential for interpreting mammal trackways. Because all mammals of the Georgia coast other than humans are quadrupedal, these gait patterns can involve the movement of four limbs in seemingly baffling combinations. Not to fear, as these gaits can be simply divided into five categories: slow diagonal walking, fast diagonal walking, trotting, loping, and galloping (Figure 8.7). Experienced trackers might say that this is oversimplifying the complicated matter of gaits, and ordinarily I would agree with them. Nonetheless, for the purposes of discerning broad behaviors and identifying mammals from their trackway patterns, these groupings will do just fine for now.

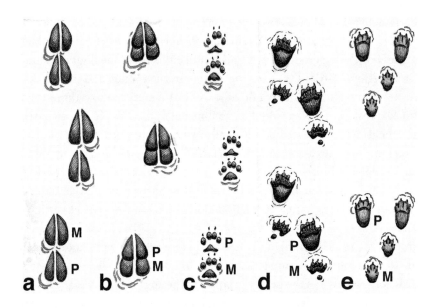

Figure 8.7. Main categories of mammal gait patterns, with manus and pes labeled accordingly (M and P). (a) Slow diagonal walking, shown by white-tailed deer (*Odocoileus virginianus*) trackway. (b) Fast diagonal walking, with indirect register of pes on manus, also by white-tailed deer. (c) Trotting, in which pes lands in front of the same-side manus, shown by coyote (*Canis latrans*) trackway. (d) Loping, where one pes exceeds the same-side manus and the other pes is beside the opposite-side manus, shown by otter (*Lutra canadensis*) trackway. (e) Galloping, where both rear feet land in front of both front feet, made by marsh rabbit (*Sylvilagus palustris*) when running.

What is the difference between slow diagonal walking and fast diagonal walking in a mammal, besides speed? This may depend on the mammal, but in general is determined by relative foot placement. The most commonly used gait by a mammal, also known as its baseline (Young and Morgan 2007), may be a slow walk, fast walk, trot, lope, or gallop, but most mammals have a baseline of a slow or fast walk. For example, wild felids—such as bobcats—tend to place their pes almost exactly into their manus impression, making a slightly smaller print inside the preceding one. In contrast, domestic cats are far too fat and lazy to do this, and their trackway patterns are often disappointing, revealing just how much we spoil them. A track-within-a-track pattern is called direct register; then if the pes intersects the manus impression, but is not perfectly inside its outline, it is indirect register (Elbroch 2003; Young and Morgan 2007). Simple enough, but what does this have to do with relative speed while walking? Look again at the indirect register: if a felid trackmaker had its pes slightly behind the manus, it was walking more slowly than if the pes was in front of the manus. As soon as the pes begins to extend beyond the manus, this felid is moving faster than its baseline. With more rapid movement, the pes will get farther in front of the manus, and the progression in speed associated with the change in trackway pattern goes something like this: slow diagonal walk (indirect register, pes behind manus) → fast diagonal walk (indirect register, pes in front of manus) → trot (pes just in front of the same-side manus) → lope (trailing pes is next to the opposite-side manus) → bounding (rear feet on either side of the front feet) → gallop (both rear feet exceed the front feet). In contrast to felids, most canids have a baseline gait that shows the pes slightly in front of the manus, indicating a trot. Yet even canids have their exceptions, as some foxes walk more like cats and their trackways often show direct register, looking at first as if a fox were walking on only two legs. Different mammals have different baselines, and it is up to a tracker to note which gait pattern is more prevalent along a given mammal trackway. Once a baseline is detected, any changes in the trackmaker's behavior can be better assessed.

Another general principle of gait patterns is that as speed increases, the straddle (trackway width) becomes narrower, and pace angulation becomes higher. One way to test this concept is to take a series of straddle measurements along any trackway that seemingly shows a transition from a slow walk to a gallop. These numbers should get smaller as the animal accelerated. The intergroup distance, or the space between a set of four tracks—all four feet touching then ground, then an airborne leap, followed by all four feet touching the ground—should also increase along a trackway of an accelerating trackmaker (Halfpenny and Bruchac 2002). Last, as a mammal changes its gait from a slow walk to a gallop in a trackway, it exerts more pressure against whatever substrate it is impacting and pushing off. Thus the pressure-release structures associated with individual tracks become more prominent with increasing speed, and sediment may even have exploded out of and behind the tracks (Brown, 1999). Are there exceptions to these generalizations? Sure! An eastern gray squirrel can perform a so-called slow gallop, in which its trackway pattern shows a gallop pattern (both rear feet exceed the front), but this movement was actually slower

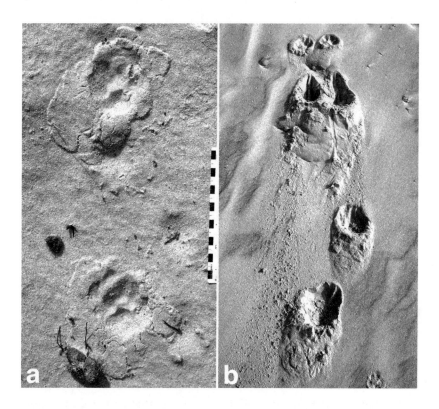

Figure 8.8. Bobcat (*Lynx rufus*) tracks on Cumberland Island. (a) Normal, slow diagonal walking pattern with claws not showing. Scale in centimeters. (b) Galloping pattern up sand dune, in which claws were extended for traction; direction of movement is toward top of photo.

than its fast diagonal walk. Take it from someone who has watched squirrels for many hours: they can do diagonal walking, and do not spend all of their time on the ground galloping. Nevertheless, gait patterns normally reveal the preferred gait of a given mammal, and variations of that gait will aid in understanding subtle changes in behavior while that mammal is in motion.

With this crash course in gait patterns in mind, let us return to mammal foot anatomy and how these details provide more information about a trackmaker. Think of claws, for example, also called unguals, such as your fingernails and toenails, which are composed of keratinized tissue on the ends of your digits. Unlike humans, though, some mammals have retractable claws, normally kept sheathed under the fleshy parts of the digits, such as in felids. Accordingly, felid tracks almost never show claw marks, unless a cat ran up a slope and needed added traction (Figure 8.8). In contrast, canids, raccoons, rodents, armadillos, deer, and every other native mammal of the Georgia barrier islands show ungual impressions because these are always exposed. The overall form of claw marks may also divulge behavioral adaptations of their tracemakers. For example, the sharp, pinpoint claw marks in the tracks of gray foxes (*Urocyon cinereoargenteus*) hint at this canid's ability to climb trees, something domestic dogs (*Canis familiaris*) only wish they could do. Nonetheless, the stouter, blunter claws of wolves and dogs are superbly adapted for traction while running, or for digging. Other anatomical details to look for in tracks are fleshy pads, calluses, dewclaws (which are vestigial digits), hair impressions around the outside of the track, and abnormalities, such as missing digits. Some of these details are important for testing or confirming the identity of a trackmaker, whereas others may even point to an individual animal.

How frequently will you see a perfectly formed manus or pes track of a mammal on a Georgia barrier island, with every digit, heel pad, claw mark, and fur impression recorded? Not as often as one might imagine, but the wealth of trackmakers and fine-grained sediments of the islands certainly makes this happenstance more likely than most places (Figure 8.9). And what if a tracker diligently follows a mammal's tracks through a back-dune meadow, then the tracks seemingly vanish into the leaf litter of a maritime forest? Well, this is where skilled trackers are separated from people who aspire to play trackers on TV. Most tracks are not exquisitely preserved, even those we just observed being made. Consequently, a tracker must look for compression shapes, which will then define a trackway pattern and aid in identifying a trackmaker and its behavior. For example, felids make circular compressions, canids make egg-shaped (oval) compressions, deer and their kin make heart-shaped compressions, and so on. Simply draw a line around the periphery of any suspected track, omitting the details of its digits, claws, and pads; measure each compression shape and the distances between them; and then note the trackway pattern. Also keep in mind that a sharp-edged, hard hoof will have a different effect on forest debris versus a soft, padded foot. Tracks made by hoofed animals thus cut through and more easily crush leaves and twigs, leaving a more sharply defined outline than a track left by a soft foot.

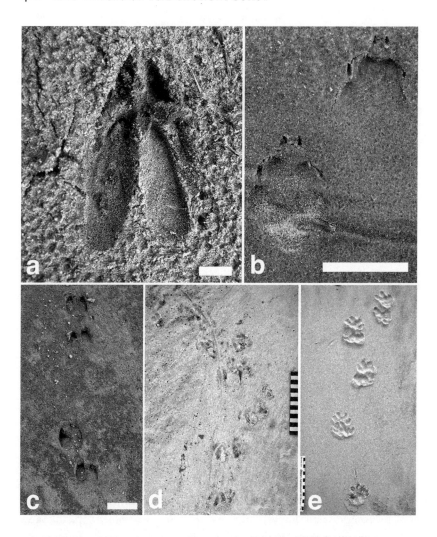

Figure 8.9. Representative terrestrial mammals tracks of the Georgia barrier islands. (a) Indirect register of left pes and manus tracks of white-tailed deer (*Odocoileus virginianus*): Sapelo Island. Scale = 1 cm (0.4 in). (b) Coyote (*Canis latrans*) tracks in upper beach, left pes (*top*) and manus (*below*): Wassaw Island. Scale = 10 cm (4 in). (c) Feral hog (*Sus scrofa*) trackway in saltpan of high salt marsh, showing slow walk: St. Catherines Island. Scale = 10 cm (4 in). (d) Armadillo (*Dasypus novemcinctus*) tracks accompanied by tail drag mark in back-dune meadow, moving from top to bottom: Sapelo Island. Scale in centimeters. (e) Opossum (*Didelphis virginiana*) trackway in back-dune meadow: Jekyll Island. Scale in centimeters.

This perspective of looking for and examining compression shapes left by mammals in less than ideal substrates, instead of only studying tracks worthy of magazine covers, is a practice that can be reinforced daily, whether on the Georgia coast or elsewhere. In my experience, using compression shapes as search images is also extremely important when scouting for fossil tetrapod tracks. All too often, I have seen paleontologists walk by imperfect vertebrate tracks just because they do not fit their ideal of what a track should look like. Such tracks were not being ignored, though: they were not even seen in the first place. In other words, every depression or sediment-filled depression in the right environment (or paleoenvironment) may be treated as a possible track, a hypothesis that should be tested first, instead of a quick dismissal based on unrealistic expectations of perfection.

To summarize the potentially lengthy subject of mammal tracks on the Georgia coast, an ichnologist, to readily identify a mammalian trackmaker and its behavior, should use a combination of digital formulae; trackway patterns; pressure-release structures; anatomical details; and ecological context. Nonetheless, the single most important descriptor is gait pattern. Sometimes a gait pattern, especially when placed in the context of the host ecosystem, will immediately reveal the identity of the mammal, rendering any further documentation as the academic equivalent of gravy.

Mammal Trails

Tracks beget trackways, and trackways beget trails. Trails are linear, worn, and slightly depressed paths in maritime forests, back-dune meadows, or through salt marshes where repeated movements of animals (trackway on trackways) have decreased or prevented the growth of vegetation. Often "trail" is also used as a verb for following a trackway (as in "trailing"), or for a trackway itself, but here will refer to habitually used paths. On the Georgia barrier islands, mammals usually make trails, and only a few are culpable: raccoons, deer, and feral species of hoofed mammals. Adult alligators, however, sometimes make trails between favorite ponds that they use for hunting or mating, and may pass through maritime forests on a regular basis (Figure 7.1c). For example, I have personally witnessed alligators going on walkabouts through the woods at night, which is rather unnerving to witness, but good to know. Hence trackers would be wise to check an overland trail for alligator tracks too. Although mammal trails are often used by more than one species of mammal, they are normally only slightly

414 LIFE TRACES OF THE GEORGIA COAST

Figure 8.10. Mammal trail indicated by repeated chipping (wear) on the trunk of a downed tree crossing trail, caused by hoofs of white-tailed deer (*Odocoileus virginianus*) and feral cows (*Bos taurus*): Sapelo Island. Scale in centimeters.

wider than the straddle of the dominant trail user. For example, deer trails seem remarkably narrow to a human trying to follow one through a maritime forest, but they faithfully reflect the straddle of a deer on the barrier islands. One factor to keep in mind with such trails, though, is the island effect on deer size, where smaller body size is selected over generations for a population restricted to an island (Sinclair and Parkes 2008). This means that deer trails on the Georgia barrier islands will be, on average, slightly narrower than those on the mainland.

Trails used habitually by raccoons, rabbits (*Sylvilagus palustris* or *S. floridanus*), armadillos (*Dasypus novemcinctus*), opossums (*Didelphis virginiana*), and other relatively short mammals run uninterrupted under branches, vines, and fence wires. Trails of smaller mammals also may go around obstacles, such as thick, fallen trees. Hence these aspects of the landscape can help define minimum heights of their tracemakers. With hoofed mammals, look for where trails are intersected by such trees, and check the trunks of downed trees for chipping and other signs of wear (Figure 8.10). These are areas where generations of durable hoofs may have dragged across a trunk, but of course also depend on how long a tree has been there lying there. Old habits are hard to die, and mammals accustomed to following a path of least resistance will only change their preferred routes through a maritime forest if coerced by a major alteration to the surrounding landscape.

Can a trail go up a tree? Yes, if it is made by arboreal mammals that also come down to the ground for foraging, such as the southern flying squirrel (*Glaucomys volans*), eastern gray squirrel (*Sciurus carolinensis*) and fox squirrel (*S. niger*). These trails are subtle, but become more visible if one thinks like a squirrel. Look at a live oak or other hardwood tree from a distance and ask yourself, "Which is the easiest way up the tree?" in which a route neatly avoids branches, knots, or other irregularities in the trunk that might impede progress. Once this possible route has been assessed, see if any slightly discolored linear patch follows from just above the ground to several meters up the trunk. If so, this is the possible squirrel trail. Be a good scientist then, and check your hypothesis by closely examining the trunk to see whether this discoloration can be attributed to bark pulled off the trunk in those places, and whether these areas have squirrel claw marks. Squirrel claw marks are small compared to those left by raccoons and other tree-climbing mammals, so this distinction should be easy to make, especially on smooth-barked trees. Yet another piece of evidence to test the identification of a vertical trail is the presence or absence of a squirrel nest in the upper reaches of the tree, giving at least two squirrels a reason to go up and down that tree regularly. Squirrel nests (dreys), which are wonders of construction, will be discussed in just a bit.

As mentioned before, scratch marks on tree trunks can be made by other mammals that either live in trees or otherwise use them, so look for these while also looking for insect borings (Chapter 5) and bird traces. Sadly, black bears (*Ursos americanus*) are no longer native to the Georgia barrier islands, but older trees may bear witness to their former presence through 4–8 deep, parallel gouges carved with the claws of their front feet. This form of territorial marking is unmistakable once seen, as these grooves are sometimes more than 2 m (6.6 ft) off the surface and longer than 50 cm (20 in). (As local Georgians might say, "That ain't no squirrel.") Bears are also accomplished climbers, and make claw marks that reflect this ability; these run well up the trunk of a tree sturdy enough to bear their weight.

Mammal Burrows

Mammal burrows may not be seen nearly as often as their tracks and trails, but are important components of terrestrial environments, and must be included in ichnological assessments of those environments. Burrows are tunnels, shafts, or chambers made by mammals that either spend much of their lives underground, or otherwise dig dens in which they raise their young or seek privacy. In this respect, shallow excavations made while foraging for insects, fungi, or roots are thus considered as the products of burrowing, but are not burrows per se. With those pedantic definitions in mind, burrowing mammals of the Georgia barrier islands include armadillos, bobcats (*Lynx rufus*), canids (coyotes and foxes), chipmunks (*Tamias striatus*), a few other small rodents (mice and voles), and small insectivores, such as the Southern short-tailed shrew (*Blarina carolinensis*). In freshwater environments, round-tailed muskrats (*Neofiber alleni*) and beavers (*Castor canadensis*) also make bank burrows in water bodies. Round-tailed muskrats (also called Florida water rats) only burrow into banks during droughts and subsequent low water levels. Beavers are likewise more prone to burrow in the absence of sufficient vegetation for making a lodge. Both species, however, are rare on the Georgia barrier islands (Byrne and Lagana 2009). So do not look too long for their bank burrows, especially when alligator burrows are so common in these same environments.

Probably the most ichnologically significant mammal of terrestrial environments on the Georgia coast is the least seen, the eastern mole (*Scalopus aquaticus*), which burrows in nearly every near-surface terrestrial substrate throughout the barrier islands. Their extensive burrows, some of which can be followed continuously for nearly 100 m (330 ft), even originate in coastal dunes and abruptly end at the intertidal zone of beaches (Figure 8.11a,b). (Did they then turn around, or go for a swim?) Part of the reason why moles burrow is to keep protected from predators, but mostly they eat. These small mammals are voracious consumers of macroinvertebrates, gobbling up subterranean insect larvae, oligochaetes, nematodes, and anything else living that might be detected by their keen sense of smell. Moles are so good at what they do, that earthworms flee in terror at any subsurface vibrations that remind them of moles (Mitra et al. 2009; Chapter 5).

In terms of behavior, mole burrows can be interpreted on the basis of depth. Shallower tunnels, easily seen as prominent, 10–15 cm (4–6 in) wide linear to meandering ridges on the surface, are the results of hunting for

TERRESTRIAL VERTEBRATES, PART II 417

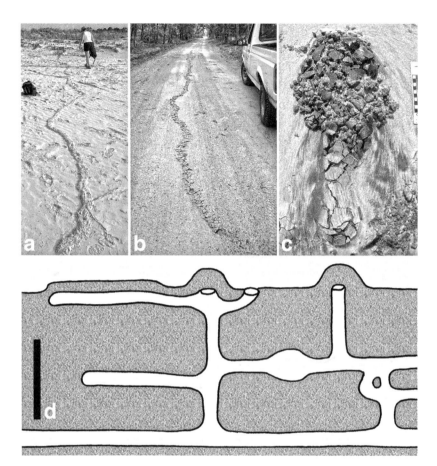

Figure 8.11. Burrows of the eastern mole (*Scalopus aquaticus*). (a) Horizontal tunnel that travels between dunes (*background*) and upper intertidal zone: Sapelo Island. (b) Surface expression of extensive horizontal tunnel in sandy road in maritime forest: Sapelo Island. (c) Molehill, in which excavated sediment from a tunnel is pushed to the surface, in back-dune sand: St. Catherines Island. Scale in centimeters. (d) Longitudinal view of typical mole burrow system, including enlarged area used for nesting. Adapted from a figure by Gorman and Stone (1990). Scale = 50 cm (20 in).

food, and thus are temporary burrows. In contrast, deeper, more permanent burrow systems are normally 15–50 cm (6–20 in) below the surface, but can be more than a meter deep (Figure 8.11c). These burrows are used for dwelling and raising young, as well as feeding (Gorman and Stone, 1990). The open parts of a burrow system normally have circular cross sections that are about 3–5 cm (1.2–2.0 in) wide. Mole dwelling burrows can also be quite complex, as these have branches off a main tunnel that split into secondary tunnels and vertical shafts. Some parts of tunnels expand to about 10 by 20 cm (4 by 8 in); these areas serve as nesting sites for nurturing young. Near the surface, moles do not actively backfill their burrows, but instead push excavated sediment upward to the surface, making small hillocks of sediment often nicknamed molehills (or mountains, depending on perspective). Mole burrowing is the stuff of mammalian legend, as they are capable of digging as far as 5 m (16 ft) in an hour, or more than 30 body lengths for the average-sized mole, which is about 16 cm (Gorman and Stone, 1990; Whitaker and Hamilton, 1998). Because of this extensive burrowing by individual moles, multiplied by many moles and over generations, the eastern mole is a keystone species through its special role in the mixing and aeration of soil in terrestrial ecosystems. The overall pattern of a mole burrow system is similar to the trace fossil *Thalassinoides*, which is more often associated with crustacean tracemakers (Chapters 5, 6, 9).

And woe to any predator that thinks of a mole as a meal that should simply slide down its gullet. In at least one documented instance, a herring gull made the mistake of snatching up a live, juvenile mole (probably one that had burrowed down to the shoreline) and swallowing it. The mole proceeded to tear open the innards of the gull, and although it died just after having nearly exited the gull's body, the gull died too. Both animals were found together, providing a nice forensic exercise for the person who encountered their corpses (Gorman and Stone, 1990). The moral of the story is to never, ever swallow a live burrowing mammal. Also, think of what a lovely body-fossil pair this would make: a small mammal stuck in the body cavity of a predatory bird or dinosaur, perhaps accompanied by claw marks on the predator's bones as trace fossils.

Other mammal burrows made in maritime forests include those of eastern chipmunks (*Tamias striatus*), which, unlike their squirrel cousins, are ground dwellers. This habit is confirmed through direct observation of chipmunks, but can also be discerned from their tracks. Their trackway patterns, which normally show a gallop, have offset manus impressions (half bound), which is more typical of ground-dwelling mammals (Halfpenny and Bruchac 2002; Elbroch 2003). This trait is shared with lagomorphs, which depend on quick, lateral changes of direction and maneuvering. Such movements done while climbing up a tree would cause the tree climber to fall off the trunk, a scene we rarely see in nature: accordingly, squirrels have full-bound patterns in their trackways. As a result of their ground-dwelling habits, chipmunks require protection while spending so much time there, which is one of the reasons why they burrow. Chipmunk burrows are open and have one or two entrances, which are perfectly circular in cross section, about 4–5 cm (1.6–2.0 cm) diameter, and their peripheries are fastidiously clean (Elbroch 2003). Unlike many other burrowing mammals, they do not leave sediment mounds, food scraps, feces, or other traces just outside burrow entrances that might attract predators. These discarded materials are instead carried or pushed out of the burrow system away from the entrance and dispersed so that they do not garner much attention. Entrances lead to inclined tunnels that go down for about 20–25 cm (8–10 in) before turning to the right or left, a burrow twist shared by many other vertebrates; most of the burrow system is at a depth of about 50–100 cm (20–40 in). Burrow systems tend to have a primary tunnel, which can be more than 5 m (16 ft) long (Thomas, 1974). Additional side tunnels branching off the main tunnel lead to expanded chambers, which are either places for caching food or nesting.

Chipmunk or similar mammal burrow systems, if preserved in the fossil record, could be easily confused with those made by crayfish (Chapter 5) or even callianassid shrimp (Chapter 6) because of their circular cross sections and complicated, three-dimensional geometry, which includes expanded chambers that connect with shafts and tunnels. Similar to mole burrow systems, the applicable ichnogenus for a trace fossil equivalent to a chipmunk burrow would be *Thalassinoides* (Chapter 9). Main distinguishing traits of chipmunk or other small mammal burrows, though, would be the presence of terrestrial nuts, seeds, and grasses (former food caches), feces, and indications of subaerial exposure, as mammals tend to burrow above the water table. Indeed, a Miocene subterranean mammal cache, thought to represent seasonal hoarding, was interpreted on the basis of more than 1,000 fossil nuts in a presumed rodent burrow (Gee et al. 2003).

Traces of meadow voles (*Microtus pennsylvanicus*) are abundant and varied in maritime forests, patchy, grassy areas between forests, and edges of salt marshes, but are difficult to characterize ichnologically. This dilemma is because most vole tunnels are actually surface trails, also called runs (Elbroch 2003). These are composed of pathways worn down by voles along the ground, but under dense, grassy cover. To observe these trails, just part some of the thick grasses in areas on the edges of maritime forests or in grassy meadows between forest, and look for 3–6 cm (1.2–2.4 in) wide, worn, linear and curved, intersecting trails in between patches of vegetation. Vole burrows are also potentially confused with those of short-tailed shrews, which are about 2.5–3.0 cm (1 in) wide, but should be directly associated with vole trail systems.

Mammal burrows vary in size corresponding to their makers, hence the creator of one of the largest of these should be mentioned, which is among the most prolific mammalian tracemakers of the Georgia coast: the nine-banded armadillo (*Dasypus novemcinctus*). This odd-looking animal, which children are sometimes surprised to find out is a mammal because of its bony plates and lack of prominent fur, is famous for its burrowing abilities, which it does for both for feeding and lodging. Armadillos prefer to burrow at the bases of large trees, which probably lend support to the roofs of their dwelling burrows (Elbroch 2003). Burrow cross sections are oval to circular and slightly larger than the body cross section of the armadillo tracemaker, about 18–25 cm (7–10 in). These burrows expand into a living or denning chamber about twice the width of the main burrow (McBee and Baker, 1982; Feldhammer et al. 2003). Their subsurface homes are normally about 50 cm (20 in) below the surface, but can be as deep as a meter, and some are as long as 7 m (23 ft). Individual armadillos often dig many burrows; on Cumberland Island, they were observed making an average of 11 burrows within a year (Bond et al. 2000). As a result, armadillos rival gopher tortoises in their displacement of sediment and alterations of habitats. They similarly leave a large sediment mound outside their burrow entrance, which often preserves fresh tracks and tail drag marks. Armadillos have also been observed stuffing their burrows with leaves and other vegetation, which is associated with denning and presumably used for insulation of offspring (Taulman, 1994). Hence a fossilized burrow and den structure made by an armadillo (or a close relative) might be expected to contain an unusual concentration of fossil plant debris too.

Armadillo tracks associated with their burrows look unusual compared to other mammal tracks. These are normally made in a slow, diagonal walking pattern showing two prominent, clawed digging toes in the manus and three obvious clawed toes in the pes, despite their overall 4–5 manus–pes digital formula (Figure 8.9d). The first time I saw armadillo tracks in southern Georgia, I puzzled over them for a few minutes because the rear foot impressions were so similar to bird tracks. Soon, though, I realized (with much embarrassment) that no bird would be walking on all four limbs while also leaving a linear impression from a stout tail. Fortunately, that mistake was corrected then and there, and I soon connected their distinctive trackways with deep burrows, shallower digging marks (Figure 1.7), wide foraging trails made by plowing through thick forest debris, feces, and the many other traces they make. Indeed, once you know what to look for, their numerous dig marks cannot be missed in a typical maritime forest, and can only be confused with those of small feral hogs (*Sus scrofa*). Armadillos are also capable of sitting back onto their hind legs to get a better whiff of the air, so their trackways sometimes contain these temporarily bipedal resting traces.

Interestingly, armadillos are not considered native animals to the Georgia barrier islands, or even the southeastern United States, as they have been migrating north and east from the southwestern United States. Then how did they get themselves to the islands? They made it through a combination of swimming, walking underwater, and floating. Yes, you read that right: armadillos not only swim, they are capable of holding their breath and walking along the bottom of a lake or stream to cross it (now those would make for some great trace fossils!). They also can float by swallowing enough air to offset their normal body density, which helps to buoy them across a water body (McBee and Baker, 1982). On the Georgia coast, strong ebb- or flood-tide currents likely aid this mode of transport. The resulting traces made by an armadillo making a beach landing, whether modern or ancient, would certainly constitute a marvelous find.

Mammal Lays, Wallows, and Nests

Lays, or temporary resting spots for mammals, are subtle but can be detected in Georgia maritime forests by looking for areas of compressed leaves or grass, then comparing the compression shape to the size and anatomy of possible tracemakers. Deer lays are by far the most commonly encountered of such traces because of their size and number. These are made by a group of deer traveling together, in which they leave multiple, closely spaced impressions that mimic the lateral outline of their torsos. Size differences then can be used to hypothesize the sexes and age structures of the group. Lays made by marsh rabbits or other small mammals are trickier to find, and are formed when their tracemakers paused or hunkered down—thus

Figure 8.12. Feral hog wallow structure in a maritime forest, in which a standing pool of water in its middle indicates the local water table: Sapelo Island. Structure is about 1.5 m (5 ft) wide.

reducing their motion and profile—while avoiding detection from predators, or simply sleeping.

Wallows are traces made by a few species of mammals, particularly feral hogs. These structures function like dust and mud baths made by some birds, helping to decrease the obnoxious effects of parasitic insects, but also leave the scent of the tracemaker (Elbroch 2003). These large depressions are 2–3 m (6.6–10 ft) wide and as much as 40 cm (16 in) deep, with irregular and uneven surfaces, a result of trampling by the primary tracemaker (Figure 8.12). Wallows also may contain tracks of other vertebrates (raccoons in particular), shallow burrows and trails of insects and oligochaetes, and the distinctive towers of crayfish burrows (Chapter 5). These structures are hard to overlook in a maritime forest, and are often easy to find near seasonal freshwater ponds or otherwise intersect the local water table; hence wallows may also contain alligator tracks. The most obvious behavioral difference between mammal wallows and mud baths taken by birds is that mammals will also use these to cool off during hot Georgia summers. Additionally, layers of mud, once dried, provide added protection against mosquitoes, ticks, and other parasites. So far, lays and wallows have not yet been interpreted from the geologic record. Nevertheless, my bet is that fossil wallows will eventually be interpreted, seeing how large some of these might have been, especially if made by the larger mammals of previous epochs in the Cenozoic Era.

Squirrel nests, or dreys, are architectural wonders, looking much like a messy, loose collection of leaves and sticks, but these actually are sturdy, weather-resistant, and well-insulated homes. I have tested their construction by taking abandoned nests, freshly fallen onto the ground, and then shaking them to see how well they stayed together. Usually nothing more than an errant dry leaf pops out; otherwise, the nest remains complete. This integrity is imparted by an interweaving of branches that the squirrels snipped off their home tree, which forms the nest framework. Squirrels then stuff dry leaves, bark, or grasses in the interstices, which completes the structure, and also keeps the interior of the nest dry, even during severe rainstorms. These nests are used for raising offspring and keeping them tucked away from predators; in more northern climates, these also are good places to keep warm during the winter (Elbroch 2003). Each branch composing the framework of a drey carries a diagnostic 45° angle bite mark at one end. Although the fossilization potential of a drey seems extremely low, they might be hypothesized on the basis of a homogenous collection of leafy material and branches in which such distinctive bite marks are also present.

Mammal Tooth Marks

Just mentioned were tooth marks, mammal traces that ordinarily are omitted from trace and trace-fossil assessments. This is a sad oversight, considering their abundance in modern terrestrial environments in durable substrates, such as nuts, other woody tissues, and bone. For example, those mammals with the proper teeth, such as chisel-like incisors on both the maxilla (top jawbone) and mandible (lower jawbone), can impart a 45° cut across a branch that is easily diagnosed as a mammal trace, rather than a random break. A little bit of knowledge about mammal ecology, occlusion (how an animal's teeth come together), and width measurements of individual tooth marks can be further applied to interpret the tracemaker. Was it a mouse, chipmunk, squirrel, eastern cottontail, marsh rabbit, or beaver? A 45° incision limits the tracemakers to rodents or lagomorphs; incisor widths further narrow the size of the animal; and the type of vegetation chewed should cinch the identification. For example, a recently snipped live oak twig, still wet with sap, green on the edges, and bearing tooth marks only a few millimeters wide, was definitely not made by a beaver or rabbit. (Also recall that both mammals do not climb trees). In such instances, the most likely tracemaker is a species of squirrel. Likewise, the gnawed base of a tree with 1-cm-wide incisor marks that are concentrated 30–40 cm (12–16 in) above the ground, come from a ground-dwelling animal with a length of at least 40 cm (16 in) when standing on its hind legs, as well as a keen interest in eating wood—in other words, a beaver.

Like all traces, though, tooth marks must be observed in the context of their habitat. For example, marsh rabbits and chipmunks are both good at clipping low-lying grasses, but rabbits are more commonly found in back-dune meadows, whereas chipmunks frequent maritime forests. How could such marks by squirrels, beavers, or similar mammals adapted for biting through wood be preserved or recognized in the fossil record? Accumulations of wood fragments from a variety of trees, in which the ends are preserved—allowing for diagnosis of incisor marks and cut angles—would do the trick. Indeed, at least one Pleistocene beaver dam has been interpreted in the Georgia coastal plain using these same criteria, providing hope that

Figure 8.13. Mammal feeding signs, large and small scale. (a) Browse line (*arrow*) in shrubbery made by fallow deer (*Dama dama*) along edge of maritime forest: Little St. Simons Island. (b) Pine cone nearly demolished by eastern gray squirrel (*Sciurus carolinensis*): Sapelo Island. (c) Tooth marks (*arrow*) on limb bone of white-tailed deer (*Odocoileus virginiana*), made by eastern gray squirrel: St. Catherines Island.

other such trace fossils might be found if geologists are made aware of their possible existence (Rich et al., 1999).

Among the more interesting but nuanced tooth marks are those evident in grasses, branches of shrubbery, or cactus, that have clean cuts on one side, but ragged, torn edges on the other. Mammals that lack teeth on their maxillas, such as deer and other ungulates, are responsible for such traces. For deer, a larger-scale manifestation of these feeding traces is a browse line, a height at which the vegetation on the edge of a maritime forest—usually at the transition between it and back-dune meadows—is clearly pruned (Figure 8.13a). This height is also representative of a moving average height for the deer, although varied by head position and industriousness when seeking food, such as when deer stand on their hind legs to browse higher than normal.

With all other tooth-mark ichnology in mind, do not forget such traces made by some of the smallest mammals evident on live-oak acorns, other nuts, and pinecones. Mammal tooth marks on nuts are distinguishable from those made by birds (such as nuthatches) by looking for paired incisor impressions, rather than the single beak marks of a bird (Elbroch and Marks 2001; Elbroch 2003). Acorns or other nuts with such tooth marks, particularly those made by squirrels, are also closely associated with caches, where rodents dug shallow pits and buried nuts to eat later. Such caches, however, are often permanently left below ground, either through a combination of poor memory or death of the tracemaker before it can excavate its hoard. Live-oak forests of the Georgia coast thus exist in part because of the digging abilities of generations of squirrels, their forgetfulness, and predation by raptors, a sort of behavioral imprint on barrier island landscapes we may take for granted. Other than the forests themselves, denuded pinecones are among the most common squirrel traces in pine-dominated forests. These are made by squirrels that pluck a fresh pinecone from a branch, sit down, and systematically chew open each seed casing surrounding the pine nuts, causing a rain of discarded casings, and eventually the core of the cone, from above (Figure 8.13b). Squirrels with greater ambition (and efficiency) chew off a pine branch containing many cones, and then descend to the ground to eat royally.

Tooth marks on bone are a result of multiple, overlapping motivations: wearing down of teeth—which grow continually in rodents—and gaining some needed calcium in diets (Figure 8.13c). Of course, not all tooth marks on bone were made by rodents, but also by predators and scavengers. Tooth marks of carnivores are circular to oval in cross section; are oriented perpendicular or otherwise at high angles to the bone surface; show lateral lineations formed by scraping of flesh or sinew from the bone; and are associated with bone fractures. All of these features are absent from tooth marks of mammals that simply scrape shallowly at a slight angle to the bone surface. Confusing matters are traces of bone-eroding insects, such as dermestid beetles (Chapter 5). Nonetheless, a skilled ichnologist, taphonomist, or other forensic scientist should be able to apply crosscutting relations to discern a typical order for marks on bone: predatory traces (made while killing the prey); scavenging traces (within a few days after the prey was dead); insect borings (weeks to months afterward); and tooth-wearing, calcium-eating traces (months or more than a year following death). Similar scenarios have been worked out for fossil bones, leading to a more complete picture of what happened to an animal's body between when it died and its discovery much later by paleontologists (Jacobsen and Bromley 2009).

Mammal Feces and Olfactory Landscapes

One cannot talk about mammal traces without also discussing their feces. Although the study of feces may seem a bit scatological, and is often the butt of jokes, it is ichnologically significant for those seeking the straight scoop on mammal behavior. Simply put, these end products of mammal digestion provide heaps of behavioral and ecological information; one might say they are full of it. Not only do feces tell us what a mammal ate, but much about the local ecology of the tracemakers, such as plant–animal and animal–animal interactions. In this respect, some mammals and flowering plants clearly coevolved, with the plants developing delicious fruits surrounding indigestible seeds. Thus what seems generous of the plant is actually a form of evolutionary bribery that assists its seed dispersal (Barlow 2000). Once a mammal has been enticed to consume a fruit, it carries the enclosed seeds to another locality that could not be reached by the plant. The mammal then dumps its load, including the seeds, all ensconced in a warm, healthy pile of fertilizer. Thus mammals, along with birds, are important dispersers for certain plants, and can impact entire ecosystems with their feces. For mammalian carnivores, the presence of prey animals in an area can be confirmed through scat, as well as the surprising items that may show up on

Figure 8.14. Examples of the wide range of sizes and forms of mammal feces (scat). (a) Pelletal scat of white-tailed deer (*Odocoileus virginiana*). (b) An impressive fecal pile left by a feral bull (*Bos taurus*), with its moistness indicating both its age (quite recent) and consumption of succulent plants by the tracemaker. (Also note the poorly defined track in the right center of the pile, doubling its ichnological value.) Both on Sapelo Island, scale in centimeters.

a mammal's menu, such as fiddler crabs or crayfish (Chapters 3, 5). Almost as important as eating for many mammals is the establishment of territory, and feces are essential for communicating who might be staking claim to a particular patch of land. Urinations, regurgitants, or other secretions often augment these obvious markers. All of these are used to announce to other mammals that someone is entering their home range, however temporary that declaration might turn out.

Like bird droppings, mammal feces are classified on the basis of shape, size, and composition of the undigested material. Once all three of these parameters are described, a preliminary identification of the tracemaker is more probable than if based on only one or two criteria. Shapes are categorized broadly as spheres (pellets), ovoids, and cords; cords are further subdivided on the basis of thickness to length ratios, and whether they are straight or folded (Halfpenny and Bruchac 2002; Elbroch 2003). Size is of course important for narrowing down the tracemaker, as scat volume can be compared to the body volume of a potential mammal for an easy reality check (Figure 8.14). For example, do you really think a chipmunk produced a 15 cm (6 in) long and 5 cm (2 in) wide chunk of feces? (If so, poor chipmunk.) Cord-like scat also can be segmented and have blunt or tapering ends. Scat composition is an essential piece of data to collect, and can lend to quick interpretations of the tracemaker dietary preferences by providing preliminary answers to simple questions: Herbivore, omnivore, or carnivore? Keep in mind, however, seasonal changes in diet and the way mammals modify their sources of food on the basis of availability. Even seemingly hardcore herbivores sometimes supplement their diets by seeking out insects or scavenging on dead animals. Conversely, carnivores will eat berries and other plant-based foods if these are abundant and easy to acquire (Halfpenny and Biesiot, 1986; Elbroch 2003). As a result, scat can hold surprises, and should be examined with the possibility of having unexpected contents.

As mentioned earlier, mammal scat also advertises the presence of a particular mammal in an area, acting much like billboards along a roadside. (This may be an unfair analogy, though, as the feces are much more attractive.) However, these signposts are not so much visual as olfactory. Think of how many mammals use their noses and nasal passages as primary sense organs that aid in their acquiring nutrition, navigation, social relations within their species, and interspecies interactions. For example, armadillos have poor vision and spend much of their time with their noses to the ground, sniffing out their food, or detecting the presence of other armadillos or predators who might make their day (mating!) or not (dying!), respectively. I have tested armadillo eyesight by standing very still and watching them make their extensive foraging traces—furrows in the forest floor punctuated by shallow dig marks—until they walk up to me, stand up on their hind legs, and take a big sniff. Only then do they run away. Admittedly, I may have contributed to such adverse reactions to my scent by spending several days in the field without bathing. Nonetheless, such behaviors still demonstrate how smell holds sway over sight in many mammals.

Some mammals also make scent posts, in which they regularly deposit scat or other secretions in the same place; urinations; wallows and lays; rubbings on trees; and random spots where they impart their scent. Even your beloved cat, rubbing against your leg while purring, is actually marking you as her or his property (but you already knew that). As a result, a tracker or other ichnologists studying mammal traces should not be surprised to find carnivore scat in prominent places, such as in the middle of a well-used trail or on top of a log lying across a trail. Canids in particular are notorious for mining trails with their fur- and bone-laden scat, announcing to rivals and anyone else who might be interested in learning, "This is my territory!" (Halfpenny and Biesiot, 1986; Elbroch 2003). Mammals are also prone to depositing their feces or urinating on other animal's waste products, lending credence to the earthy expression "pissing contest," which often can escalate to scat wars. All of these traces constitute an olfactory landscape that is mostly invisible to us, but is quite vivid and real for most terrestrial mammals, a part of their daily lives and drama. Nonetheless, such sensory abilities and behavioral interactions from the fossil record would be difficult to detect unless through inferences based on functional morphology of the skull, supplemented by trace fossils, such as tracks and coprolites.

Bats as Flying Mammalian Tracemakers

Mammalogists and other fans of furry critters have probably noticed that the preceding material omitted one of the most diverse, yet overlooked groups of mammals in North America: bats. This neglect, however, is based on ichnology in Georgia coastal environments, in which geography and geology conspire to decrease the most abundant and prominent traces normally made by bats, which are feces. Bat feces, also referred to as guano, can be quite abundant when produced by hundreds or thousands of generations of roosting bats in some long-lived caves. Such thick, nitrogen- and phosphorous-rich deposits become important economic resources, and have been mined for fertilizers (Sealey 2007). In other words, they are feeding traces that also result in people getting fed. Alas, the Georgia barrier islands lack caves or other large, subterranean environments that might serve as roosting spots for colonies of bats. In other places, such colonies are composed of hundreds of thousands of bats. Instead, we have to content ourselves with spottier finds of feces under some of their roosts, such as those left by the common brown bat (*Myotis lucifugus*). These traces are dark, cylindrical, 2–3 mm (0.08–0.1 in) wide, 5–15 mm (0.2–0.6 in) long, and filled with parts of flying insects (Elbroch 2003).

Although best adapted for incredible acrobatics in the air, a few species of bats also descend to the ground. Once on the surface, they may forage and walk on all fours, forming diagonal walking patterns from the tips of their wings—which are skin membranes stretched over elongated fingers—and rear feet. In fact, experiments done on vampire bat locomotion using small treadmills (no, I am not making this up) show that with increasing treadmill speeds, these bats (*Desmodus rotundus*) actually bound, and nearly gallop (Riskin et al. 2006). While bounding, these bats land on the ends of digits on their wings and push off with their rear feet, reaching maximum speeds of 1.2 m/s. Of course, small, bat-sized treadmills are even more rare than bat trackways in Georgia terrestrial environments, so you should not hold out much hope for seeing trackway evidence of galloping bats. Nonetheless, an identifiable landing and walking trackway made by a bat would indeed be a nice find, and its fossil equivalent would be the discovery of a lifetime. More readily available in ancient strata, though, are pterosaur tracks, whose makers also walked quadrupedally on land. Hence bats, although not directly related to these flying reptiles, may provide analogues for how some small pterosaurs moved about when on the ground (Mazin et al. 2009).

A Holistic Perspective of Mammal Traces

So rather than thinking of mere categories of mammal traces, and instead indulging in more holistic ichnological perspectives, we should remind ourselves to think about ecologically consistent assemblages of mammal traces. For example, a few mammal species of the Georgia barrier islands live in or near freshwater environments, such as river otters, marsh rabbits, muskrats, and mink, and make their respective tracks, trails, burrows, and feces there. However, taken together with the context of other ecological clues—substrates, vegetation, bones, shells—an ichnologist can put together a picture of the original ecosystem, which can serve as a model for a paleoecological analogue. Mammal traces may not be as numerous or as varied as avian traces, nor do they inspire the dinosaurian dreams—or nightmares—invoked by alligator traces. Nevertheless, their ichnological worth on the Georgia barrier islands is notable because of how some mammals so easily move across different ecosystems and substrates. This concept will be explored more in the next chapter, in which we will see that the tracks, trails, and burrows of so-called terrestrial mammals cross salt marshes, tidal-creek banks, dunes, storm-washover fans, and beaches often enough that they should be considered as marginal-marine tracemakers (Chapter 9).

A last point that should be made about mammal traces is how well they can tell stories with which we (as mammals) can so readily identify, whether in modern or fossil traces. For example, in July 2004 on Sapelo Island, I was collecting preliminary data on feral cattle (*Bos taurus*) to help with estimating their populations on the island. Unlike their dull-witted and slow-moving domestic counterparts, feral cattle seemingly revert to their Pleistocene heritage and live mostly in the maritime forests, rather than in open fields. Furthermore, they are so wily and wary of humans, one usually has to track them to detect their presence or learn much about their behavior. Fortunately, these animals are nearly as easy to track as humans, so I used this approach to learn more about them. One day while out scouting for their tracks and sign, I stopped my vehicle to look at some fresh tracks spotted on the sandy road. An adult cow had been walking along the right side of the road; those of a limping bull, who entered the scene from an intersecting trail on the left, later joined her tracks. Suddenly, in the midst of my methodically measuring and sketching their tracks, paces, straddles, and trackway patterns, I found myself bursting into laughter. The tracks clearly showed where the bull had approached the female from behind and tried to mount her, but she deftly shuffled forward just enough that he fell onto all fours and was left—shall we say?—in a frustrated state as she walked quickly away. Still, he persisted in his wooing, following behind and to her left, until she moved over to his left; soon after that, he crossed her trackway to the right and continued trailing her. I followed their tracks for nearly a kilometer, in which he kept pursuing her, but he never seemed to get close enough to attempt another coupling. Eventually I stopped tracking them because the whole scenario reminded me far too much of my dating days (minus the limp). Ichnologically speaking, though, the experience was transformational in learning that mammal traces can amuse as well as inform.

FRESHWATER AND TERRESTRIAL VERTEBRATE TRACEMAKERS: MORE NUMEROUS AND VARIED THAN YOU THOUGHT?

The noticeable lengthiness of this chapter for just two groups of tracemakers is not accidental, as the wide variety of terrestrial birds and mammals, coupled with their complex behaviors, warrant extensive treatment. As mentioned earlier, much of the neoichnology of the Georgia coast has emphasized marginal-marine invertebrate traces (Chapter 6), with only a few studies reporting vertebrate traces, but again related more to marginal-marine tracemakers (Reineck and Howard, 1978; Frey and Pemberton, 1986). Nevertheless, the majority of easily observable, macroscopic traces on the Georgia barrier islands, covering the broadest areas of maritime forests, back-dune meadows, and freshwater environments, while also holding ecological significance, are those of birds and mammals.

Perhaps upon starting this chapter, a reader also thought that the ichnology of terrestrial birds and mammals only involved the study of their tracks, with an emphasis on identifying the probable trackmakers. Now you know better. Bird and mammal traces of just the Georgia barrier islands include the following, with implied behaviors: dwelling and feeding burrows; arboreal nests made of twigs, mud, and grass; cavity nests in trees and outcrops; holes drilled into wood for acquiring insects; overland trails connecting between water bodies or food sources; trails and claw marks on trees; feces with digested remains of meals; regurgitants with undigested remains of meals; dust baths and wallows, used for health reasons; tooth marks on bone, also related to health; chew marks on plants used for food and nests; and many, many more. Of course, these traces have varying degrees of preservation potential, but their sheer number and variety increase the likelihood that similar examples exist in the fossil record (Chapter 9).

9

MARGINAL-MARINE AND MARINE VERTEBRATES

A TALE OF TURTLE TRACKS

Sea turtle trackways are perhaps the most impressive of marginal-marine vertebrate traces anyone can stumble upon on a Georgia beach, and this one was no exception. But because my wife, Ruth, and I were infrequent visitors to the coast, we had never seen one in all of its glorious three dimensions. Its presence that June morning was enhanced by its freshness, telling us that its tracemaker had been on Nannygoat Beach of Sapelo Island only four or five hours beforehand. Other prominent traces in the area were of a turtle patrol vehicle (an all-terrain vehicle, or ATV) that abruptly decelerated just after intersecting the trackway, turned sharply to the left, and looped back to stop. Based on the driver's tracks, she then spent much time at the end of the trackway, no doubt collecting data that would be analyzed later for its contribution to the larger scheme of understanding sea turtle nesting on the Georgia coast.

This particular trackway is what sea turtle researchers sometimes call a false crawl. Of course, there was nothing false about it. The sea turtle crawled out of the sea and left deep, broad, sweeping imprints of all four limbs (flippers), a partial drag mark made by the bottom of its shell, and scattered clods of moist, cohesive beach sand held together by water from the preceding tide and whatever drained off her wet body. The "false" appellation comes from the expectation that a sea turtle trackway will lead to a nest filled with eggs. In previous instances when these did not, the resulting disappointment evidently inspired such negative feelings among sea turtle researchers that they then assigned a descriptor denying the very existence of these magnificent traces left by expectant mother turtles. Incidentally, this is also one of the few instances where ichnologists can reliably identify sex from traces. After all, male turtles are not known to voluntarily crawl up on beaches, whether for exercise or otherwise; their first and last steps on land are normally done as hatchlings. So we were thrilled to find this trace, a feeling that was probably in inverse proportion to the disappointment held by the data collector, whose footprints seemed laden with melancholy.

Sea turtle conservationists who actively work along the Georgia coast throughout each nesting season are skilled ichnologists in their own realm. Their early morning patrols of beaches, typically performed by driving ATVs along the length of an assigned beach, involve rapid scanning of beach sands for evidence of fresh turtle trackways. Although spotting such sign in a fast-moving vehicle seems like a remarkable feat, it is not, because turtle trackways are large, conspicuous, and unmistakable. The real skill of these conservationists—often volunteers, who, much like writers, are working for the love of their subjects, not for the money—becomes evident when they stop the vehicle and start documenting the trackways. Recording the location of a trackway is of paramount importance, as this information is put into a GIS (geographic information system) database that later can be used to predict sea turtle nesting sites, which aids in future monitoring of nests. But the remaining data are all part of a detailed picture drawn in the sand by the mother sea turtle. A series of questions volunteers then ask might consist of the following: When did she come up on the beach? How wide and long is the trackway? How does the trackway reflect the size of the tracemaker? Where does the trackway go? Does it end at a nest in the upper beach or lower dunes, or does it turn around short of that goal? Did she stop anywhere along the way? Why did she stop at one spot versus another? What species of turtle made this trackway? All of these questions, and more, can be answered by a careful examination of her traces.

Based on our assessment of the trackway, the mother turtle had come out of the surf zone with a high (but dropping) tide in the early morning.

We were there later at low tide, which revealed the full transition of tracks made in saturated sand to those in the relatively drier, drained sand of the upper beach (Figure 9.1a). Her tracks at the high-tide mark were vaguely outlined depressions, only visible as widely separated two-by-two patterns, made more visible by darker heavy minerals on the edges of the concavities. Water had flowed off her flippers and into each track, imparting a drainage system in miniature that faded with each step up the beach and away from the waterline. Once she encountered drier sand, it was pushed back into huge, ridge-like structures with fractures around their edges. Dislodged chunks of sand that had temporarily adhered to her flippers were also scattered on the periphery of the trackway. The tracks were in pairs on each side, corresponding to manus and pes. The largest blade-like impressions in the front and outside of the trackway represented the manus, and the pes as the smaller, more medially placed one behind the manus. The front and rear flippers were separated by about 15 cm (6 in) on each side. The manus impression was boomerang shaped, with both ends pointing forward, and about 30 cm (12 in) long, whereas the pes was more of a gentle crescent moon and about 20 cm (8 in) long (Figure 9.1b). Based on the sizes of the pressure-release structures left by manus, most of the power behind the turtle's movement up the beach came from the front limbs. She alternated her steps—right, left, right—but in short increments, only 25 cm (10 in) of progress made on each side. The minimum size of her shell, 50 cm (20 in), was indicated by the central drag mark made by the shell, and with limbs outstretched and walking, she was about 75 cm (30 in) wide. Her minimum length then could be figured by measuring the distance between the opposite-side front and rear flipper tracks. This was about 60 cm (24 in), and once her head, tail, and remaining bow and stern of her shell was included, she was probably closer to a full meter (3.3 ft) long. She was a big mamma.

The trackway showed how she stopped about 80% of the way up the beach, just a few meters short of the dunes, inexplicably made an abrupt left turn, and headed back to the sea. Once we stepped back to view it in full, the trackway formed a huge, arch-like U shape. At its apex, decreased pace distances between tracks and more pronounced shell drag marks showed where she slowed down, stopped, and made her decision to leave the site. This is also where her short tail registered slightly in the middle of the trackway, little commas inside big parentheses. This would not be the place for her nest, not this time. The turtle researcher had planted a wooden

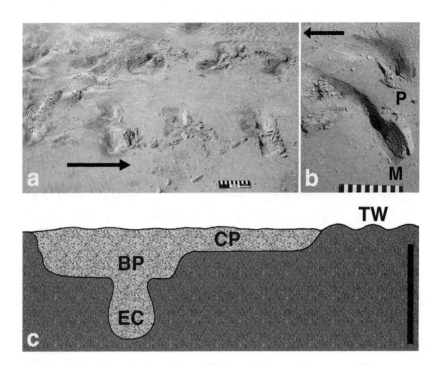

Figure 9.1. Traces of a mother loggerhead sea turtle (*Caretta caretta*) on Sapelo Island, and the nest structure she would have made, had she not changed her mind. (a) Transition in character of tracks from saturated sediments (*left*) to drained, drier sediments (*right*) as the turtle crawled onto the beach, with arrow indicating direction of movement. Scale = 15 cm (6 in). (b) Close-up of left side of trackway, showing large pressure-release structures associated with pushing motion of limbs; larger track (M) is from the front flipper (manus), smaller (P) from the rear flipper (pes), arrow indicates direction of movement. Scale in centimeters. (c) Cross-sectional view of a typical loggerhead nest structure, including trackways (TW), covering pit (CP), body pit (BP), and egg chamber (EC). Based on descriptions and figures by Brannen and Bishop (1993) and Bishop et al. (2011). Scale = 50 cm.

stake at the highest point of the turtle's path. Field data—day, number, and other brief notations—were written on one side of the stake, partly to document, but also to prevent stopping at the same trackway the next day and duplicating the data. When you're speeding by on an ATV, turtle trackways may tend to look alike, although when viewed carefully and up close, their individual personalities are soon revealed. Indeed, scientists who studied sea turtle movements were astonished to discover that some individual turtles come back to the same places to nest over the years—a form of site fidelity—and, in some instances, where they hatched (Bowen and Karl 1996; Lutz et al. 1997; Tucker 2009).

The trackway also revealed when and how long she had stayed on the beach; the part where it was preserved in drier sand and heading back to sea was slightly shorter that her ascending one; thus the tide was still rising and peaked while she was onshore. A quick look at a tide chart later bracketed when she was there to within an hour or so. In our case, she had not stayed long at all, as her tracks turned about 7–8 meters (23–26 ft) past the high tide mark before she changed her mind and turned to the left: probably less than an hour, and closer to 40 minutes.

Had she decided to nest, other distinctive traces would have been connected to the trackway. The trackway and the traces would have shown the following sequence of behavioral nuances, told in the present tense. Her body turns to the right or left well above the high tide, perhaps in the dunes, and then stops. With her front flippers, she digs a shallow depression outlining her body. Appropriately, this structure is called a body pit. In this body pit, the sand holds her in place, and makes her next efforts a little more inconspicuous to any nearby egg predators who have already smelled her in the area. She is seeking sand with the right moisture content so that the walls of the nest chamber, which she intends to make next, will remain upright. If the sand is too dry—walls would collapse—or too wet—the eggs would drown—then she leaves, returning to the sea to try again somewhere else. With her rear flippers, she digs down (vertically) in alternating, semicircular sweeps outward, and scours out an egg chamber with a shape that is best described as an upside-down light bulb (incandescent, that is: Figure 9.1c). This chamber is about 25 cm (10 in) wide, spherical to ovoid in its bottom portion, and connects with a more narrow neck-like portion, 15–20 cm (6–8 in) wide at the top. The total depth of the egg chamber is slightly less than the length of her rear flippers, about 40 cm (16 in).

Once the chamber is made, she positions her cloaca at the top of it, and fills it with as many as 120 leathery eggs. As more eggs are deposited, those on the outside of the clutch press into the sandy walls of the nest chamber, imparting crescentic impressions. Once the egg laying is complete, she pulls in the excavated piles of sand with her rear flippers, buries the chamber, and gives a few pats on the surface to help compact the sand above the nest. She makes more sweeps, but this time using the front flippers as well, making the sea turtle equivalent of a snow angel in the sand, spreading over the body pit as another attempt at concealment of the nest. She may make a complete 180° turn while covering, which maximizes bioturbation of the surface area of sand above the nest and potentially confuses any egg predators seeking the nest chamber. This area of mixed sediments can be as much as 2–3 m (6.6–10 ft) wide and 20–30 cm (8–12 in) thick, but its position on the upper beach or lower dunes means that it will be weathered quickly by wind and rain, leaving an amorphous outline within only a few days. Once finished with the covering, she crawls back to sea, off in the buoyant ocean to recover from her arduous sojourn on land.

The tracemaker of the trackway we were studying that morning on Sapelo was probably a loggerhead sea turtle. Species identification of a trackway is possible through using just a bit of knowledge about their natural history and ways they move. Although loggerheads are by far the most common sea turtles to nest on Georgia beaches, and thus become the default hypothesis for identifying their trackways, other differences in anatomy and behavior can help to distinguish different tracemakers. Five species of sea turtles live in the waters off the Georgia coast: loggerhead (*Caretta caretta*), hawksbill (*Eretmochelys imbricata*), Kemp's ridley (*Lepidochelys kempii*), green (*Chelonia mydas*), and leatherback (*Dermochelys coriacea*). Only three of these turtles are known to come up onto Georgia shores to nest—loggerhead, leatherback, and green—which further helps to narrow down the maker of a trackway (Ruckdeschel and Shoop 2006). The trackway we were studying was smaller compared to that of a nesting leatherback (discussed later) and a green turtle too. Moreover, both the leatherback and green turtle trackways have pronounced tail drag marks (missing from most loggerhead trackways) and parallel manus and pes impressions, rather than the alternating ones made by a loggerhead (Figure 9.2). Parallel flipper impressions are a result of leatherback and green turtles moving their front and rear limbs in unison, more like a butterfly stroke used by humans in swimming, instead of a freestyle crawl. This movement also causes the tail to register in the sand, where it makes a central groove as it is dragged forward, then a point as the turtle reaches its greatest height above the sand surface at the end of limb strokes. Sea turtle volunteers are taught these differences in trackways so that when they encounter one during their morning surveys, they can quickly identify which of the three species made it. This practical application of sea turtle ichnology also al-

MARGINAL-MARINE AND MARINE VERTEBRATES 441

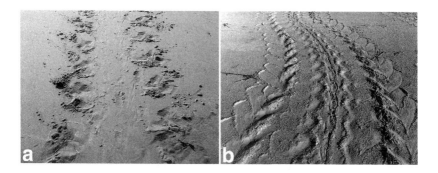

Figure 9.2. Differences between sea turtle tracks, based on gait pattern and tail drag marks. (a) Loggerhead (*Caretta caretta*) trackway, with alternating manus–pes pairs and no discernable tail drag mark: St. Catherines Island. (b) Leatherback (*Dermochelys coriacea*) trackway, with parallel (synchronized) manus–pes pairs and prominent tail drag mark in middle of trackway: Cumberland Island. In both trackways, the turtles were moving away from the viewer. Photograph in (b) by Carol Ruckdeschel.

lows for answering subsequent questions about the tracemaker. If these volunteers worked elsewhere in the world where more than a few turtle species were known to nest, then they would use a combination of trackway widths and trackway patterns to discern each species (Lutz et al. 1997).

Although we had encountered a nonnesting crawlway that morning, the next morning might have given us one that led to all of the traces associated with nesting, along with their implied behaviors. We also might see a sign that the sea turtle volunteer has already protected the nest by installing a tough, orange plastic mesh over the nest chamber, and documented its presence with a stake, which then helps with subsequent monitoring of the nest. On the other hand, if ghost crabs, raccoons, or feral hogs reached the eggs before the sea turtle volunteer, we also might have seen the traces of egg predation, which is what the mesh is meant to prevent.

Studying the ichnology of egg predators is helpful for understanding how and when sea turtle nests might be attacked, as well as what species of egg predator might be habitually looking for freshly laid eggs. Ghost crab predation (Chapter 6) would have been evidenced by freshly dug burrows on top of or on the periphery of the nest chamber that, if followed downshaft, would intersect with the chamber. Once below the sand surface, the crabs break into and consume the protein-rich contents of eggs on the outside of the nest first, then work their way inward. Although they may not eat all 100+ eggs in one day, they and other ghost crabs that discover such a buried treasure have an assured food cache for weeks afterward. The nest structure and its accompanying body pit may still be recognizable afterward, though, as a result of this precise targeting by such a small predator.

A raccoon raid of a sea turtle nest would be more obvious, having a conical pit trampled by raccoon tracks, piles of sand on the sides of the pit, and scattered empty eggshells bearing raccoon tooth marks. Not all of the eggs may be eaten in one episode: individual raccoons typically only consume about one-fourth of the eggs in a clutch (Ruckdeschel and Shoop 2006). Nevertheless, exposure of the nest chamber and remaining eggs means that it is only a matter of time before all of the embryonic turtles are dead, in which case egg predation segues into egg scavenging. Interestingly, raccoons that have previously eaten turtle eggs will link the smell of a mother sea turtle with a yummy meal. Hence raccoon trackways may follow a sea turtle's tracks up a beach, the scent leading the raccoon directly to the nest: a form of olfactory tracking. Owing to such rich opportunities for food, raccoons, which we normally associate with maritime forests (Chapter 8), are frequent visitors to Georgia beaches, as well as salt marshes at low tide, where they hunt for fiddler crabs.

The most overt of turtle egg predation traces, however, are made by feral hogs (*Sus scrofa*), which may include huge and deep pits, snout impressions, numerous tracks, and few to no eggshells, the latter situation because hogs eat the eggs whole, and all of them. In other words, they behave like, well, pigs. The impact of feral hogs on sea turtle eggs varies on the Georgia barrier islands, and is largely a function of their population size. For example, they have no effect on turtle eggs on Jekyll Island, where they are apparently absent (I have never seen feral hog tracks there), but are a constant source of concern on islands overrun by hogs, such as Ossabaw, Cumberland, or St. Catherines Islands. Hog tracks and other traces are ubiquitous in maritime forests of these islands, and common on their beaches and dunes. As a result, volunteers on ATVs on their early morning patrols of some islands are often racing against hogs, trying to beat them to a nest before the entire egg clutch is exhumed and eaten. Along those lines, I know of at least one sea turtle researcher who, while patrolling for trackways, carries an antihog and turtle-nest protection tool, known popularly as a rifle.

Other dangers to a newly made nest that leave ichnological signatures include ant nests (Chapter 5). If a mother turtle made the mistake of placing her eggs near a colony of ants—particularly fire ants (*Solenopsis invicta*)—the vertebrate nest could become a food source for the invertebrate nest. Fire ants will also attack on the basis of movement and disturbance of their nests, started by the first ant that emits alarm pheromones, which provokes an accelerating chain reaction in other soldier ants within the colony (Tschinkel 2006). A sea turtle nest placed in June thus might have a fire ant nest move in next to it in July. Then, as soon as the vibrations of hatching turtle eggs reach the colony, the ants will be perturbed enough to invade the egg chamber and sting the offending baby turtles until they stop moving. Of course, then the turtles become food for the ants and any other scavengers that are attracted to such treats. In short, unnatural selection sometimes occurs with sea turtle nesting, in which invasive species and native predators join forces, thus making sea turtle nests yet more vulnerable without human assistance.

If nesting happened early in the nesting season—May, and lasting through August—a mother turtle might return to the same beach a few weeks later. Some sea turtles make as many as seven nests in the same season (Ruckdeschel and Shoop 2006), a frequency that is seemingly dependent on fat stores of the mother (how many eggs can she produce?) and the availability of adequate beaches (how many potential good places for nests?). The latter factor seems to be the more important one now, as turtles will not just nest anywhere. The broad, mostly undeveloped beaches of the Georgia barrier islands provide ideal places for turtles to crawl onto a shore and dig a nest. Nevertheless, statistical analyses of turtle nest locations linked to proximity of human structures—buildings, street lights, piers, jetties, and so on—suggest that turtles tend to avoid locations with such items, or even beaches that have been renourished (Leong and Waller 2008; Waller and Leong 2008). Furthermore, sea turtle researchers posit that mother turtles are quite sensitive to conditions for laying their eggs. Any disturbances in the area of nesting, including bright lights, noises, human activity, and odd structures, will drive them off, and their trackways will show looping U-turns on the beach wherever they were spooked. As mentioned before, they may decide while digging a nest that the sand is too moist, too cool, or otherwise not to their liking; a fickle behavior that causes them to go back to sea, swim a bit further, and try somewhere else.

Temperature is also a factor, and warmer sand is better in this respect. For that reason alone, the vast majority of sea turtle nesting on the east coast of the United States is not on the Georgia coast, but to the south on Florida beaches. Nesting frequency is carefully recorded annually for each Georgia barrier island; since such numbers have been compiled, 2008 was a record year (1,646 nests documented).

Once safely buried by turtles and monitored by their human caretakers, eggs take 50–60 days to hatch. The temperature of the substrate (sand) becomes an important factor again, and must stay 24–32° C (75–90° F) for the eggs to incubate properly; sand temperature also relates specifically to sex determination in the brood (Spotila 2004; Ruckdeschel and Shoop 2006). Sand that is, on average, less than 28°C (82°F) causes hormonal changes in the embryos, making them males, whereas sand temperatures of more than 29°C (84°F) result in females. The reality is that sand temperatures vary around and within the nest, hence microenvironmental differences cause most egg clutches to have mixed sex ratios, but with localized clusters of male and female hatchlings. Once they are fully mature and ready to escape their leathery boundaries, hatchling turtles use a sharp projection on the tops of their beaks (called a caruncle) to slice open the eggshell. This cut leaves a hatching trace that should be distinct from those inflicted by teeth and claws in egg predation. Massed movement of hatchlings, particularly those toward the top of the nest chamber, bioturbates the sand so that it collapses around them, thus creating an exit route to the surface.

This activity pauses, though, if the sand near the surface is too hot, indicating daytime temperatures. Daylight is bad for baby sea turtles because of how quickly they would be spotted by diurnal predators. As a result, natural selection has ensured that hatchlings normally emerge under the cover of night. However, because of this heating, sand near the top of the nest is drier than below, which helps with its collapse once the baby turtles start moving upward. So once night arrives and sand temperatures decrease, hatchlings resume their upward movement and emerge through the exit hole formed by the downward collapse of sand. The clutch may hatch in stages, perhaps taking place over several days, but movement begets movement within the nest chamber and soon most of the turtles leave through the top. Once out, they are guided by light, such as moonlight and starlight reflecting off the ocean surface, as well as slope, in which a downhill direction both increases their speed and better enables getting to the surf. Consequently, their trackways may be small, numerous, and overlapping, but mostly unidirectional, heading toward the tideline at that time. The track sequences will be noticeable as 10-cm-wide (4 in) trackways composed of alternating pairs of right–left depressions and shell drag marks, although their small masses, usually about 20 g (less than an ounce for a newly hatched loggerhead), may make for faint and ephemeral traces.

All is not well upon hatching and emerging from the nest, however. This event, especially if it takes place over days, attracts predators: ghost crabs (again!), nocturnal birds, such as yellow-crowned night herons (*Nyctanassa violacea*), raccoons, and other animals that enjoy these delectable morsels, will have easy pickings from a large group of slow-moving baby turtles. As a result, traces of baby turtles may end just in front of those of their predators, with only the tracks of their consumers walking away from the scene. Additionally, hatchling trackways may indicate movement away from the ocean and toward errant light sources, such as those of nearby piers, hotels, and condominiums. In these instances, their tracks lead to impacts with motor vehicles, or otherwise tell of ignoble ends to a cycle of life based on nearly 200 million years of earth history.

Nonetheless, the huge numbers and varieties of traces left by sea turtle species today provide a window into their evolutionary history and help us better appreciate how far they have come as a lineage. Indeed, their trackways and other traces on at least one Georgia barrier island (Jekyll Island) are inspiring a rare form of revenue: ichnotourism. On Jekyll, each nesting season provides an opportunity for tourists interested in sea turtles to observe their nesting. These turtle walks involve looking for fresh trackways on the nighttime beaches of Jekyll, which then easily lead to the nesting turtles. Direct viewing of nest making and egg laying evidently leaves a lasting mental trace on the people who witness it, increasing the probability that they will support protective measures taken by conservationists to ensure reproductive success in Georgia sea turtles. With such elevated public awareness, prompted by the charismatic traces left by these marine reptiles, these endangered animals may just have a chance at surviving for a while longer.

STINGRAYS AND OTHER FISH TRACEMAKERS

Although sea turtles are clearly the most engaging of marginal-marine tracemakers on the Georgia coast, traces made by other vertebrates—various fish, alligators, birds, and mammals—are common on its tidal creek banks, beaches, and sand flats. For example, at low tide and on nearly any expanse of the lower intertidal area of the Georgia coast, large depressions, some 1–2 m (3.3–6.6 ft) across and 30 cm (12 in) deep, interrupt the rippled surfaces and draw attention because of their size and irregular outlines (Figure 9.3). When I first saw these odd structures, the hypothesis that leapt into my mind was "limulid resting trace," where a large adult limulid had settled on the sand bottom during high tide and fed on infaunal invertebrates in the sand (Chapter 6). To enable this predation, juvenile and adult limulids alike stop on sandy surfaces and move their walking legs in the saturated sand below, loosening it enough to reveal its hidden goodies. This action also causes the limulid to sink slightly into the resulting depression, making a pit surrounded by more compacted sand (Chapter 6). Inspired by that assumption, I then imagined currents—either from waves or shifting tides—that scoured the sand from around the limulid carapace, expanding the depression so it became larger than the actual tracemaker.

A wonderful and most pleasing hypothesis it was, combining my meager knowledge of limulid biology and coastal sedimentology with ichnology. Yet it was wrong. A few of these traces may indeed be as I visualized—limulid resting traces that were eroded by currents—but most are no doubt attributable to fish, specifically the southern stingray (*Dasyatis americana*), although some could also be made by other rays, such as the Atlantic (*D. sabina*) or bluntnose stingray (*D. sayi*). I have also seen these traces on exposed muddy banks of tidal channels at low tide, their identity supported by sightings of appropriately sized rays swimming nearby in the shallow, murky water.

Stingrays are cartilaginous fish (Chondricthyes) under the family Dasyatidae, and are more closely related to sharks than to bony fish, such as teleosts (Chapter 7). Normally famed for their beauty while swimming, in which they undulate broad, wing-like pectoral fins, these fish were seen in a different light after the accidental death of animal popularist Steve Irwin by a stingray in 2006. Among the few bony parts of a typical stingray is a stout, serrated, and pointed barb on the end of its tail, which it uses for defense if threatened; in some species, this barb is laced with venom. Tail barbs can be as much as 25 cm (10 in) long, which, when combined with venom, can inflict awful wounds.

MARGINAL-MARINE AND MARINE VERTEBRATES 447

Figure 9.3. Southern stingray (*Dasyatis americana*) feeding traces on rippled lower intertidal sand flat, each about 50 cm (20 in) wide, and formed by combined resting and shooting jets of water into the sand: Sapelo Island. With just a little imagination, think of how these could be perceived as large tracks, forming a trackway.

Such attacks, however, are unreported from the Georgia coast. Moreover, their barbs are not used for acquiring food. Stingrays and other rays instead use gentler means for finding their meals, as reflected by their traces. The southern stingray, for example, swims along a sandy bottom, settles down on a spot that holds promise of food items just underneath the surface, and expels a jet of water from its ventrally oriented mouth against the sand (Howard et al. 1977). This action loosens the sand, dislodging and exposing any infaunal bivalves, crustaceans, and other invertebrates, and even small vertebrates that may have been living safely there (Chapter 6). Ichnologists previously thought these traces were made by stingrays flapping their pectoral fins against the sand to put it into suspension, but direct observations of their feeding, as well as careful descriptions of the traces, have confirmed their use of water jets (Gregory et al. 1979; Hines et al. 1997). How do they sense their prey underneath the sand surface? Are stingrays actually cartilaginous ichnologists that manage to recognize invertebrate burrows? No, but they are electroreceptive, using sensory organs called ampullae of Lorenzini, which are capable of detecting weak electric fields created by infaunal animals; these organs are also in sharks (Blonder and Alevizon 1988). Once invertebrate or small vertebrate snacks are found, a stingray places its wide mouth around them and starts masticating with hard, flat plates of tightly arranged teeth, the only other bone-like tissues they have besides their barbs. Working like nutcrackers, these teeth can easily break open hard-shelled crustaceans, bivalves, and gastropods (Summers 2000). Soft parts are consumed and a stingray then spits out the crushed hard parts, sometimes leaving an accumulation of shell debris in their feeding excavations.

Do these traces sound too vaguely defined to be discerned from the fossil record? If you said "yes," you are as wrong as I was with my initial premise that such traces were made by limulids. Ray feeding traces have been interpreted from Cretaceous rocks in North America (Utah) and Europe (Spain), as well as Miocene rocks in the North Island of New Zealand (Howard et al. 1977; Gregory 1991; Martinell et al. 2001). In New Zealand, modern examples can be seen on tidal flats only about 20 km (12 mi) from an outcrop holding fossil ones. Modern ray resting traces there so perfectly outline ray bodies that some have clasper impressions of the male rays. Males use these appendages to attach to females during mating; hence their presence in a trace allows neoichnologists to identify the sex of the tracemaker. Unfortunately, the trace-fossil equivalent of these clasper impressions have not yet been interpreted, but their criteria are easy enough that some paleontologist will notice it in a ray trace fossil some day. Diagnostic criteria for ancient examples of ray traces include size and dimensions, but ideally would also include pieces of crushed and otherwise macerated molluscan shells.

Not surprisingly, such large depressions, when found in Mesozoic rocks, have been mistaken for sauropod dinosaur tracks or nests (López-Martínez et al. 2000). These hypotheses, however, were disproved once researchers realized that the structures were quite variable in size and shape; they did not define definite trackway patterns; nor were any dinosaur eggshell or embryonic bones associated with each of them (Martinell et al. 2001). Their occurrence in intertidal sediments further supported the hypothesis that these were trace fossils made by sea-dwelling animals, although some sauropods left tracks in intertidal environments too (Farlow et al. 1989). Rays then became the most likely candidates as tracemakers once these trace fossils were compared to modern examples.

Some fish in both freshwater and marginal-marine environments also burrow and live in burrows of their making. Alas, few fish in either environmental setting of the Georgia coast are documented burrowers, although many secondarily occupy invertebrate or vertebrate burrows. Among the more notable fish burrowers elsewhere are lungfish, which construct their burrows by shimmying their bodies straight downward into sediment (head first), then turning around at the burrow end and moving up the same vertical shaft. The resulting structure is shaped like an old-fashioned bulb thermometer, in which a vertical tube leads to an enlarged chamber, where the lungfish aestivates, breeds, or just hang outs (Gobetz et al. 2006). In contrast, the only documented burrowing fishes on the Georgia coast are snake eels (*Ophichthus ocellatus* or *O. gomesi*), which live in riverine estuary bottoms. These eels and their relatives start a burrow by poking a tail into the host sediment, followed by pushing the sediment aside as the rest of the body is inserted (de Schepper et al. 2007). The advantage of such a burrowing strategy is that an eel does not have to back out of its home, nor construct a turnaround chamber. Instead, it can simply stay in the burrow with its head near the entrance, ready to snatch any vulnerable fish that might swim by. Snake eel burrows are not the neat vertical or near-vertical ones made by lungfish, but are more convoluted, having mostly vertical trends but also a few horizontal turns, evocative of an eel's body movement. These burrows are nearly circular in cross section and slightly wider than the eel (a few centimeters) and can be more than 50 cm (18 in) long; burrow interiors are smooth and lined with copious amounts of fish-produced mucus (Howard and Frey 1975). As a result of their sizes and forms, such burrows could be easily mistaken for ones made by larger invertebrates, particularly those of hemichordates. Howard and Frey (1975) also speculated that the cusk eel (*Rissola marginata*) might make burrows like those of snake eels, but these structures have not yet been documented.

Traces made by other fish in Georgia coastal environments are more difficult to recognize, with the obvious exception of tooth marks inflicted by sharks. Sharks of the Georgia Bight are diverse, including nurse sharks (*Ginglymostoma cirratum*); smooth dogfish (*Mustelis canis*); Atlantic sharpnose sharks (*Rhizoprionodon terraenovae*); dusky sharks (*Carcharhinus obscurus*); blacknose sharks (*C. acronotus*); blacktip sharks (*C. limbatus*); sandbar sharks (*C. milberti*); tiger sharks (*Galeocerdo cuvier*); and scalloped hammerheads (*Sphyrna lewini*), to name a few (Dahlberg 2008). Shark diversity in offshore areas, though, is dependent on seasonal temperatures, where warmer water means more species. Shark tooth marks are not necessarily found only in the bones of prey items swimming offshore, but also may occur in those of land-dwelling animals. For example, remains of clapper rails (*Rallus longirostris*), which are salt marsh–dwelling birds, have been found in the stomachs of juvenile tiger sharks on the Georgia Bight (Carlson et al. 2002). This circumstance either indicates their predation of these birds in salt marshes, or opportunistic feeding on clapper rails that may have been washed out to sea during storms. This instance, along with scavenging of terrestrial vertebrate carcasses flushed out by rivers and tides from the mainland and barrier islands, implies that shark tooth marks also may be common traces in vertebrate remains that did not go through a shark digestive tract. Indeed, shark tooth marks have been interpreted on fossil whale bones from the Pliocene–Holocene deposits in South Carolina (Cicimurri and Knight 2009), and in dinosaur bones from Late Cretaceous deposits in Georgia (Schwimmer et al. 1997). The latter were helpfully identified in one instance by the presence of an actual shark tooth still in the trace (Chapter 9).

Other than these examples, marginal-marine fish traces of the Georgia coast are apparently understudied, as I expect many species other than rays, snake eels, and sharks are forming traces that would leave lasting marks fit for the fossil record. For example, as of this writing I could not find any published discussions on the effects of bioeroding fish on Gray's Reef (Chapter 2). Yet such fish are important tracemakers in subtropical and tropical reef environments through their munching of coral and fecal deposition of digested coral. (Think about that next time you are lying on a Caribbean beach with its beautiful, white carbonate sand, and how most of it went through the guts of parrotfish and came out their rear ends. Romantic, is it not?) Additionally, one common species of teleost fish of the Georgia Bight, the black drum (*Pogonias cromis*), was documented on the U.S. gulf coast as a major bioeroder through its feeding-related breakage of bivalve shells (Cate and Evans 1994). Nonetheless, this phenomenon has not been studied in the eastern United States. As a result, I fully expect that this relatively short section will be augmented by future research coming from many eager ichthyologically inclined ichnologists, who will show how the abundant fishes of Georgia coastal environments likewise produce many traces applicable to fossil examples.

REPTILES THAT VISIT THE BEACH AND SALT MARSH

Species of seafaring or salt marsh–dwelling reptilian tracemakers on the Georgia coast are easy to name, as there are only five, three of which are sea turtles: loggerhead turtles, leatherback turtles, green turtles, alligators, and diamondback terrapins (*Malaclemys terrapin*). Salty water excludes most amphibians and reptiles, and only a few species of reptiles worldwide have adapted to it, such as sea turtles, sea snakes, marine iguanas, and some crocodilians. In the geologic past, and just before the end of the Mesozoic Era, about 65 million years ago, this situation was dramatically different. For more than 150 million years, huge marine reptiles ruled the oceans: mosasaurs, plesiosaurs, and ichthyosaurs shared the oceans with sea turtles and sea snakes, along with massive, dinosaur-eating crocodiles that frequented shallow offshore areas when not nesting on land (Schwimmer 2002; Everhart 2005). Most marine reptiles, however, probably stayed in the sea their entire lives, giving live birth—which paleontologists have confirmed through embryonic remains in fossilized adults—instead of crawling onto land to dig nests and lay eggs. Consequently, neoichnologists working on the Georgia coast should be thankful for these five species of reptilian tracemakers as modern analogues in marginal-marine environments. However, in no way should we see these as representing anything more than mere glimpses of the past, seen between much blinking.

Sea Turtles

Sea turtle neoichnology has already been discussed in detail, but should be explained a bit more in the context of the environmental setting for this tracemaking. Georgia beaches are hot spots for sea turtle nesting, given all of the excellent reasons for turtles to be in the area and come on shore. For one, the Georgia Bight, as a wide and shallow marine area, provides superb hunting for the five species of sea turtles that inhabit its waters, where they can eat jellyfish and much other floating and swimming food. Georgia beaches are also relatively flat when approached from the water, requiring less effort for mother sea turtles to walk on their way to potential nest sites. The wide beaches of the Georgia coast, which are composed of fine-grained to very fine-grained sand and heated by summertime temperatures, supply many suitable places for digging nests and incubating eggs. Added to these factors is that most Georgia barrier island beaches remain relatively undeveloped: no skyscraping condominiums are on any of them, although a few are developed enough to send some neoichnologists away screaming. This situation means that mother turtles are more likely to crawl onto darkened beaches to make their large, beautiful trackways, and hatchlings will not be distracted by lights of nearby humans.

These advantages, however, are counterweighed by noxious and sometimes fatal challenges posed to nests by native species (raccoons, ghost crabs) and invasive species (hogs, humans) that prey on turtle eggs. Other potential dangers to sea turtle nesting are posed by plans to develop areas with beachside housing, shopping malls, and restaurants just behind the dunes, which is exactly where sea turtles prefer to nest. Even before nesting, floating plastic bags are a common form of death in turtles that mistake these for jellyfish and consume them, which then blocks their digestive tracts (Spotila 2004). Additionally, despite the best efforts of offshore shrimp trawler crews to use turtle-excluding devices (TEDs) on their nets, mother (and father) turtles frequently die traumatically when caught in nets. Consequently, their bodies wash up more often on Georgia beaches during the shrimping season, providing an exotic menu item for black and turkey vultures on Georgia beaches then.

Although sea turtle nesting strategies might seem exceedingly picky to a casual observer, they have worked well since the Late Triassic, which is when the oldest sea turtles show up in the fossil record. The oldest fossil sea turtle nest structure, trackway, and body pit, discovered in Late Cretaceous rocks of Colorado, also suggest that their onshore nesting behavior has probably always been with them too (Bishop et al. 1997, 2011). No turtles, sea dwelling or otherwise, are known to give live birth, thus the assumption is that all ancestors of present-day sea turtles laid eggs on land. Turtles were originally thought to have evolved in terrestrial and freshwater environments during the Triassic Period (more than 200 million years ago), and later underwent rapid speciation into different habitats during the remainder of the Mesozoic Era. The discovery of a primitive Late Triassic turtle in marine rocks of China, though, altered this perception, suggesting that turtles may have evolved first in marginal-marine environments (Li et al. 2008). In short, sea turtles have been conducting sea turtle business for a long time, and have followed the evolutionary equivalent of the Southern dictum, "If it ain't broke, don't fix it." Nonetheless, current human alterations of habitats, invasive species, and global climate change are forcing selection pressures on sea turtles that make their egg laying more pernicious. As I often tell my students, if you want to make a species go extinct, stop it from reproducing. Sure enough, all seven extant species of sea turtles are endangered because of the aforementioned challenges posed to their nesting and reaching sexual maturity.

A mention of the Mesozoic probably helps to provoke thoughts of reptiles of unusual size, such as the chelonian trivia question, "What is the world's most massive modern reptile?" If the recipient of such a question is not in a turtle frame of mind, he or she will probably answer "crocodile," and if pressed for a specific crocodile, "Nile crocodile" (*Crocodylus niloticus*) or "saltie" (*Crocodylus porosus*) of Africa and Australia, respectively. One that might be forgotten—yet is among the largest living reptiles, and certainly the biggest sea-dwelling one—is the leatherback turtle (*Dermochelys coriacea*). This sea turtle, the last of its genus, can reach weights of more than 700 kg (> 1,500 lb), and adults are longer than most people are tall. Additionally, I will bet that nearly no one, except for the most turtle fanatic of people, will know that they are the speediest reptiles in the world. Yes, they cheat by swimming, but they can exceed almost 40 kph (25 mph) in the water, more than four times faster than the best of Olympic swimmers.

A leatherback sea turtle is aided in its swimming by its disproportionately huge limbs. In all sea turtles, the front limbs are longer than the rear, and hence provide the power stroke when swimming. Those of leatherbacks, however, are nearly as long as their shells, bestowing an appearance of wings while swimming. With such bulk and elongated limbs, they make incredible, awe-inspiring trackways, ones that no doubt would cause neoichnologists to drop to their knees and weep with unmitigated joy. Sadly, I have never seen one, only photographs, which like those of the Grand Canyon are probably pathetic imitations of the real thing. As mentioned earlier, their trackways are also qualitatively distinctive: that is, size is not everything when identifying species from turtle trackways.

Nonetheless, when thinking about leatherback turtle trackways and nests, a paleontologist cannot help but also think about *Archelon ischyros* from the Late Cretaceous, the largest sea turtle known to have lived. Leatherback sea turtles, distant living relatives of *Archelon*, would have felt wholly inadequate swimming next to this species. This sea turtle, which swam through the midcontinental sea that covered what we now call Kansas about 70 million years ago, was more than 4 m (16 ft) long, or two to three times longer than the biggest of leatherbacks, and weighed several metric tons (Everhart 2005). Its mandible and maxilla together made a formidable beak that may have been used on hard-shelled prey, such as their ammonite contemporaries, but if they were anything like modern sea turtles, they would have eaten floating colonies of jellyfish. More importantly, at least one paleontologist, Gale Bishop, has thought about the ichnology of *Archelon* mothers during the Cretaceous, and calculated the expected dimensions of their trackways, body pits, and nests (Bishop et al. 2011). However, few paleontologists would be able to distinguish such structures because of their shear immensity.

Alligators

As discussed before, alligators are common inhabitants of salt-infused water in tidal creeks, tidal channels, and the shallow shoreface (Chapter 7). This behavior is mostly unexpected because alligators, unlike some species of crocodiles, lack the physiological means for getting rid of excess salt. Based on evolutionary analyses, modern crocodilians are descended from marginal-marine or seafaring species that later underwent habitat partitioning, which included some lineages—such as that of alligators (or alligatorids)—adapting to freshwater ecosystems (Taplin 1988; Taplin and Griggs 1989; Brochu 2003). Apparently, though, large adult alligators can tolerate the estuarine waters of the Georgia coast. This conclusion is based on sightings of both the animals and their trackways on Georgia beaches and the lower reaches of tidal creeks, as well as occasional resting traces on tidal creek banks (Chapter 7). Subadult and juvenile alligators, however, are not so flexible in their habitat ranges, and hatchlings absolutely must live in freshwater environments.

Alligator traces in marginal-marine environments, while provoking some questions, also provide important clues about their inhabitation of barrier islands. For example, during the sea-level fluctuations of the Pleistocene and Holocene epochs, did the ancestors of the present-day alligators use a combination of walking and swimming to go to the islands during sea-level lows? Or are alligator populations of the barrier islands continually renewed by more abundant and wide-ranging individuals from the mainland, even during times of rising sea level, such as now? If breeding-age individuals were—and still are—able to cross these so-called salinity barriers, their traces become even more interesting for documenting their biogeographical ranges.

Last, one of the more interesting anecdotes I have heard about possible alligator traces in nearshore waters of the Georgia barrier islands concerns their feeding on unexpected items. No, they are not eating people, or small dogs: those are expected. (Actually, just small dogs.) A naturalist on Skidaway Island, John Crawford, told me about how he has not only seen alligators in the surf zone of uninhabited Wassaw Island, he has watched them eat adult limulids there. Although this may be a predator–prey relationship that nearly no one would have predicted, it makes sense in terms of the following: first, if a large adult alligator is swimming in shallow water along a shore face, it is likely hunting for something worth eating; and second, adult limulids, which move along the sandy bottoms of these shore faces, are large enough to attract the attention of an alligator and register as prey in their reptilian consciousness. Upon hearing this tale, I started looking for such traces in dead limulids on Georgia beaches in the hope of finding evidence of this seemingly unusual (or is it common?) predation. Alas, these have not yet made themselves apparent. A quick look at the literature also confirmed that alligators have been observed feeding on limulids elsewhere (Shuster et al. 2003), but again no descriptions have been made of such feeding traces.

Diamondback Terrapins

Diamondback terrapins (*Malaclemys terrapin*), unlike their mud turtle cousins, are turtles fully adapted to brackish water, and spend much of their time in salt marshes, eating marsh periwinkles (*Littoraria irrorata*), various crabs (*Uca pugnax, Sesarma reticulatum, Callinectes sapidus*), and other marsh invertebrates. Terrapin dietary preferences are well known by studying their feces, which contain the remains of the day (Tucker et al. 1995). They also make their nests on the upper parts of sandy beaches, dunes, or other areas with sandy substrates near marginal-marine environments, but above the high tide and only during summer months (May–July). Interestingly, I have seen their 12–15 cm (4.7–6 in) wide trackways coming out of the surf and onto the beaches of Jekyll Island during winter months, thus they must have reasons for visiting areas other than marshes. Nest cavities are usually 10–20 cm (4–8 in) deep, with narrow openings and broad bases, similar to the vase-like profiles of sea turtle nests (Butler et al. 2004). Hole nests contain clutches of 4–12 eggs, which are laid in early summer, and are accompanied by crawlways coming from and going back into the salt marsh. Hatchling traces are also similar to those of sea turtles by leaving an exit hole and multiple, narrow-width trackways leading from the hole to submerged parts of a nearby salt marsh (Chapter 8); nest emergences happen in late summer (August–September). This turtle, like mud turtles, burrows into muddy banks and bottoms of tidal creeks for extended times of dormancy during the winter. Surprisingly, hibernating terrapins have been found in creek bottoms that stayed submerged through the winter, thus showing a remarkable tolerance to low-oxygen conditions (Palmer and Cordes 1988). Unfortunately for ichnologists, though, these hibernacula have not been described in detail.

Diamondback terrapin trackways are identifiable by first recognizing that they belong to turtles (Chapter 7), but also by their association with salt marshes and beaches. This species almost became extinct in the first half of the twentieth century—its flesh was evidently too tasty in soups—but is now protected and its numbers are coming back. Hence, its trackways should be studied carefully, and its nests left unbothered. Of course, hungry raccoons do not heed human rules, so their excavation traces and partially eaten eggs will sometimes mark a former terrapin egg clutch. In places where no measures are taken to guard these nests, raccoons and other egg predators, such as crows, boat-tailed grackles, and ghost crabs, can inflict 80–90% losses of nests in a single nesting season. Even plant roots can inflict damage on a diamondback terrapin or sea turtle nest: roots of saltwort (*Salicornia virginica*), sea oats (*Uniola paniculata*), and other plants pierce eggs or surround them so that hatchlings cannot move past them (Lazell and Auger 1981; Hannan et al. 2007). Hence egg-predator traces may be nearly as common as those of the adult turtles.

Naturally, seeing that marshes are the places where these terrapins feed, they are also the sites for the start of boy–girl terrapin stories. As a result, diamondback terrapin trackways are common additions to the trace assemblages of salt marsh surfaces, washover fans, or beaches near marshes, especially during mating and nesting seasons (March through July). Their earthen nests, however, are more in the ecotone between the high marsh and maritime forests or other terrestrial environments, above the high-tide mark. Unfortunately, the fossil equivalent of salt marsh turtle traces have not yet been interpreted, but will become more likely with diamondback terrapin traces as examples.

SHOREBIRDS AND THEIR MANY TRACES

As loyal readers will recall, this book started with a story of murder most foul, involving a bivalve victim that unexpectedly found itself flying through the air without aid of a suspension net, and meeting its end from the beak and wings (not hands) of a sinister laughing gull (Chapter 1). The traces made by this shorebird associated with the death of a thick-shelled clam, composed of tracks, marks made by the beak prying the bivalve out of the sand, the bounce mark of the bivalve, the broken shell, and the partially eaten remains, only represented a few of many types of fascinating shorebird traces neoichnologists can encounter on the Georgia coast. Readers will also remember how the previous introduction to bird traces (Chapter 8) mentioned that shorebird traces deserve their placement into the category of marginal-marine vertebrates because they are most varied and numerous on beaches and intertidal zones. Given the wealth of ichnological material on Georgia coast shorebirds, I will attempt succinctness but not sacrifice enthusiasm when relating basic information about these important vertebrate traces. Moreover, I will hint at how these traces could be used for interpreting ancient shorebird behaviors.

Shorebird Tracks

Shorebirds are loosely defined here as birds that are either on the water, in the water, or along a marine shoreline for the vast majority of their lives, only taking a respite from the sea for reproduction on land, which typically takes place adjacent to the shoreline. Accordingly, their tracks fall in the categories of anisodactyl (semipalmate and palmate) and totipalmate, with partial or full webbing between their digits reflecting various degrees of adaptation to aquatic environments (Table 9.1). However, numerous other forest-dwelling or forest-nesting birds frequently visit coastal dunes, foredunes, and berms. These birds are easily detected by their anisodactyl tracks and often long, continuous trackways: ground doves (*Columbina passerina*), boat-tailed grackles (*Quiscalus major*), fish crows (*Corvus ossifragus*), wild turkeys (*Meleagris gallopavo*), turkey vultures (*Cathartes aura*), and black vultures (*Coragyps atratus*), to name a few. For vultures, dead animals combined with the wide expanses of a sandy Georgia beach provide for many excellent opportunities to study their unseen behaviors (Chapter 8). Vulture tracks, beak marks, and feces are often directly associated with any sufficiently large or odorous carcass, such as a shark, sea turtle, or bottlenose dolphin that may have washed ashore with a high tide. As revealed by their traces, vultures will even eat dead adult limulids, an unexpected interaction between a maritime-forest bird and subtidal invertebrates (Figure 1.4a). Frequent avian visitors to intertidal environments are sandhill cranes (*Grus canadensis*), which stroll through high salt marshes and washover fan deposits and leave their distinctive trackways (Figure 9.4). Presumably, sandhill crane visitations to salt marshes are so they can garner some easy-to-find fiddler crab snacks, although I have not yet found further ichnological evidence (beak marks, regurgitants, feces) to support that conjecture.

Birds considered here as marginal marine but habituating upper intertidal to supratidal areas of beaches and dunes (respectively), salt marshes, and tidal creeks include green herons (*Butorides virescens*); snowy egrets (*Egretta thula*); great egrets (*Ardea alba*) bald eagles (*Haliaeetus leucocephalus*); great blue herons (*Ardea herodias*); and ospreys (*Pandion haliaetus*), among others. In contrast, the list for shorebirds that spend most of their lives looking for food in sandy intertidal areas, on the water surface, or diving underneath it, is quite long. Nonetheless, a short list includes sandpipers, plovers, terns, gulls, pelicans, and cormorants, but also a large

TABLE 9.1. TRACKS OF COMMON BIRDS IN SHORELINE ENVIRONMENTS OF THE GEORGIA COAST

Common Name	Species	Track Length (cm)
A. Anisodactyl, with prominent hallux impression		
Boat-tailed grackle	*Quiscalus major*	6.0–8.2
Fish crow	*Corvus ossifragus*	7.5–8.5
Green heron	*Butorides virescens*	8.0–9.5
Snowy egret	*Egretta thula*	10.0–11.5
Turkey vulture	*Cathartes aura*	9.5–14.0
White ibis	*Eudocimus albus*	12.0–13.5
Great egret	*Ardea alba*	16.5–18.5
Bald eagle	*Haliaeetus leucocephalus*	16.0–19.0
Great blue heron	*Ardea herodias*	16.5–21.5
B. Anisodactyl incumbent, with reduced or absent hallux impression		
Least sandpiper	*Calidris minutilla*	2.0–2.2
Sanderling	*Calidris alba*	2.0–2.2
Piping plover	*Charadrius melodus*	2.0–2.5
Spotted sandpiper	*Actitis macularius*	2.2–2.5
Semipalmated plover	*Charadrius semipalmatus*	2.2–2.7
Dunlin	*Calidris alpina*	2.2–3.0
Killdeer	*Charadrius vociferous*	2.5–3.0
Ruddy turnstone	*Arenaria interpres*	3.0–3.5
Short-billed dowitcher	*Limnodromus griseus*	3.2–3.8
Black-bellied (grey) plover	*Pluvialis squatarola*	3.5–4.0
Marbled godwit	*Limosa fedoa*	3.5–4.5
Greater yellowlegs	*Tringa melanoleuca*	4.0–4.5
Willet	*Tringa semipalmata*	4.0–5.0
Long-billed curlew	*Numenius americanus*	4.5–5.5
American oystercatcher	*Haematopus palliatus*	5.0–6.5
C. Palmate		
Least tern	*Sterna antillarum*	2.2–2.7
Forster's tern	*Sterna forsteri*	2.8–3.8
Royal tern	*Sterna maxima*	3.2–4.5
Black skimmer	*Rynchops niger*	3.8–4.5
Laughing gull	*Larus altricilla*	4.0–5.0
Ring-billed gull	*Larus delawarensis*	4.5–5.5
American avocet	*Recurvirostra americana*	5.0–5.5
Herring gull	*Larus argentatus*	6.5–7.5
Great black-backed gull	*Larus marinus*	7.5–9.5
D. Totipalmate		
Double-crested cormorant	*Phalacrocorax penicillatus*	11.5–14.0
Brown pelican	*Pelecanus occidentalis*	16.0–18.0

Note. Environments include foredune, berm, salt marsh, and intertidal. Tracks are organized on the basis of track category and ranked within each category by length. Track lengths from Elbroch and Marks (2001) and my field data. Track lengths include hallux impressions for category A, but exclude these for categories B, C, and D. See Table 8.1 for terrestrial tracks, some of which are found in both settings.

number of birds with colorful common names: sanderlings, dunlins, killdeer, marbled godwits, greater yellowlegs, willets, dowitchers, long-billed curlews, black skimmers, and American oystercatchers. These avians, so understandably beloved by birders who make pilgrimages to the Georgia coast to see them, are also prolific tracemakers, and in many instances their groupings and species can be identified by their distinctive traces.

Yet before an eager neoichnologist should bother running out to identify shorebird tracks and other traces on the Georgia coast, she or he should think ecologically. That is, keep in mind that shorebirds tend to gravitate toward certain environments and substrates, and often self-segregate by species. These habits mean the identity of the tracemaker can be narrowed down considerably by looking first at the habitat, rather than going through a list of more than a hundred possible species of shorebirds when any tracks are encountered. For example, given an idealized slope from a shallow, submerged sand flat, up a beach and berm, and to the edge of a salt marsh, shorebirds may be zoned like so: short-billed dowitcher → dunlin → western sandpiper → semipalmated plover → least sandpiper (Sibley 2001). Dunlins and dowitchers also like to forage in muddy sediments, and long-billed dowitchers prefer lower-salinity water compared to short-billed dowitchers (Sibley 2001). Thus an assessment of the ecological and sedimentological context to any given shorebird tracks can be helpful in better understanding what bird was there and why.

Sandpipers, sanderlings, dunlins, and plovers are often confused with one another by rookie birdwatchers (including me), but with enough study of their tracks, there is no mistaking them. The least sandpiper (*Calidris minutilla*), spotted sandpiper (*Actitis macularius*), and dunlin (*Calidris alpina*) have anisodactyl feet with a reduced digit I, and only proximal webbing (Elbroch and Marks 2001). This means their tracks will show a typical tridactyl arrangement with a barely noticeable nub in the rear made by the hallux. In contrast, sanderlings (*Calidris alba*), killdeer (*Charadrius vociferus*), piping plovers (*Charadrius melodus*), semipalmated plovers (*Charadrius semipalmatus*), black-bellied plovers (*Pluvialis squatarola*), and Wilson's plovers (*Charadrius wilsonia*) also have anisodactyl feet with

Figure 9.4. *Facing.* Trackway of sandhill crane (*Grus canadensis*) in a high salt marsh, a result of its hunting fiddler crabs on intertidal surfaces instead of staying in fully terrestrial environments: St. Catherines Island. Scale in centimeters.

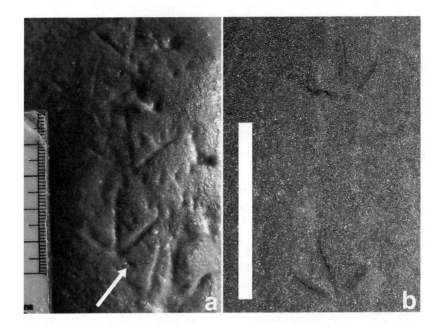

Figure 9.5. Differences between sandpiper-like and plover-like tracks. (a) Dunlin (*Calidris alpina*) tracks, in this case overlapping trackways made by two individuals, showing small digit I impression (*arrow*) and nearly symmetrical spread of other three digits. Scale in millimeters and centimeters. (b) Semipalmated plover (*Charadrius semipalmatus*) tracks, lacking digit I and asymmetrical on either side of middle digit (III). Scale = 5 cm (2 in). Both on Sapelo Island.

proximal webbing, but all lack digit I in their tracks. So based on the presence or absence of digit I, an initial quick and dirty distinction can be made between sandpiper-like and plover-like tracks in an intertidal area (Figure 9.5). This initial classification is further testable by measuring track sizes, differences in track symmetry, and noting behavioral differences indicated by individual trackways. Miscellaneous shorebirds that also leave anisodactyl tracks include ruddy turnstones (*Arenaria interpres*), short-billed dowitchers (*Limnodromus griseus*), marbled godwits (*Limosa fedoa*), greater yellowlegs (*Tringa melanoleuca*), American oystercatchers (*Haematopus palliatus*), willets (*Tringa semipalmata*), and long-billed curlews (*Numenius americanus*).

Trackways combined with feeding traces further help to place similarly sized shorebird tracemakers into categories of plovers, sanderlings, or sandpipers. Fortunately for neoichnologists, Elbroch and Marks (2001) sorted out these distinctions so that these three groups of birds are identifiable through just glancing at their feeding trails. Watch any plover and you will only see a blur where its legs should be, as this bird moves rapidly along the sandy shoreline of a Georgia beach. All the while, the plover is stabbing its beak into the sand, but just barely, because it is picking off any small invertebrates (amphipods, isopods, bivalves) that have the misfortune of being exposed by an errant wave. This superficial beak probing combined with high speed means that a plover's beak disturbs a significant amount of sand along the surface. The beak-caused disturbance shows up as a smear along a narrow path, almost perfectly defining the midline of the bird's travel, with its tracks on either side. A plover's briskness also translates into a narrow straddle for its trackway, like the bird was walking on a tightrope.

Now watch a sanderling feed along the same shoreline, and you will note that it is slower than a plover and its feeding is more punctuated, separated by periods of just running. Because a sanderling runs along for several steps without a single downward movement of its beak, this means its trackways may show no signs of feeding at all for long stretches. However, once it detects a mother lode of infaunal invertebrates, it quickly and repeatedly inserts its beak into a small area, resulting in high concentration of overlapping holes. This intermittent feeding strategy results in a trackway that shows a fluctuation between two behavioral modes: "no feeding, just running" and "stop, lots of feeding in one place." Accordingly, its straddle is slightly wider than that of a plover, and its tracks more often have the feet come together in what trackers call a "T" stop, with the feet defining either upright arm of a "T" and the midline of travel as the lower part (Figure 9.6a).

Last, watch a sandpiper, and you will soon see that it is the most meticulous of all three types of birds in its tracemaking activities. Sandpipers walk slowly, stop often, and will keep their heads down as they poke their beaks into the sand for infaunal invertebrates. The latter behavior leaves a distinctive trace that clearly links the beak marks with the trackmaker. When a sandpiper comes to a full stop, it probes once, pulls its beak up only a few millimeters, and plunges it into the sand again. This action causes a doubling of holes, which sometimes are separate—and thus show up as pairs—but will also overlap to make what looks like a single elongate hole. Look again, and you will see where the beak inserted once, then again, only

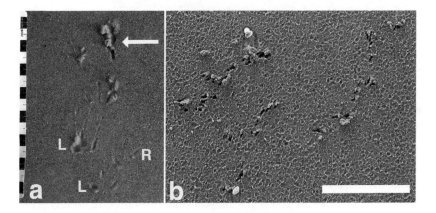

Figure 9.6. Feeding traces made by shorebirds foraging for delicious intertidal invertebrates. (a) Sanderling (*Calidris alba*) tracks and beak-probe marks (*arrow*), showing a T-stop trackway pattern, where right (R) and left (L) tracks are parallel to one another as it stopped to feed and then resumed walking forward: Jekyll Island. Scale in centimeters. (b) Beak-probe marks, probably by a least sandpiper (*Calidris minutilla*), identifiable from the irregular path and frequent double-hole pattern in probes: Sapelo Island. Scale = 10 cm (4 in).

the beak did not quite pull out of the sand after the first poke. As can be imagined, a sandpiper's straddle varies hugely along a given trackway, from tightrope-like patterns to "T" stops, with all sorts of changes in direction and looping along a trackway. Sandpipers do not like following straight lines, and it shows in their trackways (Figure 9.6b).

With regard to one special shorebird and its tracks, I recall one summer on Sapelo Island, accompanied by my wife Ruth, when we encountered a small group of birders. They had just set up a spotting scope to view a cluster of birds on an offshore sandbar at low tide, and were excited about the presence of one bird in particular. This bird made its way through all of the other, smaller birds while methodically inserting its lengthy, curved beak into the sand, probing for infaunal bivalves. It was a long-billed curlew, the largest sandpiper in North America, and a rare visitor to the Georgia coast, coming mostly during winter months (Parrish et al. 2006). Furthermore, its numbers have declined as a result of the usual litany of human-caused ills: habitat alteration, nest destruction, target practice, and so on. Accordingly, these birders were thrilled to spot one during the summer, an unusual sighting indeed.

Since then, I have looked for long-billed curlew tracks on Georgia beaches and occasionally identify them: they are about 5 cm (2 in) long and 7 cm (2.8 in) wide, with a reduced digit I and proximal webbing between digits II to IV; trackways show strides of about 20–40 cm (8–16 in). The only other tracks with which they could be mistaken are those of the much more common willets, which have the same foot configuration (anisodactyl with reduced digit I), but are noticeably smaller: 4.5 cm (1.8 in) long and 6 cm (2.4 in) wide. Willets also run faster than long-billed curlews, resulting in a longer stride, as much as 50–60 cm (20–24 in). If suspected long-billed curlew tracks also have deep (18 cm/7 in) and narrow-diameter holes next to them, some of which coincide with invertebrate burrows, these are likely probe marks made by these birds' (you guessed it) long bills. Along those lines, an ornithologist told me of watching a long-billed curlew stick its beak down a ghost shrimp burrow and pull out its tracemaker for a quick snack, making a neat composite trace (Jen Hilburn, personal communication 2010). In short, if an avian ichnologist is attempting to distinguish a willet trackway from that of a long-billed curlew, a willet trackway will show a combination of smaller tracks, longer strides, and will be one of many such trackways. In other words, in the absence of a sighting, bird tracks and other traces can be used to document the probable presence of a rare bird species.

Trackmakers with palmate (fully webbed) anisodactyl feet are a broad group, consisting of black skimmers (*Rynchops niger*), terns, and gulls. As can be intuited from their feet, these birds are fully adapted for paddling on the water, or for lending added stability (through increased surface area) when wading in saturated sediments, similar to the effect of a snowshoe. These feet, however, have their disadvantages on emergent surfaces, as increased surface area contacting firm substrates makes walking slightly more challenging, and running more so. (Think of the comic potential of a footrace between people wearing snorkeling fins, and you will see what I mean.) Consequently, trackways made by palmate birds often indicate slow walking that might lead directly into takeoff patterns. Given the choice between running and flying, the latter is a much better defense against any potential threat when you have palmate feet.

Terns are among my favorite shorebird trackmakers, partially because I think they are elegant little birds, but also because tracks of the more common ones can be separated from those of gulls, and even identified to

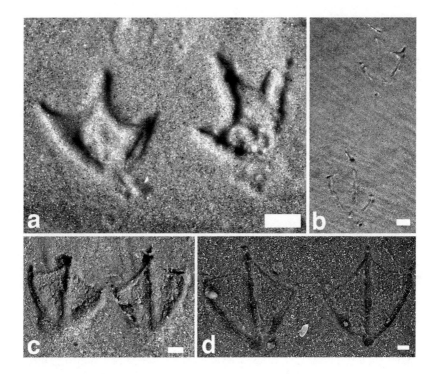

Figure 9.7. Tern and gull tracks. (a) Pair of tracks from a standing least tern (*Sterna antillarum*). (b) Normal walking trackway of royal tern (*Sterna maxima*). (c) Pair of tracks from a standing laughing gull (*Larus altricilla*). (d) Pair of tracks from a standing herring gull (*Larus argentatus*). All from Sapelo Island and with the same scale, 1 cm (0.4 in).

the species level. Least terns (*Sterna antillarum*), common terns (*Sterna hirundo*), Forster's terns (*Sterna forsteri*), and royal terns (*Sterna maxima*) all make palmate tracks, yet these are also smaller than the palmate tracks of gulls. The royal tern is largest of these, yet its tracks are all less than 5 cm (2 in) wide, whereas the smallest gull—the laughing gull—leaves wider tracks. Moreover, tern tracks are more likely to register digit I, however reduced it might be, whereas gull tracks will only show this if recorded in a soft substrate, like mud. Once tern tracks have been separated from gull tracks, they can be placed into probable species on the basis of respective narrow size ranges. Least tern tracks are tiny, typically measuring between 2–2.5 cm (0.8–1 in) wide. From there, the tracks of each species, although similar morphologically, overlap only slightly in size: least tern → common tern → Forster's tern → royal tern (Table 9.1; Figure 9.7a,b). As far as I know, no major behavioral differences between these species can be teased out of their traces, but such an investigation is certainly worth undertaking.

As an aside, and through a brief field story related to terns, I want to emphasize how careful descriptions of shorebird tracks can be used effectively for identifying species, despite a tendency to lump their tracks into a vague category of shorebird tracks. One summer day on Sapelo, I was training my binoculars on a large group of shorebirds loafing in a sandy intertidal beach, and struggling with reconciling these with their descriptions and pictures in a standard bird-identification book. I then noticed that one of the terns was considerably smaller than the others around it. This lent to wondering if it was a least tern, although I was not entirely sure. So instead of looking again at the bird and the book, back and forth, I instead memorized its position, and walked straight to that spot. Of course, by the time I got there, the whole group had flown away, not nearly as curious about me as I was of them. Nonetheless, their tracks were still there, which included the beautifully expressed tracks of a least tern. Thus my bird-identification skills were augmented by ichnological evidence, a lesson in how traces provide yet another tool for distinguishing shorebird species, and sometimes after a bird has flown away. Yes, looking at tracks may not match the thrill of seeing your favorite shorebird in person, but these can let an ardent birder know that a certain bird is in the area, and hence can serve as a predictor for future sightings of that species.

Similar to terns, gulls, which have anisodactyl tracks with full webbing, can also be identified to the species level from their tracks, although mostly on the basis of size, abundance, and ecological context. For example, tracks of laughing gulls (*Larus altricilla*), ring-billed gulls (*Larus delawarensis*), herring gulls (*Larus argentatus*), and great black-backed gulls (*Larus marinus*) are morphologically identical (Figure 9.7c–d), but in width they range in a continuum from about 5–10 cm (2–4 in). By far the most frequently encountered gull tracks on Georgia beaches, from the intertidal to back-dune meadows, are those of laughing and ring-billed gulls. Herring gull tracks, the few times I have seen them, were noticeably wider and longer than those of their smaller relatives, measuring about 8 by 7 cm (3.1 by 2.8 in) wide. The tracks of a black-backed gull seem monstrously large in comparison to those of the much more common laughing and ring-billed gull tracks; indeed, these are often twice as wide, and are more likely to be confused with tracks of Canada geese (*Branta canadensis*). Black-backed gulls, however, are rare visitors to the Georgia coast, and Canada geese almost never visit beaches, so tracks of either species in a sandy intertidal area on the Georgia coast would be unusual. Trackways in combination with track size can also be used further to identify a species, as the pace and stride increases in proportion to gull size. All gulls walk with a mildly pigeon-toed gait, as their longest toe (digit III) points toward the midline of the trackway. Gulls are of course well known for their voracious omnivory, and their tracks are nearly always connected to other traces that demonstrate their appetites and broad food choices. These traces, however, will be included later with an overview of most other shorebird feeding traces.

Totipalmate trackmakers on the Georgia coast are easy to name and distinguish from one another, as there are only two kinds: double-crested cormorants (*Phalacrocorax auritus*) and brown pelicans (*Pelecanus occidentalis*). Less common additions to this list are anhingas (*Anhinga anhinga*) and white pelicans (*Pelecanus erythrorhynchos*), which are infrequent visitors to the Georgia barrier islands. Moreover, anhingas are more commonly associated with freshwater environments; hence smaller-sized totipalmate tracks on an intertidal sand flat are far more likely to belong to cormorants. Both cormorants and brown pelicans are never going to win any foot races, but their trackways, particularly those of pelicans, can record entire sequences of their landings and takeoffs. Most often, though, tracks of these birds, especially those of pelicans, often come to a full stop, denoting where they settled down for a while. Tracks demonstrating this loafing behavior may also include a subtle impression of a bird's butt around the paired tracks, which are side by side and one slightly overlapping the other, but normally have droppings just behind the tracks. Pelican resting traces may even have pockmarks in front of them, caused by drops of water rolling off a prodigious beak (Figure 9.8a). Collectively, these traces probably indicate that the pelican was just in the water; caught and enjoyed swallowing a fish, and then got out of the water to take a nap.

Nonetheless, by far the most fascinating shorebird trackways I have encountered on Georgia intertidal areas are those that preserve the entirety of flight patterns, from landing to foraging (or loafing) to takeoff. Flying traces, termed originally *volichnia* by German neoichnologists for either arthropods or birds (Müller 1956; Reineck 1981), were summarized previ-

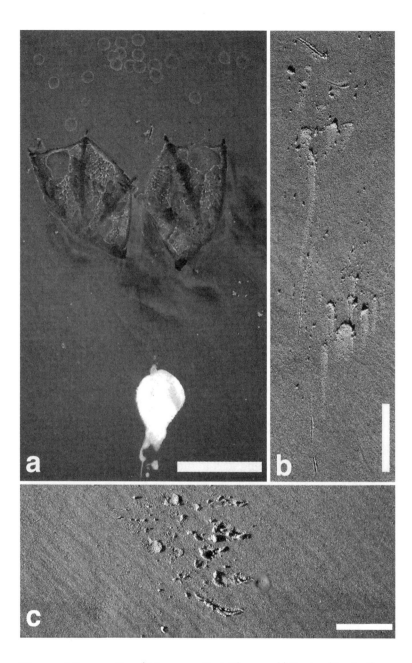

Figure 9.8. Brown pelican (*Pelecanus occidentalis*) traces. (a) Resting ("loafing") trace immediately following the pelican's immersion, based on the water-drop impressions (falling off its beak) in front of the tracks: Sapelo Island. Photograph by Stephen Henderson. (b) Landing tracks, with skid marks, piling of sand in front of digits, and long impressions from claws: Little St. Simons Island. (c) Takeoff tracks made by the same pelican as in (b); note the lack of tracks behind and in front of these. Scale in all photos = 10 cm (4 in).

ously for terrestrial birds (Chapter 8), but are worth revisiting here because of the plethora of behavioral insights bestowed by shorebird trackways.

Taken on an individual basis, shorebirds may land gracefully or awkwardly, depending on the bird and how we perceive it. For example, pelicans and larger, similarly bulky birds—bald eagles and osprey come to mind—need to slow themselves as they control their descent. They do this by running briefly after touchdown, then coming to a complete stop before walking at a normal gait. The resulting trackway pattern will be of two tracks of slightly offset feet, with gouges imparted by digit I, or other rear parts of the foot if digit I is absent; elongate skid marks from each of the digits and pressure-release structures piled up in front of the digits (Figure 8.3d; 9.8b); perhaps another offset pair or two of tracks from a long hop; alternating left–right tracks, denoting bipedal running, but with pace values decreasing with every step; side-by-side tracks, indicating a halt to forward progress; and alternating tracks, but made at a normal walking speed. In a few lucky instances, wing impressions may even be part of the pattern. I recall once seeing where a small flock of brown pelicans, just before landing, grazed the tips of their wings along the beach surface, leaving long, thin, parallel lines before their feet made contact.

In contrast, takeoff patterns for shorebirds are almost as widely varied as the birds. Nonetheless, on either end of a spectrum are trackways in which birds flew from a standing start to those where birds needed a prolonged running start (with much flapping along the way) before finally becoming fully airborne. In standing-start takeoffs, which are common in some smaller shorebirds, the feet may be parallel or slightly offset, and footprints will show lateral smears and claw impressions angled to the right or left. Take note of the directions indicated by these claw marks, and test whether they match prevailing winds coming off the open ocean. I have often watched plovers, sanderlings, and sandpipers spread their wings while still on the sand surface, and they are suddenly aloft, looking as if they jumped into the air and instantly flew. In many instances, though, what seemed like a jump is actually a wind-assisted lift, in which a strong breeze carried them away as soon as the air currents met their outstretched wings. These traces and their depictions of effortless flight are in marked contrast with those where shorebirds had to beat their long wings with frantic abandon while gawkily running or hopping across the sand flat. Yes, brown pelicans, I'm thinking of you as I describe this; although, to be fair, I have seen a few pelicans take off from standing starts (Figure 9.8c). Trackways recording a long takeoff are delightful to follow, as each set of tracks becomes farther spaced from the preceding one as the bird gradually gained more lift, and I often find myself urging them on as the stride lengths increase. Corresponding with these lengthened strides is lighter pressure exerted by the feet, which means that tracks become fainter along the flight line, just before ending with a last, offset pair.

Many smaller shorebird species, such as those of plovers and sandpipers, also travel in flocks that land and fly away in unison once approached by a large, upright biped. One difference in flocks' takeoffs, though, is that plovers more often disperse in all directions, whereas sandpipers tend to feed and fly together as a unit (Sibley 2001). Accordingly their last tracks before takeoff will show either an alignment just before flight (sandpipers) or not (plovers). Either way, these birds can become airborne quickly from a standing start, although bipedal sprinting is also common just before they decide to use their wings to full advantage. With trackways made at the same time by flocks foraging along the shore, be sure to check for evidence of group behavior and its degree of organization.

Finally, shorebird tracks do not always show what one might think, and can belie our most basic assumptions. For example, on a December afternoon on Cumberland Island, I spotted a small flock (13–14) of sanderlings loafing near the intertidal zone. A glance at them through binoculars showed them doing something curious (to me, anyway): most were standing on one leg. This stance was not because they had all suffered from the same unfortunate accident, but was done in an attempt to stay warm by keeping one leg close to the torso. As I walked toward them to take a closer look, they then started quickly hopping away. Yet it was a one-legged hop, as these birds stubbornly kept one leg tucked against their bellies while bouncing away from me. (Photographs I took of these sanderlings' pogo stick imitations later revealed that about half were hopping on their right legs, and the other half on their left legs, so leg preference seemed to be a matter of individual choice.) Several seconds passed before they remembered, "Wait a minute, I have another leg!" and these came down to assist in their escape. This reversion to bipedal motion was followed quickly by their recalling that they also had wings: only then did they take to the air as a group. As one might imagine, I was excited to see just what a one-legged sanderling trackway looked like, and indeed was rewarded by an intriguing series of tracks. Sanderling tracks are anisodactyl and lack digit I, but unlike those of plovers, are nearly symmetrical. Hence a trackway made by a one-legged sanderling or similar shorebird will, at first glance, look very much like a two-legged one. The key point of distinction between the two is the position of the middle toe (digit III) relative to the midline of the trackway. In two-legged trackways, this digit points either inward or straight forward, whereas in a one-legged trackway the digit points to the inside of the trackway, and consistently so (Fig. 9.9) In short, take a second look at every shorebird trackway, modern or fossil, and make sure it was using both legs.

Mating and Nesting Traces

Among the more complex behaviors of shorebirds, and somewhere in between flying and feeding, are their mating rituals, with their rarely seen associated trackways. Many bird species are known for their complex courtship displays, inviting much anthropomorphism or outright imitation by human admirers. Some of these displays consist of body movements approaching our concept of dancing, in which males try to convince females to mate with them by showing off their fancy footwork. Among Georgia coast shorebirds, plovers are well known for this behavior, in which male plovers perform a high-stepping display, also known as marking time. This behavior involves alternately pulling feet high above the ground and placing them down just in front of one another (Cairns 1982; Bergstrom 1988a). The resulting trackway pattern definitely stands out from all others made by plovers. It is composed of parallel and tightly spaced left–right tracks with narrow straddles about twice the width of their tracks.

Wilson's plovers in particular exhibit a wide variety of behaviors during courtship, some of which must leave an enticing suite of traces. These include horizontal and upright hunched runs, ground scraping, ground pecking, flinging of pebbles or bits of vegetation, and of course, mating itself, which results in an overprinting of trackways from the male–female pair, with the male closely following the female (Bergstrom 1988a). Perhaps the most intriguing of these traces would be those caused by the parallel run, in which two male plovers competing for the same female run next to one another in a straight line, turn around, and run back in the opposite direction along the same line. Apparently these birds are marking an imagi-

MARGINAL-MARINE AND MARINE VERTEBRATES 473

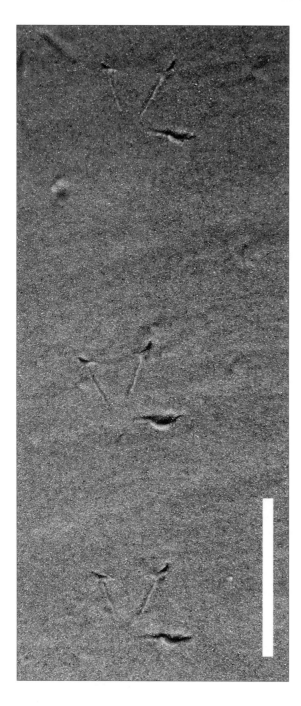

Figure 9.9. Trackway made by a sanderling (*Calidris alba*) hopping on its left leg: Cumberland Island. Scale = 5 cm.

nary territorial boundary through this behavior (Bergstrom 1988a). Think of how this would make for a fantastic interpretation if two overprinted and parallel trackways of birds or nonavian dinosaurs showed up in the fossil record!

Like trackways made by mother sea turtles, mating ritual traces are not only species specific, but also sex specific. Moreover, these trackways can be associated with a small part of a year, as plovers only mate seasonally, and the breeding ranges of some plover species are further north. For example, piping plovers do not breed as far south as Georgia, but Wilson's plovers do (Parrish et al. 2006). Although I have yet to see plover-wooing tracks, other avian-track-loving folks have been lucky enough to observe and document them (Elbroch and Marks 2001). Moreover, they related how trackway measurements by one researcher (Diane Boretos) helped to distinguish mating ritual, territorial displays, and walking in piping plovers on the basis of stride measurements. Surely the same measurements could be taken from plovers in Georgia that take time from their busy schedules to mate.

Following successful mating, fertilization of eggs, and pair bonding in some shorebird species, nests are constructed in upper parts of beaches (berms, foredunes) or in back-dune meadows. These structures are simple scrape nests made on the ground (Chapter 7), in which birds excavate sand with their feet to make a depression and a shallow sedimentary rim around the depression (Figure 9.10). For example, Wilson's plovers are scrape nesters, in which the male shows off its handiwork by tilting down, rotating its body clockwise or counterclockwise, and scraping the sand with its feet. The final nest structure is only about 8 cm (3.1 in) wide (Bergstrom 1988a), so watch out to make sure you do not step on it. Once laid, eggs in many shorebird species are left uncovered, with no vegetation or other additional nesting material. Although many shorebird eggs may have camouflaging patterns similar to Georgia dune sands, mammalian predators that rely more on smell than sight—such as raccoons and feral hogs—quickly find and consume unattended eggs, leaving their own interesting traces. As a result, parents invest much energy toward protecting their egg clutches from egg predators by staying on or near the nest. Another risk posed to eggs is from heat, as egg clutches of many species are laid in the late spring or early summer, when temperatures on the Georgia coast begin to rise and the heat is amplified by white sand surrounding the nests. Although we

MARGINAL-MARINE AND MARINE VERTEBRATES 475

Figure 9.10. Scrape nest made by American oystercatcher (*Haematopus palliatus*) containing egg clutch: Sapelo Island.

may normally think of birds keeping their eggs warm by sitting on them, in these instances, they are actually preventing their potential offspring from cooking.

However, if an active nest is approached by a potential threat, such as an upright bipedal ichnologist, some birds will leave their eggs by attempting to distract the intruder. Perhaps the most famous of protection ruses is performed by killdeer (*Charadrius vociferous*), in which a parent will emit loud alarm calls, and run away from the nest while feigning a broken wing, held at an odd angle away from the body. If the offending (or curious) animal follows the killdeer far enough away from the nest to pose less of a danger, this bird suddenly undergoes an instant healing and flies away, saying the avian equivalent of "Sucker!" as it departs. Wilson's plovers are also known to perform this same trick (Tomkins 1944), meaning that it must have worked for both species in their evolutionary history. Trackways of this behavior have not yet been described, but would lend some insights on how similar sorts of protective behavior could be interpreted from birds or their nonavian dinosaurian cousins in the geologic record.

Similar to sea turtles, many shorebird species show a tendency to nest in the same places over generations, gravitating to islands devoid of humans, such as Wassaw, Little Tybee, and Blackbeard. American oystercatchers have long-term pair bonding, and thus the same male–female pair may show up at the same site for nesting in successive years (Sibley 2001). This behavior, though, may not be so much a result of site fidelity, as it is a lack of adequate nesting sites for shorebirds. This is especially the case wherever the fatal combination of shoreline development and predatory invasive species might occur, infringing on these birds' breeding cycles. Some shorebirds also place their nests far enough away from those of their own species to prevent seeing each one another. In one study of breeding Wilson's plovers, nests were spaced minimally 35 m (115 ft) apart, to as much as 250 m (820 ft) (Bergstrom 1988b), which also limits their choices if adequate nesting habitats become scarce.

Feeding Traces

Shorebird feeding traces, as already mentioned, are normally associated directly with trackways, and fall into four broad categories: beak marks, broken shells, regurgitants, and feces. All such traces are exceedingly abundant and easily observable in the intertidal sand flats of the Georgia barrier islands, and aspiring neoichnologists will have no problem finding enough data to last a lifetime, or a dissertation, which only seems to last a lifetime. Salt marshes can also provide some of these same traces, but are more difficult for humans of considerable weight to study up close and personal. This point is not meant as an insult, but is meant to emphasis the great difference in the weight of a shorebird versus a typical human. Accordingly, a shorebird can more easily walk on a soft mud without sinking too deeply into its cool, wet, organic, and sulfurous embrace. Hence most of my sightings of avian feeding traces on or around salt marshes are in the saltpans of high marshes, or on the edges of tidal channels associated with low marshes.

Beak marks are preserved in soft substrates, such as sand or mud, and as discussed earlier can be allied with species of shorebirds based on their patterns and if found in direct association with tracks. Beak marks, however, also register in the exoskeletons and endoskeletons of invertebrates, outlined by where they pierced thin-shelled bivalves, gastropods, crabs, and sand dollars. Gulls are particularly persistent in this respect, breaking through the hard parts of many an invertebrate to get at its delicious innards. Likewise, American oystercatchers use their stout and specially adapted beaks to hammer into and pry open oysters (of course) and other bivalves. Although these feeding traces have not been described, oystercatchers must leave a wide variety of marks on intertidal bivalves in both marshes and along the edges of beaches.

On an intertidal sand flat, a complete suite of traces that involve shorebird beak marks somewhere in the mix will include the following:

- Tracks leading up to the intended meal.
- Initial beak marks that may pierce the sand to investigate a buried invertebrate (probes).
- More vigorous poke marks, some of which may reflect leveraging of the prey item out of the sand.
- Marks made by extraction of the prey with the beak and pulling it completely out of its former resting spot.
- Drip marks caused by water falling off the invertebrate, but defining a line as the shorebird walked with the prey in its beak to a more suitable place.
- A few impact marks made by the shorebird slamming its prey against the hard-packed sand.
- Beak impressions in the body of the prey, assuming the prey was sufficiently stunned or freshly killed and has given up itself for the shorebird's meal.
- Tracks that show walking away from the partially consumed remains, perhaps reflecting a temporary satisfaction and slaking of hunger.

This sequence or assemblage of traces may vary according to the predator, prey, and various factors in the local environment, but nonetheless provides a starting point for interpreting the nuances of behavior they reveal, rather than simply writing "beak marks" in a field notebook, then going off to have a cup of tea. What do these shorebirds go for? A lot: bivalves, gastropods, anemones, sea stars, sand dollars, and sea cucumbers, to name a few. I am also continually amazed at how many shorebirds perform their own form of trace recognition while walking along the edges of runnels and shoals, in which they stop to study nearly every depression, bump, or slight change in the sand surface for its possibly holding a delicious morsel. Sometimes they even pry out a buried invertebrate and recognize it as something they do not want to eat after all, leaving a gastropod, bivalve, or sea star nakedly exposed on the sandy surface, or the articulated shell of a bivalve only containing sand. So instead of a complete feeding sequence of traces as described before, the tracks pointing away from the former object of desire are instead those of rejection and disappointment (Figure 9.11a).

Broken shells, a somewhat cryptic trace of shorebird predation attributed to gulls, are often accompanied by an impact mark of the skydiving molluscan prey. Gulls most commonly apply this technique to thick-shelled bivalves, which usually are well protected against any beak-only assault, but poorly adapted for falls from heights of 5 m (16.5 ft) or more (Figures 1.1, 1.2). This type of feeding can cause seemingly random fractures of bivalve shells that might be difficult to attribute to predation, especially if shell pieces are not accompanied by other avian traces (Chapter 1). Impact sites have a slight depression caused by the shell hitting the sand, but sometimes show impressions of the bivalve's ribbing (helping to identify it if no shell fragments are nearby), or may have radiating marks caused by the rapid expulsion of water from the bivalve's body (Figure 9.11b).

An even more subtle feeding trace that indicates shorebird feeding on bivalves, but one that does not result in any shell breakage, is gaping. This method is employed by a shorebird that, rather than pulling a bivalve out of its burrow, will instead wedge its beak between the valves of a clam, then open its mouth wide. This action eventually tires the bivalve, and once its adductor muscle gives out, the bird will eat the soft parts from between the

Figure 9.11. *Facing.* More shorebird feeding traces. (a) Tracks and breakage of a bivalve by a herring gull (*Larus argentatus*), that was disappointed to find out the bivalve (long dead) only contained sand: Little St. Simons Island. Scale = 10 cm (4 in). (b) Impact mark where a bivalve was dropped and broken by an unknown bird (probably a gull) onto a sand flat: Sapelo Island. Scale = 5 cm (2 in). (c) Gaping of a disk dosinia (*Dosinia discus*) by ring-billed gull (*Larus delawarensis*): Sapelo Island. Scale in centimeters. (d) Cough pellet from a laughing gull (*Larus atricilla*) on an intertidal sand flat, yet filled with fiddler crab parts, indicating how the gull was recently in a salt marsh: Sapelo Island. Scale = 1 cm (0.4 in). (e) A three-for-one special: tracks, dropping, and cough pellet of a young laughing gull: Sapelo Island. Scale = 5 cm (2 in). (f) Fluidized trackway of a ring-billed gull where it brought its infaunal prey out of the sand, moving right to left: Tybee Island. Scale = 15 cm (6 in).

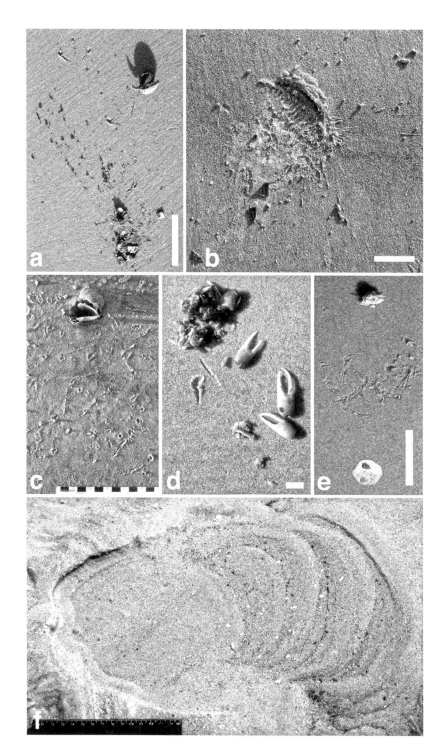

two valves. Shorebird tracks in front of these empty bivalves—still in their burrows—indicate much shuffling in a small area, where the bird stayed more or less in one place as it both attacked and ate the clam (Fig.9.11c). Some terrestrial birds use a similar gaping behavior to feed on ants by poking their beaks into the entrance of an anthill, expanding the hole by widening their mouths and letting the ants pour in (Elbroch and Marks 2001).

Regurgitants (cough pellets) are more likely linked with terrestrial avian predators, such as owls or raptors, but are also common shorebird traces. Gulls are by far the most common cough-pellet creators on Georgia sand flats; hence when I encounter these traces, I begin with gulls as the default tracemaker, but then test that identification by looking at nearby tracks (Figure 9.11d,e). Regurgitants are a natural consequence of feeding in shorebirds, and not a result of all-night avian parties gone wrong. When a gull or other shorebird is ready to look at its lunch again, this is a sign that it finished digesting its protein and is just trying to make room for more food. In making a cough pellet, a shorebird invariably walks normally, stops with its feet parallel to one another, and deposits a slightly rounded mass of invertebrate hard parts, usually about 3–5 cm (1.2–2 in) wide, but varying based on tracemaker size. Pieces of these pellets also may separate from the main mass, which can help to identify exactly what the shorebird was eating. For example, I have been surprised sometimes to find, through the cough pellets of laughing and ring-billed gulls, that they sometimes visit salt marshes, where they will chow down on abundant (and frequently victimized) fiddler crabs. This supposition is supported by the presence in the pellet of the prominent claws of male *U. pugnax* and *U. pugilator*, used only hours before to wave menacingly (but futilely) at these huge, feathered bipedal intruders.

Feces (droppings) are easily recognizable as whitish, liquid deposits that are the final waste products of digestion, consisting of the altered remains of whatever did not make it out in the cough pellet. Occasionally these will contain a few fragments of thin-shelled bivalves (especially *Mulinia* or *Donax*), crabs, or pieces of other hard-shelled invertebrates that did not quite undergo complete dissolution in a shorebird's gut. However, in my experience, shorebird droppings also can be used as indirect indicators of a bird preparing for flight within the next few seconds (Figure 9.11e). In a sizeable number of the shorebird takeoff patterns I have seen, trackways start with the bird's feet together (tracks side by side), and immediately behind these is a significant and widespread splattering of feces. This behavior is not so much attributable to shorebirds saying, "Oh, crap!" when approached by a potential threat (and thus becoming literal), as much as it is a lightening of their body weight that immediately precedes flight. Every little bit helps, especially for larger shorebirds that need more effort to get off the ground, such as brown pelicans, great egrets, and great blue herons.

Another less frequently encountered feeding trace is the skeletonized body of a vertebrate prey item in which only its head is left behind, a common trace of bald eagles. If an eagle catches a fish—which eagles are quite fond of doing—it eats all of the flesh from a fish, but leaves the bony remains and the head (Elbroch and Marks 2001). Bald eagles sometimes go for other prey, though, including their bird brothers. For example, once I encountered a rather creepy scene on Wassaw Island in which a bald eagle had fed on a ring-billed gull, leaving only scattered feathers and a gull head, neatly snipped from its former body. If a similar assemblage somehow preserved in the geologic record, it would consist of a dense accumulation of feathers and a skull with severed cervical vertebrae, clues to an example of avian-on-avian predation. Ospreys, on the other hand, will often eat their entire prey item (usually fish), leaving nothing but a cough pellet filled with fish bones a few hours afterward.

In their feeding, laughing gulls also make large, odd, and interesting sedimentary structures, features that would be almost impossible to interpret without watching the birds make them or knowing about similar examples related by scientific literature. Unexpectedly, I first witnessed the behavior that produced these enigmatic traces on Tybee Island, one of the most developed islands on the Georgia coast. While walking along the main beach of Tybee at low tide one day, I noticed that ring-billed and laughing gulls were doing their usual search-and-destroy forays, looking for anything worthy of consuming. One gull, however, stood out from the others by its, well, standing. This gull was stopped in a shallow pool of water, when it suddenly moved its feet up and down rapidly, a motion that instantly took me back to the 1980s through its clever avian imitation of Jennifer Beals in the movie *Flashdance*. This seemingly unusual behavior, however, had a purpose well beyond provoking memories of beat-filled music, sexy dancing, and big hair, as the gull occasionally stopped moving both feet and pecked at something in the water. After watching this gull for several minutes, I realized that its trampling was shaking apart the sand grains, which liquefied the sand. This fluidization in turn dislodged any infaunal invertebrates that might have been hiding in the compacted sand, such as mole crabs and various small bivalves (Chapter 6). After each instance of stationary stomping, the gull then walked forward just a bit, trampled until more food was gained, moved forward again, trampled, and so on. Once this gull had exhausted this pool and flew away, I walked up to observe the astonishing structures that had resulted: a meters-long, slightly sinuous zone of disturbance about 15 cm (6 in) wide (slightly more than two foot-widths for a laughing gull) with an apparent backfill structure composed of meniscae (Figure 9.11f). The meniscae, though, were actually caused by the progressive movement of the gull and the forward edges of its webbed feet. Had I encountered such a trace in either a modern or fossil state, I would have stupidly attributed it to a burrowing echinoid, such as a sea urchin (Chapter 6). I certainly would not have thought "bird," especially because of the apparent lack of obvious avian tracks in the trace.

Yes, I just related an anecdote, which we scientists are not supposed to do, but here is another one anyway. Later that same day on this Tybee beach, I observed several more gulls performing the same behavior, indicating a repeatability of the preceding observation, and thus providing an opportunity for other researchers to independently observe the same. Now this is starting to look a little more like science, right? A look later at peer-reviewed literature also confirmed that these structures had been described from the ichnologically famous tidal flats of the Wadden Sea, Netherlands (Chapters 1, 2), and that gulls had been observed making them (Cadée 1990, 2001). Consequently, ichnologists now have an alternative explanation for any similarly strange structures they might find in ancient intertidal deposits that overlap with the body-fossil record of shorebirds during the past 100 million years or so. In other words, ichnological anecdotes can become genuine scientific hypotheses, and then serve as predictors for future finds of trace fossils.

Overall, marginal-marine bird traces are extremely abundant and varied, bestowing finer perceptions of avian behavior on beaches and marshes than might be gained by just bird watching. Neoichnological observations of bird traces are starting to creep their way into paleoichnological studies (Genise et al. 2009), but certainly could be studied and applied more often. So with all of these Georgia shorebird tracks in mind, here are best wishes to paleoichnologists who now have the ideas and search images for interpreting walking, running, one-legged hopping, mating, feeding, and a myriad of other behaviors of shorebirds (or even nonavian dinosaurs) in the geologic record.

MAMMALS AT THE SHORE AND IN THE MARSH

Some mammals of the Georgia coast spend most or all of their time in marginal-marine or marine environments, making traces whenever they interact with substrates of salt marshes, beaches, or other intertidal areas. Examples of these mammals include river otters (*Lutra canadensis*) and mink (*Mustela vison*), which are the two most common semiaquatic mammalian tracemakers in and on the banks of tidal creeks, beaches, and intertidal zones. Nonetheless, raccoons (*Procyon lotor*) are seemingly in all terrestrial and intertidal environments of the Georgia coast, as evidenced by their tracks, trails, and feeding traces. This raccoon omnipresence is related to their sleeping, reproducing, and denning in maritime forests (Chapter 8) but also their comfort in searching for food in salt marshes or upper parts of beaches, encouraged in no small part by their love of marine crustaceans and sea turtle eggs. Raccoons are also comfortable wading in shallow water and can swim, further blurring their habitat label of maritime forest.

Less common marginal-marine mammals, but still well represented by numbers, are marsh rabbits (*Sylvilagus palustris*), which live on the edges of freshwater swamps and salt marshes but often enter and exit tidal channels. Two other small mammalian tracemakers frequently inhabit salt marshes, marsh rice rats (*Oryzomys palustris*) and star-nosed moles (*Condylura cristata*), although neither has been studied well from an ichnological perspective. Naturally, mammals that everyone agrees are fully aquatic are bottlenose dolphins (*Tursiops truncatus*) and West Indian manatees (*Trichechus manatus*), which descended from land-dwelling mammals, but are more or less helpless once out of the water today. Feral hoofed mammals, such as cattle, hogs, and horses, which we normally may not consider as marginal-marine vertebrates, also leave a significant number of traces in intertidal environments. Hence these will be discussed in terms of how their traces might give some insights on how similar hoofed mammals in the geologic past may have interacted with sediments deposited on the edge of the sea.

Otters, Mink, Raccoons

River otters, with their wide territorial ranges and year-round activity on the Georgia coast, are prolific tracemakers. These members of the weasel family (Mustelidae) form tracks, trackways, slide marks, dens, tooth marks on bivalves and other invertebrates, and feces, many of which are associated with tidal-creek banks, but also can be found on beaches and along the edges of freshwater sloughs. Most people may associate these mammals with freshwater environments, especially if they have a mental image of substantially larger sea otters (*Enhydra lutris*) on the U.S. west coast, and think the two species are separated by habitat. Indeed, I often track river otters in freshwater streams running through the middle of Atlanta, Georgia, more than 300 km (185 mi) from the coastline, which also lends to a freshwater-only vision for these mammals. Nonetheless, river otters on the Georgia coast are quite comfortable making their living in and out of the salt marshes and other environments affected by saline waters. Because of this life habit, as well as the tendency for their traces to reflect direct predator–prey relations with marine invertebrates and vertebrates, they are considered herein as marginal-marine vertebrates.

Otter tracks and trackways are difficult to confuse with those of any other mammal, considering their relatively large tracks, 5–5 digital formula on the manus and pes, distal webbing between the toes, stout ends to claw-bearing toes (as opposed to the thinner ends of raccoons), loping trackway patterns, and tail drag marks (Halfpenny and Biesiot 1986; Halfpenny and Bruchac 2002; Elbroch 2003). Of all of these criteria, though, their trackways patterns are the most distinctive and can be easily identified from afar, negating any need to count toes in individual tracks. Otter trackways typically show either 1–2–1 or 1–3 patterns (Figure 9.12a,b); rarely do they slow down enough to make 2–2 diagonal walking patterns, which should come as no surprise to anyone who has watched otters move on land. The 1–2–1 pattern is from a loping gait, in which one manus print is behind a side-by-side pair of opposite-side (right–left) manus and pes prints, which is followed by the other pes. Furthermore, if an otter performs a side lope (favoring one side or the other), the tracks may form a diagonal line that repeats with each grouping. A 1–3 pattern, however, results once an otter picks up speed and approaches a gallop. This gait causes a lone manus impression well behind a grouping of the other three tracks. Given an unim-

MARGINAL-MARINE AND MARINE VERTEBRATES   485

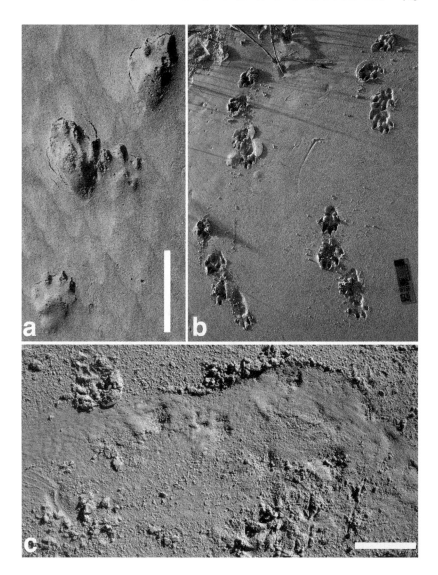

Figure 9.12. River otter (*Lutra canadensis*) traces. (a) Trackway with 1–2–1 lope pattern: Little St. Simons Island. Scale = 10 cm (4 in). (b) Parallel trackways with 1–3 lope patterns, probably made by a male–female pair moving harmoniously with one another: St. Catherines Island. Scale (*right*) = 10 cm (4 in). (c) Slide mark on beach sand near tidal creek: Sapelo Island. Scale = 10 cm (4 in).

peded line of sight along a sandy beach, an extensive otter trackway may even show mixtures of these patterns. If these patterns are not enough to convince a skeptical tracker that she or he is following an otter, look at the midline of the trackway for long, straight grooves. These are drag marks left by their heavy tails, which bounce onto and off the surface with each forward movement. Interestingly, river otter tracks on beaches may be linked to both season and sex, as male otters often go in search of mates during March–April, and consequently expand their ranges then. In my experience, early spring is when I most commonly encounter their tracks on sandy areas, such as storm-washover fans adjacent to salt marshes, or heading directly downslope on a beach, straight to the surf zone. To me, otter tracks are the most joyous and exuberant of all mammal trace I see on the Georgia barrier islands, easily inspiring an empathetic longing to likewise jump into the water and enjoy life.

Speaking of playfulness, no otter traces say "fun" more loudly than their belly-slide marks, which commonly co-occur with their tracks, especially on steep banks of tidal creeks (Figure 9.12c). These traces are irregular, slightly sinuous (low wavelength) grooves that are a little more than the width of an adult otter, about 20–25 cm (8–10 in). Lengths of slide marks vary considerably, but if associated with a high-angle slope can be several meters long. Slide marks also may have faint imprints of tracks on either side where feet helped to push the otters along on their bellies. Many animal behaviorists may have sterile interpretations of the behavioral import of such traces, such as territorial marking, in which otters are transferring their scent to the substrates, or conserving energy while moving down a bank. Nonetheless, I cannot help but think these traces are recreational, such as from male otters showing off for the ladies. Which is not so sterile after all.

River otters are known to make bank burrows, although these have not been documented from the Georgia coast. Other potential otter traces include their feces, which are often filled with fish scales and do double duty as scent-filled territorial markers. Otters tend to have site fidelity to their toilet habits, making latrines by habitually depositing feces and urine in the same places. These areas are above the banks of tidal creeks and might be spotted by looking for localized patches of dead or dying vegetation, memorably called brownouts (Rezendes 1999; Elbroch 2003). Latrines also serve as scent posts that warn competing otters. Scat is also sometimes accompanied by scratching or digging marks, as well as overt body impressions caused by sliding and rolling in the sand or mud.

Sharing the same environments with river otters on the Georgia coast are mink, which spend their time in tidal creeks and freshwater sloughs. In my experience, however, their traces are not nearly as common as those of otters. Individual mink tracks are morphologically similar to otter tracks, but much smaller. Moreover, their trackway patterns are expressed differently, showing a 1–3 bound or 2–2 lope, but also frequent 2–2 diagonal walking, with same-side manus–pes pairs (Rezendes 1999; Halfpenny and Bruchac 2002; Elbroch 2003). Of course, stride and trackway width are also proportionally shorter for a mink, reflecting their significantly smaller bodies. As proud representatives of the weasel family, mink are rapacious predators and can take on prey nearly their size, such as muskrats. Most other times, though, mink pursue fish, amphibians, crayfish, and other aquatic prey, and when on land go after small rodents, such as mice or voles, and birds. As a result, their scat contains undigested parts of these animals—crayfish parts, scales, fur, and feathers—which, like those of otters, accumulate at latrine sites or scent posts. Mink are also known to burrow for den making, and will take over old beaver or muskrat burrows (Stevens et al. 1997). Den entrances are denoted by mink tracks and well-worn trails connecting to nearby water bodies, as well as close proximity to mink latrines. Mink dens, although well studied in more northern climates, have not yet been documented on the Georgia coast.

Raccoon tracks, discussed previously (Chapter 8), are great examples of ecosystem-crossing traces, as these are found in nearly every terrestrial and semiaquatic sedimentary environment of the Georgia barrier islands (Figure 9.13a). Other raccoon traces include their well-established trails, which connect maritime forests with salt marshes and dunes, while cutting directly through ecotones separating these salinity-determined ecosystems (Figure 9.13b). Raccoon tracks and trails probably have good preservation potential in high salt marsh environments, where they are not completely submerged by daily tides, and hence are subjected to drying and hardening before burial by migrating dune sands. Raccoon predation traces frequently accompany their tracks and trails; after all, they are not just going to the marsh for the exercise, to chat, or read a novel. These traces are characterized by shallow, less than 10 cm (4 in) deep excavations that are also about 14–15 cm (5.5–6 in) wide, or a little more than the

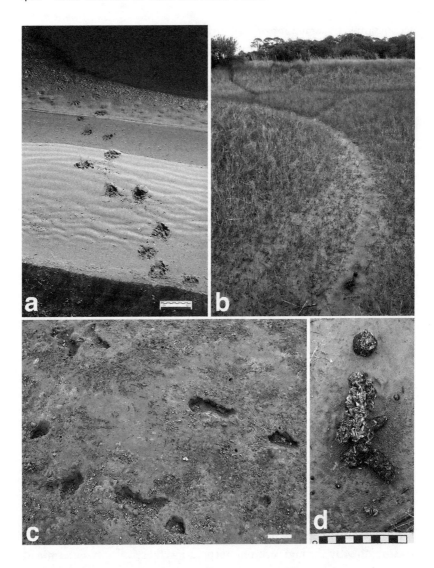

Figure 9.13. Raccoon (*Procyon lotor*) traces. (a) Typical walking trackway, with opposite-side manus and pes next to one another, made while patrolling a runnel in a sandy intertidal zone: St. Catherines Island. Scale = 10 cm (4 in). (b) Trail directly connecting salt marsh with dunes: Wassaw Island. Notice other trails in the background that connect with this one. (c) Dig marks, made with the front feet and to acquire delicious fiddler crabs: St. Catherines Island. Scale = 10 cm (4 in). (d) Feces, filled with fiddler crab parts, and placed as a territorial marker on the same raccoon trail shown in Figure 9.14b: Wassaw Island. Scale in centimeters.

width of a raccoon's two front paws put together (Figure 9.13c). Such holes are made to acquire a raccoon's favorite treat on an emergent salt marsh, fiddler crabs (Chapter 6). Fiddler crabs use burrows as their first line of defense against large vertebrates, running immediately into the nearest hole if panicked by an approaching biped or quadruped, which works well against animals that do not intend to dine on them. Raccoons do not give up so easily, though, and immediately begin digging into the burrow, a behavior they also display with crayfish burrows. For that reason, these predation traces made by the same species of vertebrate overlap with traces made by freshwater and marginal-marine crustaceans. Seeing that most fiddler crab burrows are less than 15 cm (6 in) deep (Chapter 6), a raccoon does not have to dig long before being rewarded by a crunchy treat. Additional raccoon traces that triumphantly declare successful predation of fiddler crabs are their feces, which are cylindrical with blunt ends, about 7–9 cm (2.7–3.5 in) long and 1.5–2 cm (0.6–0.8 in) wide, and filled with fine bits of undigested crab body parts (Figure 9.13d). In typical mammalian fashion, these feces are left in prominent places, such as elevated areas at trail junctions, on boardwalks, or other places where rival raccoons or other mammals will take notice of them. Raccoons also make large-scale excavations of sea turtle nests when preying on their eggs (Anderson 1981). As mentioned earlier, raccoons often associate the smell of a sea turtle with potential food sources, and track the scent of a mother turtle to a freshly made nest.

Miscellaneous Mammals: Marsh Rabbits, Marsh Rice Rats, Beach Mice, and Star-Nosed Moles

Marsh rabbits are closely related to the eastern cottontail (*Sylvilagus floridanus*), but the latter are definitely separated from marsh rabbits by habitat, and are much more common in mainland Georgia. Despite their common names, marsh rabbits are frequent inhabitants of back-dune meadows and the edges of freshwater wetlands. Nonetheless, they often stay close to water bodies and are well known for swimming. Indeed, sometimes these rabbits are seen swimming in salt marshes or tidal creeks, especially if evading landlubbing predators, such as rattlesnakes—a major predator of young rabbits—although they may also fall prey to alligators when in the water. Because they often enter and exit tidal creeks and tidal channels, these mammals, along with the less common marsh rice rats (*Oryzomys*

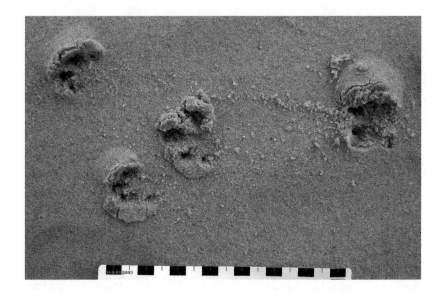

Figure 9.14. Marsh rabbit (*Sylvilagus palustris*) trackway in coastal dune showing half-bound gallop pattern, where front feet are offset and register behind both rear feet: Jekyll Island. Scale (*bottom*) in centimeters.

palustris) and star-nosed moles (*Condylura cristata*) are considered here as marginal-marine vertebrate tracemakers.

Marsh rabbit trackway patterns, once encountered, immediately tell a tracker that she or he is looking at something different from the average bunny. Although the tracks of both eastern cottontails and marsh rabbits are similarly sized and shaped, marsh rabbits spend much of their waking hours walking slowly in their preferred habitats while grazing (Chapman and Willner 1981). Eastern cottontails, in contrast, more often gallop (what most people call hopping) when getting around. Hence marsh rabbit trackways, which sometimes show a half-bound pattern caused by galloping (Figure 9.14), may reflect diagonal walking, with alternating sets of manus–pes impressions (Elbroch 2003). As a result of their preferred movement and abundance, they wear down ground vegetation along preferred pathways, thus making well-worn trails near and leading into salt marshes or bodies of water. Deposited on these trails are piles of their pelletal scat, which are subspherical and 8–10 mm (0.3–0.4 in) wide. Marsh rabbits also make nests for their young, although these have poor preservation potential compared to the underground dens made by cottontails and other mammals. Nonetheless, their nests are prominent, measuring about 35 cm (14 in) wide, 20 cm (8 in) deep, and 2–3 cm (1 in) thick, composed of fur (their own) and vegetation (Tomkins 1935; Chapman and Willner 1981). Tooth marks on vegetation are similar to those made by rodents, having 45° angles where grass stems and other plants are sheared. These traces, however, are on low-lying vegetation, rather than on branches like those snipped from trees by squirrels (Chapter 7). Marsh rabbits also dig burrows—presumably for food, not denning—that are subvertical and as deep as 30 cm (12 in) (Chapman and Willner 1981). Unfortunately, surprisingly little else is documented on the ichnology of marsh rabbits. Why do I find this surprising? Because these mammals are common on the Georgia barrier islands, even on more developed ones with many people, such as Jekyll and St. Simons. Hence neoichnologists or mammalogists would have no shortage of data if they decided to conduct a study on marsh rabbit traces.

Marsh rice rats are ecologically analogous to muskrats on the Georgia coast; the latter are more common in inland freshwater environments of Georgia and much of North America. Marsh rice rats are the only rodents known to habituate salt marshes in Georgia, hence any rodent tracks found on tidal creek margins or flat areas of salt marshes are very likely theirs. But what if a squirrel from a nearby maritime forest, dazed from eating too many live-oak acorns, loses its way and stumbles into the marsh, leaving its tracks? Not to worry, as rice rat tracks are notably smaller than those of squirrels, consisting of thin toes with reduced claws, unlike the tree-climbing claws of their arboreal cousins (Elbroch 2003). Trackway patterns also normally show 2–2 diagonal walking and a tail drag mark, although rice rats occasionally hop or bound across open areas. Food choices, based on stomach contents, include fiddler crabs, molluscans, insects, plants, and both eggs and hatchlings of marsh wrens (Sharp 1967). Hence their scat may reflect an interesting mixture of terrestrial and marginal-marine species, although their feces are quite small, only 2–3 mm (0.03–0.04 in) wide and 6–12 mm (0.25–0.5 in) long (Elbroch 2003) and would be easy to overlook. Rice rats often make nests of loosely woven vegetation, which are either positioned on high-marsh surfaces (above the high tide) or at the bottoms of shallow burrows (Wolf 1982). Unfortunately, their burrows have not been described, but probably are smaller than those of mink.

The southeastern beach mouse (*Peromyscus polionotus*), also known as the oldfield mouse, is a widespread but varied species, consisting of more than a dozen subspecies throughout the southeastern United States. Although this tracemaker occupies a large number of mostly terrestrial habitats, it is the dominant resident mammal in coastal dunes, leaving abundant trackways, trails, and burrows just shy of the upper intertidal zone (Figure 9.15). I have often been amazed by their presence so close to the shore, but now I just delight in finding their tiny footprints and hopping trackways. Their burrows consist of entrances, inclined tunnels, nest chambers, and so-called escape tunnels, which stop just short of the surface, but are used as a backdoor during times of peril, like when evading predators (Whitaker and Hamilton 1998). Tunnels are normally 2–3 cm wide and 60–90 cm long, with an average vertical depth of about 50 cm (Sumner and Karol 1929; Hayne 1936). These burrows are normally placed directly underneath sea oats or other coastal vegetation, and hence are often associated with root traces (Chapter 4). Nest chambers, which are located at the horizontal base of a burrow system, are subspherical and 4–6 cm wide, and often partially filled with pieces of sea oats and other coastal plants. If these mice do not feel like making a new burrow, they move into an old, unoccupied ghost crab burrow. Nevertheless, individual mice are known to make as many as 20 burrows within a given territory (Blair 1951). Tracks or other surface traces are usually made at night, and their nocturnal activities increase dramatically during new moons or cloudy nights, when they feel safer from owls and other nighttime predators. Why are these mice living in coastal dunes? Besides enjoying the company of other mice, they are also well adapted to expanses of easily burrowed sand, and are taking advantage of the copious food provided by seeds of seat oats and panic grass, as well as insects. Indeed, some biologists have hypothesized that burrowing

Figure 9.15. *Facing.* Southeastern beach mouse (*Peromyscus polionotus*) traces in coastal dunes. (a) Trackway, showing typical half-bound gallop pattern with offset manus impressions: Cumberland Island. Scale = 5 cm (2 in). (b) Trackway showing slow, diagonal walking pattern and tail drag marks: Little St. Simons Island. Note nearby eroded sparrow tracks (*right*). Scale = 10 cm (4 in). (c) Multiple trackways leading into and out of a ghost crab burrow, indicating the crab abandoned it and mice usurped it: Little St. Simons Island. Scale = 5 cm (2 in). (d) Trail on coastal-dune ridge, made by multiple individuals traveling the same route between burrows: Little St. Simons Island. Scale = 10 cm (4 in).

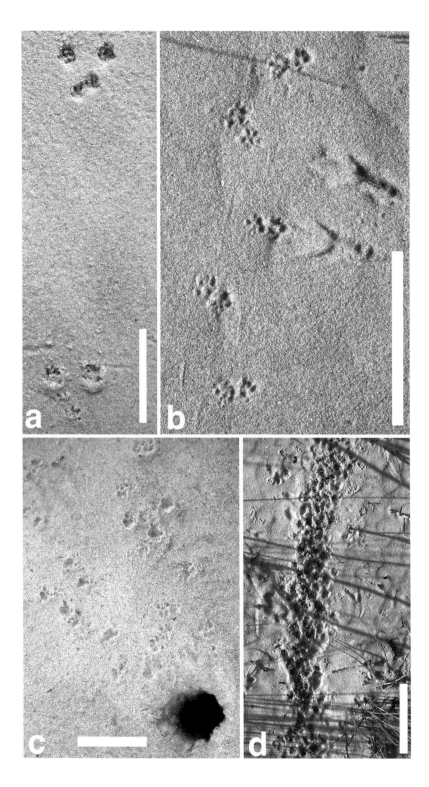

is a hardwired behavior in mice belonging to *Peromyscus,* and hence the presence of many subspecies of *P. polionotus* in coastal dunes makes sense, evolutionarily speaking (Weber and Hoekstra 2009).

Although eastern moles are well known for their swimming ability, star-nosed moles are the only moles that make a habit of being in the water, where they hunt for much of their food. The star-nosed mole's common name is derived from the starburst-like pattern of their noses, consisting of 22 (more or less) soft, fleshy projections looking much like little fingers spread out around the nose. These ray-like features, however odd looking, enable this mole to find its invertebrate prey; once a worm or insect is touched and identified as food by one of these rays, a mole quickly devours it. Some researchers have even proposed that this touch is augmented by the detection of electrical fields given off by their prey, such as earthworms (Gould et al. 1993). In other words, feeding traces left by star-nosed moles represent a sophisticated hunting method. Star-nosed mole burrows are similar in size (3–6 cm, or 1.2–2.4 in diameter) and geometry to those of eastern moles, and are as much as 60 cm (24 in) below the surface (Petersen and Yates 1980). Star-nosed mole burrows, though, more likely connect directly with water bodies, such as freshwater wetlands or salt marshes, and even have underwater entrances. In contrast, nest chambers will be above the water table, a smart strategy for keeping their offspring from drowning. Chambers are denoted as expanded zones, slightly wider than tall, such as 18 cm (7 in) versus 13 cm (5 in) along tunnel networks, accompanied by some vegetation for insulation. Mounds of excess sediment pushed to the surface out of excavated tunnels (molehills) could be potentially confused with burrow mounds made by crayfish (Chapter 5), but the latter have definite burrow openings (Whitaker and Hamilton 1998). Thus an ichnologist should not be making a crayfish burrow out of a molehill.

In short, these four relatively small mammals form a significant number and variety of traces in marginal-marine environments, particularly in coastal dunes and salt marshes. If studied from more of an ichnological perspective, each species could probably provide more evidence lending to the identification of analogous trace fossils in the geologic record. Furthermore, the proximity and direct connections of their traces to marginal-marine water bodies bode well for their potential preservation in the geologic record.

Dolphins and Manatees

One mammal commonly seen on the Georgia coast, the bottlenose dolphin, certainly comes to mind as the most fully adapted to marine environments. Yet it also comes close enough to shore that a neoichnologist, or anyone else for that matter, could easily entertain the concept of its leaving traces in shallow-marine environments. Indeed, dolphins do make traces in these environments, and their brief interactions with coastal sediments are intentional. Dolphins have been observed to go fishing through what was once considered an unusual technique, but as is often the case with such discoveries, it actually may be quite common. First of all, dolphins swim upstream in a tidal creek during low tide, but in one that still retains sufficient water to hold them. They then herd fish into dense groups by chasing them up the creek (Hoese 1971). (Just as moon snails were seen as the lions of the tidal flats, so should dolphins be viewed as the cowboys of the tidal channels.) Once properly panicked by these predators, the fish become increasingly desperate in their evasion, and eventually will do almost anything to escape, including leaving the water. Dolphins know this, and pursue the fish onto a tidal-channel bank, where entire schools of fish leap onto the exposed bank and, lacking water, promptly stop swimming. The dolphins then heave themselves onto the banks and have a feast on fresh fish. More importantly, in doing this they also leave outlines of their ventral surfaces, in addition to any imprints left by fish flopping on the bank. Sadly, I have not seen these traces, but this fishing behavior has been witnessed multiple times and briefly described by other scientists in Georgia salt marshes (Hoese 1971; Frey and Pemberton 1986), giving something to look for with future visits.

Other than the occasional wayward whale, the only other fully aquatic mammal species near and along the Georgia coast is the West Indian manatee, specifically its northern subspecies, otherwise known as the Florida manatee (*Trichechus manatus latirostris*). Manatees belong to Sirenia, a group of marine mammals that includes dugongs (*Dugong dugon*) and an extinct species, Stellar's sea cow (*Hydrodamalis gigas*). All members of this lineage reflect long-existing aquatic adaptations, such as vestigial rear legs, which, like in whales and dolphins, are not externally visible; modified forelimbs (flippers) for swimming; a spade-like tail with a circular outline; streamlined bodies; a downwardly pointing mouth (all the better for feeding on bottom-dwelling vegetation); and the ability to hold their breath and stay underwater for long periods of time. On average, these mammals are about 3 m (10 ft) long and weigh about 400 kg (880 lb), although long-lived individuals can become monstrously large, as much as 1,500 kg (3,300 lb) (Scott 2004). Furthermore, the low nutritive content of their food, much of which consists of fiber, means they must eat a large amount of vegetation to sustain themselves, about 5–10% of individual body mass per day. The Florida manatee is migratory and swims north from Florida along the Georgia coast once its water becomes warmer, during March–November. Once in Georgia, they then swim into some of the tidal channels and riverine estuaries of the coast for feeding, mating, raising calves, and conducting other manatee business.

Do these mammals leave any traces? Yes, although these have not yet been documented in an ichnological sense. Most of their interactions with substrates are likely caused by shallow-water bottom feeding, in which they pull up plants with their mouths, or excavate more firmly rooted plants by using their forelimbs. The latter traces would have paired impressions caused by the blade-like edges of the forelimb fins, with a deeper pit and mouth marks in the middle. When swimming, manatees rotate their forelimbs inward (toward the centerline of the body), so their digging traces may also show evidence of this rotational movement, although this description is speculative and should be tested with actual observations. During high tides along tidal-channel margins, manatees are known to munch on submerged patches of smooth cordgrass (*Spartina alterniflora*), which of course is wildly abundant along the Georgia coast (Baugh et al. 1988). Hence manatees may leave traces of this activity, although these may be difficult to distinguish between the normal wear and tear of tidally induced erosion on a marsh edge.

Video footage of manatees also show them walking along sedimentary surfaces, but with forelimbs nearly simultaneously stroking forward and back, similar to a galloping pattern seen in terrestrial mammals (Chapter 7). They are also likely to leave traces from their tails contacting sedimentary surfaces as they swim along the bottom. Like all aquatic mammals, manatees swim with an up-and-down motion, rather than the side-by-side movement of fish and alligators (Chapter 7). Consequently, a manatee swimming close to a sedimentary surface could have its rear paddles leave periodic impressions in a swimming trail, with undisturbed patches of sediment between marks.

Unfortunately, like manatee feeding traces, these swimming traces are undocumented, and all such traces need to be described from manatee habitats soon. This species, like other sirenians, is slow breeding and frequently killed by boat collisions; habitat alteration has also adversely affected their breeding and feeding habits. As a result, it is endangered, and only an estimated 3,000 individuals remain, with their numbers expected to decline over the next few decades (Runge et al. 2007). Consequently, neoichnologists will have fewer opportunities to study their modern traces in upcoming years, especially if they hope to compare these to possible sirenian trace fossils.

Feral Mammals

Although feral mammals on the Georgia barrier islands are primarily terrestrial, special notice is warranted for their traces in marginal-marine environments, especially those that adversely affect salt marshes, the lower parts of dunes, and upper portions of beaches (Figure 9.16). The major feral tracemakers in these environments are horses (*Equus caballus*), cattle (*Bos taurus*), and hogs (*Sus scrofa*). Feral horses are restricted to Cumberland Island; stock for the current population has been there for less than the past hundred years or so, despite romanticized claims of their descent from Spanish horses over the past 400 years (Bellis et al. 1995; Urquhart 2002). Feral cattle are only on Sapelo Island, and are the descendants of cattle owned in the 1920s by millionaire Howard Coffin (Chapter 2) that were allowed to roam free on the island. Feral cattle were also on Cumberland, but were transferred off island in 1974 (Huggett 1995). Unfortunately, feral hogs are on nearly every undeveloped barrier island, even the ones labeled in tourist brochures as pristine (Chapter 3). These animals likely have experienced a recurring presence despite several temporarily successful extirpations, owing to reintroductions and swimming swine. An uncommon feral species on the Georgia coast is the European fallow deer (*Dama dama*), which is restricted to Little St. Simons Island. Introduced originally for hunting, these deer have displaced native white-tailed deer there and impacted most of the terrestrial ecosystems on the island (Morse et al. 2009).

Feral horses are frequent visitors to coastal dunes and berms, walking across the tops of dunes and sea oats (Figure 9.16a), which they also eat by pulling these plants out of the dunes, extensive root systems notwithstanding (Chapter 4). Consequently, coastal dunes on Cumberland have un-

498 LIFE TRACES OF THE GEORGIA COAST

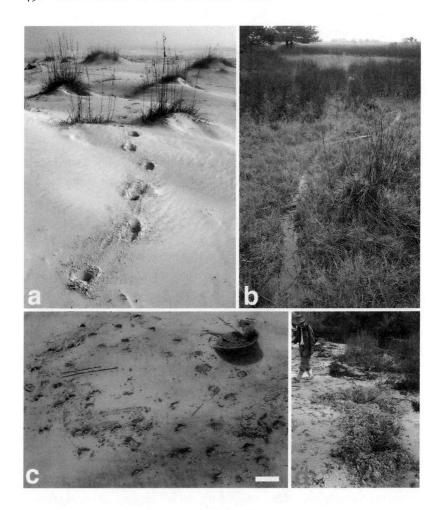

Figure 9.16. Traces of feral hoofed mammals that show their interactions with marginal-marine environments. (a) Feral horse (*Equus cabullus*) tracks crossing coastal dunes: Cumberland Island. (b) Feral cattle (*Bos taurus*) trail leading into low salt marsh: Sapelo Island. (c) Feral hog (*Sus scrofa*) tracks associated with scavenging of adult limulid in intertidal zone of beach: St. Catherines Island. (d) Excavation pits made by hogs in intertidal zone of beach: Cumberland Island. Carol Ruckdeschel (*left*) for scale.

dergone significant erosion from trampling and grazing on the plants that are largely responsible for keeping these dunes in place. These horses also overgraze salt marshes, cropping smooth cordgrass to the point where it prevents further growth, a degradation exacerbated by horse hoofs squashing roots. The long-term effects of horse-trodden and chewed landscapes have been studied on a few barrier islands in North Carolina (de Stoppelaire et al. 2004), as well as on Cumberland itself (Turner and Bratton 1987; Turner 1988) and these results predict continued poor health for shoreline ecosystems on Cumberland. The copious amounts of feces produced by horses also overwhelm the native dung beetle populations (Chapter 5), which cannot keep up with the demand for their ecological services.

Similarly, feral cattle on Sapelo often emerge from maritime forests to graze on smooth cordgrass on the edges of salt marshes, reducing tall stands to mere nubs and damaging root systems with their feet. Through a combination of sharp-edged hoofs, great weights, soft mud, and enthusiasm for eating smooth cordgrass, their tracks penetrate intertidal muds quite deeply, as much as 30 cm (12 in). Thus their tracks may be directly associated with older invertebrate traces, such as fiddler crab burrows. This sedimentary effect is further multiplied by small herds of 10–20 cattle traveling together. For example, I have seen their meter-wide (3.3 ft) trails extending hundreds of meters into marshes, indicating a habitual presence that prevents further growth of smooth cordgrass in places (Figure 9.16b).

Feral hogs, which cause significant alterations to maritime forests of the barrier islands, also cause considerable and long-lasting marks on marginal-marine environments (Figure 9.16c,d). Their ecological and sedimentological impact is largely provoked by their incredible omnivory, aided by their famed sense of smell. Hogs, which often travel in groups, leave trackways onto and over dunes, damaging dune and salt marsh vegetation. More importantly, though, they also dig up anything that hints of food in dunes, foredunes, and berms, including sea turtle nests (as mentioned earlier) and pennywort (*Hydrocotyle bonariensis*), a colonizing species of dune plant (Chapter 4). These excavations are startlingly large: some I have seen are 4–5 m (13.1–16.4 ft) across and as deep as 25–30 cm (10–12 in), depending on the depth of the food source and hog motivation. Hogs also cross high salt marshes and occasionally rip apart a marsh surface when in search of nutrition, leaving prominent concavities that raccoons can only envy.

Despite the nonnative status of these species, their traces might provide insights on hoofed mammals—perissodactyls and artiodactyls—that lived on Georgia barrier islands during the Pleistocene, or on other coastal environments throughout the Cenozoic. Some biologists have even proposed that modern-day feral mammals have ecological impacts on coastal ecosystems similar to those of large Pleistocene herbivores (Levin et al. 2002). More importantly, the encroachment of these terrestrial mammals on marginal-marine environments means their traces overlap considerably with invertebrate and other vertebrate traces, constituting a significant addition to a suite of traces. Thus we should not ignore these traces in favor of a native-species purity test, as the islands have already been sullied by a human presence for the last 5,000 years anyway (Chapter 2). Besides, the only native hoofed animal on the Georgia barrier islands, the white-tailed deer (*Odocoileus virginiana*), also crosses into these marginal-marine environments. I have often observed deer tracks in the upper intertidal zones of tidal creeks, beaches, and salt marshes, all of which they traverse during low tides. In some instances, deer may even swim long distances, especially if they have overpopulated and denuded smaller islands or other patches of land, or if they are being hunted (Rue and Rue 2004). In such situations on the Georgia barrier islands, deer tracks enter and exit subtidal environments, like deep tidal channels separating nearby islands. However improbable it might seem to urban dwellers, whose only contact with pigs may be through eating prosciutto, feral hogs also can swim, even across rivers (Mayor and Brisbin 2008). This knowledge provokes ever-increased vigilance in wildlife biologists trying to protect shorebird and sea-turtle nest sites, such as on uninhabited Holocene islands like Blackbeard and Wassaw. In short, the seemingly easy category of marginal-marine vertebrate can take on expanded meaning in the face of ecological opportunism.

MARGINAL-MARINE VERTEBRATE TRACEMAKERS OF THE GEORGIA COAST

Marginal-marine vertebrates, despite their relatively low diversity compared to their fully terrestrial counterparts (Chapter 8), comprise a significant group of tracemakers on the Georgia coast, some of which include animals thought of as exclusively land dwelling. Moreover, many of their traces are readily observable by neoichnologists without their having to get into a boat: simply walking on intertidal sand flats and along the margins of salt marshes can yield a huge number of observed tracks, trails, nests, feces, and other traces. Shorebirds in particular leave varied and rich stories of their lives on the sandy beaches and mudflats of Georgia coastal environments, waiting for further interpretations, but other vertebrates—fish, reptiles, and mammals—deserve more attention than previously received. I hope that the preceding coverage will have inspired some new insights and avenues of discovery in this respect, mixed with a lot of fun along the way.

10

TRACE FOSSILS AND THE GEORGIA COAST

YELLOW BANKS BLUFF DECODED

Yellow Banks Bluff on St. Catherines Island is one of those places where the people who named it limited themselves to pure description and allowed for no flights of fancy. It is indeed a yellowish, banked bluff, forming an outcrop about 4 m (13.1 ft) tall and 600 m (1,970 ft) long, extending much of the length of a beach on the northeast corner of the island (Figure 10.1a). The first time I saw it was with about 30 other people on a geology field trip in the spring of 2007. The trip was associated with and just before a regional meeting of the Geological Society of America, held in nearby Savannah, Georgia, that year. Our group rode in ATVs, four to five people per vehicle, zooming down the beach and weaving between downed oaks and pines, the former living parts of the maritime forest just above the bluff, but now victims of sea-level rise and coastal erosion (Chapter 2). This fast-forward way of viewing was definitely not my style—too many traces on the beach were missed along the way—but was essential for us to reach a particular part of the outcrop in a timely way. Once there, the field trip leaders would educate us about its geologic significance, with plenty of time left over to eat dinner, drink adult beverages, and mentally digest what we had seen that day.

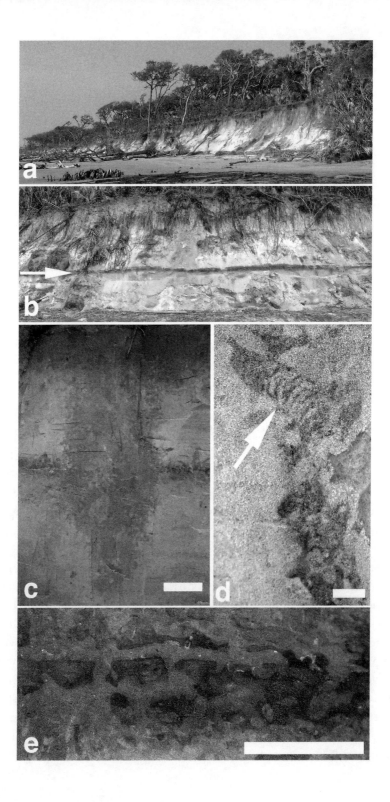

In typical field trip fashion, we crowded around the field trip leaders and strained to listen to their spiel about the sediments composing the bluff, the sedimentary structures, its trace fossils, and a few odd strata that defied identification. From these descriptions, a broad interpretation of the geologic history of the area emerged. During a previous sea-level high of the Pleistocene Epoch, about 50,000 years ago, shallow-marine shelf sediments covered this place. As sometimes happens during sea-level highs, the water level stopped going up, yet did not go down either, a stasis called a stillstand (Chapter 2). Stillstands are interesting to sedimentologists and paleoecologists because they represent times when erosion may outstrip sedimentation in a specific area. As a result, parts of the seafloor become less like loose, unconsolidated sand or mud, and more like a firmground. These firmgrounds, such as the relict marshes discussed previously (Chapters 3, 6), attract invertebrates that are well adapted to scraping away or otherwise boring into the hardened material for their domiciles, such as angelwing bivalves (Chapter 6). As a result, these animals make distinctive, discrete traces that are easy to identify and valuable for their paleoecological significance. With this idea in mind, and as we listened to the field trip leaders, we also continued to look at the bluff, which contained thin, linear and dark horizontal interruptions in the otherwise homogeneous yellow-white sand (Figure 10.1b). These were the marine firmgrounds, our leaders said, although they were not entirely sure about this interpretation because their corroborating evidence was admittedly somewhat cryptic. This is the criterion I use to I draw a line between good scientists and bad scientists: a willingness to base a degree of certainty—or uncertainty—on the weight of the evidence, and not speculation. In other words, these were good scientists.

Figure 10.1. *Facing.* Yellow Banks Bluff, an outcrop of poorly consolidated Pleistocene sandstone on St. Catherines Island, containing many interesting trace fossils. (a) Overall view of outcrop. (b) Closer view of outcrop, with a thin, dark bed (*arrow*) holding a mysterious past, begging to have its trace fossils studied. (c) Probable root trace fossil originating at dark layer in the outcrop, its outlines blurred by subsequent burrowing. Scale = 10 cm (4 in). (d) Backfilled fossil burrow (*Taenidium* isp.), indicated by arrow, probably made by an infaunal insect, cutting across a dark root trace. Scale = 1 cm (0.4 in). (e) The fossil burrows *Thalassinoides* isp., filled with lighter-colored sediment from above, and probably made by the intersecting burrows of fiddler crabs (*Uca* spp.). Scale = 10 cm (4 in).

As usually happens in such field trips, its leaders realized that it was silly to continue lecturing people about what was in front of them without allowing them to examine it, so they cut off their own rhetoric by saying, "Go take a look." My friend and colleague, Andy Rindsberg (Chapter 6), then joined me in scraping away at the weakly consolidated outcrop with putty knives in a quest to expose its secrets. As we did this, the thought came to me that we were modern-day tracemakers excavating a terrestrial firmground to investigate presumed marine firmgrounds, an iterative metaexperience that later might serve as the inspiration for some self-indulgent poetry read aloud at a local coffeehouse. Fortunately, that idea came and went, and we concentrated on examining our freshly cut surfaces of the outcrop and their newly revealed textures and structures. Rarely do geologists have this luxury with sedimentary rocks, where they can abandon their cherished rock hammers or power tools for more gently applied hand tools, which revealed more telling details with every stroke.

Perhaps the trickiest part of reconstructing ancient environments from two-dimensional vertical faces of outcrops is figuring out how to flip these surfaces 90° into a third dimension—the areas that represent their original environments—and then interpret how these environments changed under the fourth dimension of time. The vertical sequence we saw before us contained a story that began at the bottom of the outcrop, where it met the present-day beach, and ended at the top of the bluff, where it met the present-day forest. Nonetheless, it was up to us to read its details to figure out the Pleistocene cast of characters, character development, plots, leitmotivs, symbols, themes, and other such parts comprising a narrative that changed sometimes within the vertical distance of a centimeter. Moreover, like any scientists should, we did not necessarily accept the story given to us by the field trip leaders, which would have been like taking the word of a reviewer for a book's contents before reading the book for yourself. Instead, we were testing the hypothesis—this was a marine firmground, formed during a sea-level high and stillstand—by examining the veracity of the stated evidence, adding any previously undocumented observations, and asking whether this new evidence supported, modified, or rejected the original hypothesis.

Thus it was no surprise to any of us that a few strokes of a putty knife across a brownish structure in the outcrop instantly supplied new evidence that rendered the previous explanation in doubt, and instead favored a radically different alternative. That is how science works, or as evolutionary scientist and noted Darwin supporter Thomas Huxley (Chapter 6) would have said, "Science is organized common sense where many a beautiful theory was killed by an ugly fact." The structure was vertically oriented, about 30 cm (12 in) wide at its top and about 80 cm (31 in) long, with its top above a dark firmground layer, and cutting through it and a similar layer about 40 cm (15 in) below it, tapering to a point (Figure 10.1c). Other vertical structures near it branched and tapered to points near their distal ends, toward the bottom of the outcrop. Andy and I looked carefully at its oddly blurred margins, which were rendered that way by small, macaroni-like structures filled with meniscae (Figure 10.1d). I had seen a similar relationship between such structures in Pleistocene rocks at Raccoon Bluff on Sapelo Island, which in turn were echoed by nearly identical ones in slightly younger rocks on the North Island of New Zealand (Chapter 4; Gregory et al. 2004). The larger structures were tree (or shrub) root traces, and the smaller ones were backfilled invertebrate burrows of the ichnogenus *Taenidium*, which matched those made by infaunal insects (Chapter 7). These facts—the in situ trace fossils of plant roots and insects—had neatly slain the previous hypothesis: the sea was not responsible for the traces we saw before us, but the land.

I have seen and participated in several outcrop conversions, in which field trip leaders bring a group of geologists or paleontologists to an outcrop with a story in hand, only to have the story changed or completely demolished within 10–20 minutes of arriving. This shift often causes either a minor or more heated verbal tussle, followed by some of the flock becoming born again by the new story, and others sticking with their previous faith. This was yet another example of that necessary part of science—peer review—in which we do not necessarily agree to the prevailing story until we see its underpinnings, while also keeping in mind that new evidence can always change the status quo. Now Andy and I were in the midst of upsetting the apple cart (or the ATV, as it were), as we gave our assessment of what was here. I suddenly became an impromptu field trip leader, a role I normally shun because it usurps attention from the people who put many hours of work into planning and carrying out the field trip. Nevertheless, science is science, and I am a teacher too. So I related our preliminary identifications, pointed to the evidence and explained it, cited previous peer-reviewed works that backed up these interpretations, and stated how all of this collectively contradicted the previous interpretations. The field trip was enlivened, discussion and debate ebbed and flowed, and some people tried to replicate (and thus test) our results right then and there, scraping their own spots on the outcrop and looking for the same criteria I had just described. Fortunately, the field trip leaders were both bemused and amused by all of this activity, and based on discussions with them later, they were happy to have had such a fruitful and respectful exchange between the participants. In other words, everyone won.

As a result of our finds, Andy and I were encouraged to pursue our hypothesis about the firmgrounds, and we made plans with one of the field trip leaders—Gale Bishop—to come back to St. Catherines several months later to study this outcrop further. Trace fossils had already proven their worth in our first visit to Yellow Banks Bluff, so we expected these would be our allies in a second visit too. We were much more thorough in our analysis this time, measuring, sketching, describing, photographing, and discussing our results over the course of several days. The dark, thin horizontal strata were clearly the key to discerning the rest of the story, and because we had seen trace fossils in them during our previous visit, these received our undivided attention. Our intensive study paid off. Within two days, we identified the trace fossils in the firmgrounds as the ichnogenera *Psilonichnus* and *Thalassinoides* (Figure 10.1e), and their sizes and shapes matched burrows made by fiddler crabs (*Uca* spp.; Chapter 6, Figure 6.10). Yet these layers were also intersected by tree shrub root traces and insect burrows. Hence we had a snapshot of short-term ecological succession in our geological section.

But what kind of succession? What situations could allow fiddler crabs to be followed so quickly by insects and terrestrial plants in the same place? Our questions were answered by spending several afternoons walking across a few sandy deposits that covered former salt marshes, some of which were only a few hundred meters south of Yellow Banks Bluff. These deposits were in various stages of ecological maturity, and their trace assemblages gave us an inkling of how they might translate into the fossil record. We found ourselves shuffling back and forth—mentally and physically—between the modern traces and the trace fossils in the outcrop, before a clear picture began to emerge.

Applying these neoichnological examples, we proposed that Yellow Banks Bluff had preserved layers made by storms, specifically storm-wash-

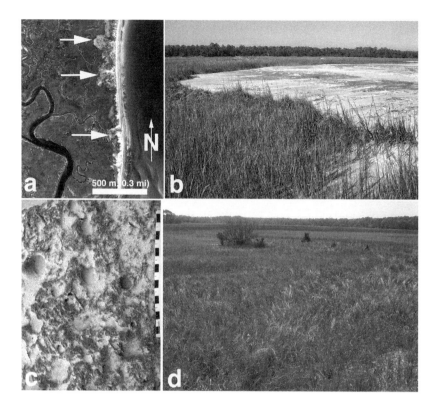

Figure 10.2. Modern storm washover fans and preservation of traces. (a) Aerial view of washover fans (*arrows*) on St. Catherines Island, less than a kilometer (0.6 mi) south of Yellow Banks Bluff: from Google Earth (source U.S. Geological Survey), image taken May 31, 2008. (b) Relatively new washover fan on Sapelo Island that has not yet been colonized by terrestrial plants. (c) Abandoned modern mud fiddler crab (*Uca pugnax*) burrows in top surface of relict marsh being filled with windblown sand from a nearby beach. The differing sediments provide a sedimentary contrast that would be easily observable in the geologic record: Sapelo Island. Scale in centimeters. (d) Well-vegetated washover fan adjacent to high salt marsh, with extent of fan denoted by tall grasses and shrubs: Wassaw Island.

over fans (Figure 10.2a; Chapter 3). Such deposits are made when storm surges, caused by hurricanes or other tropical storms, inundate a shoreline, breach the lowest parts of the dunes, and flood a nearby salt marsh (Figure 10.2b) or back-dune meadow with sea water and transported sand (Deery and Howard 1977). Sand layers deposited by these surges are often suffused with finely organic material, making for a dark, coffee-grounds-like mixture that settles onto the marsh surface once the storm waves have waned. Soon after the storm subsides, fiddler crabs move into these freshly paved surfaces and begin feeding, burrowing, mating, and making more crabs, which also start burrowing once they reach maturity. With continued exposure and not receiving daily tides, though, the ecosystem may not sustain a fiddler crab population, especially if other storm surges and windblown sand build the height of the area above normal high-tide levels. Accordingly, substrates then become more firm and less occupied by fiddler crabs, and the open holes of their vacant burrows are filled with white windblown sand that contrasts nicely with their darker host sediments (Figure 10.2c). With enough time and freshwater input, terrestrial plants from ecotones of nearby maritime forests start colonizing along the edges of the washover fans, then eventually in the fan deposits themselves (Figure 10.2d). Burrowing insects associated with these plants are the last of the colonizers, mixing the formerly distinct sediments and cutting across traces made by roots and fiddler crabs. The succession is then complete until the next burial or erosion, either of which might also arrive with another storm surge.

Almost a year later, in March 2008, I delivered a presentation about Yellow Banks Bluff in a coastal geology session at the same regional meeting of the Geological Society of America, but this time held in Charlotte, North Carolina. This was peer review again, only this time the roles were reversed, as our hypothesis was open to skeptical scrutiny, and if someone else noted an annoyingly contradictory fact, it might wipe out our explanation as neatly as a hurricane-caused storm surge. In my presentation, I laid out the evidence, showed how it falsified the previous hypothesis—the sediments, most certainly, were not deposited offshore—and related the trace fossil evidence supporting the new scenario. Afterward, two geologists who had put together the previous hypothesis came up and congratulated me (with gusto) about how Andy and I had finally figured out Yellow Banks Bluff.

However gratifying such positive feedback might have felt at the time, though, I kept in mind how easily some future geologists might revisit the outcrop and demonstrate the foolishness of the hypothesis proposed by Martin and Rindsberg, especially if these geologists took a more critical look at the trace fossils. Along those lines, I was asked the following year (2009) to talk about Yellow Banks Bluff and its trace fossils yet again, but this time in front of an audience of experts on the geology of St. Catherines, many of whom had been working there for more than 20 years. Even more daunting, this conference was on St. Catherines Island, and a field trip to Yellow Banks Bluff was scheduled for the afternoon of the same day of my talk. All trepidation notwithstanding, the talk was well received, a quick visit to the outcrop clarified the concepts further, and a research paper solicited on the same subject passed peer review with the proverbial flying colors (Martin and Rindsberg 2011).

In short, we had successfully applied neoichnology to solving a Pleistocene paleoichnological problem on the Georgia coast. This may not be all too surprising, though: after all, one of the fundamental principles of geography is Tobler's Law, which states, "Everything is related to everything else, but near things are more related than distant things" (Tobler 1970). Appropriately enough, we were proximal to modern storm-washover fans on St. Catherines and, geologically speaking, quite close in time. Then perhaps a better follow-up question to ask was this: How applicable was this neoichnological knowledge to trace fossil assemblages from more geographically and temporally distant places? More specifically, how well would such neoichnological models hold up as we traveled away from these modern life traces in both time and space, back into the Mesozoic or Paleozoic Eras, and on continents far removed from Georgia? This question is tested, and answered, by trace fossils.

TRACEMAKERS AND THEIR TRACES: ONE ICHNOLOGY

Uniformitarianism, also known as actualism (Chapter 2), has its limits in ichnology, as most species of plants and animals that made trace fossils are extinct. Hence we cannot directly observe these organisms making traces and unequivocally compare their vestiges to trace fossils in the geologic record. For an extreme example, we will never have the opportunity to study how a 7 tonne (7.7 ton) bipedal dinosaur moved and made tracks, no matter how well Hollywood digitally recreates such animals. Along those lines, observing a large bipedal bird today is a poor substitute for directly observing a living theropod dinosaur in the Mesozoic, however thrilling it might be to watch a sandhill crane or great blue heron walk gracefully and leave beautiful, theropod-like tracks in a salt marsh (Chapter 9). Instead, we scrutinize the behaviors of analogous tracemaking species that we think approach the size and anatomy of past species, preferably in the context of similar environments, while accepting, with a shrug, that it just will not be the same. We also understand that all tracemakers and traces we observe today are under the influence of a recent human presence, which has altered ecosystems on local and global scales so that these are also dissimilar from the prehuman past (Chapter 2).

Paramount to observations of trace fossils, however, is making sure the paleoecological context of a trace fossil is hypothesized on the basis of all available information. This requires thorough descriptions of the substrate preserving the trace fossil and accompanying physical sedimentary structures, then asking the right questions. For example, for a given substrate, we should pose the following inquiries: Is a trace fossil normally preserved in a mudstone or sandstone? How well sorted is the sediment? Was the sediment wet or dry when the trace was made? How much oxygen was in the original sediment? Was the substrate consolidated, either as a firmground or a hardground? Moreover, with physical sedimentary structures, we might ask: Was the trace formed underwater or in emergent conditions? Was the trace made in spring, summer, fall, or winter? Were waves and tides affecting the sediment and the tracemaker? Did any storms punctuate quieter conditions before or after the trace was made?

With these abiotic clues in mind, an ichnologist should also examine the entire assemblage of trace fossils in the given stratum—appropriately called an ichnoassemblage—and test the hypothesis of whether or not it represents an ichnocoenose, which is an ecologically consistent grouping of traces, modern or fossil. However, sometimes an ichnoassemblage and ichnocoenose do not coincide, such as in firmgrounds where younger tracemakers overprint traces of an entirely different and much older ecological community, analogous to a modern person carving his initials next to ancient petroglyphs. Describing and interpreting these easy-to-see macroscopic traits of a sedimentary rock are essential first steps before resorting to geochemical tests or anything that may involve intricate lab equipment and otherwise cost considerable amounts of money. In fact, I like to think this cost-effectiveness is yet another example of the utility and facility of ichnology as a science, and why it is a science that is open and available to all who might be interested.

Regardless of whether an ichnologist is just starting her or his career, or is a seasoned veteran of many trace campaigns, another important theme to remember is that traces beget traces, and many modern traces can be complex (made over multiple generations of tracemakers), compound (strikingly different types of traces made by the same species of organism), and composite (more than one species of organism contributed). Does this sound complicated? Well, it is. As an attentive reader may have noticed in the preceding chapters, the vast majority of traces are not singular alterations of a substrate in an environment with definite parameters, caused by an organism's response to a simple stimulus. Instead, they are more like puzzles that, once deconstructed and put back together, create a picture bearing shades of meaning, uncertain settings, and a cast of characters who all require elaborate development to better understand. For example, a deer track encountered in a maritime forest is much more than just that simplistic label, "deer track." Instead, it is only one small part of a string of tracks that tell of that deer's day and how it was affected by (or affected) its surroundings. Furthermore, the deer's behavior, and even the tracks themselves, caused changes in the behaviors and traces of other organisms that in turn influenced one another, with each track like a pebble dropped in a pond that imparts intersecting waves. In other words, traces in nature are not made in isolation under strictly controlled laboratory conditions with all variables taken into account. Instead, traces are shaped by many external and interconnected factors that should be viewed holistically, including whatever might not be observable today.

Thus a philosophy of "One Ichnology," a rallying cry often issued by esteemed ichnologist Richard Bromley to other ichnologists, needs to be adopted when examining any modern and ancient trace left by an organism interacting with its environment. This, in turn, is a worldview aided by asking the right questions. One could start in this ichnological quest for enlightenment by following a journalistic outline, and formulate the most basic of questions: Who (was the tracemaker), what (was the tracemaker's behavior), when (did the trace get made), where (was it made, ecologically speaking), why (did the tracemaker behave that way), and how (did the tracemaker and its traces affect other aspects of its environment)? Through such inquiry, minds can be opened and trace fossils can be viewed as much more than mere curios in a cabinet of paleontology, but more as media that bring back their makers in our imaginations, living and breathing.

Additionally, ichnologists often divide themselves into neoichnologists, paleoichnologists, invertebrate ichnologists, vertebrate ichnologists, marine invertebrate ichnologists, and even deep-sea ichnologists. Sadly, there is still no one who self-identifies simply as a plant ichnologist, although this may change some day (Chapter 4). Unfortunately, this sort of specialization and self-identification contributes to communication breakdowns and the maintenance of separate realms among a relatively small group of scientists. However, one step taken to alleviate this division is a meeting, held once every four years (sort of the Olympics of ichnology), in which all ichnologists are invited to participate and exchange information about their subdisciplines. The first of these meetings, the International Ichnological Congress—also called *Ichnia*—was held in Trelew, Argentina, in 2004 and the second in Krakow, Poland, in 2008. The second meeting resulted in two volumes of peer-reviewed papers published in separate Polish journals, with one dealing with marine ichnology (Uchman and Rindsberg 2010) and the other with continental (nonmarine) ichnology (Pienkowski et al. 2009). These gatherings and their published findings were great successes in pulling together ichnologists from all six of the inhabited continents and of all ichnological persuasions, thus better approaching the ideal of working together under the banner of "One Ichnology."

WHAT'S IN A NAME? ICHNOTAXONOMY AND ITS USES

Oddly enough, if a trace became a part of the geologic record, when these trace fossils are seen by human eyes thousands or millions of years later, a few people care enough about them to assign official scientific names to them. This methodology is called ichnotaxonomy. One might think that the tracemaker itself is what should determine the name of a trace fossil: in other words, a track made by a theropod dinosaur could just be identified as a theropod dinosaur track, and leave it at that. Yet names are applied to theropod tracks perceived as distinctive from one another. For example, *Eubrontes*, *Grallator*, *Carmelopodus*, and *Megalosauripus* are ichnogenera given to theropod tracks, which through their different names implies a splitting of them into distinctive categories, even if some of these clearly have affinities to living theropods, namely birds (Figure 10.3). An alert reader will have noticed ichnogenera names sprinkled throughout the preceding chapters, in which modern traces were compared to trace fossils, as in, "This callianassid shrimp burrow is nearly identical in form to the ichnogenus *Ophiomorpha*" (Chapter 6). Although a few readers may have already known how and why paleoichnologists identify trace fossils with formal names, I apologize for any opacity, and I will now divulge a much-owed explanation for the basis of these names to those who gamely struggled along until now.

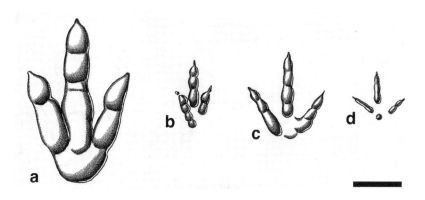

Figure 10.3. Ichnogenera applied to theropod dinosaur tracks, and a modern sandhill crane (*Grus canadensis*) track from St. Catherines Island (Georgia), drawn to the same scale for easier comparison. Which ichnogenus most closely resembles that of a large bird? (a) *Eubrontes*, from the Late Triassic to Early Jurassic Periods, about 200 million years ago. (b) *Grallator*, also from the Late Triassic to Early Jurassic Periods. (c) *Carmelopodus*, from the Middle Jurassic Period, about 165 million years ago. (d) Modern sandhill crane. All tracks drawn as natural casts (positive relief) and depicting the right pes. Scale = 10 cm.

Ideally, the naming of trace fossils as ichnogenera and ichnospecies is meant to serve a dual purpose: (1) succinctly communicate scientific information about the trace fossil, while also (2) acknowledging the scientific reality that good descriptions can last forever, whereas interpretations are always subject to change. In conveying information, a name should reflect the form of the trace itself, rather than any ascriptions conferred on it. In other words, the ichnogenus name *Celliforma*, for those who are familiar with this trace fossil, should immediately trigger an image of a small-diameter, cylindrical fossil burrow that expands into a smooth-walled and vase-like cell at one end, similar to the brooding cells of bees (Chapter 5). Unfortunately, such a sensible practice has not always been followed, and a few decidedly nondescriptive names have crept in from various sources— usually using non-Greek or -Latin roots—as well as through well-meaning tributes to people who are in some way associated with these trace fossils. For example, I actually was part of such a conspiracy, in which my coauthors and I named a beautifully intricate and multifaceted Eocene bivalve trace fossil *Hillichnus lobosensis* (Figure 10.4). In this study, we named the ichnogenus after Robert Hill, the geologist who first recognized the trace

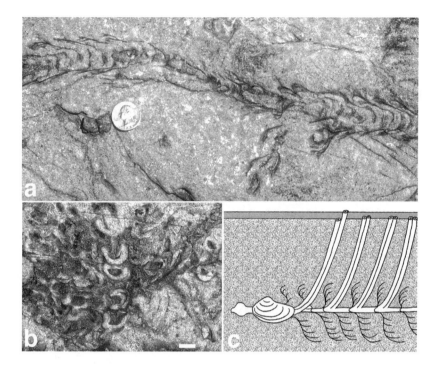

Figure 10.4. The beautiful and multifaceted trace fossil *Hillichnus lobosensis* from the Paleocene Epoch (about 55 million years ago) of California, interpreted as the combined locomotion and feeding trace of an infaunal bivalve. (a) Lower part of the trace fossil, made by the bivalve foot movement (central part of the trace) and its feeding on nearby sediments (lateral marks). (b) Upper part of the trace fossil, with paired circular rings caused by its siphons intersecting the sediment surface on the original seafloor. Scale = 10 cm (.4in). (c) Interpretive sketch showing how these very different-looking traces are actually connected to one another and made by the same tracemaker. Based on figures by Bromley et al. (2003).

fossil and its type locality in Point Lobos State Park, California, which then inspired the ichnospecies name (Bromley et al. 2003).

Nonetheless, the naming of ichnogenera and ichnospecies is helped considerably by neoichnology, which through modern examples can lend insights to paleoichnologists on the complete geometry and architecture of comparable trace fossils. Do these names necessarily reflect the identity of the tracemaker? The answer is "sometimes," especially with vertebrate ichnologists who were absolutely certain about what animal made a specific type of track. For instance, the previously mentioned fossil theropod track *Megalosauripus* is named after the theropod dinosaur *Megalosaurus*. This was done despite the poor definition of this dinosaur genus (Lockley et al. 1998). This sort of hubris, however, is easily deflated by facts, which, as we learned previously, can be pesky things. Along those lines, I have seen fossil vertebrate trackways, made by the same individual animal, that contain at least three ichnogenera along the course of the trackway. Why the need for different names? Remember the holy trinity of ichnology: substrate, anatomy, and behavior (amen!; Chapter 1), and ask yourself what changed for the tracemaker while it was walking, as well as what stayed the same, then the explanation for this nominal multiplicity is simple. Rarely does an animal's anatomy alter radically in midstep (Monty Python scenarios aside), so an ichnologist can assume this factor stays constant. Thus variations in the tracemaker's behavior and its affected substrate must be at play, perhaps in combination, such as when an animal slowed down once it stepped into a soft mud, and the digits of the foot splayed in response to this unstable ground. Such factors could have caused each track in a trackway to look different enough from one another that, if viewed in isolation and not as part of a trackway, their receipt of separate ichnotaxonomic designations becomes more understandable.

Another common source of confusion about ichnogenera and ichnospecies is that most of the nonichnological world (or, rather, most of the world) associates italicized words with actual organisms, especially those words using an uppercase name followed by a lowercase name, such as *Homo sapiens*. Hence when an ichnologist says that she or he has specimens of *Thalassinoides suevicus* and *Ophiomorpha nodosa* in a sedimentary deposit, a well-meaning science writer may then report how the rock contains "*Thalassinoides suevicus* and *Ophiomorpha nodosa* animals," which is utterly wrong. Even a subsequent explanation of how the trace fossils are not actual bodily remains of an animal may fall onto deaf ears, as the mesmerizing effects of italics lead people astray. Moreover, the modern analogues of these trace fossil genera are not trace fossils (yet); hence a geologist also would be amiss in stating that a recently made callianassid shrimp burrow on a Georgia beach is *Ophiomorpha nodosa*. Instead, this ichnologist should say the traces are *Ophiomorpha*-like (however awkward that might sound) or are incipient *Ophiomorpha*, recognizing the full future potential of a trace that seemingly aspires for fossilization.

Probably the best instance of where modern traces helped to better understand a trace fossil is provided by the well known, easily recognized, and just mentioned trace fossil *Ophiomorpha nodosa*. S. A. B. Lundgren (1891) originally named this ichnogenus, which was based on a representative type specimen of the trace fossil, otherwise known as a holotype. After his work, though, many paleontologists assigned their own names to what was, morphologically speaking, the same trace fossil. Such duplications are termed synonymies, and although paleontologists may have really enjoyed naming new ichnogenera and ichnospecies in the past, synonymies are frowned upon nowadays because they tend to muck up an already difficult naming system. Synonymies often provoke paleontologists to use the rule of priority, in which the oldest name for the same taxon is considered as the valid one. For example, the rule of priority is why the sauropod dinosaur *"Brontosaurus"* is now known as *Apatosaurus*, because *Apatosaurus* is the older (and hence more valid) name for the same dinosaur (Gould 1992).

Anyway, *Ophiomorpha nodosa* is a descriptive name, albeit with a twist of imagination, using the Greek roots, *ophio*, "snake"; *morpha*, "form," and *nodosa*, "[having] nodes [bumps]." The ichnogenus name reflects the often sinuous, tubular geometry of this trace fossil, even if it neglects its common branching habit. The ichnospecies name also acknowledges the nodes composing its outer wall (Figure 10.5). *Ophiomorpha nodosa* is a common trace fossil, and has been reported worldwide in rocks from the Permian Period through the Pleistocene Epoch, covering about 250 million years (Frey et al. 1978; Netto et al. 2007). However, nearly 75 years passed after Lundgren's (1891) description before sedimentologists noted that the eroded, lower parts of modern callianassid shrimp burrows they saw on Georgia coast beaches were identical to nearby Pleistocene *Ophiomorpha nodosa* in outcrops of the Georgia barrier islands (Weimer and Hoyt 1964; Howard and Scott 1983). Hence the incipient *Ophiomorpha* of modern environments were linked directly to their fossil counterparts, a relationship strengthened all the more by Tobler's Law. Since this recognition of Georgia callianassid burrows, similar modern burrow systems made by related shrimp in different parts of the world have also been recognized, lending to a more uniform acceptance that such burrows are consistent in their form and tracemakers. As a result, neoichnology enabled us to reasonably hypothesize animals that had anatomies and behaviors very much like

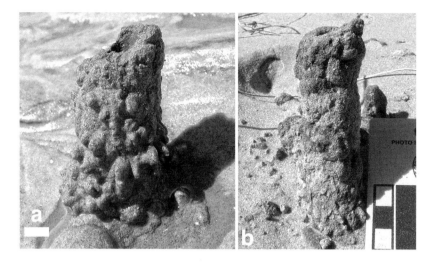

Figure 10.5. Comparison of callianassid shrimp burrow to the ichnospecies *Ophiomorpha nodosa* on the Georgia coast. (a) Modern burrow of the Carolina ghost shrimp (*Callichirus major*), in which lower part of the vertical burrow shaft was exposed by shoreline erosion: Sapelo Island. Scale = 1 cm (0.4 in). (b) Specimen of *Ophiomorpha nodosa*, exposed through erosion of an outcrop of the Raccoon Bluff Formation (Pleistocene): Sapelo Island. Scale in centimeters.

modern callianassid shrimp must have made *Ophiomorpha nodosa* (Frey et al. 1978, 1984; Goldring et al. 2007).

A bit more exploration of *Ophiomorpha*, ichnotaxonomically speaking, is further demonstrated by looking at its ichnospecies other than *O. nodosa*. These are considered morphological variants of the ichnogenus, and their distinctions are based on the character of the pelletal wall. For example, *O. annulata* is the ichnospecies name assigned to those *Ophiomorpha* with nodes arranged in definite layers (annulations), whereas if the nodes are more pointed, then it is given the name *O. rudis*. Does this seem like unnecessary splitting? In my opinion, yes, but in the opinions of other ichnologists, no, especially for those who even find reasons for naming varieties of ichnospecies as subichnospecies (no, I am not making that up). Hence paleoichnologists, just like other paleontologists, zoologists, and botanists, often find themselves dividing into two groups based on their taxonomic philosophy, lumpers and splitters. (Which I suppose makes them all splitters, unless they all agree one day, which would then make them all lump-

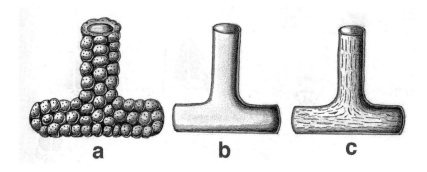

Figure 10.6. Comparison in form and preservation between three ichnogenera that may have been made by the same species of (or individual) decapod tracemaker: (a) *Ophiomorpha*; (b) *Thalassinoides*; (c) *Spongeliomorpha*. Note how the interior diameter is identical for each burrow. Based on a figure by Ekdale et al. (1984).

ers.) Out of sheer practicality, though, when in the field, most ichnologists and other paleontologists are lumpers, using ichnogenera when naming trace fossils. In other words, they let the experts later quibble over whether, say, a trace fossil resembling a single sine wave, made by a fish swimming along a sedimentary surface (Chapter 7), is *Undichna britannica, U. unisulca, U. simplicitas*, or at least 10 other ichnospecies under the ichnogenus *Undichna* (Minter and Braddy 2006). In fact, my saying that the trace fossil consisted of a single sine wave just gave reason for the five people in the world who know better to inform me, with smug satisfaction, that this trace fossil cannot possibly be *U. britannica, U. simplicitas*, or most other ichnospecies of *Undichna*, but must be *U. unisulca*. Thus proving my point.

Nonetheless, the use of ichnogenera names is not always straightforward and can result in complications. These problems ensue when the same tracemaker made similar trace fossils, such as the previously mentioned fossil trackway or burrow system, but vary morphologically along their parts owing to behavioral, substrate, or preservational factors. For example, take a typical *Ophiomorpha*-like burrow from the geologic record (Figure 10.6a), fill its tunnels and shafts with sediment to form an internal mold of the burrow system, then strip away the outer wall as it preserves in the geologic record. Viola! You now have the ichnogenus *Thalassinoides*, which retains the overall geometry of what could have been (and should have been) *Ophiomorpha* (Figure 10.6b). *Thalassinoides* (*Thalassi*, "seagrass," and *noides*, "like") is another commonly identified trace fossil, but has an even broader geologic range than *Ophiomorpha*. This ichnogenus shows up in rocks from the earliest part of the Ordovician Period, nearly 450 million years ago, and persists up through the Holocene, mostly in strata formed originally in marine environments (Goldring et al. 2007), but occurs in continental facies too (Martin et al. 2008). The most common ichnospecies of *Thalassinoides* is *T. suevicus*, although the justification for this ichnospecies is badly defined and may explain why the ichnogenus name is applied more often.

What then can be terribly confusing to a budding paleoichnologist, now having nearly graduated from neoichnological schooling on the Georgia coast, is discerning such intergradations between ichnogenera in the same fossil burrow system. This preservation-dependent situation means that you can follow part of a specimen of *Ophiomorpha nodosa* with your eyes and see it shift over to *Thalassinoides*, and vice versa. So just to be consistent with the purpose of ichnotaxonomy—that is, communicating to others the forms of trace fossils through a name—both names need to be applied to the burrow system. Nonetheless, a paleoichnologist also should helpfully note that these are part of the same burrow system, and hence were likely made by the same species of tracemaker. This lack of correlation between the number of ichnotaxa and tracemakers is an important concept in ichnology, in that ichnodiversity does not necessarily translate into biodiversity. In other words, a long and impressive list of ichnogenera and ichnospecies might very well represent the lifeworks of only one species of tracemaker (Rindsberg and Martin 2003).

Perhaps then, with the help of the preceding explanation, you thought ichnotaxonomy was becoming more approachable and understandable. Alas, this hope is quickly dashed, like a bivalve dropped on a sand flat by a hungry seagull (Chapters 1, 9). Instead, it gets more convoluted. Let us say, for example, that the just-discussed *Ophiomorpha–Thalassinoides* complex has part of its internal sediment fill exposed, and you note that this fill bears short, narrow, longitudinally arranged ridges. Now the ichnogenus is no longer *Thalassinoides*, but switches to another one, *Spongeliomorpha* (Figure 10.6c). You see, the ridges correspond to scratch marks made on the wall of the burrow—now preserved in positive relief because of the sediment fill—and the presence of these scratch marks indicates a possible firmground, which has paleoecological significance. But why not just say, it is *Thalassinoides* with a few scratch marks in places, and leave it at that? Or, even simpler, call it an *Ophiomorpha* in which the interior was filled with sediment, and note that parts of the *Ophiomorpha* lack the wall, and its fill has scratch marks here and there? Will we have communicated more or less scientific information by reducing the number of names in such an exercise?

By now, the gentle reader is probably wondering whether this discussion will naturally pass into a treatment of the traces left by angels dancing on the head of a pin; the sexes of the angels; their ages (and sizes); whether they were fallen angels, still serving in heaven, or somewhere in between; the types of dances used, from solo ballet performances, to square dancing, to the much-dreaded "Electric Slide"; and what ichnogenera and ichnospecies names we should assign to all of the preceding categories. Not to worry, it stops here. Suffice it to say that ichnotaxonomy, like any science, has a lot of problems to solve yet, but has progressed to a state where it mostly works. In fact, an International Ichnotaxonomy Workshop, composed of a small group of ichnologists keenly interested in all facets of how trace fossils are named, meets once every four years to contemplate such matters. These ichnologists usually gather in a place where museum or university collections contain many trace fossil specimens over which they can argue. Most importantly, though, this place also must have plenty of good beer, which is necessary if they want to accomplish anything meaningful. For example, in the first of these meetings, the participants spent several days going through instance after instance of purported categories of traces, and asked the seemingly simple question, "Is this a trace?" Ah, this is too easy, you say: I want to go to that meeting, where all I have to do is drink beer and argue! OK then, it's test time. Answer whether the following items are traces or not traces, and give your reasons why: insect cocoons, eggs, skin impressions in tracks, feather impressions in resting traces, healed bones, and cavities in teeth (to name a few). Even root traces still provoke spirited discussion among people who insist that plants do not behave like animals, thus we are terribly amiss to say our vegetative friends made trace fossils (Chapter 4). In other words, assumptions underpinning basic premises—this is a trace, this is not a trace—must be examined first before considering whether ichnotaxonomy needs to be applied in the first place. This is how science has been and should be done, and ichnologists are still upholding this tradition, fueled by passionate curiosity and adult beverages.

Also, beware of composite trace fossils, in which more than one species of tracemaker contributed to parts of what is seemingly a single trace fossil. These are instead Frankenstein's monsters; or, more properly, chimeras, because of their species-crossing portions. Composite trace fossils need each of their disjunctive aspects to be recognized as the works of their respective tracemakers, each of which warrant separate ichnogenera (Martin 2006b; Martin and Varricchio 2008). These will be explored further once we examine assemblages of traces and how organisms from completely different environments, but living in the same place at different times, can add traces upon traces.

One last lesson in the potential usefulness of ichnotaxonomy is provided by the ichnogenus *Oichnus* (Figure 10.7). This ichnogenus, like *Ophiomorpha*, is an easily recognized and common trace fossil, but is simple enough in its form that children can be taught to spot it instantly. Its modern equivalents cannot be missed on any Georgia beach, or most beaches worldwide, as these are the perfectly circular, beveled holes bored into bivalve and gastropod shells by predatory marine gastropods, such as naticids and muricids (Chapter 6). Immediately after this ichnogenus was named, it was applied to trace fossils fitting its prescribed form and occurrence in a molluscan shell, and the interpretation—predation trace made by a marine gastropod—became almost implicit to its identification. Two developments then followed. First, a few ichnologists realized that some octopi were capable of making a similar trace in bivalve shells, with the only major difference being relative size and a slight difference in shape: the octopus traces were often much smaller than the gastropod ones, and more cylindrical (Bromley 1993). Second, other ichnologists started noticing *Oichnus*-like borings in freshwater bivalves, which provoked the question, "What did this?" (Lawfield and Pickerell 2006). As of this writing, we are still not sure about the answer to that question, a good example of why more neoichnology is needed in freshwater environments (Chapters 5, 7). Yet with these interpretive exceptions noted, the ichnogenus stayed the same, beautifully and simply expressive of its form, and recognizable by novice and expert alike.

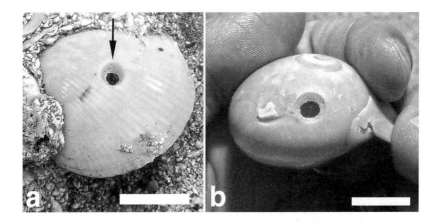

Figure 10.7. The ichnogenus *Oichnus* and its comparison to a modern boring made by the common moon snail (*Neverita duplicata*). (a) *Oichnus* in an unidentified Pleistocene bivalve (about 125,000 years old), San Salvador, Bahamas. (b) Predation boring made by common moon snail, and preserved in the shell of another moon snail, thus doubling as a trace of cannibalism: Sapelo Island. Scale = 1 cm (.4in).

ICHNOFACIES, ICHNOFABRIC, AND THE SHIFTING PARADIGMS OF SCIENCE

In the 1950s, the widely acknowledged founder of modern ichnology, Dolf Seilacher, came up with behavioral categories for trace fossils, most of which still are used today as basic ways to classify and interpret these (Seilacher 1953). In the 1960s, he took this idea one step further by grouping trace fossils with similar behaviors and rock types, which he then posited were directly related to specific environments and their physical conditions (Seilacher 1967). These trace fossil assemblages were (and still are) called ichnofacies (*ichnos*, "trace"; *facies*, "face"; Chapter 3). Probably more than any single concept in ichnology, the ichnofacies paradigm was the most significant advance in the use of trace fossils for interpreting ancient environments. For this and other accomplishments, ichnologists revere Seilacher's cognitive traces, which have lasted for more than 50 years and will likely outlive many of us.

Ichnofacies were organized mostly around distinctive suites of trace fossils that represented organisms behaving in the context of their host environments. However, they were named after a single ichnogenus, such as *Skolithos, Cruziana,* or *Psilonichnus,* that represent their collective quali-

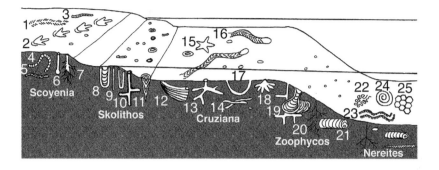

Figure 10.8. The ichnofacies concept, as originally defined by Seilacher (1967) on the basis of recurring ichnoassemblages that correspond with softground substrates and environments, from continental to deep-marine environments (*left to right*). Compare these to ichnoassemblages of the Georgia coast discussed in Chapters 4–9. Notice the gap between the *Scoyenia* and *Skolithos* ichnofacies, just above the high tide mark, which was later filled by the *Psilonichnus* ichnofacies. Key: 1 = *Hexapodichnus*; 2 = theropod dinosaur tracks; 3 = *Ancorichnus*; 4 = *Taenidium*; 5 = *Scoyenia*; 6 = "*Skolithos*,"; 7 = root traces; 8 = *Diplocraterion*; 9 = *Skolithos*; 10 = *Ophiomorpha*; 11 = *Rossellia*; 12 = *Teichichnus*; 13 = *Thalassinoides*; 14 = *Planolites*; 15 = *Asteriacites*; 16 = *Cruziana*; 17 = *Arenicolites*; 18 = *Asterosoma*; 19 = *Zoophycos*; 20 = *Chondrites*; 21 = *Scolicia*; 22 = *Cosmoraphe*; 23 = *Nereites*; 24 = *Spiroraphe*; 25 = *Paleodictyon*.

ties, a sort of avatar that speaks for every trace in its assemblage. Perhaps confusingly, this nominal ichnogenus does not need to be present in every occurrence of its ichnofacies, but the ichnofacies must have representatives of its signature assemblage. For example, the *Skolithos* ichnofacies was characterized as an assemblage of mostly vertically oriented burrows, such as *Skolithos* and *Arenicolites* (explained later), formed in shifting, soft, sandy substrates of shallow intertidal to shallow-marine environments. With that said, the ichnofacies could be dominated by *Arenicolites* with only rare or no specimens of *Skolithos*, just as long as it reflects the environmental parameters of that ichnofacies and the behaviors of tracemakers in the context of those parameters.

As originally conceived, ichnofacies were also arranged in an order that more or less reflected relative water depth, and from the perspective of water depth in marine environments. For instance, the order of ichnofacies from shallowest to deepest marine environments was originally *Skolithos* → *Cruziana* → *Zoophycos* → *Nereites* (Seilacher 1967; Figure 10.8). These were also so-called softground ichnofacies, comprised primarily of burrows and surface trails made in unconsolidated sediments. The only substrate-dependent ichnofacies named originally was the *Glossifungites* ichnofacies, which recognized that some trace fossil assemblages were made in firmgrounds that were well on their way to becoming rock (Chapter 3), but this was later followed by hardground ichnofacies (*Trypanites:* Frey and Seilacher 1980) and woodground ichnofacies (*Teredolites:* Bromley et al. 1984), explained later. The only nonmarine ichnofacies in this scheme was the *Scoyenia* ichnofacies, a sad oversight that fortunately was later rectified. Nevertheless, it represented the bias that then favored marine ichnology as the preset approach for the study of all trace fossils. This prejudice is even reflected in the term "nonmarine ichnofacies" used by some ichnologists, when a marine ichnofacies might just as well be "noncontinental." Seilacher further noted that ichnoassemblages and their host rocks, in order to qualify as ichnofacies, must be laterally widespread—in other words, they could be followed continuously for a ways—and recur in the geologic record. For example, a one-time occurrence of a few trace fossils in a 10 cm (4 in) thick and 2 m (6.6 ft) wide stratum is insufficient to warrant ichnofacies status. Or, as Seilacher famously once said, delivered with his characteristic Teutonic bombastic aplomb, "A cavity in my tooth is not an ichnofacies!"

Several decades passed after Seilacher's ichnofacies concept was proposed, and this model was tested repeatedly until it became the default paradigm for describing ichnoassemblages in the context of their environments. The ichnofacies model was also modified, tweaked, and expanded to include a few other substrate-dependent ichnofacies, such as the *Trypanites* and *Teredolites* ichnofacies, which were named after trace fossil assemblages formed in marine hardgrounds and woodgrounds, respectively (Frey and Seilacher 1980; Bromley et al. 1984). Yet other ichnofacies were proposed for transitional zones (ecotones of a sort) between other established ichnofacies. For example, the *Psilonichnus* ichnofacies was named for trace fossils formed in beaches and marine-coastal dunes, in between the continental *Scoyenia* ichnofacies and marine *Skolithos* ichnofacies. Relevantly, the *Psilonichnus* ichnofacies was based largely on examples from the Georgia coast (Frey and Pemberton 1987).

This testing of ichnofacies and their utility did not just take place in the esoteric vacuum of academe, though. During the 1960s through the 1980s, energy companies were exploring for petroleum and natural gas by using ichnofacies as one of their interpretive tools. This ichnological concept helped exploration geologists decide whether potential source rocks of hydrocarbons represented environments adjacent to or underneath rocks storing those hydrocarbons (reservoirs), each of which were formed in different sedimentary environments. Once depositional environments became easier to define through trace fossils—whether observed in surface outcrops or subsurface cores drilled by these companies—petroleum geologists, government geologists, and academics alike applied and tested the ichnofacies concept more often (Pemberton 1992). Once properly described and interpreted, ichnoassemblages thus became consistent, reliable, and predictable in their reflecting the original environments of the tracemakers, such as shallow-marine sandy coasts, muddy tidal flats, fluvial floodplains, and so on. The labeling of *Scoyenia, Skolithos, Cruziana, Zoophycos,* and other ichnofacies from the geologic record became an orthodox practice not just for ichnologists, but also for geologists in general. In fact, oil company geologists began attending field trips on the Georgia coast led by Robert Frey and James Howard (Chapter 2) to learn more about ichnofacies and their actualistic framework. (As an amusing aside, I recall stories from Frey about how some of these desk-bound petroleum geologists hesitated or balked completely at following him when he walked straight into a salt marsh and his feet promptly sank into the gooey, black, sulfurous-smelling mud.) Incidentally, Frey, Howard, George Pemberton, and other ichnologists published papers about their modern traces and trace fossils in peer-reviewed journals with the names *American Association of Petroleum Geologists Bulletin* and *Canadian Journal of Petroleum Geology.* Moreover, ichnological studies are still funded by the Petroleum Research Fund through the American Chemical Society, in recognition of how neoichnology and paleoichnology frequently connect directly with petroleum geology.

As often happens in science, though, things change. Although the ichnofacies approach worked well enough for several decades, an attempt was made in the mid-1980s through the 1990s to refine this methodology by studying groups of trace fossils and quantifying the amounts of biological alteration of substrates. This approach was termed ichnofabric analysis, whose proponents noted that traces might change the fabric of a substrate sufficiently to impart a distinctive overall appearance and other properties (Bromley and Ekdale 1986; Droser and Bottjer 1986). At first, this approach applied only to softground trace fossils and their modern equivalents, as the original definition only referred to bioturbation, but later included firmground and hardground ichnoassemblages too (Lewis and Ekdale 1992; Savrda et al. 1993).

As a result of this innovation, a schism ensued in the small, normally close-hewn ichnological community during the 1990s that continued into the early part of the twenty-first century. This mild conflict, resulting mostly in the utterance of unflattering adjectives, occasional scorn, and quite rare loathing, took place between traditional adherents to the ichnofacies paradigm and the new acolytes of ichnofabric analysis. Spirited arguments, both in print and delivered in presentations at meetings, became the academic equivalent of "tastes great, less filling" dichotomies that begged for some sort of reconciliation. In that respect, a few ichnologists wrote and published articles that attempted to find a middle ground by pointing out the strengths and flaws of both systems (Taylor and Goldring 1993; Taylor et al. 2003; McIlroy 2008). Otherwise, ichnologists found themselves falling into one of the two camps. Fortunately, this rift has mostly healed by the time of this writing, which was partially facilitated by International Ichnofabric Workshops, held biannually in a differing host country containing great trace fossils and large quantities of beer. These workshops consist of weeklong meetings and field trips composed of about 30–50 ichnologists, in which many misunderstandings were explored and sorted out. As a result, a sort of scientific détente in the relatively small ichnological community has become the norm, in which ichnologists have more or less realized that each approach has its own time and place, or can be combined to still yield satisfactory results.

Which side am I on, you ask? Both and neither. I am more of a traditionalist who likes the intuitiveness of the ichnofacies approach, but who also recognizes its shortcomings. I also acknowledged some of the advantages of ichnofabric analysis, particularly its emphasis on quantification. In other words, I am a pragmatist. Additionally, I will not look at a crossbedded sandstone containing a few specimens of *Skolithos* and label it as a "*Skolithos* ichnofabric" or "*Skolithos* ichnofacies," but will simply call it a cross-bedded sandstone with a few specimens of *Skolithos*. In my experience, slavish adherence to a paradigm tends to put on ideological blinders and hampers perceptions, particularly while attempting to describe modern traces and trace fossils in the field, where all ichnological hypotheses are subject to instant testing and revision.

Along these lines, I will never forget the response of one of my professors in graduate school, Dr. Wayne Martin (no relation), to a question posed in class. When asked by one of my fellow students why some field-based evidence cited by Dr. Martin in class contradicted what was in our textbook, he replied, with a slow, West Virginia drawl, "What the rock record and the literature say are often two different things." I hold this statement close to my heart whenever I am in the field, whether studying modern or fossil traces, and whenever my observations run counter to what has been previously published, I think of Dr. Martin, smile, and start describing what is there in front of me, literature be damned.

Consequently, the following coverage of Georgia coast ichnocoenoses and their comparison to ichnofacies is made with the full acknowledgement that these, like all things in science, are subject to critique and change. For example, if someone else is willing to revise the following coverage from an ichnofabric perspective and can explain it so that future generations of readers (many of whom are not specialists) will be engaged and enlivened by such an approach, be my guest, have fun, and invite me to have a beer (or two) with you so it can be discussed and debated properly. In the meantime, the following treatment should provide readers with a few insights on how direct comparisons can be made between modern ichnocoenoses of the Georgia barrier islands and ichnoassemblages preserved in the geologic record. In many instances, these modern ichnocoenoses are so strikingly similar to their fossilized counterparts that they inspire us to neatly toss aside any disagreements about what terms we may use to describe them. After all, once all is said and done, the organisms do not care what we call them or their traces.

TRACE FOSSILS OF TERRESTRIAL AND FRESHWATER FACIES

Ichnocoenoses of terrestrial and freshwater environments, along with their substrates, reflect three ichnofacies: (1) *Coprinisphaera;* (2) *Mermia;* and (3) *Scoyenia* (Figure 10.9; Table 10.1). Of these, the *Scoyenia* ichnofacies is one of the originals named by Seilacher, whereas the others were proposed much later: the *Mermia* ichnofacies in the mid-1990s (Buatois and Mángano 1995) and the *Coprinisphaera* ichnofacies five years later (Genise et al. 2000). The *Coprinisphaera* ichnofacies, named after a trace fossil interpreted as a dung beetle brooding chamber, is applied to ichnoassem-

TABLE 10.1. TERRESTRIAL AND FRESHWATER ICHNOFACIES AND REPRESENTATIVE ICHNOGENERA APPLICABLE TO THE GEORGIA BARRIER ISLANDS

Ichnogenus	Description	Interpretation
A. *Coprinisphaera* Ichnofacies		
Attaichnus	Vertical multichambered burrow	Ant nest
Cellicalichnus	Vertical burrow, connected cells	Compound bee nests
Celliforma	Cylindrical burrow with cell	Bee brooding chamber
Chubutholithes	Burrow, multiple cells	Wasp brooding chambers
Coprinisphaera	Spherical burrow, walled	Beetle brooding chamber
Eatonichnus	Ovoid burrow, external spirals	Beetle brooding chamber
Ellipsoideichnus	Ovoid structure, spiraled exterior	Bee brooding chamber
Fontanai	Spherical burrow	Beetle brooding chamber
Monesichnus	Elongate burrow	Beetle brooding chamber
Pallichnus	Spherical burrow	Insect brooding chamber
Palmiraichnus	Burrow with cell	Bee brooding chamber
Parowanichnus	Interconnected galleries, chambers	Ant nest
Rebuffoichnus	Ovoid cell, external weave	Wasp cocoon
Rosellichnus	Cluster of closely spaced cells	Bee brooding chambers
Syntermesichnus	Complex, multichambered structure	Termite nest
Tacuruichnus	Complex, multichambered structure	Termite nest
Teisseirei	Horizontal burrow, rounded ends	Beetle brooding chamber
Termitichnus	Complex, multichambered structure	Termite nest
Uruguay	Cluster of cells	Bee brooding chambers
B. *Scoyenia* Ichnofacies		
Beaconites	Meniscate horizontal burrow	Arthropod burrow
Camborygma	Vertical to branching burrow	Crayfish burrow
Cochlichnus	Meandering trail	Worm, larval insect trail
Cylindrichnus	Vertical burrow	Insect or arachnid burrow
Hexapodichnus	Trackway with six impressions	Insect, scorpion trackway
Octopodichnus	Trackway with eight impressions	Spider trackway
Palaeophycus	Simple horizontal burrow, lining	Earthworm burrow
Planolites	Simple horizontal burrow, no lining	Insect burrow
Root traces	Branching, tapering structures	Various terrestrial plants
Scoyenia	Meniscate burrow, braided exterior	Larval beetle burrow
Skolithos	Simple vertical burrow	Spider or insect burrow
Taenidium	Meniscate horizontal burrow	Backfilled insect burrow
Umfolozia	Series of impressions	Arthropod trackway
Undichna	Sinusoidal trails (can be multiple)	Fish locomotion trace
Vertebrate tracks	2–4 alternating impressions	Tetrapod locomotion
C. *Mermia* Ichnofacies		
Circulichnus	Ring-like surface trail	Nematode, annelid trail
Cochlichnus	Sinusoidal surface trail	Nematode, annelid trail
Gordia	Overlapping looping trail	Nematode, annelid trail
Helminthoidichnites	Meandering to straight trail	Annelid, larval insect trail
Mermia	Complex looping trail	Nematode, annelid trail
Planolites	Horizontal unlined burrow	Nematode, annelid burrow
Treptichnus	Horizontal branched burrow	Insect larval burrow
Undichna	Sinusoidal trail or offset trails	Fish swimming trace
Vertebrate traces	Various forms	Fish nests, feeding traces, tracks

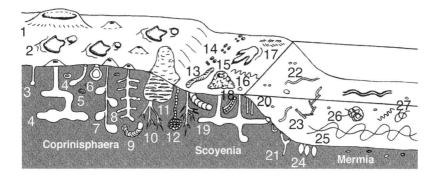

Figure 10.9. Continental ichnofacies, as defined by Frey et al. (1984), Buatois and Mángano (1995), and Genise et al. (2000), with *Coprinisphaera* (*left*), *Scoyenia* (*middle*), and *Mermia* ichnofacies (*right*), each with representative trace fossils. Key: 1 = dinosaur ground nest; 2 = ornithopod dinosaur tracks; 3 = "*Skolithos*"; 4 = *Celliforma*; 5 = *Rebuffoichnus*; 6 = *Coprinisphaera*; 7 = *Attaichnus*; 8 = *Cellicalichnus*; 9 = *Naktodemasis*; 10 = root traces; 11 = *Termitichnus*; 12 = *Castrichnus*; 13 = *Scoyenia*; 14 = hopping mammal tracks; 15 = theropod dinosaur track; 16 = *Cochlichnus*; 17 = *Hexapodichnus*; 18 = *Ancorichnus*; 19 = *Camborygma*/*Thalassinoides*; 20 = *Skolithos*; 21 = *Polykladichnus*; 22 = *Aulichnites*; 23 = *Treptichnus*; 24 = *Lockeia*; 25 = *Undichna*; 26 = *Mermia*; 27 = *Gordia*.

blages from terrestrial environments with the right ecological conditions to have permitted the formation of insect nests, such as ancient soils (paleosols). In contrast, the *Mermia* ichnofacies consists of traces made in low-energy (with minimal currents or waves) and submerged freshwater environments, such as lakes and river channels. Trace fossils include fish trails, aquatic insect burrows, nematode trails, and other traces made on these subaqueous bottoms. The namesake of the *Mermia* ichnofacies is a presumed grazing trail made by a legless invertebrate, such as a nematode worm (Buatois and Mángano 1995). The *Scoyenia* ichnofacies is more of an assemblage of traces made in between the other two ichnofacies. This ichnofacies normally consists of vertebrate tracks, invertebrate burrows, and root traces formed in moist sediments, either on the peripheries of aquatic environments, such as fluvial (river) floodplains and channels, or otherwise close to the water table (Frey et al. 1984; MacEachern et al. 2007a). Thus the traces of the *Scoyenia* ichnofacies are made in more emergent conditions than those of the *Mermia* ichnofacies, yet not quite in the high-and-dry state of the *Coprinisphaera* ichnofacies.

A few cautionary notes should be made, however, before discussing these ichnofacies any further, and how they relate to the Georgia coast ichnocoenoses. In my experience, talking about traces and trace fossils from terrestrial and freshwater environments is the intellectual equivalent of diving headfirst into a 30-cm-deep (12 in) pool of freshwater that turns out to be an alligator pond, and one where a hungry mother alligator is watching over her hatchlings.

So here goes. As mentioned before, when Seilacher originally devised the ichnofacies scheme in the 1960s—no, he is not the metaphorical alligator in this tale, nor is anyone else—he placed all trace fossil assemblages not formed in marine environments into one category, the *Scoyenia* ichnofacies. The namesake of this ichnofacies, *Scoyenia*, is a beautiful little burrow with an intricately braided exterior, looking very much like a rope, and with a meniscate interior. *Scoyenia* is likely a burrow made by beetle larvae (or similar arthropods) in firm but moist sediments of fluvial floodplain deposits, in which the ropy external texture was imparted by short, pointed legs, and the meniscate structures were formed by its active backfilling (Chapter 5). The problem with Seilacher's ichnofacies designation was its extreme lumping, as opposed to the easy splitting of marine-related ichnofacies into *Skolithos*, *Cruziana*, *Zoophycos*, *Nereites*, and *Trypanites*. As any terrestrial or freshwater ecologist can tell you, this is an ineffective one-size-fits-all solution for taking into account a huge number and variety of terrestrial and freshwater environments with plant, invertebrate, and vertebrate tracemakers. Accordingly, paleoichnologists should take heed of this biodiversity and its potential ichnodiversity.

In following Seilacher's lead, paleoichnologists soon began dividing ichnoassemblages into marine and nonmarine, with the latter dropped unceremoniously into the *Scoyenia* ichnofacies that, once named, ended most additional inquiry. Later investigations into continental trace fossils and modern traces revealed this simplistic solution to be greatly wanting, and the *Scoyenia* ichnofacies began to tear asunder under this new scrutiny, starting with Frey and others (1984), but continuing through the 1990s and up to the time of this writing. A small sampling of this explosion of subsequent research, cited pedantically (so you now have fair warning), includes works dealing with the paleoichnological significance of crayfish burrows; insect burrows, nests, and their interactions with plants; root traces; and vertebrate traces, such as tracks, burrows, nests, tooth marks, and coprolites

(Lockley 1991; Hasiotis et al. 1993, 2004; Hasiotis and Mitchell 1993; Lockley and Hunt 1995; Chin and Gill 1996; Jacobsen 1998; Lockley and Meyer 1999; Genise et al. 2000, 2004; Gregory and Campbell 2003; Chiappe et al. 2004; Gregory et al. 2004, 2006; Melchor et al. 2006; Chin 2007; Bedatou et al. 2008; Martin et al. 2008; Sarzetti et al. 2008; Smith et al. 2008; Varricchio et al. 2007; Martin 2009b; Jacobsen and Bromley 2009; Martin and Varricchio 2011). These studies and many others collectively resulted in an improved understanding of the multitudinous traces made in lakes, rivers, floodplains, forests, grasslands, and other areas well above sea level, as well as the behaviors of their tracemakers in those environments (Figure 10.10).

However, some of the best science behind terrestrial traces and tracemakers is not only fresh, but also somewhat raw, provoking some strong disagreements among ichnologists. Dinosaur trace fossils, which mostly occur in continental settings, also are the disproportionate focus of such studies. This bias certainly reflects a response by ichnologists to an area with high public interest (Martin 2006c), but can thus overshadow studies of trace fossils made by more ecologically important tracemaking organisms that lived alongside dinosaurs, such as insects (Genise et al. 2000, 2004, 2007; Chin 2007; Martin and Varricchio 2011). This uncomfortable situation will likely be alleviated as the science of continental ichnology matures and its practitioners get mellower, but will also benefit greatly from the inclusion of more studies of modern continental tracemakers, such as insects and terrestrial vertebrates other than dinosaurs (Chapters 5, 7, 8).

Another potentially confusing point about continental environments and their ichnofacies is that the *Skolithos* ichnofacies also has been used to describe ichnoassemblages in high-energy sandy river deposits (Buatois and Mángano 2007). In a few instances, these ichnoassemblages may have been associated with rivers, but were also affected by marine salinity, such as in the upper reaches of a riverine estuary (Martin 1992; Buatois et al. 2005). After all, marine organisms living in the denser, saline waters hugging a river bottom do not need to be told they are in a river; for all practical purposes, they are in a marine or brackish water environment. In other instances, though, ichnologists equated a fast-flowing, sandy river with the high-energy qualities of a shallow-marine sandy shelf battered daily by waves (Buatois and Mángano 2007). Both environments experience much erosion, deposition, and movement of sand and other sediments, thus giving tracemakers a reason to burrow vertically, whether to quickly make

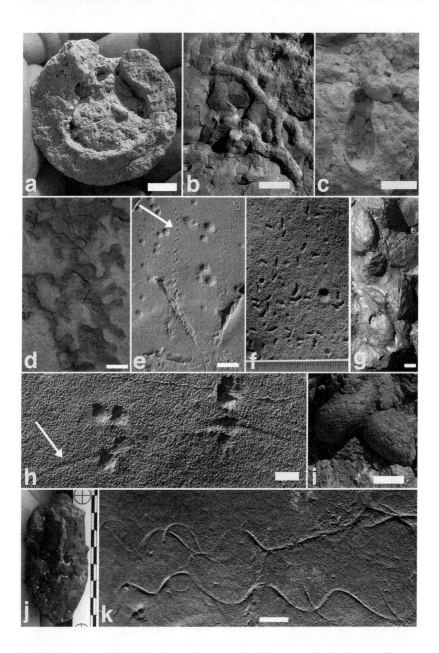

Figure 10.10. Trace fossil ichnogenera characteristic of continental environments that are typical of the *Coprinisphaera*, *Scoyenia*, or *Mermia* ichnofacies. (a) *Coprinisphaera*, with cross section showing brooding chamber, probably of a dung beetle, in a paleosol: Miocene, Argentina; (b) *Scoyenia*, a probable insect burrow in a paleosol: Late Devonian, Pennsylvania (USA); (c) *Celliforma*, a bee brooding chamber in a paleosol: Late Cretaceous, Argentina; (d) *Thalassinoides*, freshwater decapod (crayfish) burrow system in a fluvial deposit: Early Cretaceous, Australia; (e) *Cochlichnus* (*arrow*), a probable nematode trail, and an anisodactyl bird track (*bottom*) from a lakeshore:

domiciles (burrow down!) or to escape from a layer on sand dumped on top of their domiciles (burrow up!). More properly, though, the *Skolithos* ichnofacies was defined specifically for marine environments. Thus its retroactive application to continental environments would be similar to later applying a continental ichnofacies to a marine setting because they have a few paleoecological parameters in common, such as oxygen, organics, and traces reflecting feeding behaviors. In short, the *Skolithos* ichnofacies will be referred to here as marine, while also acknowledging that a few traits of that ichnofacies may apply to ichnoassemblages in continental environments.

With regard to the Georgia barrier islands, their limited areas of terrestrial and freshwater environments compose both a blessing and a curse, ichnologically speaking. The "blessing" part is related to the fantastic diversity of plant, invertebrate, and vertebrate tracemakers and their respective behaviors in those environments, most of which can be examined by simply walking around island interiors, and in relatively short jaunts. No research vessels, box cores, radiographs, or other expensive, complicated accouterments of offshore marine neoichnology are needed to conduct such studies. On the other hand, the "curse" part is that these snippets of information may not properly represent terrestrial and freshwater ichnocoenoses of barrier island deposits preserved in the geologic record. Barrier islands are, by definition, narrow strips of land separating the open ocean from the mainland, and are often bordered on their landward sides by lagoons or salt marshes (Pilkey and Fraser 2003). Moreover, barrier islands are ephemeral, shifting seaward, landward, down a coastline, or vanishing completely in the aftermath of high-intensity storms. What traces could get preserved from these slivers of land sandwiched between broader marine-related ichnofacies, especially when an entire softground ichnofacies (or more) can be wiped clean with one well-placed hurricane? Fortunately, modern analogues for

Eocene, Utah (USA); (f) *Hexapodichnus,* two intersecting insect trackways on a floodplain surface: Middle Jurassic, Argentina; (g) *Lockeia,* bivalve resting traces in a fluvial deposit preserved as natural casts: Late Devonian, Pennsylvania (USA); (h) *Ameghinichnus,* mammal tracks, preserved as natural casts, showing full bound pattern (*from right to left*) and tail drag mark (*arrow*) on a floodplain surface: Middle Jurassic, Argentina; (i) *Palmiraichnus,* two probable wasp cocoons in a paleosol: Late Cretaceous, Montana (USA); (j) Vertebrate coprolite, likely from a large archosaur: Late Triassic, Arizona (USA); (k) *Undichna,* here represented by two subparallel fish trails from an freshwater-dominated estuary: Late Carboniferous, Alabama (USA). Scale bars = 1 cm (0.4 in) in all except (d), which is 10 cm (4 in).

terrestrial and freshwater ichnofacies exist on the Georgia barrier islands and are worth describing for their possible future diagnosis in barrier island or stratigraphic deposits closely associated with ancient coastlines.

Coprinisphaera *Ichnofacies*

Of the continental ichnofacies, the *Coprinisphaera* ichnofacies is best exemplified by similar ichnocoenoses on the Georgia barrier islands, and has the best preservation potential (Figure 10.9). The well-drained upland soils of maritime forests, grasslands, and back-dune meadows are all conducive to the formation of insect nests, such as those of ants, termites, beetles, bees, and wasps, as well as abundant root traces and cicada nymph burrows. In contrast, the *Scoyenia* and *Mermia* ichnofacies are much less likely to form and preserve, mostly owing the relative paucity of freshwater environments on island interiors. Nonetheless, the *Scoyenia* ichnofacies in particular is well represented by seasonal alligator ponds, and the *Mermia* ichnofacies may be represented in larger freshwater environments that might form between relict dunes and above underlying strata that confine the groundwater, such as Lake Whitney on Cumberland Island.

Tracemaking organisms characteristic of the *Coprinisphaera* ichnofacies inhabit subaerial environments, which might be occasionally submerged by seasonal runoff from tropical storms, but otherwise are well above the local water table (Genise et al. 2000). As mentioned previously, insect nesting traces dominate this ichnofacies, particularly the complex traces of social insects (Chapter 5). Trace fossil ichnogenera include purported brooding structures from a variety of insects, such as beetles (*Coprinisphaera*, etc.), bees (*Celliforma*), wasps (*Chubutolithes*), ants (*Attaichnus*), and termites (*Termitichnus*), to name a few. Analogous terrestrial Georgia tracemakers and their respective traces overlap considerably in form and function to ichnogenera of the *Coprinisphaera* ichnofacies, consisting of beetle, bee, wasp, ant, and termite structures, but also include the burrows of cicada nymphs, ant lions, earthworms, reptiles, and mammals. As the neoichnology of Georgia coast environments has taught us, some vertebrate structures, such as gopher tortoise burrows, are also particularly important as microhabitats for other tracemaking species. Additionally, from the perspective of neoichnologists, perhaps the most noteworthy exception in the description of the *Coprinisphaera* ichnofacies is how it deals only with trace fossils in soils, but neglects traces above soil surfaces, such as those made by insects, birds, and mammals in the tissues of trees and other plants. Hence if the *Coprinisphaera* ichnofacies were some day expanded, it would likely also pull in these traces as part of the assemblage, an insight provided by observing traces from the Georgia maritime forests. Interestingly, many holometabolous and hemimetabolous insect tracemakers of the *Coprinisphaera* ichnofacies spend separate parts of their life cycles underground and above ground, resulting in a vertical distribution of their traces that is perhaps underappreciated by previous researchers.

Numerous ancient examples of the *Coprinisphaera* ichnofacies were compiled in its initial diagnosis (Genise et al. 2000), with none so far in Georgia, but many in South America. This geographical bias is not surprising, as the people who defined it are from South American countries. Some ichnologists, though, have regarded the limited geologic range of this ichnofacies as a hindrance in its application. Ground-nesting insects were apparently absent during the Paleozoic Era, and only a few (and controversial) social insect nests have been interpreted from Triassic and Jurassic rocks (Hasiotis and Dubiel 1995; Bordy et al. 2004; Hasiotis 2004; Genise et al. 2005; Bromley et al. 2007). Nevertheless, strata formed in terrestrial environments from the Cretaceous to the Pleistocene are common enough that this ichnofacies can be tested further with trace fossil assemblages worldwide. For example, at the time of this writing, a *Coprinisphaera* ichnofacies had been interpreted in Late Cretaceous rocks closely associated with dinosaur nest sites in Montana. This demonstrates the usefulness of ichnofacies as a paleoenvironmental indicator in vertebrate paleontology, even in the study of dinosaurs (Martin and Varricchio 2011).

Scoyenia *Ichnofacies*

In contrast to the *Coprinisphaera* ichnofacies, the *Scoyenia* ichnofacies has its traces rendered by plants and animals living in close association with the margins of freshwater environments, such as floodplains or river channels, which are absent from the Georgia barrier islands. Nonetheless, this ichnofacies also relates well to freshwater pond or lake margins, or wherever sediments can be both wetted and emergent, which, however limited, occur on the islands (Figure 10.9). Traces of the *Scoyenia* ichnofacies include, most obviously, terrestrial and aquatic vertebrate tracks, as well as invertebrate trackways and trails. These traces, however, may not preserve as well as the deeper burrows of invertebrates and vertebrates, thus the

Scoyenia ichnofacies is often identified on the basis of these, even though tracks and trails are certainly not ignored. Among the more common and diagnostic trace fossil ichnogenera are *Ancorichnus, Beaconites, Naktodemasis, Scoyenia,* and *Taenidium,* which all comprise a group of similar-looking backfilled meniscate burrows that were likely formed by insect larvae or nymphs, or other burrowing arthropods that imitated them (Frey et al. 1984; MacEachern et al. 2007a; Smith et al. 2008). Additional trace fossils are of arthropod trackways made by insects or arachnids (*Hexapodichnus, Octopodichnus*); horizontal oligochaete or nematode burrows or trails (*Planolites* and *Cochlichnus,* respectively); vertical and U-shaped burrows attributed to insects, arachnids, myriapods, crayfish, and similar burrowing arthropods (*Skolithos, Arenicolites, Cylindrichnus, Camborygma, Thalassinoides*); root traces; and a wide range of vertebrate tracks. This ichnofacies has been around as long as plants and animals have been in or near freshwater environments, probably since the Ordovician Period (MacEachern et al. 2007a). As a result, it can be applied to strata from the Ordovician Period and younger, covering well over 400 million years.

On the Georgia barrier islands, modern ichnocoenoses evoking the *Scoyenia* ichnofacies are severely restricted compared to those reflecting the *Coprinisphaera* ichnofacies, but nonetheless manifest on the edges of freshwater alligator ponds or broad swales between dunes. In that respect, alligator tracks and trackways, trails worn down by their overland movements, den structures, and alligator ponds are neoichnological gold mines, ripe for application to the study of similar traces made by large archosaurs in the geologic past. Additionally, the more surprising tracemakers associated with freshwater wetlands on the Georgia barrier islands, but perhaps those most diagnostic of such environments, are crayfish. Although their prominent surface towers are readily eroded, the deep vertical shafts and branching tunnels of their burrow systems frequently make it into the geologic record, with some excellent examples from the Mesozoic (Bedatou et al. 2008; Martin et al. 2008). Other burrowers include moisture-loving insects, such as beetles and mole crickets (Chapter 5), and amphibians (Chapter 7), although the neoichnology of these tracemakers in such habitats is scantily known. So far, the *Scoyenia* ichnofacies has not been identified in any Paleozoic or Mesozoic strata of Georgia, although it is partially present in Pleistocene formations cropping out on the islands and other parts of the coast, such as the previously described Yellow Banks Bluff and Sapelo

Island (Gregory et al. 2004; Martin and Rindsberg 2011). Moreover, its long geologic range, the widespread nature of such environments (think of the breadth of some river valleys and floodplains), the well-established history of the ichnofacies, and its common association with vertebrate track sites, ensured that many ichnologists have interpreted the *Scoyenia* ichnofacies in ancient strata well outside of Georgia (Frey et al. 1984; MacEachern et al. 2007a).

Mermia *Ichnofacies*

The *Mermia* ichnofacies, as mentioned earlier, is even more limited in area and significance on the Georgia coast compared to the preceding continental ichnofacies, owing to the general lack of long-lasting and low-energy bodies of freshwater. Nevertheless, larger and more permanent freshwater ponds, such as Lake Whitney on Cumberland Island, should in principle contain traces matching or otherwise resembling this ichnofacies (Figure 10.9). The *Mermia* ichnofacies, because it reflects traces made at the sediment–water interface of submerged freshwater environments, will necessarily include a large number of surface trails, trackways, and shallow burrows, although the latter may be absent if bottom waters get too low in oxygen. In fact, without oxygen, most tracemaking ceases, unless you are like the marine polychaete worm *Heteromastus filiformis* and can use your rear end like a snorkel (Chapter 6). Interestingly, though, some midge larvae can get their oxygen from photosynthetic algae in oxygen-poor sediments on lake bottoms (Gingras et al. 2007).

For the most part, trace fossils of the *Mermia* ichnofacies thus reflect surface trails or other marks made by locomotion or bottom feeding made by mobile worms or arthropods (*Mermia, Gordia, Cochlichnus*); traces left by fins of swimming fish (*Undichna*); shallow, horizontally oriented arthropod burrows (*Treptichnus, Circulichnus*); or bivalve resting traces (*Lockeia*). Shallow-water areas also may incorporate root traces of plants adapted for submerged conditions, as well as trackways made by amphibians, reptiles, or mammals formed on lake bottoms. Tracks, however, have low preservation potential because they are lightly impressed by buoyant animals, which do not sink their feet into bottom sediments unless they really need to.

Sadly for neoichnologists, the *Mermia* ichnofacies is terrible to study from the perspective of modern analogues in the coastal plain and other parts of Georgia because of its lack of natural lakes. Yes, almost all extant lakes in Georgia—the locations of many an inebriated boat ride—were formed artificially by the damming of formerly free-flowing streams or rivers. Nonetheless, former examples of the *Mermia* ichnofacies in the southeastern part of North America were likely much more common when beavers were widespread and abundant. As noted earlier (Chapters 2, 3, 8), these keystone species can make extensive ponded freshwater areas throughout North America, and their ancestors must have done this even more so in the absence of human predation. With regard to the rest of the world, Buatois and Mángano (1995) and others (Bromley et al. 2007) have supplied many occurrences of the *Mermia* ichnofacies from the geologic past, dating from the Carboniferous Period through the Pleistocene. Hence these many ancient examples of the ichnofacies can be used for comparison and testing to any in Georgia.

One last point with regard to terrestrial and freshwater ichnoassemblages of the Georgia barrier islands is how, in their placement so close to the sea, they will be necessarily influenced by ecological factors and tracemakers associated with marine environments. In essence, tracemakers from seemingly separate realms should readily cut across and blend their traces with one another, sometimes within hours. This situation is especially acute when taking into account the twice-daily effects of tides bringing marine salinities closer to land, or storm surges that dump huge amounts of salty water and marine sediments into maritime forests or freshwater ponds. Time and again, we have noted how tracemakers from the terrestrial environments encroach on marine environments (mole burrows in surf zones, anyone?) and vice versa. Hence this reality must be taken into account when describing and interpreting continental ichnofacies in the context of a barrier island setting. In other words, on a barrier island, no terrestrial barrier island ichnoassemblage is truly free from the sea.

TRACE FOSSILS OF MARGINAL-MARINE FACIES

The Georgia coast is probably most famous, ichnologically speaking, for its modern marginal-marine traces and the way trace fossils in the Georgia coastal plain can be related directly to these contemporary counterparts. For example, coastal plain ridges containing abundant *Ophiomorpha* help to identify every former barrier island of ancient Georgia coasts (Weimer and Hoyt 1964; Hoyt and Hails 1967), as well as estuarine marine environments of the Eocene Epoch, from about 35 million years ago (Martin 2009b). Also in Georgia, marine-related ichnofacies are readily identifiable in Late Cretaceous deposits (about 70–75 million years old) in the southwestern coastal plain of Georgia, as well as Paleozoic rocks in northwest Georgia. Outside of Georgia, marine ichnofacies from the Paleozoic through the Pleistocene are often compared to the modern traces of the Georgia coast, testing how well such models can be applied. In this treatment, the modern ichnocoenoses of the Georgia coast will be covered in approximate order from onshore to offshore, with the caveat that such seemingly facile divisions do not always represent reality. Additionally, examples of ancient ichnofacies that mimic some of the properties of modern Georgia ichnocoenoses are duly noted.

Psilonichnus *Ichnofacies*

The *Psilonichnus* ichnofacies is associated with ecotones between the land and sea, with tracemakers living in coastal dunes, back-dune meadows, storm-washover fans, and high salt marshes (Figure 10.11). This ichnofacies, defined by Frey and Pemberton (1987), was based primarily on Georgia coast examples, and its namesake (*Psilonichnus*) is the trace-fossil equivalent of the Y- or J-shaped vertical burrows often made by modern ocypodid crabs, such as ghost crabs (*Ocypode quadrata*), fiddler crabs (*Uca* spp.), or other crabs that can live out of the water for extended periods of time. Other traces of this ichnofacies reflect its mixture of marine-related and terrestrial tracemakers, as it includes insect and arachnid burrows, vertebrate tracks and burrows, and root traces of halophytes and terrestrial plants. For example, the previously discussed case study of Yellow Banks Bluff was of the *Psilonichnus* ichnofacies, where it was succeeded vertically by the *Scoyenia* ichnofacies.

So how is this ichnofacies different from the *Scoyenia* or *Coprinisphaera* ichnofacies? The main distinction revolves around crab tracemakers that need the nearby sea for their physiology and reproduction. However much we might think of ghost crabs and fiddler crabs as terrestrial animals, they need to wet their gills in saline water and to lay their eggs in the sea. Even terrestrial crabs that live in areas south of Georgia, such as the Bahamas and the Caribbean, still must go to the ocean to lay their eggs (Burggren and McMahon 1988). Furthermore, burrowing is an essential survival strategy for these crabs, a habit that, when multiplied by their great numbers,

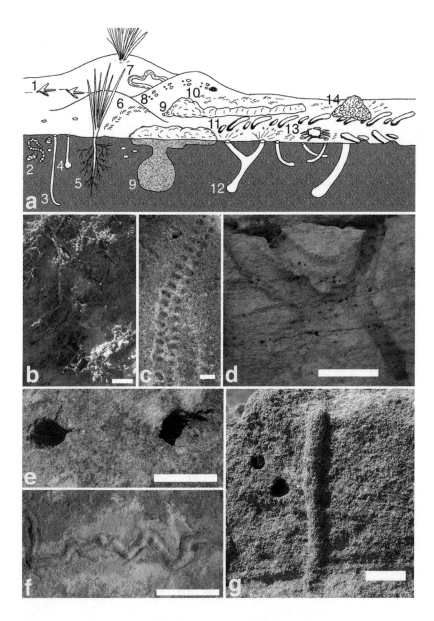

Figure 10.11. The *Psilonichnus* ichnofacies, betwixt the land and sea, but mostly influenced by the sea. (a) Diagram showing representative trace fossils, as defined by Frey and Pemberton (1987) and augmented by modern analogues described in this book. Key: 1 = Avian tracks; 2 = *Taenidium*; 3 = "*Skolithos*"; 4 = "*Skolithos*"; 5 = root traces; 6 = *Hexapodichnus*; 7 = *Aulichnites*; 8 = mammal tracks; 9 = sea turtle nest; 10 = decapod tracks; 11 = sea turtle tracks; 12 = *Psilonichnus*; 13 = decapod resting trace; 14 = decapod burrow mound. Note that some of these trace fossils (e.g., avian and mammal tracks, decapod resting traces and burrow mounds) are predicted on the basis of Georgia coast analogues. (b) Root traces associated with a paleosol in coastal

makes their burrows and other traces more likely to preserve in the fossil record. The importance of this ichnofacies as a paleoenvironmental indicator is key, because it signifies the landward side of a shoreline, whereas the *Skolithos* ichnofacies points toward the seaward side. Thus the use of both ichnofacies in conjunction can orient a geologist or paleontologist to the former position of an ancient shoreline.

On the Georgia coast, the *Psilonichnus* ichnocoenose is well represented in all of the environments one would expect, such as coastal dunes, back-dune meadows, storm-washover fans, and high salt marshes. Beautiful examples of modern Y- and J-shaped burrows made by *O. quadrata*, identical to the trace fossil *Psilonichnus*, are easily viewed in vertical cross sections of coastal dunes wherever a dune face has eroded. Smaller-diameter examples of such burrows, some of which branch into *Thalassinoides*-like burrow systems or intersect with one another (making false branches), likewise occur in salt marshes and storm-washover fans. Epoxy resin casts made of *Uca* burrows are strikingly similar to their fossil counterparts, and as mentioned at the start of this chapter, Pleistocene examples from St. Catherines Island match the forms and sizes of modern fiddler crab burrows. Also noteworthy components of the *Psilonichnus* ichnofacies are vertical burrows made by predatory arthropods, such as wolf spiders and tiger beetle larvae, which are common in back-dune areas. Furthermore, ant nests may be built adjacent to ghost crab burrows (Chapter 5). With so many delicious invertebrates in the sediment, moles dig their extensive feeding burrows in these areas too, churning all of the way through the dunes and into the intertidal zone, where they might pick up some more salty snacks (Chapter 8). Vertebrate tracks and shallow excavations might also be part of the ichnoassemblage, especially those made by sea turtles and any egg predators that happen upon sea turtle nests (Chapter 8). Sea

---

dune deposit: Pleistocene, Bahamas. Scale = 10 cm (4 in). (c) *Coenobichnus* (horizontal view) a hermit crab trackway: Holocene, Bahamas. Scale = 1 cm (0.4 in). (d) *Psilonichnus* (vertical view), a ghost crab burrow. Note the overall Y shape with dual branches, but also how a new burrow branch (*left*) crosscuts an older branch: Holocene, Bahamas. Scale = 5 cm (2 in). (e) *Psilonichnus* (horizontal view), showing dual entrances to Y-shaped burrow: Holocene, Bahamas. Scale = 5 cm (2 in). (f) *Aulichnites* (horizontal view), a probable gastropod trail: Pleistocene, Bahamas. Scale = 10 cm (4 in).
(g) *Skolithos* (vertical view), an insect or spider burrow: Holocene, Bahamas. Scale = 1 cm (0.4 in). All trace fossils preserved in carbonate coastal dune deposits.

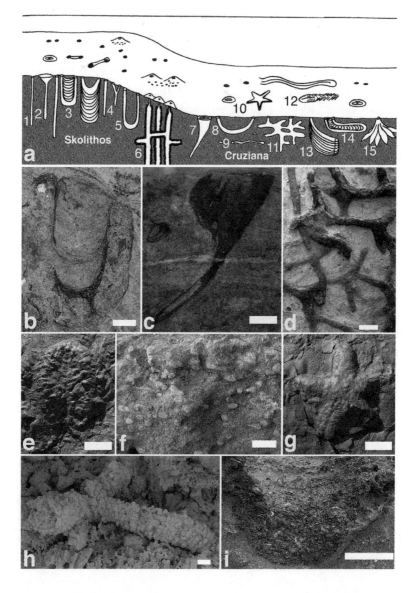

Figure 10.12. The shallow-marine soft-bottom *Skolithos* and *Cruziana* ichnofacies. (a) Diagram showing representative trace fossils of the *Skolithos* (*left*) and *Cruziana* (*right*) ichnofacies. Note that some trace fossils are present in both ichnofacies, but each ichnofacies is defined on the basis of entire ichnoassemblages. Key: 1 = *Skolithos*; 2 = *Monocraterion*; 3 = *Diplocraterion*; 4 = *Polykladichnus*; 5 = *Arenicolites*; 6 = *Ophiomorpha*; 7 = *Rosselia*; 8 = *Arenicolites*; 9 = *Planolites*; 10 = *Asteriacites*; 11 = *Thalassinoides*; 12 = *Cruziana*; 13 = *Teichichnus*; 14 = *Rhizocorallium*; 15 = *Asterosoma*. (b) *Diplocraterion* (vertical view), a probable polychaete worm burrow, in shallow-marine (nearshore) clastic deposits: Middle Jurassic, U.K. (c) *Rosselia*, a probable polychaete worm burrow, in offshore clastic deposits: Late Permian, Australia.

turtle nests in particular may impart minor ecological changes to shoreward-facing dunes, especially wherever failed nests have provided food for ants, ghost crabs, or other infaunal egg scavengers that might burrow in and around nest structures, thus making composite traces. Occasional additions to the ichnocoenose are adult limulid tracks, some of which I have seen just behind and in front of coastal dunes, left by wayward horseshoe crabs stranded by high tides and seeking the sea.

Well, of course this works, one might say. After all, it is a tautology, in which modern Georgia coast ichnocoenoses defined the ichnofacies, and then are used to support the interpretation of the same ichnofacies, in a redundant sort of way. Not really, I reply. Yes, modern Georgia traces comprised the basis for the ichnofacies, using many of the handy examples of modern traces and the known ecology of their tracemakers. Fortunately, this scientific form of self-affirmation—often expressed through self-citation in the literature—ends when one goes back to the place where the originators of the ichnofacies (Frey and Pemberton 1987) openly encouraged other researchers to apply and test this ichnofacies by comparing it to fossil examples. Indeed, paleontologists took up this challenge, and have since refined the definition of *Psilonichnus* as an indicator ichnogenus for the ichnofacies, as well as affirmed its applicability to many geological examples (Nesbitt and Campbell 2006).

Skolithos *Ichnofacies*

The *Skolithos* ichnofacies differs from the *Psilonichnus* and all other ichnofacies discussed thus far, because it typically represents traces made in the shallowest of submerged marine environments (Figure 10.12). An important aspect of marine ichnofacies, however, is that water depth is not always responsible for their physical factors. For example, the *Skolithos* ichnofacies

(d) *Thalassinoides* (horizontal view), a crustacean burrow complex, in shallow offshore carbonate deposits: Late Cretaceous, U.K. (e) *Cruziana* or *Rusophycus* (horizontal view), a trilobite resting trace, in shallow estuarine deposits: Late Ordovician, Georgia (USA). (f) *Macaronichnus* (vertical view), probable polychaete worm burrows in shallow nearshore clastic deposits: Pliocene, New Zealand. (g) *Asteriacites* (horizontal view), a sea star resting trace in shallow estuarine deposits: Late Ordovician, Georgia (USA). (h) *Ophiomorpha* (oblique view), a callianassid shrimp burrow complex in shallow nearshore carbonate deposits: Pleistocene, Bahamas. (i) *Piscichnus* (vertical view), a ray feeding structure in shallow nearshore deposits: Miocene, New Zealand. Scale bars = 1 cm (0.4 in) in all photos except (d) and (i); these are 10 cm (4 in).

is more reflective of high energy, in that waves, tides, and other high-energy movement of water causes sand and other sediments to erode, deposit, and otherwise shift, which causes animals to burrow in a certain manner. These sedimentary conditions are admittedly more common in shallow, subtidal environments just offshore, but can also take place in deeper water if storm waves reach a normally undisturbed sediment surface (Frey et al. 1990). Regardless, energetically moving water that regularly impinges on sediments causes unstable bottom conditions for any animal attempting to live there, requiring certain adaptations and behavioral strategies to cope. Consequently, the most efficient ways for any given animal to live in such conditions is to burrow, and most of the burrows are made straight down or straight up, and hence are preserved as vertical structures.

The namesake of the *Skolithos* ichnofacies is the simple, thin, unadorned vertical fossil burrow *Skolithos*, which has many analogues on the Georgia coast in shallow marine environments, such as the mucus-lined vertical burrows of the polychaete worm *Onuphis microcephala* (Figure 6.4b). Lining or otherwise reinforcing burrows is another way that animals deal with recurring attrition of their sandy foundations. As noted earlier, though, *Skolithos* is such a simple, effective burrow plan that it has been imitated by many tracemakers in a wide range of environments. Hence it should not in itself indicate the *Skolithos* ichnofacies, and in fact may be entirely absent from a *Skolithos* ichnocoenose. Instead, think vertically. A key trait of the *Skolithos* ichnofacies is an assemblage of trace fossils with a distinctive up-and-down orientation cutting across more or less horizontal bedding, usually cross-bedded and rippled sand. On the Georgia coast then, not just the burrows of *O. microcephala* represent archetypical burrows of the *Skolithos* ichnofacies, but also the pellet-lined burrows of the ghost shrimps *Callichirus major* and *C. biformis*; U-shaped burrows of the acorn worm *Balanoglossus aurantiactus* and sea cucumbers; and equilibrium and escape burrows of burrowing anemones and bivalves, to name a few tracemakers. Other than *Skolithos*, trace fossil equivalents of such burrows are *Arenicolites, Diplocraterion, Monocraterion,* and *Ophiomorpha,* among others (Table 10.2).

Cruziana *Ichnofacies*

The *Cruziana* ichnofacies is traditionally regarded as an ichnofacies that embodies a reprieve from the constantly shifting substrates of the *Skolithos* ichnofacies, representing quieter places in which sand and mud do not move so much, but not necessarily in much deeper water (Figure 10.12). The ichnogenus representing this ichnofacies, *Cruziana,* is a characterized by a horizontally to obliquely oriented, bilobed, ribbon-like structure with a medial furrow and an internal series of chevrons. This ichnogenus and its numerous variants are nearly always interpreted as trilobite burrows, and are preserved as sandstone casts of the originally hollow, open burrows that must have connected with the seafloor. Trilobites, though, "only" lived during the Paleozoic Era (the quotations marks in recognition that a time span of 300 million years should not deserve such an adverb), which may lend to the idea that this ichnofacies can be applied simply to Paleozoic examples. Fortunately, this idea is false, as the behavioral responses of tracemakers represented by *Cruziana* continued well after the Paleozoic in their absence and were taken up by other animals. In other words, similar modifications of subtidal substrates are manifested in trace fossils made by organisms other than trilobites, such as limulids, decapods, molluscans, polychaetes, and a host of other mobile animals, also exemplified by modern ichnocoenoses of the Georgia coast.

Environments reflecting the parameters of the *Cruziana* ichnofacies are realms for surface trails and subsurface burrows, the latter of which would have more horizontal orientations when compared to burrows in the *Skolithos* ichnofacies. Moreover, surface trackways and trails, such as those made by arthropods and gastropods, are mainly made while either deposit feeding or grazing. Ichnogenera that represent such behaviors are *Diplichnites, Gyrochorte,* and *Helminthopsis,* among many others. Burrows also may be the result of deposit feeding or predation, rather than serving more as domiciles. Fossil burrows may also include networks, such as *Ophiomorpha* and *Thalassinoides,* jungle-gym complexes made originally by decapods. Other ichnogenera are more representative of resting, which as discussed before actually may reflect behaviors that include saving energy, feeding, hiding from predators, or predation itself (Chapters 6, 9). For example, trace fossils associated with stationary behaviors that are likely in the *Cruziana* ichnofacies are *Lockeia* (a bivalve trace), *Rusophycus* (trilobite trace), and *Piscichnus* (ray feeding trace).

As hinted previously, estuarine environments can be a source of ichnological conundrums or seeming contradictions, because of the common mixing of components from *Skolithos* and *Cruziana* ichnofacies in such

TABLE 10.2. MARGINAL-MARINE ICHNOFACIES AND REPRESENTATIVE ICHNOGENERA APPLICABLE TO THE GEORGIA BARRIER ISLANDS

A. *Psilonichnus* Ichnofacies

Ichnogenus	Description	Interpretation
Aulichnites	Bilobed horizontal trail	Snail, worm, insect trail
Lockeia	Almond-shaped impression	Bivalve resting trace
Planolites	Horizontal unlined burrow	Nematode, annelid burrow
Psilonichnus	Y- or J-shaped burrow	Ocypodid crab
Root traces	Branching, tapering structures	Various coastal plants
Vertebrate tracks	Variably shaped indentations	Tetrapod locomotion
Vertebrate burrows	Horizontal tunnels, chambers	Tetrapod dwelling

B. *Skolithos* Ichnofacies

Ichnogenus	Description	Interpretation
Arenicolites	Vertical U-shaped burrow	Invertebrate dwelling burrow
Bergaueria	Stout (plug-like) vertical burrow	Anemone dwelling burrow
Conichnus	Conical burrow with spreite	Anemone, bivalve trace
Cylindrichnus	Vertical, tapering burrow	Polychaete dwelling burrow
Diplocraterion	Vertical U-shaped burrow, spreite	Invertebrate dwelling burrow
Gyrolithes	Spiraling vertical burrow	Crustacean dwelling burrow
Macaronichnus	Sinuous, lined horizontal burrow	Polychaete feeding burrow
Ophiomorpha	Branching burrow, pelleted wall	Crustacean dwelling burrow
Palaeophycus	Lined, simple burrow	Various invertebrates
Polykladichnus	Y-shaped burrow	Bivalve dwelling burrow
Piscichnus	Oval, shallow depression	Ray feeding trace
Rosselia	Conical burrow, concentric fill	Polychaete feeding burrow
Schaubcylindrichnus	Lined, curved vertical burrow	Acorn-worm dwelling burrow
Siphonichnus	Vertical burrow, wider below	Bivalve dwelling burrow
Taenidium	Meniscate burrow	Invertebrate feeding burrow
Vertebrate traces	Various sizes, shapes	Fish burrows, feeding traces

C. *Cruziana* Ichnofacies

Ichnogenus	Description	Interpretation
Asterosoma	Star-like bulbous cluster	Invertebrate feeding burrow
Chondrites	Downwardly branching burrow	Polychaete feeding burrow
Cruziana	Bilobed burrow, chevrons	Arthropod burrow
Cylindrichnus	Tapering burrow, concentric fill	Polychaete feeding burrow
Gyrochorte	Upraised, bilobed trail	Arthropod locomotion trace
Helminthopsis	Tightly meandering burrow	Invertebrate feeding burrow
Lockeia	Almond-shaped impression	Bivalve resting trace
Ophiomorpha	Branching burrow, pelleted wall	Crustacean dwelling burrow
Phoebichnus	Central shaft, meniscate burrows	Invertebrate feeding burrow
Phycodes	Shaft, meniscate branches	Invertebrate feeding burrow
Phycosiphon	Looping meniscate burrow	Invertebrate feeding burrow
Planolites	Horizontal unlined burrow	Invertebrate feeding burrow
Rhizocorallium	Oblique U-shaped burrow, spreite	Invertebrate dwelling burrow

C. *Cruziana* Ichnofacies (*cont.*)

Ichnogenus	Description	Interpretation
Rosselia	Conical burrow, concentric fill	Polychaete feeding burrow
Rusophycus	Bilobed horizontal burrow	Arthropod resting trace
Schaubcylindrichnus	Lined, curved vertical burrow	Acorn-worm dwelling burrow
Taenidium	Meniscate burrow	Invertebrate feeding burrow
Teichichnus	Horizontal burrow with spreite	Invertebrate feeding burrow
Zoophycos	Complex U-shaped burrow, spreite	Invertebrate feeding burrow

D. *Glossifungites* Ichnofacies

Ichnogenus	Description	Interpretation
Arenicolites	U-shaped vertical burrow	Invertebrate dwelling burrow
Bergaueria	Stout (plug-like) vertical burrow	Bivalve dwelling burrow
Conichnus	Conical burrow with spreite	Anemone dwelling trace
Diplocraterion	U-shaped vertical burrow, spreite	Invertebrate dwelling burrow
Gastrochaenolites	Vase-shaped vertical burrow	Bivalve dwelling burrow
Palaeophycus	Lined horizontal burrow	Polychaete dwelling burrow
Psilonichnus	J-or Y-shaped vertical burrow	Arthropod dwelling burrow
Skolithos	Simple vertical burrow	Polychaete dwelling burrow
Spongeliomorpha	Branching burrow, scratchmarks	Crustacean dwelling burrow
Rhizocorallium	U-shaped horizontal burrow, spreite	Invertebrate dwelling burrow
Thalassinoides	Branching tunnels, shafts	Crustacean dwelling burrow

E. *Trypanites* Ichnofacies

Ichnogenus	Description	Interpretation
Entobia	Spherical borings, connected	Sponge dwelling trace
Gastrochaenolites	Vase-like boring	Bivalve dwelling trace
Gnathichnus	Surface scrape marks	Echinoid feeding trace
Oichnus	Beveled boring, in shells	Gastropod predation trace
Rogerella	Long, cylindrical boring	Barnacle dwelling trace
Trypanites	Cylindrical boring	Polychaete dwelling trace

E. *Trypanites* Ichnofacies

Ichnogenus	Description	Interpretation
Diplocraterion	U-shaped boring, spreite	Polychaete dwelling trace
Arenicolites	U-shaped boring	Isopod dwelling trace
Teredolites	Vase-like boring, ridges	Bivalve dwelling trace
Thalassinoides	Branching tunnels, shafts	Decapod dwelling trace

environments (Buatois et al. 2005). For example, the seemingly idyllic, low-energy subtidal environments inhabited by surface dwellers and burrowers may be punctuated by the full-fledged terror of tropical storms, which rip up sandy–muddy bottoms and unload suspended sediment as thick, suffocating layers. In such instances, the former inhabitants of these environments are temporarily replaced by suspension-feeding tracemakers that burrow straight down into these new surfaces, forming the mostly vertical traces of the *Skolithos* ichnofacies. These ichnoassemblages are often referred to as poststorm colonization suites (Savrda and Nanson 2003), and their opportunistic settling on freshly made surfaces in the wake of such storms certainly has some real estate parallels in human society. Continuing this analogy further, animals that were living in the shallow sediments, but were entrained along with the eroded sediment, displaced, and deposited in onshore or more offshore environments, may make traces for just a generation or two before dying out. Thus their traces reflect the efforts of what have been called "doomed pioneers." Regardless, the prevailing, so-called average low-energy conditions of these subtidal environments resumed at some point, so more of a mix of mud and sand was introduced, and the tracemakers more typical of the *Skolithos* ichnofacies were ecologically pushed aside by the *Cruziana* ichnofacies tracemakers. In the geologic record, such variations of ichnofacies in subtidal environments might then show up as alternations of *Cruziana* and *Skolithos* ichnofacies, in which normally low-energy bottom conditions were interrupted by high-energy events, such as storms or river floods that eroded previous sediments and pumped in large amounts of new sediment in short time spans.

Substrate-Dependent Marine Ichnofacies

Nearshore substrate-dependent marine ichnofacies include the *Glossifungites*, *Trypanites*, and *Teredolites* ichnofacies (Figure 10.13). Of these three, excellent examples of the *Glossifungites* ichnofacies are well studied on the Georgia coast, whereas the other two are not so well expressed in any significant way. For the *Trypanites* ichnofacies, only isolated bivalve or gastropod shells, shell beds (such as oyster reefs), and offshore clusters of bioeroded rock can be used as models for some of its traces and tracemakers. For example, clionaid sponge borings resembling the trace fossil *Entobia* are common in individual molluscan shells that were exhumed and bored. Furthermore, relict marshes containing well-developed oyster colonies may provide limited areas in which elements of the *Trypanites* ichnofacies might occur. Likewise, modern analogs to the *Teredolites* ichnofacies, which involves the deposition, compaction, and submergence of woodgrounds (Bromley et al. 1984; Gingras et al. 2004b), are spotty at best on the Georgia barrier islands. Nonetheless, plenty of borings made by the wedge piddock, a marine bivalve (*Martesia cuneiformis*), can be observed in driftwood or in-place trees drowned by the rising sea on Georgia beaches (Chapter 6). These borings, along with those of wood-boring isopods, hint at what sorts of traces to expect in a marine woodground. Of course, borings in trees made by marine animals crosscut (and thus postdate) borings made by wood-chewing insects or birds that consumed these insects. All of these traces formed while the tree was still firmly rooted on the landscape of island interiors, above the equivalent of the *Coprinisphaera* ichnofacies.

As mentioned previously (Chapter 3), examples of the *Glossifungites* ichnofacies on the Georgia coast are exposed along the sandy shorelines of the composite Pleistocene–Holocene islands, and represented mostly by newly emergent relict salt marshes. These relict marshes are rightfully compelling for the glimpses they give us of taphonomy in action, but they also supply insights on how organisms adapt to and colonize older (but newly exposed) surfaces. On the Georgia coast, bioeroding bivalves are responsible for most of the traces described from these firmgrounds, and the identification of these traces is aided considerably by the presence of bivalve shells in their domiciles. Some of these borings, though, are also carved out by fiddler crabs that have no idea they are occupying the same salt marsh surfaces of their ancestors. While walking across these relict marshes, I have seen fiddler crabs ducking in and out of borings of their own making, as well as bivalve borings they modified slightly for their needs. How are these firmground traces then preserved and later noticed in the fossil record? Once a drifting layer of windblown sand migrates from a nearby coastal dune or beach, borings are filled with whitish sediment, providing easy contrast against the dark brown mud and a ready taphonomic model for the preservational style of similar trace fossils. Mixed in with these secondary firmground traces may be primary invertebrate burrows or root traces from halophytes. Amazingly, modern clumps of smooth cordgrass can grow on these older substrates, contributing new, modern

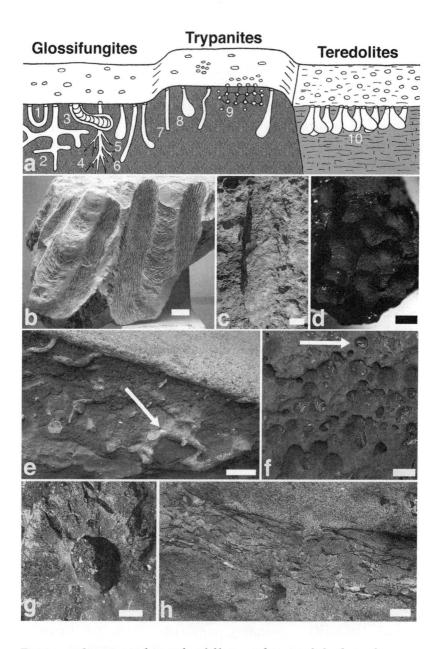

Figure 10.13. Some tracemakers prefer solid bottoms: firmgrounds, hardgrounds, and woodgrounds, defined by Seilacher (1967), Frey and Seilacher (1980), and Bromley et al. (1984). (a) Diagram of the *Glossifungites* (*left*), *Trypanites* (*middle*) and *Teredolites* (*right*) ichnofacies and their representative trace fossils. Note that some trace fossils are present in all ichnofacies, but each ichnofacies is defined on the basis of entire ichnoassemblages, and in these instances, substrates. Key: 1 = *Arenicolites*; 2 = *Thalassinoides*; 3 = *Rhizocorallium*; 4 = root traces; 5 = *Gastrochaeonolites*; 6 = *Psilonichnus*; 7 = *Trypanites*; 8 = *Gastrochaenolites*; 9 = *Entobia*; 10 = *Teredolites*.

roots alongside those of 500–1,000-year-old root systems. This implies that root-trace assemblages in the fossil record must be viewed with caution for their paleoecological significance too (Chapter 4).

Similar in principle to these *Glossifungites* ichnofacies are terrestrial firmgrounds adjacent to marine environments. These are represented in modern Georgia coastal environments by Pleistocene outcrops next to tidal channels or beaches where some birds, such as northern rough-winged swallows (*Stelgidopteryx serripennis*), have bored horizontal tunnels for nesting chambers (Chapter 7). An important aspect of these firmgrounds is how the Pleistocene sandstones composing these outcrops are poorly cemented and hence semiconsolidated; with these rocks, geologists can easily impress others by pulverizing samples with their bare hands. The aforementioned Yellow Banks Bluff, as well as Raccoon Bluff and outcrops on the western side of Cumberland Island, are all composed of similarly soft rock, readily bioeroded by determined birds, or even nesting hymenopterans, such as solitary wasps and bees. As originally defined, however, the *Glossifungites* ichnofacies is restricted to marine firmgrounds. Hence no ichnofacies equivalent for terrestrial situations has been noted: until now, that is.

Zoophycos *and* Nereites *Ichnofacies*

The Georgia Bight is either not deep enough, or otherwise lacks qualities that would mimic most ecological conditions of deep-marine environments, and thus it has no modern ichnocoenoses that approach the

(b) *Rhizocorallium* (vertical view) in chalk firmground, probably made by a crustacean, judging from scratchmarks on the burrow walls: Miocene, Ukraine. Specimen on display in Geological Museum of Jagiellonian University, Krakow, Poland. (c) *Gastrochaenolites*, (vertical view) in reef limestone, boring made by a marine bivalve: Pleistocene, Bahamas. (d) *Teredolites* (horizontal view) in top of lignite coal bed, made by wood-boring marine bivalves: Late Cretaceous, South Carolina (USA). (e) *Thalassinoides* (*arrow*) and other burrows (vertical view) preserved in a former clastic firmground, probably made by marine arthropods; note how burrows are filled with sediment from overlying bed: Miocene, New Zealand. (f) Abundant *Gastrochaeonolites* at omission surface of sandstone, some containing the boring marine bivalves that made them (*arrow*): surface between the Early Miocene and Early Pliocene, New Zealand. (g) *Entobia* (horizontal view), in top of Late Jurassic limestone, made by a large boring sponge; note lines radiating from central depression: boring was made during Late Cretaceous, Poland. (h) *Teredolites* (vertical view) in fossil log preserved in sandstone, made by marine wood-boring bivalves: Miocene, New Zealand. Scale bars = 1 cm (0.4 in) in all photos except (e), where is 5 cm (2 in).

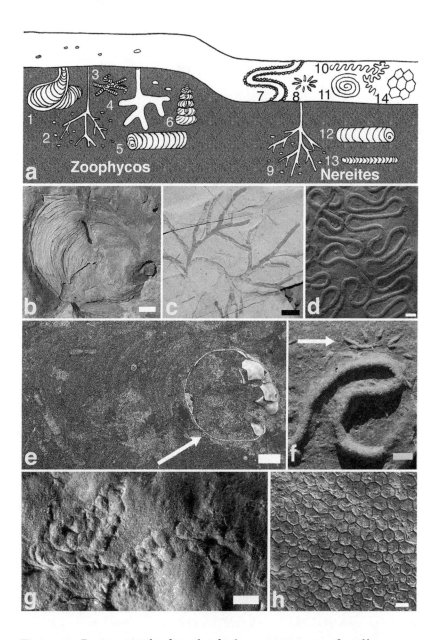

Figure 10.14. Deep-marine ichnofacies that, for the most part, are not reflected by Georgia coast traces. (a) Diagram of *Zoophycos* (*left*) and *Nereites* (*right*) ichnofacies. Note that some trace fossils are present in both ichnofacies, but each ichnofacies is defined on the basis of entire ichnoassemblages. Key: 1 = *Zoophycos*; 2 = *Chondrites*; 3 = *Phycosiphon*; 4 = *Thalassinoides*; 5 = *Scolicia*; 6 = *Spirophyton*; 7 = *Nereites*; 8 = *Lorenzinia*; 9 = *Cosmoraphe*; 11 = *Spiroraphe*; 12 = *Scolicia*; 13 = *Zoophycos*; 14 = *Paleodictyon*. (b) *Zoophycos* (horizontal view), a feeding burrow made by an unknown worm-like animal, in a prodelta sandstone: Lower Carboniferous, Kentucky (USA).

Zoophycos or *Nereites* ichnofacies (Figure 10.14). For instance, the *Zoophycos* ichnofacies, aptly represented by the beautiful and complex trace fossil *Zoophycos*, is normally associated with continental slopes or deep-sea deposits called turbidites. Turbidites are sedimentary deposits formed by turbidity currents, which are powerful, high-velocity flows that result from slope failures. Turbidity currents are sometimes triggered by earthquakes, which cause submarine avalanches that suddenly dump thick layers of sediment onto formerly quiescent environments. Accordingly, the *Zoophycos* ichnofacies, which consists of complex, deeply penetrating feeding traces (*Zoophycos, Phycosiphon, Spirophyton*), can be used to discern these alternations of seafloor quietude and reigns of depositional chaos. Additionally, and similar to the situation with the *Mermia* ichnofacies—especially in deep lakes with poor circulation—traces in this assemblage can be sensitive indicators of oxygen. This ichnofacies is typified by deposit-feeding invertebrate traces that increase in both abundance and depth with elevated oxygen, both at the sediment surface and within the sediment, or correspondingly vanish in the face of anoxia.

The *Nereites* ichnofacies, which is associated with muddy abyssal plains on the deep-ocean floor, contrasts with the *Zoophycos* ichnofacies by its ichnoassemblage of meandering or spiraling surface trails, shallow horizontal burrows, and a near absence of vertical burrows. As mentioned before, both the *Zoophycos* and *Nereites* ichnofacies were predicted on the basis of reverse uniformitarianism, in which trace fossils were used to predict that similar modern traces would be found on continental slopes and abyssal

(c) *Chondrites* (horizontal view), a feeding burrow probably made by a polychaete worm, preserved in a deep-water mudstone: Eocene, Switzerland. (d) *Cosmoraphe* (horizontal view), a feeding burrow made by a worm-like animal, in a deep-water sandstone: Paleocene, Poland. Specimen in Geology Museum of Jagiellonian University. (e) *Scolicia*, a feeding and locomotion burrow with its tracemaker, a burrowing sea urchin, partially preserved at the end of the burrow (*arrow*), and preserved in a deep-water sandstone: Miocene, Argentina. (f) *Lorenzinia* (*arrow*), a feeding burrow probably made by a worm-like animal, crosscut by *Cosmoraphe*; Eocene, Switzerland. (g) *Nereites* (horizontal view), a trilobite burrow and trail in a sandstone in this instance, but could have been made by worms in deeper water by worms in other examples: Early Silurian, Alabama (USA). (h) *Paleodictyon,* a systematic feeding burrow made by a worm-like animal, in a deep-water sandstone: Oligocene, Poland. Specimen in Geology Museum of Jagiellonian University. Scale = 1 cm (4 in) in all photos.

plains at depths of hundreds and thousands of meters (Chapter 2). Because this ichnofacies is so unlike anything represented by modern ichnocoenoses of the Georgia coast, it will not be explored further here, despite its fascinating suite of trace fossils.

What is interesting to note, though, is how individual modern traces analogous to trace fossils from the *Nereites* ichnofacies can pop up in the shallower environments of the Georgia coast. For example, at low tide during the summer through fall, millions of small (< 1 cm wide) juvenile limulids plow through wet sand, looping and meandering throughout the rippled sand flats and runnel edges in the intertidal zones of Georgia beaches. These combined shallow burrows and trails, which occasionally show tiny, stitching-like tracks and drag marks left by their telsons, bear a superficial resemblance to the trace fossil *Nereites* (Martin and Rindsberg 2007). *Nereites* is a trace fossil with a medial furrow flanked by strings of beaded sediment, often making a ribbon-like burrow that also meanders or loops on horizontal bedding planes. This ichnogenus, however, has always been interpreted as an infaunal feeding burrow, and is associated with worm-like tracemakers and deep-water environments, well away from any shoreline (Wetzel 2002; Martin and Rindsberg 2007). This supposition was based partially on the frequent occurrence of *Nereites* in deep-sea trace fossil assemblages, so often that it became the namesake of an ichnofacies. Unlike a few other deep-sea traces, though, no modern tracemakers have been caught in the act forming *Nereites*-like traces, and only one instance of a modern analogue in deep-sea sediments has been found thus far (Wetzel 2002). Thus with these juvenile limulid traces from the Georgia coast, ichnologists now have an explanation for seemingly anomalous incursions of *Nereites* in shallow-water settings, as well as the idea that small, short-bodied arthropods—such as trilobites—may have made similar traces. The lateness of this realization, made nearly twenty years after Frey, Howard, and Pemberton stopped studying the neoichnology of the Georgia coast, hints at other important insights that might be gained through further study of the life traces of the Georgia coast.

TRACE FOSSILS IN CARBONATE SEDIMENTS AND LIMESTONES

A subtle bias runs through this study of Georgia coast traces and their application to the fossil record, and it deals directly with one of the three pillars of ichnology: substrate. Nearly all Georgia coast traces, especially burrows, tracks, trails, and some feces, are preserved in substrates composed of silicate sediments, especially quartz sand and clay minerals (Chapter 2). How easily, then, do the ichnological lessons learned from silicate-rich settings apply to places with nonsilicate substrates, such as those composed of carbonate sand and mud, which eventually would form limestones? The short answer is "very well," although a few important differences should be kept in mind when looking at trace fossils in limestones and other carbonate rocks (Curran 1992; Curran and Martin 2003; Martin 2006c).

Nearly all modern carbonate sediments are made of two minerals—aragonite and calcite—both of which have the same chemical formula ($CaCO_3$). Of this sediment, organisms manufacture nearly all of it, a result of biomineralization. This process has been around for the past 550 million years and is showing no signs of subsiding any time soon, as it is essential for the survival and reproduction of a huge variety of marine organisms (and a few freshwater ones), from planktonic algae to echinoderms. In subtropical and tropical coastal areas, organisms are responsible for producing the vast majority of sediments in any coastal environment, including coastal dunes. Indeed, as mentioned before, much of the sand seen in the Bahamas and other subtropical–tropical areas are the digestive by-products of parrotfish and other bioeroders (Chapter 9). Carbonate sediments are thus produced in house, so to speak, then later processed and otherwise reworked by organisms and physical processes, such as waves, tides, and wind, although still within a relatively limited area. In contrast, the Georgia barrier islands, as well as many other barrier islands worldwide, depend on erosion of nearby highlands (e.g., the Appalachian Mountains) and rivers carrying that eroded sediment to the coast (Pilkey and Fraser 2003).

When ichnologists first started testing the ichnofacies concept by studying modern analogues, they did not limit themselves to environments with silicate sediments, but also considered the ichnology of carbonate sedimentary systems. Much to their surprise, they found that some modern marine tracemakers in southern Florida, the Bahamas, and other subtropical–tropical carbonate settings were constructing burrows identical to those of the same or similar species in Georgia coast environments (Shinn 1968; Curran 1984, 1992). Because these tracemakers, such as ghost crabs and ghost shrimp (Chapter 6), had been studied so well on the Georgia barrier islands, this knowledge was readily transferred to the study of traces and trace fossils in carbonate sediments and rocks. Making matters even easier for ichnologists, the Bahamas has extensive outcrops of Holocene and Pleistocene rocks, most of which were less than 200,000 years old and many of which contain trace fossils (Curran 1984 1992; Curran and Martin 2003; Martin 2006b, 2006c; Curran 2007). Hence ichnologists could do direct comparisons between modern traces and their trace fossil equivalents, and modern ichnocoenoses could be connected directly to ancient ichnofacies. For example, ichnocoenoses corresponding to the *Psilonichnus, Skolithos,* and *Trypanites* ichnofacies have been well defined from Holocene and Pleistocene limestones of the Bahamas (Martin 2006c; Curran 2007), and have a probable presence of the *Coprinisphaera* and *Cruziana* ichnofacies. Of these, the *Psilonichnus* ichnofacies is well represented (Fig. 10.11); indeed, the type specimen of the ichnogenus *Psilonichnus* is in an outcrop on San Salvador Island.

Another instance of how environments dominated by carbonate sediments hold ichnological similarities to Georgia coast environments is through their development of composite ichnological landscapes, in which biodeposition and burrowing combine to form entire ecosystems (Chapter 2). For instance, lagoons in the Bahamas are low-energy, muddy–sandy environments where wave action is minimal. This allows for extensive burrowing by a callianassid shrimp (*Glypturus acanthochirus*) and upogebiid shrimp (*Upogebia vasquezi*), each of which makes distinctive burrow systems (Curran and Martin 2003). The callianassid shrimp construct deep, interconnected shafts and tunnels, all reinforced by pellets, and are nearly identical to those made by ghost shrimp of the Georgia barrier islands (Chapter 6; Figure 6.13). Where the Bahamian species differ ichnologically from their American cousins is through their forming of 20–30 cm (8–12 in) high burrow mounds, which look much like small stratovolcanoes on the lagoon bottom. These mounds, formed by shrimp excavation of subsurface sediment and copious feces, overlap with one another and make for a rolling topography. Algal mats stabilize these mounds, which allow for upogebiid shrimp to burrow into the mounds and make their own distinctive domiciles. These are doubled, U-shaped burrows that intersect one another at their bottoms; thick (2–5 cm, or 0.8–2 in) pelleted walls encase these burrows. In slightly shallower water along the edges of the lagoons, mangrove roots and fiddler crab burrows, the latter made by *Uca major*— a close relative of the fiddler crabs in Georgia—penetrate sediments and make simpler traces that lack pelleted walls, setting them apart from traces made by the burrowing shrimp. All of these lagoonal traces have fantastic preservation potential, especially when considering how quickly carbonate sediments can cement to become rock. Sure enough, following a prognostication by Curran and Williams (1997) that upogebiid shrimp burrows should be in the Bahamian fossil record, I found numerous examples in a Pleistocene formation on San Salvador Island (Bahamas). These trace fossils had thick pelleted walls, doubly intersecting U-shaped burrows, and connections with probable fossil callianassid shrimp burrows, almost perfectly matching their predicted forms (Martin 1999; Curran and Martin 2003). This is yet another example of neoichnology triumphing, regardless of substrate composition.

Similar to the Bahamian lagoons, Georgia salt marshes have abundant upogebiid shrimp burrows, fiddler crab burrows, and root traces, the last not made by mangroves but smooth cordgrass (*Spartina alterniflora*). Furthermore, organisms are also responsible for depositing nearly all of the mud of a Georgia salt marsh, albeit most coming from the feces of filter-feeding mussels and oysters. These marshes, though, lack deep-burrowing ghost shrimp, which are restricted to sandy shorelines and subtidal environments (Chapter 6). These ichnological traits, combined with the respective sedimentary differences of Bahamian lagoons and Georgia salt marshes, demonstrate some of the limits of using Georgian examples for interpreting trace fossils, especially those preserved in carbonate rocks.

Oddly enough, an idea held sway for a brief time that trace fossils in limestones were less common than those in silicate rocks, or were thought to be altered so much by diagenesis that these would be unrecognizable (Kennedy 1975). Fortunately, this negating principle has been thoroughly flogged, sacked, and relegated to the rubbish heap of ichnological concepts gone wrong. The fine preservation of trace fossils in Cretaceous chalks was noted early on by Bromley (1967), then a few years later by Frey (1969, 1970a), and many exquisitely expressed trace fossils have been recognized since in limestones from the Cambrian to the Holocene (Bromley 1996; Curran 1992, 2007; Goldring et al. 2007). An exciting new direction with traces in carbonates now deals with those of more terrestrial environments, such as ichnocoenoses of the *Coprinisphaera, Scoyenia,* and *Psilonichnus* ichnofacies (Martin 2006b, 2006c). Limestones are not necessarily always associated with marine environments, either. Some inland lake deposits are composed of limestone, and often can preserve suites of trace fossils corresponding to the *Scoyenia* and *Mermia* ichnofacies (Buatois and Mángano 2007), including fish trails (Martin et al. 2010). Terrestrial soils can also have high concentrations of calcium carbonate, thus potentially becoming limestones too. Trace fossils preserved in calcareous paleosols include those of the *Coprinisphaera* ichnofacies, which can have abundant insect nests, cocoons, and vertebrate nests (Melchor et al. 2002; Martin and Varricchio 2011). In short, modern traces of the Georgia coast can provide some means for comparison for carbonate rocks, but neoichnologists should pay further attention to places with carbonate sediments for better fitting traces with trace fossils.

TRACE FOSSILS AND SEA-LEVEL CHANGE

One of the traditional uses of trace fossils since the unifying concept of ichnofacies was erected is that of discerning sea-level fluctuations, and this utility has not changed. As mentioned previously, the presence of specific ichnogenera or ichnoassemblages can be diagnostic of relative shoreline positions. To paraphrase Shakespeare, and to quote James Howard (Chapter 2), an ichnologist or geologist can simply ask, "The sea, or not the sea?" and be answered quite handily by trace fossils. Once the approximate positions of shorelines are identified, vertical successions of ichnofacies can further test whether sea level was going up or down in a particular stratigraphic sequence, in which the oldest strata are on the bottom and the youngest on top. For example, if a geologist is looking at a vertical sequence of strata, and observes that a *Coprinisphaera* ichnofacies at the base of the outcrop is succeeded by a *Psilonichnus* ichnofacies, which in turn is overlain by *Skolithos* and then *Cruziana* ichnofacies, a geologist can reasonably hypothesize this sequence reflects a rise in sea level over time, or a transgression (Chapter 2). Appropriately, the opposite sequence suggests a lowering of sea level, or regression. Could this same interpretation be made without trace fossils, using just physical sedimentary structures, body fossils, and other geologic clues? Sure, but trace fossils supply an independent check on these. Moreover, they are often among the first tools a geologist should reach for when trying to unravel a sea-level history recorded by sediments. To use a simple analogy, ichnology is both the corkscrew and the bottle opener on a Swiss Army knife, serving multiple functions while also making everyone happy that they, in critical moments, chose to have these tools included with their knives.

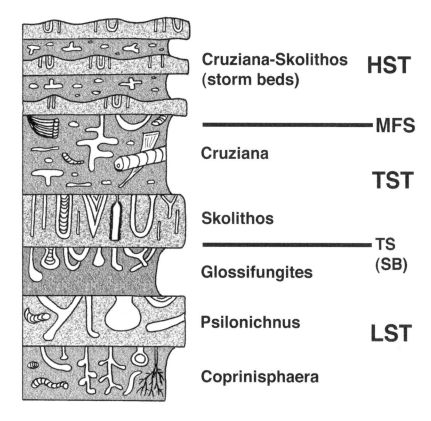

Figure 10.15. Hypothetical stratigraphic section in which trace fossils save the day, or at least help to interpret sea-level fluctuations. In this case, a transgression happened, which led from terrestrial environments to marginal-marine environments in a lowstand systems tract (LST); to a transgressive surface (TS), which is also a sequence boundary (SB); to a transgressive systems tract (TST); to a highstand systems tract (HST), in which the ichnofacies alternated because of storms affecting the shallow marine shelf. The maximum flooding surface (MFS) indicates the time when the transgressive systems tract switched to a highstand systems tract.

With that said, the application of trace fossils to discerning changes in sea level in the geologic past can become even more sophisticated. For instance, substrate-related ichnoassemblages, such as the *Glossifungites*, *Trypanites*, and *Teredolites* ichnofacies, can pinpoint exactly when and where a transgression began after a regression, or vice versa, as well as when sea level paused in between these changes. Such precision stands out in stark contrast to broad, arm-waving pronouncements that an outcrop contains evidence of a transgression, regression, both, or neither, and instead pulls together a more refined story of oceanic change over time in a given area (Figure 10.15).

The science of interpreting sea-level changes in the geologic past has a long history in geology, starting minimally with Leonardo da Vinci in the sixteenth century observing fossil bivalves in outcrops of northern Italy and concluding they must record the former position of the sea. He also noted, though, that the rocks containing the fossils, occurring in the Alps, must have been uplifted far above normal oceanic height—which was correct—and that a global flood could not have been responsible for depositing these shells. Had I mentioned that he also accurately identified some trace fossils in the same area as the former remains of organismal behavior (Baucon 2010)? (Smart guy, this Leonardo.) In the late nineteenth and first half of the twentieth century, as stratigraphic sequences and fossil assemblages from around the world became better documented, a framework emerged that began to define global sea-level change throughout the past 600 million years or so. Along these lines, geologist Laurence (Larry) Sloss published the best and most influential summary of such works, defining what he called cratonic sequences (Sloss 1963). These sequences comprised predictable orders of strata on the North American craton, which is the stable interior of a continent. His organizational method for defining these sequences was simple but brilliant. Each sequence represented a sea-level cycle in a given place, in which the rocks recorded how the sea went up (transgression) and then down (regression). These rocks were then sandwiched between obvious lower and upper unconformities, which are surfaces of nondeposition or erosion. With this framework in mind, geologists then tested its applicability outside of North America and brought in many other methods, including ichnology, to better define such sequences.

In the 1970s, geologist Peter Vail, a former student of Sloss, along with some colleagues pushed these concepts one big step forward by publishing a new and improved take on cratonic sequences (Vail et al. 1977). These researchers, who were employed by what was then called Exxon Oil Company, took full advantage of technological advances—namely, computers and subsurface geophysical methods—to map the relationships of strata below the earth's surface. As expected, these relationships were connected directly to changes in sea level, so with these new details, Vail and his colleagues were able to construct the most detailed global sea-level curve yet proposed. Like Sloss, the sequences they defined were bounded by unconformities. Unconformities are the stratigraphic equivalent of tearing pages out of a book (please, not this one). Geologists knew something was missing from the geologic record when they encountered these surfaces, as the narrative of earth history recorded by the rocks jumped far too abruptly. All strata in between two unconformities were then considered as tracts (or packages) related in origin, whether from transgressions, regressions, or stillstands (mentioned earlier with regard to Yellow Banks Bluff on St. Catherines Island). These bounding surfaces were then called sequence boundaries, and tracts were given unwieldy names, such as transgressive system tract, regressive system tract, highstand system tract, and lowstand system tract. Thus was born the science of sequence stratigraphy, which matured into more of a well-tested and workable hypothesis, and less of a scientific bandwagon, which is what it was in the early to mid-1980s when it was fueled by copious hydrocarbon-derived dollars. Other geologists since the late 1970s have added new insights to clarify the basic principle of sequence stratigraphy, which is that the origins of strata can be defined on the basis of globally significant transgressions, regressions, and everything that happens in between these.

What does this have to do with trace fossils and ichnology? Well, as an attentive reader might have divined by now, ichnology often precedes other fields of science in its usefulness, and only later do people from outside of ichnology become enlightened enough to accept its utility. With regard to sequence stratigraphy, ichnologists, without prompting from their more geologically inclined colleagues, immediately recognized stratigraphically important surfaces that marked changes in sea level (Bromley 1975; Kobluk et al. 1977; Pemberton et al. 1980; Fürsich et al. 1981; Pemberton and Frey

1985; and many others). For example, an unconformity in which the sea dropped to its lowest level would have exposed vast areas of previously submerged sediments and rocks near or on a coast. This situation decreased sedimentation to a minimum and lent to the erosion of firmgrounds, including bioerosion by bivalves or other tracemakers. Given enough time and exposure of this surface, it formed a *Glossifungites* ichnofacies. This ichnofacies may even cut across a previous softground ichnofacies, such as a *Skolithos* or *Cruziana* ichnofacies (formed during higher sea level) or a *Psilonichnus* or *Scoyenia* ichnofacies (formed during lower sea level). If the exposed surface was more of a rocky one, such as a hardground, then the softground and firmground traces might be intersected by a *Trypanites* ichnofacies. Using this logic, ichnologists who applied ichnofacies as stratigraphic tools began to pinpoint where and when sea level dropped, rose, or stayed the same. They also easily identified unconformities on the basis of ichnoassemblages and their interpreted paleoenvironments, as well as the relative amount of marine or continental processes affecting a given area through time. In other words, ichnologists were picking out sequence boundaries and system tracts before they were called such by sequence stratigraphers (MacEachern et al. 2007b; Martin 2009b).

Rather than going into more detail about the connection between ichnology and sequence stratigraphy, suffice it to say that ichnologists quickly integrated their perspectives into the latter science, and now the use of ichnofacies or ichnofabric analysis in sequence stratigraphy is standard practice (MacEachern et al. 2007a, 2007b). Starting in the 1980s, ichnologists even began saying phrases such as "basal surface forced regression," whether they liked it or not. (As an irritated aside, the lingo of sequence stratigraphy can become somewhat daunting for geological novices. In my experience, a slightly wrong use of "maximum flooding surface" and "transgressive systems tract" in a conversation with sequence stratigraphers often yields scornful stares and self-satisfied snorts of derision.) In Georgia, ichnology is also part of the tool kit that geologists use for figuring out the sequence stratigraphy of Cenozoic coastal-plain deposits, particularly economically important Eocene clay-bearing formations (Martin 2009b).

Other than improving our sciences, and occasionally demonstrating the power of words in establishing social hierarchies in science, why does this all matter? For one, as mentioned earlier, sequence stratigraphy and ichnology are fully integrated into economic geology as well-tested methods for figuring out the spatial relations of hydrocarbon source and reservoir rocks, thus aiding in exploration for such strata (MacEachern et al. 2007b). In other words, if you drive an internal combustion vehicle—hybrids included—or even use natural gas for heating or cooking, thank an ichnologist or geologist today. More importantly, however, is the application of sequence stratigraphy for figuring out earth history, especially for unraveling the long relationship of the world's oceans and atmospheres with life and how this relates to present conditions. Between the time I originally wrote this sentence and your reading these words, sea level went up just a little more, a rise resulting in transgressive systems tract deposits on top of a transgressive surface, migrating landward on marine-flooding surfaces. Will these surfaces become maximum-flooding surfaces sooner, rather than later? Trace fossils and sequence stratigraphy can give a few tentative answers, in which the past helps us to predict the future of changing sea levels (Chapter 11).

THE NEO–PALEO DANCE OF ICHNOLOGICAL IDEAS

In the preceding material, I explained how trace fossils and ichnofacies, when integrated with other geological information, can be used not just to interpret ancient environments from the land to the deep sea, but to figure out how the biota of these environments reacted to changing environments through time. This may seem like quite a claim, especially if the ancient deposits lack what most people associate with "real" paleontology, such as shells, bones, and other evidence of bodily remains that form the bulk of museum displays, or sell for millions of dollars at auctions. Nonetheless, the guiding principles supplied by neoichnology make such necromancy more than just possible, and much more robust than if geologists described body fossils alone from a sedimentary deposit, or just the sediments and their physical sedimentary structures.

For example, what if a sedimentary sequence like the Georgia barrier islands became preserved intact in the geologic record, but lacked any body fossil evidence, such as carbonized plant debris, shells, or bones? This is not such a flight of the imagination, as many sedimentary rocks lack body fossils, which were victims of either their original depositional environments not preserving them, or acidic groundwater that later dissolved any buried shells and bones. In the case of the Georgia barrier islands, their paleoecology could still be interpreted for the most part, as many of the keystone species are also, not coincidentally, prolific tracemakers. In terrestrial and freshwater environments, alligators and gopher tortoises would clearly leave evidence of their presence through their large den structures (Chapter 7), and the deep burrows of crayfish would make it into the fossil record (Chapter 5). In fully terrestrial environments, ants and other important burrowing insects would have made nests with excellent preservation potential, although evidence for the ecological impacts of wood-boring and leaf-eating insects would require preservation of plant body fossils (Chapter 5). In coastal-dune environments, ghost crabs, along with sea oats, ants and other insects, mice, lizards, and sea turtles would all have left abundant and prominent subsurface traces, although invertebrate and vertebrate tracks would be more rare (Chapters 5, 7, 9). Intertidal and subtidal environments would likewise hold a record of infaunal burrows, although much of the predatory behavior, such as shell drilling, chipping, and breaking, would be more difficult to interpret if no molluscan shells were preserved (Chapter 6). Among the more challenging groups of animals to interpret in a paleoecological context without their body fossils would be most birds and mammals that lived in terrestrial, freshwater, and marginal-marine environments. For example, the most common traces of mammals are tracks, trails, and feces, which would have a lower chance of becoming trace fossils than, say, burrows. Nonetheless, the sheer number of these traces would ensure that at least a few would win the fossilization lottery, although a paleoichnologist would have to already hold a good mental picture of how such tracks might look once in the geologic record.

How will we better refine definitions of ichnofacies, ichnofabrics, and other holistic concepts of using trace fossils in toto for interpreting ancient communities and environments? I submit that a better understanding of paleoichnology will require going back to the basics and studying neoichnology, especially the holy trinity—substrates preserving traces, anatomies of tracemakers, and behaviors of tracemakers—as well as how small and cumulative processes resulting in individual traces and ecologically consistent assemblages might unite with the preservation of those traces. Once we have studied all of these in earnest, we can sneak a peek at what is in the rock record and compare our new search images with those of the past. Sounds basic and boring, but it is only one of those two if you let yourself go to the dance with the trace fossils. Just put yourself in the place of the tracemakers, and ask, "How could I make that trace?," given the right combination of substrate, anatomy, and behavior in the setting of its original environment, with its winds, waves, tides, temperatures, and plant and animal cohorts.

11

FUTURE STUDIES, FUTURE TRACES

LOOKING FOR TRACES IN ALL THE WRONG PLACES

A self-deprecating realization struck me in the middle of the fall of 2008, as startling as looking in a mirror and suddenly noticing a large, unsightly growth projecting from the side of my head. I had been spending too much time inside lately, devoted largely to teaching classes; grading exams (or, more likely, putting off their grading); committee meetings; reading, writing, and sending e-mails; and, once in a great while, in between all of these activities, reading about traces, trace fossils, and ichnology in general. In contrast, too little had been devoted to fieldwork, especially tracking. Tracking and neoichnology in general are similar to the use of another language, where skills atrophy with disuse, become tattered around the edges, and otherwise become traces themselves of knowledge and wisdom once held so keenly. Yes, some of that previous experience can be put in the bank, so to speak, but if left alone, it does not gain any interest.

So it was that my daily scanning of the ground, often engaged wherever I went, faded into a past version of itself. A walk across campus normally yielded a bounty of noticed animal sign, much of it done by eastern gray squirrels (*Sciurus carolinensis*): small, shallow, circular dig marks in pine straw; broken pecans and acorns with distinctive tooth marks; branches with leaves still green and fresh, diagnostic 45° angle shear marks on their ends; four-by-four bounding trackway patterns across piles of oak leaves; subtle trails leading up tree trunks from habitual wear on the bark, caused by up-and-down travel; leafy dreys secured high up in the trees; and so on. Occasionally these everyday squirrel traces were accompanied by something a little more unusual, such as small piles of songbird feathers—either left by a feral cat on the ground or a hawk in a tree—as well as the distinctive five-toed tracks of raccoons (*Procyon lotor*) or opossums (*Didelphis virginiana*), denoting a midnight amble across the campus. I would also smile whenever I saw a perfectly round and meticulously maintained hole in a flower bed, as I knew a chipmunk burrow was not so welcome to a university employee who might be tending to that tiny patch of earth. Human traces sometimes amused, too, like the parallel trails left on a campus sidewalk by a student's skateboard, preserved by thin, orange residues of Georgia clay, imparted as the student hurried to class after it rained the previous day.

Instead of noticing any or all of these, though, inner reflections that were so often prompted by the question "What traces did you see?" were summarized by "Oops, forgot to look!" In tracking, what is also analogous to learning a foreign language is the tenet of immersion. My best and most significant leaps of learning in tracking and neoichnology happen during uninterrupted spans of field time, whether over days or weeks. Sadly, this sort of raptness had been replaced with distraction. Consequently, traces went unnoticed, and my attitude suffered.

Hence my receiving approval for a proposal on a book about ichnology was a needed tonic, having arrived via e-mail in the middle of that semester. A further anticipatory boost came from planning fieldwork for December–January, between semesters filled with teaching in classrooms and committee meetings: the former personally rewarding, the latter something other than that. Now I had incentive to leave behind artificial classroom and office environments, with their constant attachment to electronic communication methods and attendant responsibilities, for the environments of the Georgia coast and the myriad of traces being produced there daily, all ripe for study. My wife Ruth and I made plans to visit five islands—Cumberland, Jekyll, Ossabaw, St. Catherines, and Sapelo—all of which were composite islands, made of Pleistocene and Holocene sediments (Chapter 2). Each island was idiosyncratic in its own way, holding similar traces

Figure 11.1. Learning neoichnology on the Georgia coast without just reading about it in a book (no offense). (a) The author with his students on Sapelo Island, teaching about ecological succession in storm-washover fans and their expected traces, and using the beach sand as a whiteboard. (b) Students using their bodies and minds to make educational traces in the upper intertidal zone, Sapelo Island. In this exercise, they were instructed to draw the idealized sequence of ichnofacies that would result from a sea-level rise affecting a maritime forest, again using the beach sand to aid in their expression of these ideas. Both photographs by Ruth Schowalter.

but with different accents or preservational modes. We experienced these islands and their traces in a glorious, whirlwind trip that took just under three weeks, with both of us wishing it could have lasted longer. Photographs, field notes, and friendships remain, though, and as is typical for any of my visits to the Georgia coast, I saw traces that were either new to me or otherwise previously unnoticed, a continual learning that fed into curiosity and made us want to go back again.

Indeed, nearly one year later, when the same melancholy caused by a dearth of nonichnological activities (both physical and cerebral) intruded again during the fall of 2009, I knew the cure was to go to the coast again. This time it was with undergraduate students in tow, intended mainly to prepare them for an upcoming field course, but also to teach them how to start viewing the world as an ichnological patchwork (Figure 11.1a,b). Even a mere weekend trip was a shot in the arm that helped us to gain a shared perspective, all through using an off-campus and outdoor classroom teeming with life and its traces.

This seeking of renewal via traces and nature is not unusual. Famed ecologist, entomologist, and naturalist E. O. Wilson has written extensively about the concept of biophilia, which he explained in depth in a book with the same title, although with a noteworthy subtitle: *The Human Bond with Other Species* (Wilson 1984). In this book and another he coedited (Kellert and Wilson 1993), Wilson proposed that humans have a hardwired, intrinsic love of nature, a deep-seated, lingering, and inheritable adaptation residing in all of us that is often suppressed (or subconscious) in everyday lives spent in urbanized environments. He even gathered enough evidence about this idea to propose it as a hypothesis, albeit a difficult one to test or disprove on a broad scale. After all, think about the challenges of representative sampling for this phenomenon when the species you are studying has more than 7 billion individuals with a worldwide distribution, all overlain with cultural variations that overlap or conflict with one another. Yet a few studies in the health sciences have supported the basic tenet that "nature is good for you." For example, in one such study, researchers found that sick people in hospitals who had a view of some natural scene from their windows experienced healing rates significantly faster those who looked onto parking lots or other artificial vistas, or they otherwise just felt better (Frumkin 2001).

As a result of these investigations and additional studies, other researchers from a variety of disciplines have joined Wilson in the chorus of people who see a connection between our evolutionary history and a profound fondness for plants, animals, and places filled with rocks, sediments, water, and life. Along these lines, one of the more well-known and influential works was Richard Louv's (2005) *Last Child in the Woods: Saving Our Children from Nature-Deficit Disorder.* In this book, he documented how children in the United States, increasingly detached from natural settings during the past few decades, were suffering from maladies that manifested as decreased student engagement in learning, poor mental and physical health, and societal problems in general. This condition is easily diagnosed when a child can identify more commercial brand names than species of plants or animals in her or his surroundings. One of Louv's key assumptions, which he supports with a wealth of data and anecdotes—albeit mostly the latter—is that "nature is good for kids." Of course, this implies that being outside and learning lessons from the natural world should benefit adults too. Louv's book was followed up by one where another author encouraged cultivating nature awareness in children by teaching them to pay attention to changes in the seasons (Van Noy 2008). Tracking—and by default neoichnology—has entered this worldview too, as naturalists now often incorporate observations and interpretations of animal tracks and other traces as a means for honing a broader awareness of nature in younger people and adults (Rezendes 1999; Elbroch and Marks 2001; Elbroch 2003; Young and Morgan 2007). Some nature programs designed for children, such as one hosted at the University of Michigan (BioKids), even take advantage of technological tools that trick kids into observing nature by recording data on handheld computers (Parr et al. 2004). On the other hand, attempts to show how tracking is innate and part of human history (Liebenberg 2001) have met with varying degrees of success. Nonetheless, this idea is supported by the common occurrence of ancient petroglyphs in many parts of the world that accurately illustrate vertebrate tracks and other traces, as well as a few indigenous cultures that still retain well-honed tracking skills, passed down through many generations (Chapter 1).

But is this just a form of tribalism, in which the lovers of an ideology become advocates who imagine the same love is submerged inside all others? Like all human experiences, the individual situation should be examined before the collective. In my case, I was lucky compared to many of my contemporary urban dwellers, having been encouraged to learn about natural history at an early age and having access to real, living examples for regular study. While growing up in a relatively small town in Indiana (Terre Haute) in the 1960s, I had a backyard that served as an ecological microcosm of interacting insects and plants that supplied good, curiosity-based reasons to go outside every day. Ants were a source of constant fascination, as I watched them moving along invisible routes, systematically touching antennae or otherwise acting like a well-behaved network. Only much later would I learn about pheromone trails and the physiological complexity of ant nests, and how this way of life for ants based on their traces was yet another facet of ichnology (Chapter 5). Praying mantises, not lions, were my favorite predators, because I could watch them snatch hapless insects and consume them with their complicated mouthparts from mere centimeters away. Wolf spiders, with their instant reactions to any insect touching their funnel-like ground webs, often competed for my affections in this wild kingdom of the backyard and mind. No matter how closely I sat near the television, which I did too often—a harbinger of this malaise for the next few generations of American children—I could not actually touch or directly observe lions, grizzly bears, crocodiles, sharks, tyrannosaurs, or all of the other killers featured in programs so much more often than arthropods. I also began to learn about the mutualism of plants and animals by watching monarch butterflies alight on milkweed, and honeybees and bumblebees taking flight from flowers, their hind legs laden and yellowed by pollen. Likewise, galls in leaves and stems became sources of fascination, as I tried to figure out which insects caused such wound reactions in their host plants. In short, this nascent affinity for nature provided the spark of interest that continued into adulthood, when I finally got to see my first ocean, and later led to frequent trips to the Georgia barrier islands, a circle that keeps closing on itself, yet starting anew.

My parents, neither of whom had gone to college, were bemusedly supportive of my zoological and botanical obsessions, and we made frequent trips to the public library, my favored place for answering questions provoked by my field observations. This was also my first taste of becoming stymied by some of the contradictions between field-gained knowledge and published literature, and I soon learned that the only sureness about natural history was that no one was really sure. On weekends, my father frequently took my siblings and me on brief day trips to the nearby Indiana countryside for hunting or fishing, forays that imbued a comfort in forests, lakes, and streams not evident in most young people I teach or otherwise encounter today. Crayfish in particular caught my attention whenever we went to local streams. The bottom-dwelling ones, which I now presume were nonburrowing, could be spotted in the clear waters rather easily if one sat patiently on a nearby bank. At the slightest hint of danger, a flick of the tail sped them away, a marvelously simple display of predator avoidance. More than 40 years later, when I completed a study of the oldest fossil crayfish from the Southern Hemisphere, discovered in Australia (Martin et al. 2008), I thought back fondly to my earliest memories of those benthic crayfish in Indiana and their behaviors. Interestingly, I now do a similar sort of hunting and fishing with neoichnology, which I like to think is returning to the previously mentioned human roots of pattern recognition that had practical purposes in our evolutionary history. So again, am I just projecting an inherent love of nature on others because of my special experiences, or is this feeling really there inside others? I like to think the two ideas are not mutually exclusive, and that both could be right.

With this background in mind, tradition can dictate what to do next in the neoichnology of the Georgia coast: get into the field; observe organismal traces; meticulously describe these traces; and (if at all possible) watch organisms make traces. Only through this simple outline can we better understand tracemakers in the context of their environments, which will make for more accurate assessments of trace fossils. Initially, this appeal to such a fundamental approach may seem retrogressive and curmudgeonly, and perhaps suggests a rejection of modern technological aids available to someone practicing neoichnology in the early twenty-first century. Indeed, am I a modern ichnologically oriented version of a Luddite, advocating a return to the ways of our forebears by leading a rebellion that smashes the machines that have enslaved us? Not quite, however much a recalcitrant computer may tempt. Moreover, the popular perception of Luddites as irrational antitechnologists should be corrected. The original Luddites, named after their fictional leader, Ned Ludd, destroyed mechanized looms because these devices were replacing their jobs (and hence their livelihoods) in the textile mills of early nineteenth century Great Britain (Thompson 1965; Fox 2004). Accordingly, the new tools we have available now to augment or enhance our ichnological studies should not be ignored, but we would be wise to ensure fieldwork ("dirt time") is not being replaced entirely with machine work ("screen time"), thus taking away the very lifeblood of our curiosity-driven inquiries into the natural world.

What are some of these suspiciously regarded new tools, and how do they relate to ichnology? Perhaps the most important are GPS (global positioning systems) and GIS (geographic information systems), which are used routinely to document where organismal traces occur. These devices are considerably more accurate and reliable than the previous use of magnetic compasses and paper maps, and when used in conjunction with computer programs, such as CyberTracker, can add place-based ecological and paleoecological context to traces and trace fossils, respectively. This sort of technology can prevent problems related to when traces and trace fossils were treated like postage stamps or coins, curios to be collected—photographed, measured, and identified—and then cataloged, with no further analysis or geographical context. Digital cameras represent another technological boon to neoichnologists, who can now instantly check their photographs while still standing in front of their subjects, erase and retake photos, or record high-definition digital videos of organisms making traces that can be later studied image by image. (Ironically, soon after I started recording digital video clips on the Georgia coast, my students, born and bred in the indoors generation of the post-1980s computer age, had to endure in-class showings of lovingly taken, minutes-long, pixelated digital videos of moon snails moving slowly across the Georgia sand-flat surfaces. Meanwhile, there they sat in their Atlanta classroom, most no doubt wishing they were on that Georgia beach instead.) For terrestrial traces, infrared sensors connected to digital cameras can now be set up across trails to record pictures of the animals that traverse them, capturing mug shots that confirm the identities of suspected trackmakers, including rare and endangered species. The pictures can be downloaded and sent across the world in seconds through satellite transmission. Multipurpose handheld devices, such as PDAs (personal digital assistants) of varying brands that will remain unnamed (no unconscious advertising here), can be used to send and receive e-mail; locate your position with the built-in GPS; plot it on digital maps stored on the PDA; and take a digital photograph linked to the location on the map.

So thanks to modern technology, you can now take a picture or video of an alligator trackway on a Georgia beach, record its location, e-mail the picture to your friends on the other side of the world, upload it to your personal Web site (accompanied by much braggadocio, of course), and, time zones permitting, receive feedback from your friends on it, all while taking in the sweet sounds of fado, bossa nova, or other world music on your earpiece, yet not leave the beach. Sounds idyllic, scientifically advanced, and downright civilized, does it not?

In a sense, though, if all of these actions are occurring, you are not fully in that place. As a result, good science is probably not happening, either: a divided mind that does not settle down in the here and now is one that will gather an incomplete picture. For example, often on the Georgia coast, particularly when I am leading a weekend field trip with my students, the limited time there creates a sense of haste, in which we hurriedly take pictures of a trace and tracemaker, replay the photo on the view screen of the camera, and once satisfied with the quality of the picture, move on. Later, we may or may not download it onto a computer, and may or may not look at it again once it is stored on a hard drive. Furthermore, if we do glance at an image later in the comfort of our homes or offices on a computer screen, it omits (minimally) the briny, organic smells of the nearby ocean or marsh, wafted by a cool breeze from the east; the high-pitched calls of laughing gulls and the croaks of fish crows; the feel of the fine-grained beach sand underfoot; and the scurrying motion of the ghost crabs just upslope on the beach, caught in the periphery of your widely dispersed vision. Likewise, questions that could have been asked about the previously mentioned alligator tracks become difficult to answer when taken out of their ecological setting. For example, where did the alligator go after it walked onto the beach? When did it walk through the area? Did it pause or otherwise slow down during its journey, whether to rest or when encountering an obstacle or another animal? What was the obstacle or animal? Why was the alligator there at that time—was it looking for food, a mate, or simply out for a stroll? Did its walking through that area impart any changes in the behaviors of other tracemakers? How did the weather or time of day affect its activities? Imagine your e-mailed picture of the trackway provoking these questions from friends and colleagues, and ask yourself whether you could answer any of them, even with the actual alligator trackway directly in front of you.

This is why I recommend simpler tools for the sake of better science and teaching. In my experience, most good science stems from slowing down and observing, which can be accomplished by using old-fashioned, low-tech, and completely unsexy field notebooks and pencils. Drawing and taking notes in a journal requires deliberate thought and repeated looking, hence leading to new insights, discoveries, and other "aha!" moments that may never be discerned from a 12-megapixel image taken on a whim, or recorded by a speeding finger on a PDA screen or miniature keyboard. This is not to say that GPS, PDAs, GIS, and other acronym-laden software and hardware should be eschewed completely, as these can gather an enormous amount of data in short spans of time, which accordingly can be processed much more quickly than written notes in a book. Nonetheless, to truly observe and learn from field observations, repeated and thoughtful looking, listening, smelling, and other sensory means of gathering information are essential. In other words, twenty-first-century tools are not substitutes for taking the time to detect the phenomena of your surroundings, including the ubiquitous traces of organismal behavior that simply should not be missed if they are sought.

Accordingly, whenever teaching my students in the field, and particularly on the Georgia barrier islands, awareness is emphasized from start to finish: cell phones are turned off, computers are put away, and idle chatter about other places and people is discouraged so they can concentrate on receiving the sights, sounds, smells, and tactile sensations of the here and now. Some of my field discussions on ichnology will then employ the ready-made canvases of recently washed beach sands, where I will draw in the sand with my fingers to illustrate some concepts. (In other words, my presentation hardware is actually soft sand, and the software is my mind, even if the latter requires constant upgrades.) This method is by no means original, though, as I am imitating the elders of indigenous clans in Australia, who have probably used this mode of teaching for more than 30,000 years. Yet there is a good reason why this way of passing on information has persisted in humanity since the Pleistocene. Recently, I instructed students to try the same technique of peer teaching while we are on a Georgia beach, causing a lively, active scene as they separated into small groups and explained concepts to one another, using only the sand and their fingers (Figure 11.1b). This is done partly so they learn better how to teach in the absence of technology, employing whatever materials may be immediately available around them, but also how to be more place based in their teaching and learning. For example, the ecological zones between ghost crab and ghost shrimp burrows (Chapters 3, 6, 9) are much easier to describe and explain if they are next to one another at that time and place, rather than waiting (and hoping) that their significance is somehow discerned later in an enclosed room, far removed from where these tracemakers live.

Last, keep in mind that the extensive discoveries and documentation of life traces on the Georgia coast from the 1960s through the 1980s (Chapter 2) all took place before the computer revolution had seeped into and dominated all aspects of modern society, including academe. In fact, the most productive and successful of Georgia coast ichnologists, Robert (Bob) Frey, whose output of papers are still the envy of most modern ichnologists, never used a computer, and many of his predecessors did not either. The most accomplished of all ichnologists, Dolf Seilacher, likewise shunned computers for the vast majority of his life, and only succumbed to them for e-mail and word processing when he was in his 80s. In fact, the most sophisticated instrument Seilacher used in most of his studies was a camera lucida, an instrument invented in the early nineteenth century. A camera lucida consists of an arrangement of mirrors that projects images of viewed objects onto paper that are then easily traced. Seilacher used one to draw traces and trace fossils, which resulted in his exquisitely rendered illustrations that embody his sharp observations (Seilacher 2007). In short, our highly evolved brains and senses are always our primary tools, and putting ourselves in "the places with the traces" is also essential for making new discoveries, even in the well-studied sands, muds, shells, and other substrates of the Georgia coast.

Interestingly, one of the most common questions I received from people when telling them the setting of this book was, "Why such a specific place?" Skeptical, literal-minded scientists were perhaps the most critical, saying that an area representing a fraction of a percent of world coastlines, and with traces documented only from the last 50 years or so, would comprise slices of space and time too narrow to properly represent neoichnology as a science. Hopefully, the preceding hundreds of pages describing traces, tracemakers, and their natural history likely provided a small bit of evidence to refute such naysayers. Nonetheless, at a more visceral and somewhat heretical level, my choosing of the Georgia coast for the study of neoichnology does not really matter. Perhaps the most important realization a reader of this book may gain is that neoichnology is everywhere, and cannot be ignored if one knows what to observe. There is no need to go to the barrier islands of Georgia, South Carolina, Florida, or even the east coast of the United States to see organisms behaving and interacting with substrates. Ichnology can be seen, appreciated, and studied wherever life leaves its marks, including college campuses and urban settings (Rindsberg and Martin 2009). The principles just learned, and the search images just made, are universally applicable, and can be applied wherever sand, mud, wood, shells, or bones might be. Just put down this book, switch off your computer, hide your cell phone, and step outside.

NEW TOOLS, NEW EYES, NEW TRACES

Apocalyptic scenarios aside, technology will not suddenly disappear in the foreseeable future, so neoichnology, whether it is done on the Georgia barrier islands or elsewhere, will likely incorporate tools that augment our well-evolved senses. Furthermore, if the past 30 years can be used as a guide for prediction, everything technological mentioned in this section will be terribly outdated in just a few years, and any specific references will be as chuckle inducing as a punch card formerly used for programming computers. (The probability that many of you have no idea what a punch card might be is an example of what I mean.) Hence I will only discuss tools in a general sense that will likely persist in one form or another, while omitting details of their operations so their applications become more apparent. An old acronym used in computer programming is applicable here: GIGO, which translates to "garbage in, garbage out." An updated translation is "garbage in, gospel out," reflecting how faith in a computer's infallibility can also result in faulty reasoning. So as new tools become available for studying Georgia coast traces and their tracemakers, we should be aware of how science is not automatically improved through the use of updated technology, especially if accompanied by the descriptors "cutting edge" and "state of the art." Nonetheless, our wariness can be tempered by knowing that technology can certainly help to advance the science of ichnology in some instances; a few instruments may eventually become as standard as rock hammers and hand lenses are for geologists.

Film cameras and videotape, replaced rapidly by digital photography and video (respectively) in the first decade of the twenty-first century, are now nearly extinct as items ichnologists used to routinely include in their lists of field equipment. As a result, digital cameras are more typical, and will certainly continue to offer detailed depictions of traces and tracemaking into the near future. Field-oriented ichnologists now have fewer worries about their photos making it home with them: no more lost film canisters, film fogged by x-ray machines at airports, entire rolls of film shot with the lens cap on, or otherwise wondering if pictures came out, a worry that extended from the time a picture was taken until after the film was processed. Now a photographer quickly reviews images on her or his camera, deletes the bad ones, retakes shots, and backs up files that same day on a laptop computer or other external storage device. Many digital cameras also include the ability to take audio memos, in which the operator attaches an audio file to a specific photograph or video and thus provides added information. In other words, an ichnologist can now whisper lovingly into her or his camera, "callianassid shrimp burrow, actively pumping out feces," and unlike previous times, that information is now stored in the camera for future reference. This combination of audio and visual information also helps to prevent later viewings that provoke quizzical expressions and earthy questions varying on the theme of, "What the hell is that?" And yes, this combination of video and audio also opens up further possibilities for performance art, an all too common circumstance in ichnology.

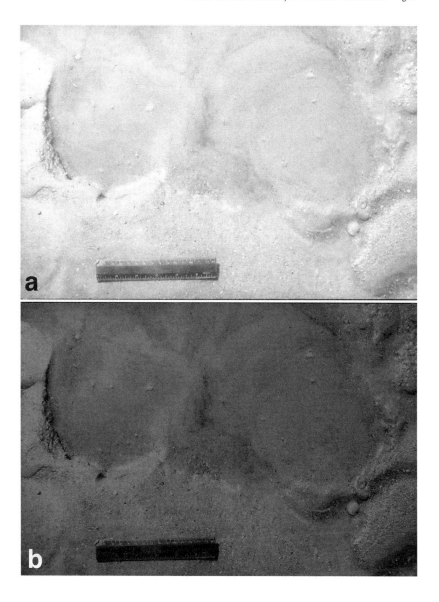

Figure 11.2. A badly taken photograph of a modern trace on the Georgia coast is improved through just a little bit of digital processing, providing some information that might not have otherwise have been observed. (a) Original image from a 35-mm slide, taken in 1998 on Tybee Island and showing the feeding traces made by a ring-billed gull (see Figure 9.13g). The original photo was overexposed and slightly out of focus. (b) After digital scanning of the slide in 2010, the image was sharpened and its contrast and brightness adjusted, pulling out details of the trace and thus requiring less squinting and imagination.

Yet another advantage of digital photography is the ease with which an image can be improved later on a computer to help communicate important information. Image processing programs can sharpen photos that were taken slightly out of focus, increase their contrast, decrease their brightness, adjust their colors to more of their original state, or otherwise pull out details that our eyes saw in the field, but may not be conveyed by the initial image (Figure 11.2). The disadvantage of this ability to manipulate an image, of course, can lead to inaccurate or otherwise misleading representations, as well as dangling temptations in front of scientists that most resist, but a few may use to cheat. Although such a problem has not yet been detected in ichnology or paleontology, a team of cellular biologists who published a paper in a prestigious journal were later found to have done just that: they altered digital images to better fit their preferred hypothesis (Snyder and Loring 2006). As a result, I tell my students to never believe what they see in a photograph nowadays, and keep in mind how these images, like any other forms of scientific data removed from their original sources, might be later proved wrong. Nonetheless, this advice also would have been just as valid in the late nineteenth century, as film photographers began using technical tricks, such as double exposures, to fool otherwise smart people, among them Sir Arthur Conan Doyle, amateur ichnologist and author of the Sherlock Holmes stories (Stashower 2001). The moral of the story is to stay moral, while being aware of the possibility of scientific immorality.

One of my favorite forms of digital processing, and far less controversial than the preceding for its scientific value, involves combining successive photos for panoramic images. As a practical application, I have applied this technique to depict outcrops or vertebrate trackways that could not be fitted into a single frame, and labeled on these where trace fossils, traces, or other forms of data were found by using the ensuing composite image (Figure 11.3). Through this technology, geologists who were once infamous for taking multiple photographs of long, tall outcrops, then pasting them together into odd-looking collages—thus impressing their art school colleagues—can now better represent what they actually saw in the field. These photographs are also well augmented by high-resolution digital video, in which sweeping, continuous pans help to document the view of an outcrop. This is a more important form of documentation than one might think, as outcrops that once held beautiful and easily accessible trace fossils can all too quickly vanish under the blades of bulldozers and

Figure 11.3. Digitally merged panorama arranged from two successive photos of raccoon (*Procyon lotor*) tracks on a beach showing back-and-forth movement by raccoons parallel to the shore: Cumberland Island. Scale bar = 10 cm.

backhoes, only to be replaced by anonymous strip malls. I also sometimes use sequential photographs and video for visual documentation of modern trackways, providing a way to later double-check measurements and maps made of these, in which the primary data might have been recorded in paper notebooks. Paper, pencils, and tape measures, though, are still the best tools for studying trackways: if a tracker wants to actually learn anything from the tracks, that is.

Much of the visual and auditory information gathered by ichnologists, however, will be linked increasingly to GIS databases. This practice ensures that organismal traces or trace fossils are more directly connected to their geographical and ecological context, and also aids follow-up research. One of the more powerful techniques for preserving a place-based context to traces or trace fossils involves taking large numbers of overlapping, high-resolution digital photographs of a locality, which are later stitched together into 360° panoramas that can be viewed as if the person is standing at the site. An added benefit of this digital technology will be the ability to zoom in and out of specific features in the panoramas, which allows for varied perspectives on large- to small-scale features. For example, a photograph of ghost crab burrows on a beach could be seen for its details—the width of the burrow, the sand pile outside the burrow, trackways leading to and out of it. Then these details can be compared to those of other burrows within the same virtual stroll, all while placing them within their relative ecological zones on a beach: foredunes, berm, upper intertidal, and intertidal. This sort of visual documentation, when integrated with high-precision GPS-linked data, provides archival information of a sort, especially as sea level continues to rise during the twenty-first century and changes local environments, while simultaneously displacing their tracemakers into adjacent habitats (Chapter 2).

Measurements of traces in the field normally can be taken quickly with a ruler or tape measure, and hopefully aided by a field assistant who can write down measurements as they happen. Nevertheless, precision, which affects accuracy, can be a problem with this tried and almost true method, and I suggest that some such work, especially for smaller traces, is best done with calipers. This might seem retrogressive in the face of recent technological advances, but calipers have been upgraded considerably in the last few years too. For example, when taking measurements that I know will later require statistical analyses, I use digital calipers, which, like a typical pocket calculator, have a small-screen display of its numbers. Numbers are measured to the nearest $1/100$ of a millimeter, which often constitutes accuracy overkill, but these can be rounded up at the site or later if necessary. More sophisticated (and expensive) calipers will even download measurement data directly to a computer, taking out the middle man of writing down the information and later transcribing it, a process that can produce unwanted errors. Once measurement data are placed in spreadsheets, these can be analyzed statistically, but of course should always be viewed through the previously mentioned awareness of an ichnologist's mortal enemy, GIGO.

What about batteries, you say? Yes, battery life is a primary concern of every field-oriented ichnologist, paleontologist, or ecologist, and can quickly become a topic of lively conversation among those scientists who have embraced the use of technology in the field. (On the other hand, laboratory-bound scientists would never be caught too far from their precious electrical outlets; thus they never have such discussions.) Solar cells, however, provide power for the just-mentioned digital calipers; so as long as I have light, I can take measurements of, say, a 7.83 mm wide wasp burrow (Chapter 5). This example hints at some of the big changes that will take place with ichnological fieldwork in the twenty-first century: unplugging of field computers and other electronic instruments from power grids.

Wireless communication networks have already taken out a few of the tethers that used to keep scientists leashed to nearby phone line or cable outlets, but the omission of outlets will surely change some aspects of how fieldwork is planned and conducted. Solar and battery technologies are becoming more advanced every day, and increased battery life combined with more efficient solar panels will probably be one of the most liberating aspects of scientists spending more time in the field. This is said with the hope, however, that field scientists will somehow still find the excuse to stay disconnected from e-mail, a luxury that keeps them out of the reach of administrators hungering for more unpaid committee work from their faculty during field seasons.

Laser scanning is becoming more prevalent as a technique with dinosaur tracks and other trace fossils (Petti et al. 2008; Platt et al. 2010; Belvedere and Mietto 2010), but also can be applied to modern tracks and other traces. In this technique, a handheld or mounted scanner emits a laser beam across the surface of a track or other trace. The laser beam reflects off that surface, calculates distances from the source to points on the trace, and these data are stored. Data points are then sent to a computer that translates these differences into surface relief, and reconstructs the trace, rendered as a three-dimensional model on a computer screen and analyzed using standard spatial analysis software. These three-dimensional models can even be imported into manufacturing software that programs a machine to fabricate an exact, solid replica of the trace. More importantly, though, tracks or other traces imaged by laser scans suddenly appear as the equivalent of topographic maps, showing all of the low and high points, including pressure-release structures (Chapters 7–9), and otherwise aiding in communicating differences in relief. I imagine many trackers, frustrated by people who represent tracks as simple cartoon outlines, will nod their heads in approval at this method for illustrating the wrinkles, dimples, and moles composing the personality of individual tracks. Using this same technique, sediment mounds made by burrowing crayfish (Chapter 5), ghost crabs, fiddler crabs, or ghost shrimp (Chapter 6) can be reproduced in startling, three-dimensional detail, looking very much like volcanoes or mountains if viewed without accompanying scales. Consequently, the precision of such devices will probably also aid in better defining invertebrate traces and trace fossils, and not just tracks. Indeed, as of this writing, laser scanning was being used to describe both invertebrate and vertebrate trace fossils. In some instances, it was used to create archival, digital files of dinosaur tracks, which if left in the field will surely erode or otherwise degrade from their original state (Bates et al. 2008).

For those people who do not have ready access to a laser scanner (in other words, nearly everyone), a more readily available tool for people interested in accurately imaging a trace or trace fossil is a so-called poor man's version: a flatbed document scanner. These devices, which normally act like a photocopier and make a digital scan of a document to reproduce it as an image, can also be used to scan traces. How is this done, especially with softground traces, such as tracks and burrows? The short answer is "not often," and neoichnologists normally would be limited to scanning surfaces of, say, plaster casts of tracks with little surface relief (such as shorebird tracks) or traces impregnated with epoxy resin that were later sliced to show two-dimensional surfaces. I have also known paleoichnologists to use it as a technique for quick, high-resolution digital imaging of trace fossils on bedding planes in small hand samples (Richard Bromley, personal communication, 2008). However, beware of how rocks might scratch or otherwise damage the glass on these scanners. I also do not recommend that vertebrate feces or other more squishy traces be placed on such scanners, unless done as some part of an elaborate practical joke (and that idea definitely did not come from me).

What about enhanced study of burrows that go underground for long distances, such as those made by burrowing owls, gopher tortoises, or alligators (Chapter 7)? Subsurface details of smaller, invertebrate burrows in aquariums (or terrariums) can be examined through computer tomography (CT) scans. CT scans have even been used on coprolites, which can help to define undigested materials inside a coprolite and thus better identify the tracemaker and its diet (Farlow et al. 2010). Of course, these techniques are impractical for burrows (and their surrounding sediment) that cannot be fitted into a CT scanner, not to mention the difficulty of convincing medical personnel to use such equipment for ichnological investigations in the first place. Smaller burrows, though, can be cast, whether through dental plaster, epoxy resin, or, in the case of ant nests, molten zinc (Chapter 5), then mapped by laser scanners. More conventionally, large burrows, like those of gopher tortoises, have been examined since the 1990s through combinations of fiber optics, infrared light, and cameras (Kent et al. 1997). With this technique, researchers manually insert these visual probes into the burrow, which sends a live feed to a viewer and is recorded for future review. (People who have endured colonoscopies or other forms of endoscopy are probably simultaneously nodding their heads and cringing in recognition at this description.) Such systems will likely become more popular with ichnologists and field biologists interested in further learning about the still-mysterious world of burrowing animals living underneath our feet. This technique also holds many advantages over casting with epoxy resin or similar liquids that harden after pouring into a hole. For example, the small camera and its mount constitute a minimal invasion of a burrow, do not harm the animals (how many epoxy resin casts have inadvertently included the tracemaker, frozen in perpetuity?), and are especially valuable for detecting any commensal species in larger burrows.

A potentially effective and noninvasive technique for imaging large burrows is ground-penetrating radar (GPR). GPR uses an emitter to send high-frequency electromagnetic radiation (microwaves, specifically) underground. These waves bounce off objects, open spaces, or other areas with different electrical properties, and are received by a unit that translates them into two-dimensional profiles, which are then compiled and rendered as three-dimensional models. At the time of this writing (2011), some colleagues and I had just begun successfully applying this method to gopher tortoise burrows on St. Catherines Island, and are hoping to test its effectiveness further. GPR also could be useful for studying alligator burrows (Chapter 7). With these structures, researchers can observe the confines of the burrow from its outside, with a greater degree of personal safety than sending in a human as a (very) temporary burrow-commensal species.

Less connected to present-day field research in neoichnology is the matter of previous studies on the ichnology and natural history of the Georgia coast, some going back into the nineteenth century. In this respect, information technology, through digital scans of older documents and their online distribution, is helping researchers become more aware of who preceded them and what was studied. Original documents of these field naturalists, including field notebooks, are often held in libraries or museums far away from researchers and the general public alike, and may not be readily accessible, especially as university and public libraries continue to cut budgets and personnel. Furthermore, some older literature is becoming too fragile to handle or otherwise does not need to be subjected to more wear. I have sometimes held original copies of classic natural history books with trembling hands, hoping I would not sneeze or otherwise add my traces to their physical states.

How to avoid such dilemmas? The gentle reader may have noticed literature cited in this book dating from the first half of the twentieth century, as well as more recent studies. Most of the older references I found through online databases and search engines, which then connected to electronic files. Most of these files are digital scans of original papers, saved in portable document format (PDF). Thus I was able to read the original words of wisdom of people who not only lived well before the technology that enabled this miraculous access, but preceded me in my personhood and consciousness. Yes, this serves the same purpose as going to a library and looking up the references in paper volumes, which all good naturalists should do anyway. In practice, however, it is potentially more powerful in its capacity to quickly pull together, cross reference, and synthesize a wide range of literature from over several centuries. As tireless library scientists and work study students (no doubt fueled by caffeine) scan more of this literature, then upload files and link them to online databases, I predict that ichnologists and nonichnologists will become increasingly aware of, enthralled, and inspired by these older papers, some filled with superb descriptions of plants, animals, behaviors, and landscapes of the recent past. Digital versions of these documents also allow for word searches, which further streamline any quests for the written origin of ichnological ideas. This ease helps ichnologists to see how sometimes a "new" ichnological insight was actually noted by someone else a long time ago, thus giving credit where it is due and providing a historical context to the development of ichnological ideas.

For example, Han Reineck first defined categories for relative amounts of bioturbation per area in vertical exposures of sediment, such as those sampled through box cores (0 percent bioturbation, 1–15 percent bioturbation, and so on), in the early 1960s in a German-language journal (Reineck 1963). His categories, however, did not become better known until their inclusion in an English-written textbook (Reineck and Singh 1975). This meant that many English-only speakers missed the evidence for his birthing of this important concept. As a result, some practitioners of the ichnofabric concept in the 1980s (Chapter 10) often failed to cite Reineck's original 1963 paper, even if they managed to mention his later publications. In turn, Charles Darwin preceded Reineck's works by about 80 years, as he conducted long-term experiments on the burrowing of earthworms and meticulously measured their effects (Darwin 1881). Thus Darwin was very likely the first quantitative neoichnologist (Pemberton and Frey 1990). Such oversights of historical literature were completely understandable in the 1980s, especially if professors were exhorting their graduate students to ignore the dim, dark past of the 1960s–1970s. Of course, the current availability of such literature as PDFs or other electronic documents, all distributed through the World Wide Web, and aided by online searching tools and translation programs for nonnative speakers of some languages, makes such oversights less common, or at least less excusable. Translation programs, however clunky in their production of amusing word choices and sentence structures, also can assist monolingual folks to get something out of this literature, as opposed to nothing. Best of all, works of many other nineteenth-century naturalists are freely available online, including Darwin's book on earthworms, *The Formation of Vegetable Mould through the Action of Worms* (1881).

In essence, a basic principle in the history of science I emphasize to my students is that newer is not always better, no matter how much recent authors may hype their published results. Furthermore, I urge that my students open their eyes to observations and ideas written well before they were born. These pregenerational traces of naturalist thoughts can often enlighten a reader on how a paper or book on neoichnology published in 2012 (such as this one) may just be replicating the results of other researchers from 50–150 years ago. Granted, the modern scientists may be using really cool gadgets that make the research look a lot sexier, trendier, and hip, and thus more worthy of accolades from colleagues and increased research funding. (Remember that the descriptors "cutting edge" and "state of the art" are crucial for convincing someone about the importance of the research.) How many times have neoichnological studies been repeated without knowledge of who may have left their footprints in the same places? As Isaac Newton famously once said, "If I have seen a little further, it is by standing on the shoulders of giants," an aphorism that reminds us how progress in science stems from the discoveries of the past. (To be fair, though, Newton may have meant this in a mocking way, aimed at a rival scientist—Robert Hooke—and not so much as one holding others in awe [Keyes

2007]. Thus your metaphorical mileage may vary.) Reading the works of the past improves our ability to learn from the present, which in neoichnology is again projected into the far more distant past when applied to paleoichnology.

THE ALIENS AMONG US: THE NEOICHNOLOGY OF INVASIVE SPECIES

Aside from technological tools, and on a more pessimistic note, the increasing effects of invasive species will likely play havoc with neoichnologists documenting traces they think are from native species. The most prominent examples of much-reviled exotic tracemakers on the Georgia barrier islands are feral hogs (*Sus scrofa*) and fire ants (*Solenopsis invicta*), both of which have adverse impacts on charismatic and loved species of tracemakers, such as sea turtles (Chapter 8). Invasive species, however, also include mostly benign animals, such as mole crickets, despised only by golfers and other advocates of monotonously manicured grassy lawns. Moreover, feral cattle, normally seen as innocuous (but wild) versions of their milk- and beef-producing brethren, are known to excessively tromp onto and chomp coastal salt marshes (Chapter 9).

Does this mean that such traces should be ignored, disregarded as unnatural, and treated with disdain? By no means! Descriptions of tracks, trails, burrows, feces, and other traces made by invasive species not only have intrinsic scientific value, but more practically can be used to detect the presence of feral species without depending on visual sightings. For example, people rarely spot feral cats on Jekyll Island. Hence a nonichnologically inclined casual visitor may react with disbelief if told that hundreds of these cats are spread throughout the most pristine parts of the island. Nonetheless, their traces tell the tales, such as piles of feathers accompanied by tracks, speaking of the devastating effects cats have on ground-nesting birds. Feral cats are exceedingly common in back-dune meadows on the southern end of the island, and their traces then blend in with those of domestic cats on the more human-populated northern end (Figure 11.4a). Furthermore, ichnology is already being used as a research tool in both conservation and the management of invasive species on the Georgia coast with regard to sea turtles. For instance, volunteers trained in recognizing turtle trackways and nest sites may also record evidence of hogs sniffing out these nests, conveyed by tracks and depredation traces at nest sites. These

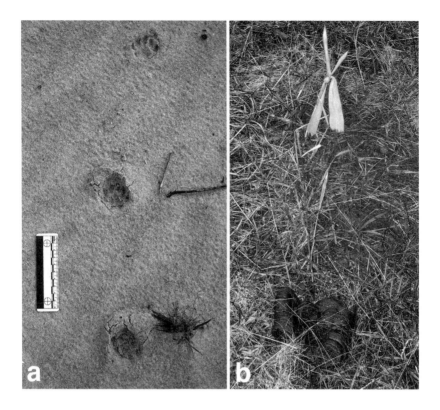

Figure 11.4. Traces made by feral animals that likely will be better understood through neoichnological approaches and applied in conservation biology along the Georgia coast. (a) Feral cat (*Felis domestica*) tracks: Jekyll Island. (b) Feral cow (*Bos taurus*) scat, its position marked by wire topped by bright-colored flagging: Sapelo Island.

same volunteers also use ichnology to determine whether sea turtle nests are in danger from nearby fire ant nests, as well as potential hazards posed by native species, such as burrowing ghost crabs and raccoons.

Feral cattle, which live on Sapelo Island, are likewise hard to find in maritime forests there. They have seemingly reverted to behaviors shared by their Pleistocene ancestors, and, once spotted, quickly vanish into the thick understory of a maritime forest. However, their tracks, trails, grazing sign, and dung piles all lend to firmly establishing their presence, numbers, sexes, and population structures (Chapters 8, 9). In an exercise done with my students on Sapelo, we found numerous, separate and fresh (less than

24 hours old) dung piles left by feral cattle in an open, grassy field. Moreover, these were clearly made within an hour of one another. Dung-eating insects had barely touched them, and raindrops from the morning had just started to register on the still-soft patties, which exuded an earthy, grassy perfume. We hypothesized that these piles represented a herd, grazing together and staying more or less stationary in the field. Without the use of GPS units, CyberTracker, GIS, or other technological tools, we then proceeded to map the distribution of the cows with what we had: our senses, and stiff wires with brightly colored flagging on one end, which we planted upright next to each patty (Figure 11.4b). A quick and dirty field version of peer review happened, as I told them they could not mark the location of a patty unless they all agreed it matched their already established criteria, thus excluding older, desiccated droppings. Once completed, we stood back to admire our map, a field spotted by bright blue flags denoting a former bovine presence. These visual aids revealed that the herd had 23–25 cattle, although a few cattle had likely defecated twice in succession along a trackway, making the determination of their numbers less certain; they were spaced apart by at least one or two cow lengths (which most assuredly is not a formal international unit of measurement); they were mostly adults, but included a few subadults, in which variously sized patties were accompanied by appropriately sized tracks; and the herd mostly faced the north, slowly moving in that direction as it grazed. Later that morning, we proudly showed our handiwork to a Georgia state-employed conservation biologist, and the students related their analysis to him. With a mixture of amusement (at our methods) and surprise (at our results), he told us that we had just described a specific herd he and others had seen moving through this same area during the last few days. Thus our dung-mapping exercise, accomplished with youthful enthusiasm, minimal equipment, and ichnological know-how, had succeeded in defining the group behavior, distribution, and population composition of a feral species. We also effectively demonstrated the potential for how some simply and frugally applied neoichnological methods can answer difficult questions about invasive species, which then may result in responsible policy decisions affecting their management.

Accordingly, as conservation biologists and ecologists assess the feasibility of ecosystem management and restoration on some Georgia barrier islands, no doubt they will use neoichnology as a means for better detecting feral species, documenting their impacts, and understanding their unseen behaviors. Basic tracking techniques, which include track identification, track aging, interpretations of behavior, and so on, can be combined with other neoichnological concepts and facilitated by GPS devices using tracking programs or other means of combining spatial data with field observations. For example, CyberTracker compiles downloadable databases of ichnological and ecological information that can be analyzed through spreadsheet programs, GIS, and other spatial analysis software (Young and Morgan 2007). Professional biologists and trained volunteers, some of who include much maligned plugged-in youths who were well trained by adults to fear the outdoors, are already doing such studies worldwide in wilderness areas, urban centers, and everywhere in between those extremes. Hence these developments reflect hope that the ways of the past can be melded with those of the present to enrich the future.

Alas, as of this writing (and as far as I know), CyberTracker and tracking methods have not been attempted for assessing feral species on the Georgia coast. Nevertheless, I hope the preceding discussion will help to inspire such future cooperation between students, teachers, local residents, and wildlife biologists, thus combining ichnology, education, and research in ways that meaningfully contribute to improving ecosystems and communities on the Georgia coast.

THE GEORGIA COAST: A TRACE ODYSSEY

For anyone who wants to understand the ichnological landscapes of the Georgia barrier islands, as well as the individual traces and their tracemakers, you simply have to go there. All of the dazzling, gee-whiz technological applications just discussed, including those that create beautifully colorful digital maps, high-resolution and three-dimensional renderings of traces, and sweeping panoramas of trace-bearing sites that can be viewed at the click of a button, are no substitute for the sensory cornucopia of field experiences. Similarly, visualizing a Georgia barrier island from space through satellite photos that are then translated as digital images on a computer screen, however convenient and informative, chokes the joy out of self-discovery and exploration. Exploration sometimes means just getting lost on the sandy roads or trails of island interiors or a lengthy and expansive beach, and not seeing another human being the remainder of the day, however frightening that might seem to most people, especially if cut off from e-mail or cell phones. On such glorious days, traces are your companion wherever you go, filling your eyes and mind with scripts that tell stories of constantly evolving landscapes, seascapes, and their biota.

If we could go back to the Georgia coast for one last indulgence of fantasy, we might consider which traces and tracemakers of the barrier islands will gain more of our attention in future years. Much like the single word of advice—"Plastics"—received by Dustin Hoffman's character in the 1967 movie *The Graduate,* I will say to present and future ichnologists who have graduated by getting this far into the book: "Insects." These animals were the most neglected of tracemakers in all of the classic neoichnological studies done by Frey, Howard, and other ichnologists in the 1960s–1980s (Chapter 2), and only the traces of a few species were mentioned briefly. As readers may recall, this remarkably diverse group of arthropods receive a decidedly large amount of attention in this book, and I overtly encouraged follow-up research on their traces and behavioral ecology (Chapter 5). Such endeavors will require more cooperative studies between ichnologists and entomologists, but the results will be immediately applicable for those scientists. Additionally, considering how termite mounds, ant nests, wasp brooding chambers, cicada burrows, and other insect traces can serve as precise environmental indicators, their better definition should reap immediate rewards for geologists and paleontologists. The clearest benefit of such study is the easier discernment of whether trace fossils in an outcrop, such as Yellow Banks Bluff on St. Catherines Island or Raccoon Bluff on Sapelo Island, represent marine or continental conditions (Chapter 9). Another area of unplumbed knowledge in insect ichnology concerns the many traces they form in leaves, wood, and other plant tissues. Previous studies done on modern plant–arthropod interactions and their applicability to the fossil record have emphasized tropical regions (Currano et al. 2008), rather than subtropical or temperate zones. A bonus of such research on the Georgia barrier islands would be a better understanding of how plants react or protect against insect herbivory, as well as more precise identifications of similar tracemakers from fossil leaves and wood, done recently with leaf-cutter bees (Sarzetti et al. 2008).

Of course, more work is needed on the ichnology of terrestrial vertebrates (Chapters 7–8), marginal-marine invertebrate and vertebrates (Chapters 6 and 9, respectively), and plants (Chapter 4). For example, the ichnology of alligators, done in a more comprehensive way, would benefit legions of vertebrate paleontologists who would like to know how the behavioral ecology of these archosaurs compares to that of their long-lost Mesozoic and Cenozoic cousins (Farlow and Elsey 2010). Related to this is the lack of intensive study on large vertebrate burrows in general. Although gopher tortoises (*Gopherus polyphemus*) are well studied, other sizeable burrow makers that dig deep dens, such as coyotes (*Canis latrans*) and armadillos (*Dasypus novemcinctus*), are less documented for their ichnological prowess. Armadillos in particular should be examined from the standpoint of their potential ecological impact as an invasive species, having made their way to the Georgia barrier islands only in the 1970s (Laerm et al. 1999). Likewise, the ichnology of shorebirds on the Georgia coast would make for a fine study, documenting how these birds leave their marks and the profusion of behavioral nuances these reflect. Such a study would require multiple years and the seasons they hold, so it would be for those who love working in the field and with birds.

Additionally, although some marginal-marine vertebrates and invertebrates are the best-studied tracemakers of the Georgia barrier islands, they certainly should not be ignored for their contributions to future research, either. For example, studies others and I have done in recent years have demonstrated how ichnologists Frey and Howard (Chapter 2) may have overlooked a few marine invertebrate traces and their potential importance in paleoichnology (Martin 2006a; Martin and Rindsberg 2007, 2011). A few other recent investigations conducted by former students of George Pemberton have also examined burrowing rates of some marine invertebrates and other ichnological parameters in marginal-marine environments of the Georgia coast (Gingras et al. 2008a, 2008b).

By devoting an entire chapter to plant traces (Chapter 4), I hope to spur some interest in this terribly neglected area of ichnology that, much like a wounded tree sending out chemotransmitters, cries out for more attention. Similar to how cooperative work between entomologists and ichnologists will greatly advance the ichnology of insects, so will botanists combining forces with ichnologists make plant traces better understood. For example, I predict that species-level recognition of root traces can be achieved, especially if aided by laser scanning and computer analyses. Such methods can more easily place root structures and architectures into morphological categories, which will lend to more accurate diagnoses of root trace fossils (Gregory et al. 2004, 2006).

Figure 11.5. The predicted future of a coastal dune on St. Simons Island as expressed by traces and their substrates, in which a vertical sequence shows how dunes containing ghost crab, insect, and root traces (*below*) are succeeded vertically by balanoglossid, polychaete, sea cucumber, and ghost shrimp burrows; then mud urchin, mantis shrimp, bivalve burrows, stingray feeding traces, and various marine invertebrate burrows (*above*). Human traces in the form of artifacts are not included, although the sea-level rise itself is a human trace.

With all of that said, the neoichnology of the Georgia coast is expected to undergo noticeable alterations as global climate change affects the lives and distributions of tracemakers (Figure 11.5). In my personal inquiries of the islands' longtime residents and nearby mainland, they have confirmed what is now a certainty in scientific circles: sea level has gone up noticeably during their lifetimes along the coast. This transgression has radically altered coastal environments, causing a lateral migration of salinity to travel up Georgia coastal rivers and otherwise invade formerly terrestrial and freshwater habitats. When combined with the effects of invasive species, encroaching human development, and other modifications of ecosystems, modern traces become moving targets that may vary daily in small ways, but will undergo huge changes within another human generation. The greater likelihood of severe tropical storms hitting the Georgia Bight and other parts of the southeastern U.S. coast is another prediction of global climate change models (Michener et al. 1997; Greenland et al. 2003; Fraser 2006). Hence the seemingly slow, incremental adjustments of ecosystems associated with rising sea level may be punctuated by high-impact events that impart instantly disastrous effects, affecting both tracemaking communities and human residents alike.

This is not to be all doom and gloom, a common public relations pitfall of many modern environmentalists and a trait gleefully seized on by their everything-is-just-dandy critics, some of whom (not coincidentally) are advocates of further coastal development. Traces will continue to be made, albeit as ever-changing ichnocoenoses as sea level shifts more landward and as humans adjust to whatever such alterations might bring. Live oaks will send their roots outward, as pine trees grow their roots downward. Crayfish will burrow in freshwater wetlands of island interiors, regardless of whether people notice their prominent pelletal towers popping up behind shopping malls and condominiums. Alligators, armadillos, and gopher tortoises will construct their dens, just as insects will build their nests, whether as individuals or through communal activity. Callianassid shrimp will dig straight down before moving off into their intricate subsurface tunnels, reinforcing their burrow walls and ensuring at least some of their former homes will become fossils some day. Seagulls will continue to fly across tidal flats, searching for the outlines of burrowing bivalves and gastropods, using their own avian form of ichnology. Species will come and species will go, but very likely, long after these words have faded, either on paper or as digital media, tracemakers will continue to leave their marks on the Georgia coast. The trace fossils of the future are being made in the present as we contemplate, imagine, and celebrate them. How fortunate we are to see these life traces being made before our eyes, now.

Appendix

Modern tracemakers of the Georgia barrier islands, including recently extirpated organisms and nonnative species. List is organized alphabetically by genus and species, and sorted according to broad taxonomy (plants, invertebrates, vertebrates) and habitats (terrestrial–freshwater and marginal–marine).

Terrestrial–Freshwater Plants

Acer rubrum—red maple
Aristida stricta—wiregrass
Cladium jamaicense—giant sawgrass
Equisetum spp.—scouring rushes
Gordonia lasianthus—loblolly bay
Ilex opaca—American holly
Ilex vomitoria—yaupon holly
Magnolia grandiflora—southern magnolia
Magnolia virginiana—sweet bay
Myrica cerifera—wax myrtle
Nymphaea stellata—water lily
Peltandra virginica—green-arrow arum
Persea borbonia—red bay
Pinus echinata—short-leaf pine
Pinus elliottii—slash pine
Pinus palustris—long-leaf pine
Pinus serotina—pond pines
Pinus taeda—loblolly pine
Polypodium polypodioides—resurrection fern
Pontederia cordata—pickerel weed
Quercus laurifolia—laurel oak
Quercus virginiana—live oak
Sagittaria latifolia—arrowhead
Sarracenia spp.—pitcher plants
Scirpus spp.—bulrushes
Taxodium ascendus—pond cypress
Tillandsia usneoides—Spanish moss
Typha angustifolia—cattail
Woodwardia virginica—Virginia chain fern
Zizania aquatica—wild rice
Zizaniopsis miliacea—cutgrass

Terrestrial–Freshwater Invertebrates

Agapostemon virescens—halictid bee
Andricus spp.—oak gall wasps
Aphaenogaster miamiana—funnel ant
Aphodius spp.—dung beetles
Archodontes melanopus—live-oak stump borer
Bombus spp.—bumblebees
Calcaritermes nearcticus—nearctic spur-legged termite

Camponotus floridanus—Florida carpenter ant
Canthon spp.—dung beetles
Centruroides hentzi—striped scorpion
Centruroides vittatus—striped scorpion
Ceratophaga vicinella—gopher tortoise moth
Chalybion californicum—blue mud dauber
Cicindela dorsalis—tiger beetle
Cicindela hirticollis—tiger beetle
Cicindela marginata—tiger beetle
Cicindela striga—tiger beetle
Cicindela trifasciata—tiger beetle
Copris gopheri—gopher tortoise beetle
Coptotermes formosanus—Formosan termite (nonnative)
Crematogaster clara—acrobat ant
Cryptotermes brevis—rough-headed drywood termite (nonnative)
Dendroctonus frontalis—southern pine beetle
Diplocardia mississippiensis—earthworm
Disholcaspis cinerosa—gall wasp
Dolichovespula maculata—bald-faced hornet, or yellow jacket
Dorymyrmex bureni—pyramid ants
Eutrichota gopheri—gopher tortoise fly
Forelius pruinosus—lawn ants
Geolycosa spp.—wolf spiders
Hesperotettix floridensis—marsh grasshopper
Hogna spp.—wolf spiders
Hydrocanthus spp.—burrowing water beetles
Idia gopheri—gopher tortoise moth
Incisitermes minor—termite (nonnative)
Incisitermes snyderi—southern drywood termite
Kalotermes approximatus—dark southeastern drywood termite
Lumbricus rubellus—red worm, or earthworm
Machimus polyphemus—gopher tortoise fly
Magicicada septendecim—periodical cicada
Mischocyttarus mexicanus—paper wasp
Myrmeleon crudelis—ant lion
Onthophagus spp.—dung beetles
Phanaeus spp.—dung beetles
Pogonomyrmex badius—Florida harvester ant
Polyergus lucidus—slave raider ant
Procambarus lunzi—hummock crayfish
Procambarus talpoides—mole crayfish
Pseudosuccinea columella—American ribbed-fluke snail
Reticulitermes flavipes—eastern subterranean termite
Scapteriscus abbreviatus—short-winged mole cricket (nonnative)
Scapteriscus borellii—southern mole cricket (nonnative)
Scapteriscus vicinus—tawny mole cricket (nonnative)
Sceliphron caementarium—black-and-yellow mud dauber
Solenopsis invicta—imported red fire ant (nonnative)
Sphecius speciosus—cicada killer
Stictia carolina—Carolina sand wasp
Tibicen auletes—dog-day (annual) cicada
Trachymyrmex septentrionalis—leaf-cutter ant
Trypoxylon politum—organ-pipe mud dauber
Vejovis carolinianus—Southern devil scorpion
Vella americana—ant lion
Vespula maculifrons—eastern yellow jacket
Vespula squamosa—southern yellow jacket
Xyleborus glabratus—red-bay ambrosia beetle (nonnative)

Xylocopa micans—southern carpenter bee
Xylocopa virginica—eastern carpenter bee

Terrestrial–Freshwater Vertebrates

Agkistrodon piscivorus—cottonmouth
Alligator mississippiensis—American alligator
Ambystoma talpoideum—mole salamander
Ambystoma tigrinum—tiger salamander
Amphiuma means—congo eel
Anhinga anhinga—anhinga
Anolis carolinensis—green anole
Apalone ferox—Florida softshell turtle
Athene cunicularia—burrowing owl
Bison bison—bison (extirpated)
Blarina carolinensis—southern short-tailed shrew
Bos taurus—cattle (nonnative)
Branta canadensis—Canada geese
Bubo virginianus—great horned owl
Bufo terrestris—southern toad
Campephilus principalis—ivory-billed woodpecker (extirpated)
Canis latrans—coyote
Canis lupus—gray wolf (extirpated)
Castoroides canadensis—beaver
Cemophora coccinea—northern scarlet snake
Centropomus pectinatus—tarpon snook (nonnative)
Ceryle alcyon—belted kingfisher
Chelonia mydas—green turtle
Chelydra serpentina—snapping turtle
Cnemidophorus sexlineatus—six-lined racer
Colinus virginianus—bobwhite
Coluber constrictor—southern black racer
Columbina passerina—ground dove
Crotalus adamanteus—eastern diamondback rattlesnake
Cyprinodon variegatus—sheepshead minnow
Dasypus novemcinctus—nine-banded armadillo
Deirochelys reticularia—chicken turtle
Didelphis virginiana—Virginia opossum
Drymarchon couperi—eastern indigo snake
Dryocopus pileatus—pileated woodpecker
Elaphe obsoleta quadrivittata—yellow rat snake
Equus cabullus—horse (nonnative)
Eucinostomus lefroyi—mottled mojarra
Eumeces egregious—mole skink
Eumeces fasciatus—five-lined skink
Eumeces inexpectatus—skink
Eumeces laticeps—broad-headed skink
Farancia abacura—eastern mud snake
Fundulus majalis—striped killifish
Gambusia affinis—western mosquitofish (nonnative)
Glaucomys volans—southern flying squirrel
Gopherus polyphemus—gopher tortoise
Grus canadensis—sandhill crane
Hirundo rustica—barn swallow
Homo sapiens—human
Hyla spp.—tree frogs
Kinosternon baurii—striped mud turtle
Kinosternon subrubrum—eastern mud turtle
Lemur catta—ring-tailed lemur (nonnative)
Lepomis gulosus—warmouth
Lepomis macrochirus—bluegill
Lutra canadensis—river otter
Lynx (Felis) rufus—bobcat
Masticophis flagellum—coachwhip
Melanerpes carolinus—red-bellied woodpecker
Meleagris gallopavo—wild turkey
Microtus pennsylvanicus—meadow vole
Myotis lucifugus—common brown bat
Nerodia fasciata—Florida water snake
Notophthalmus viridescens—eastern newt

Odocoileus virginianus—white-tailed deer
Ophisaurus spp.—glass lizards
Otus asio—eastern screech owl
Peromyscus floridanus—Florida mouse
Picoides pubescens—downy woodpecker
Procyon lotor—raccoon
Pseudacris spp.—peepers
Puma concolor—cougar (extirpated)
Rana capito—gopher frog
Rana sphenocephala—southern leopard frog
Riparia riparia—bank swallow
Scalopus aquaticus—eastern mole
Scaphiopus holbrookii—eastern spadefoot toad
Sceloporus undulatus—northern fence lizard
Scincella lateralis—ground skink
Sciurus carolinensis—eastern gray squirrel
Sciurus niger—southern fox squirrel
Sphyrapicus varius—yellow-bellied sapsucker
Stelgidopteryx serripennis—northern rough-winged swallow
Sus scrofa—hog (nonnative)
Tamias striatus—eastern chipmunk
Thamnophis sauritus—peninsular ribbon snake
Thamnophis sirtalis—eastern garter snake
Turdus migratorius—American robin
Ursus americanus—black bear
Zenaida macroura—mourning dove

Marginal-Marine Plants

Batis maritima—saltwort
Borrichia frutescens—sea-oxeye daisy
Distichlis spicata—salt grass
Hydrocotyle bonariensis—pennywort
Ipomoea pes-caprae—railroad vine
Ipomoea stolonifera—fiddle-leaf morning glory
Iva frutescens—marsh elder
Iva imbricata—beach elder
Juncus roemerianus—black needle rush
Juniperus virginiana—red cedar
Panicum amarum—bitter panic grass
Salicornia virginica—glasswort
Sabal palmetto—cabbage palm
Spartina alterniflora—smooth cordgrass
Spartina patens—salt-meadow cordgrass
Tamarix gallica—salt cedar (nonnative)
Uniola paniculata—sea oats
Yucca alifolia—Spanish bayonet

Marginal-Marine Invertebrates

Acanthohaustorius millsi—amphipod
Albunea paretii—beach mole crab
Alpheus heterochaelis—snapping shrimp
Amphipholis gracillima—burrowing brittle star
Anadara brasiliana—incongruous ark
Ancinus depressus—isopod
Armases (Sesarma) cinereum—squareback marsh (or wharf) crab
Astropecten articulatus—margined (or royal) sea star
Balanoglossus aurantiactus—golden acorn worm
Barnea truncata—Atlantic mud piddock, or fallen angelwing
Biffarius biformis—Georgia ghost shrimp
Brachidontes recurvus—hooked mussel
Busycon carica—knobbed whelk
Callichirus major—Carolina ghost shrimp
Callinectes sapidus—blue crab
Cerebratulus lacteus—silky ribbon worm
Ceriantheopsis americanus—North American tube anemone
Chaetopterus variopedatus—parchment worm
Chiridotea caeca—isopod
Clibanarius vittatus—striped hermit
Cliona celata—red boring sponge
Crassostrea virginica—eastern oyster
Cyathura polita—slender isopod
Cyrtopleura costata—angelwing

Dinocardium robustum—giant Atlantic cockle
Diopatra cuprea—tube worm
Donax variabilis—coquina clam
Dosinia discus—disk dosinia
Emerita talpoida—common mole crab
Ensis directus—Atlantic jackknife
Eupleura caudata—thick-lipped oyster drill
Eurytium limosum—white-clawed mud crab
Exosphaeroma diminutum—isopod
Geukensia demissa—ribbed mussel
Glycera dibranchiata—bloodworm
Haustorius spp.—amphipods
Hemipholis elongata—blood brittle star
Heteromastus filiformis—capitellid thread worm
Ilyanassa obsoleta—mud snail
Lepidactylus dytiscus—amphipod
Lepidopa websteri—square-eyed mole crab
Limnoria tripunctata—wood-boring gribble, or isopod
Limonoria lignorum—wood gribble (isopod)
Limulus polyphemus—horseshoe crab
Littoraria irrorata—marsh periwinkles
Luidia clathrata—lined sea star
Martesia cuneiformis—wedge piddock
Melampus bidentatus—coffee-bean snail
Mellita isometra—keyhole sand dollar
Menippe mercenaria—stone crab
Mercenaria mercenaria—southern quahog
Microphiopholis gracillima—brittle star
Moira atropos—mud heart urchin
Mulinia lateralis—dwarf surf clam
Nassarius vibex—common mud snail
Neohaustorius schmitzi—amphipod
Nephtys bucera—shimmy worm
Nereis succinea—clam worm
Neverita duplicata—common moon snail
Ocypode quadrata—ghost crab
Oliva sayana—common olive shell
Onuphis microcephala—parchment worm
Ophiophragmus filograneus—brittle star
Pagarus annulipes—hairy hermit crab
Pagarus longicarpus—long-wristed hermit crab
Pagarus pollicaris—flat-clawed hermit
Panopeus herbstii—Atlantic mud crab
Parahaustorius longimerus—amphipod
Paranthus rapiformis—sea onion anemone
Pectinaria gouldii—trumpet worm
Petricola pholadiformis—false angelwing
Renilla reniformis—sea pansy
Saccoglossus kowalevskii—helical acorn worm
Scolelepis squamata—palp worm
Sesarma reticulatum—heavy (or purple) marsh crab
Sinum perspectivum—baby's ear gastropod
Solen viridis—green jackknife clam
Sphaeroma destructor—isopod
Sphaeroma quadridentatum—isopod
Squilla empusa—mantis shrimp
Synapta (Leptosynapta) inhaerens—apodous sea cucumber
Tagelus divisus—purplish tagelus
Tagelus plebeius—stout razor clam
Tagelus plebeius—razor clam
Terebra dislocata—Atlantic auger
Thyone briareus—sea cucumber
Uca minax—red-jointed fiddler
Uca pugilator—sand fiddler
Uca pugnax—mud fiddler
Upogebia affinis—coastal mud shrimp
Urosalpinx cinerea—eastern oyster drill

Marginal-Marine Vertebrates

Actitis macularius—spotted sandpiper
Ardea alba—great egret
Ardea herodias—great blue heron
Arenaria interpres—ruddy turnstone
Calidris alba—sanderling
Calidris alpina—dunlin
Calidris canutus—red knot

Calidris minutilla—least sandpiper
Carcharhinus acronotus—blacknose shark
Carcharhinus limbatus—blacktip shark
Carcharhinus milberti—sandbar shark
Carcharhinus obscurus—dusky shark
Caretta caretta—loggerhead sea turtle
Charadrius melodus—piping plover
Charadrius semipalmatus—semipalmated plover
Charadrius vociferous—killdeer
Charadrius wilsonia—Wilson's plover
Cistothorus palustris—marsh wren
Condylura cristata—star-nosed mole
Corvus ossifragus—fish crow
Dasyatis americana—southern stingray
Dasyatis sabina—Atlantic stingray
Dasyatis sayi—bluntnose stingray
Dermochelys coriacea—leatherback sea turtle
Galeocerdo cuvier—tiger shark
Ginglymostoma cirratum—nurse sharks
Haematopus palliatus—American oystercatcher
Haliaeetus leucocephalus—bald eagle
Larus altricilla—laughing gull
Larus argentatus—herring gull
Larus delawarensis—ring-billed gull
Larus marinus—great black-backed gull
Limnodromus griseus—short-billed dowitcher
Limosa fedoa—marbled godwit
Malaclemys terrapin—diamondback terrapin
Mugil cephalus—striped mullet
Mustela vison—mink
Mustelis canis—smooth dogfish
Neofiber alleni—round-tailed muskrats
Numenius americanus—long-billed curlew
Nyctanassa violacea—yellow-crowned night heron
Ophichthus gomesi—shrimp eel
Ophichthus ocellatus—palespotted eel
Oryzomys palustris—marsh rice rat
Pandion haliaetus—osprey
Pelecanus erythrorhynchos—white pelican
Pelecanus occidentalis—brown pelican
Peromyscus polionotus—southeastern beach mouse
Phalacrocorax auritus—double-crested cormorant
Pluvialis squatarola—black-bellied (grey) plover
Poecilia latipinna—sailfin molly
Pogonias cromis—black drum
Quiscalus major—boat-tailed grackle
Rhizoprionodon terraenovae—Atlantic sharpnose shark
Rynchops niger—black skimmer
Sphyrna lewini—scalloped hammerhead
Sterna antillarum—least tern
Sterna forsteri—Forster's tern
Sterna hirundo—common tern
Sterna maxima—royal tern
Sylvilagus palustris—marsh rabbit
Trichechus manatus—West Indian manatee
Tringa melanoleuca—greater yellowlegs
Tringa semipalmata—willet
Tursiops truncates—bottlenose dolphin

Bibliography

Abe, T., D. E. Bignell, and M. Higashi. 2000. Termites: Evolution, Sociality, Symbioses, Ecology. University of Chicago Press, Chicago, Illinois, 466 pp.

Albers, G., and M. Alber. 2003. A vegetative survey of back-barrier islands near Sapelo Island, Georgia; pp. 1–4 in J. K. Hatcher (ed.), Proceedings of the Georgia Water Resources Conference, Institute of Ecology, University of Georgia, Athens, Georgia.

Alber, M., and J. E. Sheldon. 1999. Trends in salinities and flushing times of Georgia Estuaries; pp. 528–531 in K. Hatcher (ed.), Proceedings of the 1999 Georgia Water Resources Conference, Institute of Ecology, University of Georgia, Athens, Georgia.

Alexander, C., and V. J. Henry Jr. 2007. Wassaw and Tybee Islands: comparing undeveloped and developed barrier islands; pp. 187–198 in F. J. Rich (ed.), Guide to Fieldtrips, 56th Annual Meeting, Southeastern Section of the Geological Society of America, Georgia Southern University Department of Geology and Geography Contribution Series.

Alford, D. V. 1975. Bumblebees. Davis-Poynter, London, 352 pp.

Allen, E. A., and H. A. Curran. 1974. Biogenic sedimentary structures produced by crabs in lagoon margin and salt marsh environments near Beaufort, North Carolina. Journal of Sedimentary Research 44:538–548.

Allen, J. R. L. 1982. Sedimentary Structures: Their Character and Physical Basis. Developments in Sedimentology 30: Elsevier, Amsterdam, Netherlands, 593 pp.

Anderson, S. 1981. The raccoon (*Procyon lotor*) on St. Catherines Island, Georgia: nesting sea turtles and foraging raccoons. American Museum Novitates 2713:1–9.

Andrews, P. 1990. Owls, Caves, and Fossils. University of Chicago Press, Chicago, Illinois, 239 pp.

Andrus, C. F. T., and V. D. Thompson. 2008. Oxygen isotope analysis of midden mollusks to assess site formation processes and subsistence strategies at the Sapelo Island shell rings. Geological Society of America Abstracts with Programs 40(6):521.

Aresco, M. J. 1999. Habitat structures associated with juvenile gopher tortoise burrows on pine plantations in Alabama. Chelonian Conservation and Biology 3:507–509.

Ash, S. R., and G. T. Creber. 2000. The Late Triassic *Araucarioxylon arizonicum* trees of the Petrified Forest National Park, Arizona, U.S.A. Palaeontology 43:15–28.

Aspey, W. P. 1978. Fiddler crab behavioral ecology: burrow density in *Uca pugnax* (Smith) and *Uca pugilator* (Bosc) (Decapoda Brachyura). Crustaceana 34:235–244.

Bailey, C., and C. Bledsoe. 2000. God, Dr. Buzzard, and the Bolito Man: A Saltwater Geechee Talks about Life. Doubleday, New York, 334 pp.

Baker, L. A., R. J. Warren, D. R. Diefenback, W. E. James, and M. J. Conroy. 2001. Prey selection by reintroduced bobcats (*Lynx rufus*) on Cumberland Island, Georgia. American Midland Naturalist 145:80–93.

Barlow, C. 2000. The Ghosts of Evolution: Nonsensical Fruit, Missing Partners, and Other Ecological Anachronisms. Basic Books, New York, 291 pp.

Barker, G. M. 2001. The Biology of Terrestrial Molluscs. CABI Publishing, Oxon, U.K., 558 pp.

Basan, P. B., and R. W. Frey. 1977. Actual-palaeontology and neoichnology of salt marshes near Sapelo Island, Georgia; pp. 41–70 in T. P. Crimes and J. C. Harper (eds.), Trace Fossils 2. Seel House Press, Liverpool, U.K.

Bates, K. T., F. Rarity, P. L. Manning, D. Hodgetts, B. Vila, O. Oms, A. Galobart, and R. L. Gawthorpe. 2008. High-resolution LiDAR and photogrammetric survey of the Fumanya dinosaur tracksites (Catalonia): implications for the conservation and interpretation of geological heritage sites. Journal of the Geological Society 165: 115–127.

Baucon, A. 2010. Leonardo da Vinci: the founding father of ichnology. Palaios 25: 361–367.

Baugh, T. M., J. A. Valade, and J. A. Zoodsma. 1988. Manatee use of *Spartina alterniflora* in Cumberland Sound. Marine Mammal Science 5:88–90.

Beck, B. B. 1980. Animal Tool Behavior: The Use and Manufacture of Tools by Animals. Garland STPM Press, New York, 307 pp.

Bedatou, E., R. N. Melchor, E. Bellosi, and J. F. Genise. 2008. Crayfish burrows from Late Jurassic–Late Cretaceous continental deposits of Patagonia, Argentina: their palaeoecological, palaeoclimatic and palaeobiogeographical significance. Palaeogeography, Palaeoclimatology, Palaeoecology 257:169–184.

Beddington, S. D., and J. B. McClintock. 1993. Feeding behavior of the sea star *Astropecten articulatus* (Echinodermata: Asteroidea): an evaluation of energy-efficient foraging in a soft-bottom predator. Marine Biology 115:669–676.

Bell, C. M. 2004. Asteroid and ophiuroid trace fossils from the Lower Cretaceous of Chile. Palaeontology 47:51–66.

Bellis, V. J., and J. R. Keough. 1995. Ecology of the Maritime Forests of the Southern Atlantic Coast: A Community Profile. Biological Report 30, National Biological Service, U.S. Department of the Interior, 95 pp.

Belvedere, M., and P. Mietto. 2010. First evidence of stegosaurian *Deltapodus* footprints in North Africa (Iouaride'ne Formation, Upper Jurassic, Morocco). Palaeontology 53:233–240.

Benner, J. S., J. C. Ridge, and N. K. Taft. 2008. Late Pleistocene freshwater fish (Cottidae) trackways from New England (U.S.A.) glacial lakes and a reinterpretation of the ichnogenus *Broomichnium* Kuhn. Palaeogeography, Palaeoclimatology, Palaeoecology 260:375–388.

Bennett, D. H. 1972. Notes on the terrestrial overwintering of mud turtles (*Kinosternon subrubrum*). Herpetologica 28:345–347.

Bergstrom, P. W. 1988a. Breeding displays and vocalizations of Wilson's plovers. Wilson Bulletin 100:36–49.

———. 1988b. Breeding biology of Wilson's plovers. Wilson Bulletin 100:25–35.

Bertone, M., J. Green, S. Washburn, M. Poore, C. Sorenson, and D. W. Watson. 2005. Seasonal activity and species composition of dung beetles (Coleoptera: Scarabaeidae and Geotrupidae) inhabiting cattle pastures in North Carolina (U.S.A.). Annals of the Entomological Society of America 98:309–321.

Bishop, G.A., and E. C. Bishop. 1992. Distribution of ghost shrimp, North Beach, St. Catherines Island. American Museum Novitates 3042:1–17.

Bishop, G. A., and N. A. Brannen. 1993. Ecology and paleoecology of Georgia ghost shrimp; pp. 19–29 in K. M. Farrell, C. W. Hoffman, and V. J. Henry Jr. (eds.), Geomorphology and Facies Relationships of Quaternary Barrier Island Complexes Near St. Mary's Georgia. Georgia Geological Society Guidebook 13.

Bishop, G. A., N. B. Marsh, J. Barron, F. L. Pirkle, and R. S. Smith. 1997. A Cretaceous sea turtle nest, Fox Hills Formation, Elbert Co., CO. Geological Society of America, Abstracts with Programs 29(6):104.

Bishop, G. A., R. H. Hayes, B. K. Meyer, H. E. Rollins, F. J. Rich, D. H. Thomas, and R. K. Vance. 2007. Transgressive barrier island features of St. Catherines Island, Georgia; pp. 39–85 in F. J. Rich (ed.), Guide to Fieldtrips: 56th Annual Meeting, Southeastern Section of the Geological Society of America. Georgia Southern University, Department of Geology and Geography Contribution Series 1.

Bishop, G. A., F. L. Pirkle, B. K. Meyer, and W. A. Pirkle. 2011. The foundation for sea turtle geoarchaeology and zooarchaeology: morphology of recent and ancient sea turtle nests, St. Catherines Island, Georgia, and Cretaceous Fox Hills Sandstone, Elbert County, Colorado; pp. 247–269 in G. A. Bishop, H. B. Rollins, and D. H. Hurst (eds.), Geoarchaeology of St. Catherines Island, Georgia. Anthropological Papers of the American Museum of Natural History 94.

Blair, W. F. 1951. Population structure, social behavior, and environmental relations in a natural population of the beach mouse (*Peromyscus polionotus leucocephalus*). Contributions of the Laboratory of Vertebrate Biology, University of Michigan 48:1–47.

Blonder, B. I., and W. S. Alevizon. 1988. Prey discrimination and electroreception in the stingray *Dasyatis Sabina*. Copeia 1988:33–36.

Booth, R. K., F. J. Rich, and G. A. Bishop. 1999. Palynology and depositional history of Late Pleistocene and Holocene coastal sediments from St. Catherines Island, Georgia USA. Palynology 23:67–86.

Booth, R. K., F. J. Rich, and S. T. Jackson. 2003. Paleoecology of mid-Wisconsinan peat clasts from Skidaway Island, Georgia. Palaios 18:63–68.

Bond, B. T., M. I. Nelson, and R. J. Warren. 2000. Home range dynamics and den use of nine-banded armadillos on Cumberland Island, Georgia. Proceedings of the Annual Conference of the Southeastern Association of Fish and Wildlife Agencies 54:415–423.

Bordy, E. M., A. J. Bumby, O. Catuneanu, and P. G. Eriksson. 2004. Advanced Early Jurassic Termite (Insecta: Isoptera) nests: evidence from the Clarens Formation in the Tuli Basin, Southern Africa. Palaios 19:68–78.

Bost, K. C., and D. D. Gaffin. 2004. Sand scorpion home burrow navigation in the laboratory. Euscorpius: Occasional Publications in Scorpiology 17:1–5.

Botton, M. L. 1984. Diet and food preferences of the adult horseshoe crab *Limulus polyphemus* in Delaware Bay, New Jersey, U.S.A. Marine Biology 81:199–207.

Botton, M. L., B. A. Harrington, N. Tsipoura, and D. Mizrahi. 2003a. Synchronies in migration: shorebirds, horseshoe crabs, and Delaware Bay; pp. 5–32 in C. N. Shuster Jr.,

R. B. Barlow, and H. J. Brockmann (eds.), The American Horseshoe Crab, Cambridge, Massachusetts, Harvard University Press.

Botton, M. L., R. E. Loveland, and A. Tiwari. 2003b. Distribution, abundance, and survivorship of young-of-the-year in a commercially exploited population of horseshoe crabs *Limulus polyphemus*. Marine Ecology Progress Series 265:175–184.

Bowen B. W, and S. A. Karl. 1996. Population structure, phylogeography, and molecular evolution; pp. 29–50 in P. L. Lutz and J. A. Musick (eds.), The Biology of Sea Turtles. CRC Press, Boca Raton, Florida.

Bowers, A. K., L. D. Lucio, D. W. Clark, S. P. Rakow, and G. A. Heidt. 2001. Early history of the wolf, black bear, and puma in Arkansas. Journal of the Arkansas Academy of Science 55:22–27.

Braddy, S. J. 1998. An overview of the invertebrate ichnotaxa from the Robledo Mountains ichnofauna (Lower Permian), southern New Mexico. New Mexico Museum of Natural History and Science Bulletin 12:93–98.

Brandenburg, R. L., Y. Xia, and A. S. Schoeman. 2002. Tunnel architectures of three species of mole cricket (Orthoptera: Gryllotalpidae). Florida Entomologist 85:383–385.

Brannen, N. A., and G. A. Bishop. 1993. Nesting traces of the loggerhead turtle (*Caretta caretta* [Linne]), St. Catherines Island, Georgia: implications for the fossil record; pp. 30–36 in K. M. Farrell, C. W. Hoffman, and V. J. Henry Jr. (eds.), Geomorphology and Facies Relationships of Quaternary Barrier Island Complexes Near St. Mary's, Georgia. Georgia Geological Society Guidebook 13.

Bretz, D. D., and R. V. Dimock Jr. 1983. Behaviorally important characteristics of the mucous trail of the marine gastropod *Ilyanassa obsoleta* (Say). Journal of Experimental Marine Biology and Ecology 71:181–191.

Briggs, D. E. G., W. D. I. Rolfe, and J. Brannan. 1979. A giant myriapod trail from the Namurian of Arran Scotland. Palaeontology 22:273–291.

Bright, D. B., and C. L. Hogue. 1972. A synopsis of the burrowing land crabs of the world and list of their arthropod symbionts and burrow associates. Contributions in Science 220, Natural History Museum, Los Angeles, California, 58 pp.

Britt, B. B., R. D. Scheetz, and A. Dangerfield. 2008. A suite of dermestid beetle traces on dinosaur bone from the Upper Jurassic Morrison Formation, Wyoming, U.S.A. Ichnos 15:59–71.

Brochu, C. A. 2003. Phylogenetic approaches to crocodilian history. Annual Review of Earth and Planetary Sciences 31:357–397.

Brockmann, H. J. 1990. Mating behavior of horseshoe crabs, *Limulus polyphemus*. Behaviour 114:206–220.

———. 2004. Variable life-history and emergence patterns of the pipe-organ mud-daubing wasp, *Trypoxylon politum* (Hymenoptera: Sphecidae). Journal of the Kansas Entomological Society 77:502–527.

Bromley, R. G. 1967. Some observations on burrows of thalassinidean Crustacea in chalk hardgrounds. Quarterly Journal of the Geological Society of London 123:157–182.

Bromley, R. G. 1970. Borings as trace fossils and *Entobia cretacea* Portlock, as an example; pp. 49–90 in T. P. Crimes and J. G. Harper (eds.), Trace Fossils. Seel House Press, Liverpool, U.K.

———. 1975. Trace fossils at omission surfaces; 399–428 in R. W. Frey (ed.), The Study of Trace Fossils. Springer-Verlag, New York.

———. 1992. Bioerosion: eating rocks for fun and profit; pp. 121–129 in C. G. Maples and R. R. West (eds.), Trace Fossils. Paleontological Society Short Course 5.

———. 1993. Predation habits of octopus past and present and a new ichnospecies, *Oichnus ovalis*. Geological Society of Denmark Bulletin 40:167–173.

———. 1996. Trace Fossils: Biology and Taphonomy. Routledge, New York, 384 pp.

Bromley, R. G., and U. Asgaard. 1975. Sediment structures produced by a spatangoid echinoid: a problem of preservation. Bulletin of the Geological Society of Denmark 24:261–281.

Bromley, R. G., and A. D'Alessandro. 1984. The ichnogenus *Entobia* from the Miocene, Pliocene, and Pleistocene of southern Italy. Rivista Italiana Paleontologia e Stratigrafia 90:227–290.

Bromley, R. G., and A. A. Ekdale. 1986. Composite ichnofabrics and tiering of burrows. Geological Magazine 123:59–65.

Bromley, R. G., S. G. Pemberton, and R. A. Rahmani. 1984. A Cretaceous woodground: the *Teredolites* ichnofacies. Journal of Paleontology 58:488–498.

Bromley, R. G., A. Uchman, M. R. Gregory, and A. J. Martin. 2003. *Hillichnus lobosensis* igen. et isp. nov., a complex trace fossil produced by tellinacean bivalves, Paleocene, Monterey, California, U.S.A. Palaeogeography, Palaeoclimatology, Palaeoecology 192:157–186.

Bromley, R. G., L. A. Buatois, J. F. Genise, C. C. Labandeira, M. G. Mángano, R. N. Melchor, M. Schlirf, and A. Uchman. 2007. Comments on the paper "Reconnaissance of Upper Jurassic Morrison Formation ichnofossils, Rocky Mountain Region, USA: Paleoenvironmental, stratigraphic, and paleoclimatic significance of terrestrial and freshwater ichnocoenoses" by Stephen T. Hasiotis. Sedimentary Geology 200:141–150.

Brown, C. A. 2002. Surface density and nocturnal activity in a west Texas assemblage of scorpions. The Southwestern Naturalist 47:409–419.

Brown, C. R. 1993. Origin and history of the potato. American Journal of Potato Research 70:363–373.

Brown, T., Jr. 1999. The Science and Art of Tracking. Berkley, New York, 240 pp.

Brundrett, M. C. 2002. Coevolution of roots and mycorrhizas of land plants. New Phytologist 154:275–304.

Bryan, J., V. S. Wood, and M. R. Bullard. 1996. Journal of a Visit to the Georgia Islands of St. Catherines, Green, Ossabaw, Sapelo, St. Simons, Jekyll, and Cumberland, with Comments on the Florida Islands of Amelia, Talbot, and St. George, in 1753. Mercer University Press, Macon, Georgia, 103 pp.

Buatois, L. A., and M. G. Mángano. 1995. The palaeoenvironmental and palaeoecological significance of the *Mermia* ichnofacies: an archetypal subaqueous non-marine trace fossil assemblage. Ichnos 4:151–161.

———. 2007. Invertebrate ichnology of continental freshwater environments; pp. 285–323 in W. M. Miller III (ed.), Trace Fossils: Concepts, Problems, Prospects. Elsevier, Amsterdam, Netherlands.

Buatois, L. A., M. K. Gingras, J. MacEachern, M. G. Mángano, J.-P. Zonneveld, S. G. Pemberton, R. G. Netto, and A. J. Martin. 2005. Colonization of brackish-water systems through time: evidence from the trace-fossil record. Palaios 20:321–347.

Buczkowski, G., and G. Bennett. 2008. Behavioral interactions between *Aphaenogaster rudis* (Hymenoptera: Formicidae) and *Reticulitermes flavipes* (Isoptera: Rhinotermitidae): the importance of physical barriers. Journal of Insect Behavior 21:296–305.

Buhs, J. B. 2004. The Fire Ant Wars: Nature, Science, and Public Policy in Twentieth-Century. University of Chicago Press, Chicago, Illinois, 216 pp.

Bullen, R. P. 1961. Radiocarbon dates for south-eastern fiber-tempered pottery. American Antiquity 27:104–106.

Burggren, W. W., and B. R. McMahon (eds.). 1988. Biology of the Land Crabs. Cambridge University Press, Cambridge, U.K., 479 pp.

Burke, R. L., M. A. Ewert, J. B. McLemore, and D. R. Jackson. 1996. Temperature-dependent sex determination and hatching success in the gopher tortoise (*Gopherus polyphemus*). Chelonian Conservation and Biology 2:86–88.

Burrison, J. A. 2007. Roots of a Region: Southern Folk Culture. University Press of Mississippi, Jackson, Mississippi, 236 pp.

Burrows, M., and G. Hoyle. 1973. The mechanism of rapid running in the ghost crab *Ocypode ceratophthalma*. Journal of Experimental Biology 58:327–349.

Bush, D. M., O. H. Pilkey, and W. J. Neal. 1996. Living by the Rules of the Sea. Duke University Press, Durham, North Carolina, 179 pp.

Buta, R., A. K. Rindsberg, and D. Kopaska-Merkel. 2005. Pennsylvanian Footprints in the Black Warrior Basin of Alabama. Alabama Paleontological Society Monograph No. 1, 390 pp.

Butler, J. A., C. Broadhurst, M. Green, and Z. Mullin. 2004. Nesting, nest predation, and hatchling emergence of the Carolina diamondback terrapin *Malaclemys terrapin centrata* in northeastern Florida. American Midland Naturalist 152:145–155.

Byrne, M. W., and D. M. Lagana. 2009. Live-specimen Key for the Mammals of Southeast Coast Network Parks, Natural Resource Report NPS/SECN/NRR-2009/122, U.S. Department of the Interior, National Park Service, Fort Collins, Colorado, 13 pp.

Cadée, G. C. 1989. Size-selective transport of shells by birds and its palaeoecological implications. Palaeontology 32:429–437.

———. 1990. Feeding traces and bioturbation by birds on a tidal flat, Dutch Wadden Sea. Ichnos 1:23–30.

———. 1991. The history of taphonomy; pp. 3–21 in S. Donovan (ed.), The Processes of Fossilization. University of Chicago Press, Chicago, Illinois.

———. 2001. Sediment dynamics by bioturbating organisms; pp. 127–148 in K. Reise (ed.), Ecological Comparisons of Sedimentary Shorelines. Springer-Verlag, Berlin.

Cadée, G. C., and R. Goldring. 2007. The Wadden Sea, cradle of invertebrate ichnology; pp. 3–13 in W. M. Miller III (ed.), Trace Fossils: Concepts, Problems, Prospects. Elsevier, Amsterdam, Netherlands.

Cain, M. L., H. Damman, and A. Muir. 1998. Seed dispersal and the Holocene migration of woodland herbs. Ecological Monographs 68:325–347.

Cairns, W. E. 1982. Biology and breeding of piping plovers. Wilson Bulletin 94:531–545.

Callaway, R. M., K. O. Reinhart, G. W. Moore, D. J. Moore, and S. C. Pennings. 2002. Epiphyte host preferences and host traits: mechanisms for species-specific interactions. Oecologia 132:221–230.

Cane, J. H., T. Griswold, and F. D. Parker. 2007. Substrates and materials used for nesting by North American *Osmia* Bees (Hymenoptera: Apiformes: Megachilidae). Annals of the Entomological Society of America 100:350–358.

Capinera, J. L. 2008. Encyclopedia of Entomology (2nd Edition 3). Springer, Berlin, 3218 pp.

Capinera, J. L., and N. C. Leppla. 2008. Shortwinged mole cricket, *Scapteriscus abbreviatus* Scudder; Southern mole cricket, *Scapteriscus borellii* Giglio-Tos; and tawny mole cricket, *Scapteriscus vicinus* Scudder (Insecta: Orthoptera: Gryllotalpidae). Publication EENY-235 (IN391), Florida Cooperative Extension Service, University of Florida, 8 pp.

Capon, B. 2005. Botany for Gardeners (2nd Edition). Timber Press, Portland, Oregon, 239 pp.

Carlson, J. K., M. A. Grace, and P. K. Lago. 2002. An observation of juvenile tiger sharks feeding on clapper rails off the southeastern coast of the United States. Southeastern Naturalist 1:307–310.

Carney, J. A. 2001. Black Rice: The African Origins of Rice Cultivation in the Americas. Harvard University Press, Cambridge, Massachusetts, 240 pp.

Carriker, M. R. 1951. Observations on the penetration of tightly closing bivalves by *Busycon* and other predators. Ecology 32:73–83.

———. 1969. Excavation of boreholes by the gastropod, *Urosalpinx*: an analysis by light and scanning electron microscopy. American Zoologist 9:917–993.

———. 1977. Ultrastructural evidence that gastropods swallow shell rasped during hole boring. Biological Bulletin 152:325–336.

Castro, G., and J. P. Meyers. 1993. Shorebird predation on eggs of horseshoe crabs during spring stopover on Delaware Bay. Auk 110:927–930.

Cate, A. S., and I. Evans. 1994. Taphonomic significance of the biomechanical fragmentation of live molluscan shell material by a bottom-feeding fish (*Pogonias cromis*) in Texas coastal bays. Palaios 9:254–274.

Cate, M. D. 1930. Our Todays and Yesterdays: A Story of Brunswick and the Coastal Islands. Glover Brothers, Brunswick, Georgia, 243 pp.

Chalmers, A. G. 1997. The Ecology of the Sapelo Island National Estuarine Research Reserve. National Oceanic and Atmospheric Administration, Office of Coastal Resource Management, Sanctuaries and Reserves Division, Washington, D.C., 129 pp.

Chang, S.-C., H. Zhang, P. R. Renne, and Y. Fang. 2009. High-precision ^{40}Ar/^{39}Ar age for the Jehol biota. Palaeogeography, Palaeoclimatology, Palaeoecology 280: 94–104.

Chapman, J. A., and G. R. Willner. 1981. *Sylvilagus palustris*. Mammalian Species 153:1–3.

Chiappe, L. M., J. G. Schmitt, F. D. Jackson, A. Garrido, L. Dingus, and G. Grellet-Tinner. 2004. Nest structure for sauropods: sedimentary criteria for recognition of dinosaur nesting traces. Palaios 19:89–95.

Chin, K. 2007. The paleobiological implications of herbivorous dinosaur coprolites from the Upper Cretaceous Two Medicine Formation of Montana: why eat wood? Palaios 22:554–566.

Chin, K., and J. R. Bishop. 2004. Exploited twice: bored bone in a theropod coprolite from the Jurassic Morrison Formation of Utah, U.S.A.; pp. 379–387 in R. G. Bromley, L. A. Buatois, G. M. Mángano, J. F. Genise, and R. N. Melchor (eds.), Sediment–Organism Interactions: A Multifaceted Ichnology. Society for Sedimentary Geology, Special Publication, 88.

Chin, K., and B. D. Gill. 1996. Dinosaurs, dung beetles, and conifers: participants in a Cretaceous food web. Palaios 11:280–285.

Choate, P. M. 2003. Illustrated Key to Florida Species of Tiger Beetles (Coleoptera: Cicindelidae), http://entomology.ifas.ufl.edu/choate/tigerbeetle_key.pdf, 21 pp.

———. 2009. Tiger beetles of Florida, *Cicindela* spp., *Megacephala* spp. (Insecta: Coleoptera: Cicindelidae). Publication EENY-005 (IN131), Florida Cooperative Extension Service, University of Florida, 6 pp.

Cicimurri, D. J., and J. L. Knight. 2009. Two shark-bitten whale skeletons from Coastal Plain deposits of South Carolina. Southeastern Naturalist 8:71–82.

Clark, R. B. 1962. Observations on the food of *Nephtys*. Limnology and Oceanography 7:380–385.

Clark, R. B., and M. E. Clark. 1960. The ligamentary system and the segmental musculature of *Nephtys*. Quarterly Journal of Microscopical Science 101:149–176.

Clayton, D. 2001. Acoustic calling in four species of ghost crabs: *Ocypode jousseaumei*, *O. platytarsis*, *O. rotundata* and *O. saratan* (Brachyura: Ocypodidae). Bioacoustics 12:37–55.

Cobb, W. R. 1969. Penetration of calcium carbonate substrates by the boring sponge, *Cliona*. American Zoologist 9:783–790.

Coelho, J. R., and C. W. Holliday. 2008. The effect of hind-tibial spurs on digging rate in female eastern cicada killers. Ecological Entomology 33:1–5.

Coelho, J. R., and K. Weidman. 1999. Functional morphology of the hind tibial spurs of the cicada killer (*Sphecius speciosus* Drury). Journal of Hymenoptera Research 8: 6–12.

Coleman, K. 1991. A History of Georgia. University of Georgia Press, Athens, Georgia, 461 pp.

Colquhoun, D. J., M. J. Brooks, and P. A. Stone. 1995. Sea-level fluctuations: emphasis on temporal correlations with records from areas with strong hydrologic influences in the southeastern United States. Journal of Coastal Research, Special Issue 17: 191–196.

Cornilessen, T. G., and G. W. Fernandes. 2001. Induced defences in the neotropical tree *Bauhinia brevipes* (Vog.) to herbivory: effects of damage-induced changes on leaf quality and insect attack. Trees: Structure and Function 15:236–241.

Counts, J. W., and S. T. Hasiotis. 2009. Neoichnological experiments with masked chafer beetles (Coleoptera; Scarabaeidae): implications for backfilled continental trace fossils. Palaios 24:74–91.

Coward, S. J., H. C. Gerhardt, and D. T. Crockett. 1970. Behavioral variation in natural populations of two species of fiddler crabs (*Uca*) and some preliminary observations on directed modifications. Journal of Biological Psychology 12:24–31.

Craige, B. J. 2002. Eugene Odum: Ecosystem Ecologist and Environmentalist. University of Georgia Press, Athens, Georgia, 226 pp.

Crawford, C. S. 1992. Millipedes as model detritivores. 8th International Congress of Myriapodology, Innsbruck, Austria, Supplement 10:277–288.

Croker, R. A. 1968. Distribution and abundance of some intertidal sand beach amphipods accompanying the passage of two hurricanes. Chesapeake Science 9:157–162.

Crook, M. R. 1980. Archaeological indications of community structures at the Kenan Field site; pp. 89–100 in D. P. Juengst (ed.), Sapelo Papers: Researches in the History and Prehistory of Sapelo Island, Georgia. West Georgia College Studies in the Social Sciences 19.

Crook, M. R., C. Bailey, N. Harris, and K. Smith. 2003. Sapelo Voices: Historical Anthropology and the Oral Traditions of Gullah. State University of West Georgia, Carrolton, Georgia, 293 pp.

Curran, H. A. 1984. Ichnology of Pleisteocene carbonates on San Salvador, Bahamas. Journal of Paleontology 58:312–321.

———. 1992. Trace fossils in Quaternary, Bahamian-style carbonate environments: the modern to fossil transition; pp. 105–120 in G. G. Maples and R. R. West (eds.), Trace Fossils, Short Courses in Paleontology 5, Paleontological Society.

———. 2007. Ichnofacies, ichnocoenoses, and ichnofabrics of Quaternary shallow-marine to dunal tropical carbonates: a model and implications; pp. 232–247 in W. M. Miller III (ed.), Trace Fossils: Concepts, Problems, Prospects. Elsevier, Amsterdam, Netherlands.

Curran, H. A., and A. J. Martin. 2003. Intertidal mounds of tropical callianassids provide substrates for complex upogebiid shrimp burrows: modern and Pleistocene examples from the Bahamas. Palaeogeography, Palaeoclimatology, Palaeoecology 192:229–245.

Curran, H. A., and B. White. 1999. Ichnology of Holocene carbonate eolianites on San Salvador Island, Bahamas: diversity and significance; pp. 22–35 in H. A. Curran and J. E. Mylroie (eds.), Proceedings of the 9th Symposium on the Geology of the Bahamas and other Carbonate Regions, Bahamian Field Station, San Salvador, Bahamas.

Curran, H. A., and A. B. Williams. 1997. Ichnology of an intertidal carbonate sand flat: Pigeon Creek, San Salvador Island, Bahamas; pp. 33–46 in J. L. Carew (ed.), Proceedings of the 8th Symposium on the Geology of the Bahamas, Gerace Research Centre, San Salvador (Bahamas).

Currano, E. D., P. Wilf, S. L. Wing, C. C. Labandeira, E. C. Lovelock, and D. L. Royer. 2008. Sharply increased insect herbivory during the Paleocene-Eocene thermal maximum. Proceedings of the National Academy of Sciences 105:1960–1964.

Curtis, L. A. 1987. Vertical distribution of an estuarine snail altered by a parasite. Science 235:1509–1511.

———. 2004. Ecology of larval trematodes in three marine gastropods; pp. 843–856 in R. Poulin (ed.), Parasites in Marine Systems. Cambridge University Press, Cambridge, U.K.

Dahlberg, M. D. 2008. Guide to Coastal Fishes of Georgia and Nearby States. University of Georgia Press, Athens, Georgia, 208 pp.

Dame, R. F. 1996. Ecology of Marine Bivalves: An Ecosystem Approach. CRC Press, Boca Raton, Florida, 254 pp.

———. 2009. Shifting through time: oysters and shell rings in past and present southeastern estuaries. Journal of Shellfish Research 28:425–430.

Danchin, E., L.-A. Giraldeau, and F. Cézilly. 2008. Behavioural Ecology: An Evolutionary Perspective on Behaviour. Oxford University Press, Oxford, U.K., 688 pp.

Dangerfield, A., B. B. Britt, R. Scheetz, and M. Pickard. 2005. Jurassic dinosaurs and insects: the paleoecological role of termites as carrion feeders. Geological Society of America, Abstracts with Programs 37(7):44.

Darwin, C. 1881. The Formation of Vegetable Mould through the Action of Worms, with Observations on their Habits. John Murray, London, U.K., 326 pp.

Davis, R. B., N. J. Minter, and S. J. Braddy. 2007. The neoichnology of terrestrial arthropods. Palaeogeography, Palaeoclimatology, Palaeoecology 255(3–4):284–307.

de Bruxelles, G. L., and M. R. Roberts. 2001. Signals regulating multiple responses to wounding and herbivores. Critical Reviews in Plant Sciences 20:487–521.

de Gilbert, J. M. 1996. *Diopatrichnus odlingi* n.isp. (annelid tube) and associated ichnofabrics in the White Limestone (M. Jurassic) of Oxfordshire: sedimentological nd palaeoecological significance. Proceedings of the Geologists' Association 107: 189–198.

de Schepper, N., B. de Kegel, and D. Adriaens. 2007. *Pisodonophis boro* (Ophichthidae: Anguilliformes): specialization for head-first and tail-first burrowing? Journal of Morphology 268:112–126.

de Stoppelaire, G. H., T. W. Gillespie, J. C. Brock, and G. A. Tobin. 2004. Use of remote sensing techniques to determine the effects of grazing on vegetation cover and dune elevation at Assateague Island National Seashore: impact of horses. Environmental Management 34:642–649.

Deery, J. R., and J. D. Howard. 1977. Origin and character of washover fans on the Georgia coast, U.S.A. Transactions of the Gulf Coast Association of Geological Societies 27:259–271.

DePratter, C. B., and J. D. Howard. 1981. Evidence for a sea-level lowstand between 4500 and 2400 years B.P. on the southeast coast of the United States. Journal of Sedimentary Research 51:1287–1295.

Deyrup, M., N. D. Deyrup, M. Eisner, and T. Eisner. 2005. A caterpillar that eats tortoise shells. American Entomologist 51:245–248.

Dietl, G. P. 2004. Origins and circumstances of adaptive divergence in whelk feeding behavior. Palaeogeography, Palaeoclimatology, Palaeoecology 208:279–291.

Dolch, R., and T. Tscharntke. 2000. Defoliation of alders (*Alnus glutinosa*) affects herbivory by leaf beetles on undamaged neighbours. Oecologia 125:504–511.

Donovan, S. K., W. Renema, and R. K. Pickerell. 2005. The ichnofossil *Scolicia prisca* de Quatrefages from the Paleogene of eastern Jamaica and fossil echinoids of the Richmond Formation. Caribbean Journal of Science 41:876–881.

Doonan, T. J., and I. J. Stout. 1994. Effects of gopher tortoise (*Gopherus polyphemus*) body size on burrow structure. American Midland Naturalist 131:273–280.

Dörjes, J. 1972. Georgia coastal region, Sapelo Island, U.S.A: sedimentology and biology, VII. Distribution and zonation of macrobenthic animals. Senckenbergiana Maritima 4:183–216.

———. 1977. Marine macrobenthic communities of the Sapelo Island, Georgia Region; pp. 399–421 in B. C. Coull (ed.), Ecology of Marine Benthos. University of South Carolina, Columbia, South Carolina.

Dörjes, J., and G. Hertweck. 1975. Recent biocoenoses and ichnocoenoses in shallow-water marine environments; pp. 459–491 in R. W. Frey (ed.), The Study of Trace Fossils. Springer-Verlag, New York.

Dörjes, J., and J. D. Howard. 1975. Estuaries of the Georgia coast, U.S.A.: sedimentology and biology. IV. Fluvial-marine indicators in an estuarine environment, Ogeechee River-Ossabaw Sound. Senckenbergiana Maritima 7:137–179.

Doyle, P., J. L. Wood, and G. T. Gareth. 2000. The shorebird ichnofacies: an example from the Miocene of southern Spain. Geological Magazine 137:517–536.

Driese, S. G., C. I. Mora, and J. M. Elick. 1997. Morphology and taphonomy of root and stump casts of the earliest trees (Middle to late Devonian), Pennsylvania and New York, U.S.A. Palaios 12:524–537.

Droser, M. L., and D. J. Bottjer. 1986. A semiquantitative field classification of ichnofabric. Journal of Sedimentary Research 56:558–559.

Duncan, G. A. 1986. Burrows of *Ocypode quadrata* (Fabricus) as related to slopes of substrate surfaces. Journal of Paleontology 60:384–389.

Duncan, P. B. 1987. Burrow structure and burrowing activity of the funnel-feeding enteropneust *Balanoglossus aurantiacus* in Bogue Sound, North Carolina, U.S.A. Marine Ecology 8:75–95.

Edwards, C. A., and P. J. Bohlen. 1996. Biology and Ecology of Earthworms (3rd Edition). Springer, Berlin, 426 pp.

Eastwood, E. B., and C. Eastwood. 2006. Capturing the Spoor: An Exploration of Southern African Rock Art. New Africa Books, Glodserry, South Africa, 216 pp.

Edwards, J. M., and R. W. Frey. 1977. Substrate characteristics within a Holocene salt marsh, Sapelo Island, Georgia. Senckenbergiana Maritima 9:215–259.

Eickwort, G. C. 1981. Aspects of the nesting biology of five Nearctic species of *Agapostemon* (Hymenoptera, Halictidae). Journal of the Kansas Entomological Society 54: 337–351.

Ekdale, A. A. 1980. Graphoglyptid burrows in modern deep-sea sediment. Science 207: 304–306.

Ekdale, A. A., and R. G. Bromley. 2001. A day and a night in the life of a cleft-foot clam: *Protovirgularia–Lockeia–Lophoctenium*. Lethaia 34:119–124.

Ekdale, A. A., R. G. Bromley, and S. G. Pemberton. 1984. Ichnology: The Use of Trace Fossils in Sedimentology and Stratigraphy. SEPM Publication, Tulsa, Oklahoma, 317 pp.

Elbroch, M. 2003. Mammal Tracks and Sign: A Guide to North American Species. Stackpole Books, Mechanicsburg, Pennsylvania, 779 pp.

Elbroch, M., and Marks, E. 2001. Bird Tracks and Sign of North America. Stackpole Books. Mechanicsburg, Pennsylvania, 456 pp.

Ellers, O. 1995a. Discrimination among wave-generated sounds by a swash-riding clam. Biological Bulletin 189:128–137.

———. 1995b. Behavioral control of swash-riding in the clam *Donax variabilis*. Biological Bulletin 189:120–127.

Emerson, S. B. 1976. Burrowing in frogs. Journal of Morphology 149:437–458.

Emery, K. O., C. A. Kaye, D. H. Loring, and D. J. G. Nota. 1968. English Cretaceous Flints of North America. Science 160:1227.

Engel, M. S., D. A. Grimaldi, and K. Krishna. 2009. Termites (Isoptera): their phylogeny, classification, and rise to ecological dominance. American Museum Novitates 3650:1–27.

Epperson, D. M., and C. D. Heise. 2003. Nesting and hatchling ecology of gopher tortoises (*Gopherus polyphemus*) in southern Mississippi. Journal of Herpetology 37: 315–324.

Erickson, G. M., A. K. Lappin, and K. A. Vliet. 2003. The ontogeny of bite-force performance in American alligator (*Alligator mississippiensis*). Journal of Zoology 260: 317–327.

Erickson, G. M., O. Rauhut, Z. Zhou, A. Turner, B. Inouye, D. Hu, and M. Norell. 2009. Was dinosaurian physiology inherited by birds? Reconciling slow growth in *Archaeopteryx*. PLoS One 4:1–9: doi:10.1371/journal.pone.0007390.

Ernst, C. H., and J. E. Lovich. 2009. Turtles of the United States and Canada (2nd Edition). Johns Hopkins University Press, Baltimore, Maryland, 827 pp.

Erwin, D. M. 1984. Growth of the rhizomorph of *Paurodendron* and homologies among the rooting organs of Lycopsida. American Journal of Botany 71:112–113.

Etheridge, K., L. C. Wit, and J. C. Sellers. 1983. Hibernation in the lizard *Cnemidophorus sexlineatus* (Lacertilia: Teiidae). Copeia 1983:206–214.

Evans, H. E., and K. M. O'Neill. 2007. The Sand Wasps: Natural History and Behavior. Harvard University Press, Cambridge, Massachusetts, 340 pp.

Everhart, M. J. 2005. Oceans of Kansas: A Natural History of the Western Interior Sea. Indiana University Press, Bloomington, Indiana, 321 pp.

Ezzo, J. A., C. S. Larsen, and J. H. Burton. 2005. Elemental signatures of human diets from the Georgia Bight, American Journal of Physical Anthropology 98:471–481.

Farlow, J. O., and R. M. Elsey. 2010. Footprints and trackways of the American alligator, Rockefeller Wildlife Refuge, Lousiana; pp. 31–40 in M. G. Lockley, S. G. Lucas, J. Milàn, J. D. Harris, M. Avanzini, J. R. Foster, and J. A. Spielmann (eds.), The Fossil Record of Crocodilian Tracks and Traces: An Overview. New Mexico Museum of Natural History Bulletin 51.

Farlow, J. O., J. G. Pittman, and J. M. Hawthorne. 1989. *Brontopodis birdi*, Lower Cretaceous sauropod footprints from the U.S. Gulf Coastal Plain; pp. 371–394 in D. D. Gillette, and M. G. Lockley (eds.), Dinosaur Tracks and Traces. Cambridge University Press, Cambridge.

Fasano, J. M., G. D. Massa, and S. Gilroy. 2002. Ionic signaling in plant responses to gravity and touch. Journal of Plant Growth Regulation 21:71–88.

Feldhamer, G. A., B. C. Thompson, and J. A. Chapman. 2003. Wild Mammals of North America: Biology, Management, and Conservation. Johns Hopkins University Press, Baltimore, Maryland, 1216 pp.

Fellendorf, M., C. Mohra, and R. J. Paxton. 2004. Devastating effects of river flooding to the ground-nesting bee, *Andrena vaga* (Hymenoptera: Andrenidae), and its associated fauna. Journal of Insect Conservation 8:311–322.

Fiebel, C. S. 1987. Fossil fish nests from the Koobi Fora Formation (Plio-Pleistocene) of northern Kenya. Journal of Paleontology 61:130–134.

Fierstien, J. F., IV, and H. B. Rollins. 1987. Observations on intertidal organism associations of St. Catherines Island, Georgia. II. Morphology and distribution of *Littorina irrorata* (Say). American Museum Novitates 2873:1–31.

Fincher, G. T. 1975. Dung beetles of Blackbeard Island (Coleoptera: Scarabaeidae). Coleopterists Bulletin 29:319–320.

———. 1979. Dung beetles of Ossabaw Island, Georgia. Georgia. Journal of Entomological Science 14:330–334

Fincher, G. T., and R. E. Woodruff. 1979. Dung beetles of Cumberland Island, Georgia (Coleoptera: Scarabaeidae). Coleopterists Bulletin 33:69–70.

Fisher J. B., and K. Jayachandran. 1999. Root structure and arbuscular mycorrhizal colonization of the palm *Serenoa repens* under field conditions. Plant and Soil 217:229–241.

Fitzpatrick, J. W., M. Lammertink, M. D. Luneau Jr., T. W. Gallagher, B. R. Harrison, G. M. Sparling, K. V. Rosenberg, R. W. Rohrbaugh, E. C. H. Swarthout, P. H. Wrege, S. B. Swarthout, M. S. Dantzker, R. A. Charif, T. R. Barksdale, J. V. Remsen Jr., S. D. Simon, and D. Zollner. 2005. Ivory-billed woodpecker (*Campephilus principalis*) persists in continental North America. Science 308:1460–1462.

Fletcher, W. O., and W. A. Parker. 1994. Tree nesting by wild turkeys on Ossabaw Island, Georgia. Wilson Bulletin 106:562.

Flood, J. 1997. Rock Art of the Dreamtime: Images of Ancient Australia. Angus & Robertson, Sydney, Australia, 384 pp.

Fox, N. 2004. Against the Machine: The Hidden Luddite Tradition in Literature, Art, and Individual Lives. Island Press, Washington, D.C., 424 pp.

Fraaije, R. H. B. 2003. The oldest in situ hermit crab from the Lower Cretaceous of Speeton, U.K. Palaeontology 46:53–58.

Fraedrich, S. W., T. C. Harrington, R. J. Rabaglia, M. D. Ulyshen, A. E. Mayfield III, J. L. Hanula, J. M. Eickwort, and D. R. Miller. 2008. A fungal symbiont of the redbay ambrosia beetle causes a lethal wilt in redbay and other Lauraceae in the southeastern United States. Plant Disease 92:215–224.

Frank, J. H., and M. C. Thomas. 2009. Rove beetles of the world. University of Florida/IFAS Extension, EENY-114 (IN271): http://edis.ifas.ufl.edu/.

Frankenberg, D., and K. L. Smith Jr. 1967. Coprophagy in marine animals. Limnology and Oceanography 12:443–450.

Franks, S. J. 2003. Facilitation in multiple life-history stages: evidence for nucleated succession in coastal dunes. Plant Ecology 168:1–11.

Freeman, J., D. F. Gleason, R. Ruzicka, R. W. M. van Soest, A. W. Harvey, and G. McFall. 2007. A biogeographic comparison of sponge fauna from Gray's Reef National Marine Sanctuary and other hard-bottom reefs of coastal Georgia, U.S.A; pp. 319–325 in M. R. Custódio, G. Lôbo-Hajdu, E. Hajdu, and G. Muricy (eds.), Porifera Research: Biodiversity, Innovation and Sustainability. Série Livros 28, Museu Nacional, Rio de Janeiro.

Fraser, W. J., Jr. 2006. Lowcountry Hurricanes: Three Centuries of Storms at Sea and Ashore. University of Georgia Press, Athens, Georgia, 319 pp.

Frey, R. W. 1968. The lebensspuren of some common marine invertebrates near Beaufort, North Carolina. I. Pelecypod burrows. Journal of Paleontology 42:570–574.

———. 1969. Stratigraphy, ichnology, and paleoecology of the Fort Hays Limestone Member of the Niobrara Chalk (Upper Cretaceous) in Trego County, Kansas. Unpublished PhD dissertation, Indiana University, Bloomington, Indiana, 368 pp.

———. 1970a. Trace fossils of Fort Hays Limestone Member, Niobrara Chalk (Upper Cretaceous), west-central Kansas. University of Kansas Paleontological Contributions, Article 53 (Cretaceous 2), 41 pp.

———. 1970b. The lebensspuren of some common marine invertebrates near Beaufort, North Carolina. II. Anemone burrows. Journal of Paleontology 44:308–311.

———. 1987. Hermit crabs: neglected factors in taphonomy and paleoecology. Palaios 2:313–322.

——— (ed.). 1975. The Study of Trace Fossils. New York: Springer-Verlag, 562 pp.

Frey, R. W., and P. B. Basan. 1978. Coastal salt marshes; pp. 101–169 in R. A. Davis Jr. (ed.), Coastal Sedimentary Environments, Springer-Verlag, New York.

———. 1981. Taphonomy of relict Holocene salt marsh deposits, Cabretta Island, Georgia. Senckenbergiana Maritima 13:111–155.

Frey, R. W., and J. D. Howard. 1969. A profile of biogenic sedimentary structures in a Holocene barrier island-salt marsh complex, Georgia. Transactions of the Gulf Coast Association Geological Society 19:427–444.

———. 1972. Georgia coastal region, Sapelo Island, U.S.A.—sedimentology and biology. VI. Radiographic study of sedimentary structures made by beach and offshore animals in aquaria. Senckenbergiana Maritima 4:169–182.

———. 1982. Trace fossils from the Upper Cretaceous of the Western Interior; potential criteria for facies models. The Mountain Geologist 19:1–10.

———. 1985. Trace fossils from the Panther Member, Star Point Formation (Upper Cretaceous), Coal Creek Canyon, Utah. Journal of Paleontology 59:370–404.

———. 1986. Mesotidal estuarine sequences: a perspective from the Georgia Bight. Journal of Sedimentary Petrology 56:911–924.

———. 1988. Beaches and beach-related facies, Holocene barrier islands of Georgia. Geological Magazine 125:621–640.

———. 1990. Trace fossils and depositional sequences in a clastic shelf setting, Upper Cretaceous of Utah. Journal of Paleontology 64:803–820.

Frey, R. W., and T. V. Mayou. 1971. Decapod burrows in Holocene barrier islands, beaches and washover fans, Georgia. Senckenbergiana Maritima 3:53–77.

Frey, R. W., and S. G. Pemberton. 1986. Vertebrate lebensspuren in intertidal and supratidal environments, Holocene barrier island, Georgia. Senckenbergiana Maritima 18:97–121.

———. 1987. The *Psilonichnus* ichnocoenose, and its relationship to adjacent marine and nonmarine ichnocoenoses along the Georgia coast. Bulletin of Canadian Petroleum Geology 35:333–357.

Frey, R. W., and A. Seilacher. 1980. Uniformity in marine invertebrate ichnology. Lethaia 13:183–207.

Frey, R. W., P. B. Basan, and R. M. Scott. 1973. Techniques for sampling salt marsh benthos and burrows. American Midland Naturalist 89:228–234.

Frey, R. W., M. R. Voorhies, and J. D. Howard. 1975. Estuaries of the Georgia coast, U.S.A.: sedimentology and biology. VIII. Fossil and recent skeletal remains in Georgia estuaries. Senckenbergiana Maritima 7:257–295.

Frey, R. W., J. D. Howard, and W. A. Pryor. 1978. *Ophiomorpha:* its morphological, taxonomic, and environmental significance. Palaeogeography, Palaeoclimatology, Palaeoecology 23:199–229.

Frey, R. W., H. A. Curran, and S. G. Pemberton. 1984. Tracemaking activities of crabs and their environmental significance: the ichnogenus *Psilonichnus.* Journal of Paleontology 58:333–350.

Frey, R. W., J. D. Howard, and J. S. Hong. 1986. Naticid gastropods may kill solenid bivalves without boring: ichnologic and taphonomic consequences. Palaios 1:610–612.

———. 1987a. Prevalent lebensspuren on a modern macrotidal flat, Inchon, Korea: ethological and environmental significance. Palaios 2:571–593.

Frey, R. W., J.-S. Hong, J. D. Howard, B.-K. Park, and S.-J. Han. 1987b. Zonation of benthos on a macrotidal flat, Inchon, Korea. Senckenbergiana Maritima 19:295–329.

Frey, R. W., S. G. Pemberton, and J. A. Fagerstrom. 1984. Morphological, ethological, and environmental significance of the ichnogenera *Scoyenia* and *Ancorichnus.* Journal of Paleontology 58:511–528.

Frey, R. W., S. G. Pemberton, and T. D. A. Saunders. 1990. Ichnofacies and bathymetry: a passive relationship. Journal of Paleontology 64:155–158.

Frick, E. A., M. B. Gregory, D. L. Calhoun, and E. H. Hopkins. 2002. Water Quality and Aquatic Communities of Upland Wetlands, Cumberland Island National Seashore, Georgia, April 1999 to July 2000. U.S Geological Survey Water-Resources Investigations Report 02-4082, 72 pp.

Frumkin, H. 2001. Beyond toxicity: human health and the natural environment. American Journal of Preventive Medicine 20:234–240.

Fürsich, F. T., W. J. Kennedy, and T. J. Palmer. 1981. Trace fossils at a regional discontinuity surface: the Austin/Taylor (Upper Cretaceous) contact in central Texas. Journal of Paleontology 55:537–551.

Gamble, J. R., and D. A. Cristol. 2002. Drop-catch behaviour is play in herring gulls, *Larus argentatus*. Animal Behaviour 63:339–343.

Gans, C. 1986. Locomotion of limbless vertebrates: pattern and evolution. Herpetologica 42:33–46.

Garrison, E. G., W. Weaver, and M. Mitchell. 2003. Geoarchaeology and paleontology of Gray's Reef National Marine Sanctuary. Geological Society of America Abstracts with Programs 35(6):100.

Garrison, E. G., G. McFall, and S. E. Noakes. 2008. Shallow marine margin sediments, modern marine erosion and the fate of sequence boundaries, Georgia Bight (U.S.A.). Southeastern Geology 45:127–142.

Gee, C. T., P. M. Sander, and B. E. M. Petzelberger. 2003. A Miocene rodent nut cache in coastal dunes of the Lower Rhine embayment, Germany. Palaeontology 46: 1133–1149.

Genise, J. F., and G. Cladera. 2004. *Chubutolithes* and other wasp trace fossils: breaking through the taphonomic barrier. Journal of the Kansas Entomological Society 77:626–638.

Genise, J. F., M. G. Mángano, L. A. Buatois, J. H. Laza, and M. Verde. 2000. Insect trace fossil associations in paleosols: the *Coprinisphaera* ichnofacies. Palaios 15:49–64.

Genise, J. F., E. S. Bellosi, and M. G. Gonzalez. 2004. An approach to the description and interpretation of ichnofabrics in palaeosols; pp. 355–382 in D. McIlroy (ed.), The Application of Ichnology to Palaeoenvironmental and Stratigraphic Analysis. Geological Society of London, Special Publication 228.

Genise, J. F., E. S. Bellosi, R. N. Melchor, and M. I. Cosarinsky. 2005. Comment—Advanced Early Jurassic termite (Insecta: Isoptera) nests: evidence from the Clarens Formation in the Tuli Basin, Southern Africa (Bordy et al., 2004). Palaios 20:303–308.

Genise, J. F., R. N. Melchor, E. S. Bellosi, M. G. González, and M. Krause. 2007. New insect pupation chambers (Pupichnia) from the Upper Cretaceous of Patagonia, Argentina. Cretaceous Research 28:545–559.

Genise, J. F., R. N. Melchor, M. Archangelsky, L. O. Bala, R. Stranecke, and S. de Valais. 2009. Application of neoichnological studies to behavioural and taphonomic interpretation of fossil bird-like tracks from lacustrine settings: the Late Triassic–Early Jurassic? Santo Domingo Formation, Argentina. Palaeogeography Palaeoclimatology, Palaeoecology 272:143–161.

Gerling, D., and H. R. Hermann. 1978. Biology and mating behavior of *Xylocopa virginica* L. (Hymenoptera, Anthophoridae). Behavioral Ecology and Sociobiology 3:99–111.

Gibbons, J. W., and J. W. Coker. 1978. Herpetofaunal colonization patterns of Atlantic coastal barrier islands. American Midland Naturalist 99:219–233.

Gielazyn, M. L., S. E. Stancyk, and W. W. Piegorsch. 1999. Experimental evidence of subsurface feeding by the burrowing ophiuroid *Amphipholis gracillima* (Echinodermata). Marine Ecology Progress Series 184:129–138.

Gilbert, M. T. P., D. L. Jenkins, A. Götherstrom, N. Naveran, J. J. Sanchez, M. Hofreiter, P. F. Thomsen, J. Binladen, T. F. G. Higham, R. M. Yohe II, R. Parr, L. S. Cummings, and E. Willerslev. 2008. DNA from pre-Clovis human coprolites in Oregon, North America. Science 320:786–789.

Gingras, M. K., C. Mendoza, and S. G. Pemberton. 2004a. Fossilized worm-burrows influence the resource quality of porous media. AAPG Bulletin 88:875–883.

Gingras, M. K., J. A. MacEachern, and R. K. Pickerill. 2004b. Modern perspectives on the *Teredolites* ichnofacies: observations from Willapa Bay, Washington. Palaios 19: 79–88.

Gingras, M. K., S. V. Lalond, L. Amskold, and K. O. Konhauser. 2007. Wintering chironimids mine oxygen. Palaios 22:433–438.

Gingras, M. K., S. Dashtgard, S. G. Pemberton, and J. A. MacEachern. 2008a. Biology of shallow marine ichnology: a modern perspective. Aquatic Biology 2:255–268.

Gingras, M. K., S. G. Pemberton, S. Dashtgard, and L. Dafoe. 2008b. How fast do marine invertebrates burrow? Palaeogeography, Palaeoclimatology, Palaeoecology 270: 280–286.

Glass, M. L., J. Amin-Naves, and G. S. F. da Silva. 2009. Aestivation in amphibians, reptiles, and lungfish; pp. 179–189 in M. L. Glass and S. C. Wood (eds.), Cardio-Respiratory Control in Vertebrates. Springer, Berlin.

Gobetz, K. E., S. G. Lucas, and A. J. Lerner. 2006. Lungfish burrows of varying morphology from the Upper Triassic Redonda Formation, Chinle Group, eastern New Mexico; pp. 140–147 in J. D. Harris, S. G. Lucas, J. A. Spielmann, M. G. Lockley, A. R. C. Milner, and J. I. Kirkland (eds.), The Triassic–Jurassic Terrestrial Transition. New Mexico Museum of Natural History and Science Bulletin 37.

Godfray, H. C. J. 1994. Parasitoids: Behavioral and Evolutionary Ecology. Princeton University Press, Princeton, New Jersey, 473 pp.

Goldring, R., G. C. Cadée, and J. E. Pollard. 2007. Climatic control of marine trace fossil distribution; pp. 159–171 in W. M. Miller III (ed.), Trace Fossils: Concepts, Problems, Prospects. Elsevier, Amsterdam, Netherlands.

Gore, R. H. 1966. Observations on the escape response in *Nassarius vibex* (Say), (Mollusca: Gastropoda). Bulletin of Marine Science 16:423–434.

Gorman, M. L., and R. D. Stone. 1990. The Natural History of Moles. University of Chicago Press, Chicago, Illinois, 138 pp.

Gould, E., W. McShea, and T. Grand. 1993. Function of the star in the star-shaped mole. Journal of Mammalogy 37:223–231.

Gould, S. J. 1992. Bully for Brontosaurus. W.W. Norton and Company, New York, 540 pp.

Goulson, D. 2003. Bumblebees: Their Behaviour and Ecology. Oxford University Press, Oxford, U.K., 235 pp.

Graça, M. A., S. Y. Newell, and R. T. Kneib. 2000. Grazing rates of organic matter and living fungal biomass of decaying *Spartina alterniflora* by three species of salt-marsh invertebrates. Marine Biology 136:281–289.

Grant, J. 1981. A bioenergetic model of shorebird predation on infaunal amphipods. Oikos 37:53–62.

Greenberg, C. H., and G. W. Tanner. 2004. Breeding pond selection and movement patterns by eastern spadefoot toads (*Scaphiopus holbrookii*) in relation to weather and edaphic conditions. Journal of Herpetology 38:569–577.

Greenland, D., D. G. Goodin, and R. C. Smith. 2003. Climate variability and ecosystem response at long-term ecological research sites. Long-Term Ecological Research Network Series. Oxford University Press, Oxford, U.K., 459 pp.

Gregory, M. R. 1991. New trace fossils from the Miocene of Northland, New Zealand, *Rosschachichnus amoeba* and *Piscichnus waitemata*. Ichnos 1:195–206.

Gregory, M. R., and K. A. Campbell. 2003. A *"Phoebichnus* look-alike": a fossilised root system from Quaternary coastal dune sediments, New Zealand. Palaeogeography, Palaeoclimatology, Palaeoecology 192:247–258.

Gregory, M. R., P. F. Ballance, G. W. Gibson, and A. M. Ayling. 1979. On how some rays (Elasmobranchia) excavate feeding depressions by jetting water. Journal of Sedimentary Petrology 49:1125–1130.

Gregory, M. R., A. J. Martin, and K. A. Campbell. 2004. Composite trace fossils formed by plant and animal behavior in the Quaternary of northern New Zealand and Sapelo Island, Georgia (U.S.A.). Fossils and Strata 51:88–105.

Gregory, M. R., K. A. Campbell, R. Zuraida, and A. J. Martin. 2006. Plant traces resembling *Skolithos*. Ichnos 13:205–216.

Grimaldi, D. A., and M. S. Engel. 2005. Evolution of the Insects. Cambridge University Press, Cambridge, U.K., 755 pp.

Gross, M. R., and A. M. MacMillan. 1981. Predation and the evolution of colonial nesting in bluegill sunfish (*Lepomis macrochirus*). Behavioral Ecology and Sociobiology 8:163–174.

Guyer, C., and S. M. Hermann. 1997. Patterns of size and longevity for gopher tortoise burrows: implications for the longleaf pine–wiregrass ecosystem. Bulletin of the Ecological Society of America 78:254.

Hafemann, D. R., and J. I. Hubbard. 2005. On the rapid running of ghost crabs (*Ocypode ceratophthalma*). Journal of Experimental Zoology 170:25–31.

Halfpenny, J. C., and E. A. Biesiot. 1986. A Field Guide to Mammal Tracking in North America. Johnson Books, Boulder, Colorado, 164 pp.

Halfpenny, J. C., and J. Bruchac. 2002. Scats and Tracks of the Southeast: A Field Guide to the Signs of Seventy Wildlife Species. Globe Pequot Press, Guilford, Connecticut, 192 pp.

Hagadorn, J. W., and A. Seilacher. 2009. Hermit arthropods 500 million years ago? Geology 37:295–298.

Hager, S. B., and B. J. Cosentino. 2006. An identification key to rodent prey in owl pellets from the northwestern and southeastern United States: employing incisor size to distinguish among genera. American Biology Teacher 68:135–144.

Halloran, M. M., M. A. Carrel, and J. E. Carrel. 2000. Instability of sandy soil on the Lake Wales Ridge affects burrowing by wolf spiders (Araneae: Lycosidae) and Antloins (Neuroptera: Myrmeleontidae). Florida Entomologist 83:48–55.

Hannan, L. B., J. D. Roth, L. M. Ehrhart, and J. M. Weishampel. 2007. Dune vegetation fertilization by nesting sea turtles. Ecology 88:1053–1058.

Hansell, M. H. 1993. The ecological impact of animal nests and burrows. Functional Ecology 7:5–12.

———. 2000. Bird Nests and Construction Behavior. Cambridge University Press, Cambridge, U.K., 280 pp.

Hansen, T. A., P. H. Kelley, and J. C. Hall. 2003. Moonsnail Project: a scientific collaboration with middle school teachers and students. Journal of Geoscience Education 50:35–38.

Harding, J. M., P. Kingsley-Smith, D. Savini, and R. Mann. 2007. Comparison of predation signatures left by Atlantic oyster drills (*Urosalpinx cinerea* Say, Muricidae) and veined rapa whelks (*Rapana venosa* Valenciennes, Muricidae) in bivalve prey. Journal of Experimental Marine Biology and Ecology 352:1–11.

Hasiotis, S. T. 2004. Reconnaissance of Upper Jurassic Morrison Formation ichnofossils, Rocky Mountain Region, U.S.A.: paleoenvironmental, stratigraphic, and paleoclimatic significance of terrestrial and freshwater ichnocoenoses. Sedimentary Geology 167:177–268.

Hasiotis, S. T., and D. F. Dubiel. 1995. Termite (Insecta: Isoptera) nest ichnofossils from the Upper Triassic Chinle Formation, Petrified Forest National Park, Arizona. Ichnos, 4:119–130.

Hasiotis, S. T., and C. E. Mitchell. 1993. A comparison of crayfish burrow morphologies: Triassic and Holocene fossil, paleo- and neo-ichnological evidence, and the identification of their burrowing signatures. Ichnos 2:291–314.

Hasiotis, S. T., C. E. Mitchell, and R. F. Dubiel. 1993. Application of morphologic burrow architects; lungfish or crayfish? Ichnos 2:315–333.

Hasiotis, S. T., R. W. Wellner, A. J. Martin, and T. M. Demko. 2004. Vertebrate burrows from Triassic and Jurassic continental deposits of North America and Antarctica: their paleoenvironmental and paleoecological significance. Ichnos 11:103–124.

Hastings, J. 1986. Provisioning by female western cicada killer wasps *Sphecius grandis* (Hymenoptera: Sphecidae): influence of body size and emergence time on individual provisioning. Journal of the Kansas Entomological Society 59:262–268.

Havill, N. P., and K. F. Raffa. 2000. Compound effects of induced plant responses on insect herbivores and parasitoids: implications for tritrophic interactions. Ecological Entomology 25:171–179.

Hayes, M. O. 1975. Morphology of sand accumulation in estuaries: an introduction to the symposium; pp. 3–22 in L. E. Cronin (ed.), Estuarine Research II. Academic Press, New York.

Hayes, M. O., and J. Michel. 2008. A Coast for All Seasons: A Naturalist's Guide to the Coast of South Carolina. Pandion Books, Columbia, South Carolina, 286 pp.

Hayne, D. W. 1936. Burrowing habits of *Peromyscus polionotus*. Journal of Mammalogy 17:420–421.

Hefetz, A. 2008. The role of Dufour's gland secretions in bees. Physiological Entomology 12:243–253.

Hembree, D. I. 2009. Neoichnology of burrowing millipedes: linking modern burrow morphology, organism behavior, and sediment properties to interpret continental ichnofossils. Palaios 24:425–439.

Hembree, D. I., S. T. Hasiotis, and L. D. Martin. 2005. *Torridorefugium eskridgensis* (new ichnogenus and ichnospecies): amphibian aestivation burrows from the Lower Permian Speiser Shale of Kansas. Journal of Paleontology 79:583–593.

Hendrix, P. F. 1995. Earthworm Ecology and Biogeography in North America. CRC Press, Boca Raton, Florida, 244 pp.

Hendström, A. 2002. Aerodynamics, evolution and ecology of avian flight. Trends in Ecology and Evolution 17:415–422.

Henry, V. J., Jr., K. M. Farrell, and S. V. Cofer-Shabica. 1993. A regional overview of the geology of barrier complexes near Cumberland Island, Georgia; pp. 2–10 in K. M. Farrell, C. W. Hoffman, and V. J. Henry Jr. (eds.), Geomorphology and Facies Relationships of Quaternary Barrier Island Complexes Near St. Mary's Georgia. Georgia Geological Society Guidebook 13.

Hermann, H. R., and J.-T. Chao. 1984. Nesting biology and defensive behavior of *Mischocyttarus* (Monocyttarus) *mexicanus cubicola* (Vespidae: Polistinae). Psyche 91:51–66.

Hertweck, G. 1972. Georgia coastal region, Sapelo Island, U.S.A.; Sedimentology and biology; V. Distribution and environmental significance of lebensspuren and in-situ skeletal remains. Senckenbergiana Maritima 4:125–161.

Hertweck, G., A. Werhmann, and G. Liebezeit. 2007. Bioturbation structures of polychaetes in modern shallow marine environments and their analogues to *Chondrites* group traces. Palaeogeography, Palaeoclimatology, Palaeoecology 245:382–389.

Hester, M. W., and I. A. Mendelssohn. 1989. Water relations and growth responses of *Uniola paniculata* (sea oats) to soil moisture and water-table depth. Oecologia 78: 289–296.

———. 1991. Expansion patterns and soil physiochemical characterization of three Louisiana populations of *Uniola paniculata* (sea oats). Journal of Coastal Research 7:387–401.

Hill, G. E., D. J. Mennill, B. W. Rolek, T. L. Hicks, and K. A. Swiston. 2006. Evidence suggesting that ivory-billed woodpeckers (*Campephilus principalis*) exist in Florida. Avian Conservation and Ecology—Écologie et Conservation des Oiseaux 1(3): 2–15.

Hill, G. W., and R. E. Hunter. 1973. Burrows of the ghost crab *Ocypode quadrata* (Fabricus) on the barrier islands, south-central Texas Coast. Journal of Sedimentary Research 43:24–30.

Hill, J. R. 1988. Nest-depth preferences in pipe-nesting rough-winged swallows. Journal of Field Ornithology 59:334–336.

Hillier, R, D. Edwards, and A. N. Other. 2008. Sedimentological evidence for rooting structures in the Early Devonian Anglo–Welsh Basin (U.K.), with speculation on their producers. Palaeogeography, Palaeoclimatology, Palaeoecology 270:366–380.

Hines, A.H., R. B. Whitlach, S. F. Thrust, H. J. E. Hewi, J. Cunnings, P. K. Dayton, and P. Legendre. 1997. Nonlinear foraging response of a large marine predator to benthic prey: eagle ray pits and bivalves in a New Zealand sand flat. Journal of Experimental Marine Biology and Ecology 216:191–210.

Ho, C.-K., and S. C. Pennings. 2008. Consequences of omnivory for trophic interactions on a salt marsh shrub. Ecology 89:1714–1722.

Hobbs, H.H., Jr. 1981. The Crayfishes of Georgia. Smithsonian Institute Press, Washington, D.C., 549 pp.

———. 1988. Crayfish distribution, adaptive radiation and evolution; pp. 52–82 in D. M. Holdich and R. S. Lowery (eds.), Freshwater Crayfish: Biology, Management and Exploitation. Croom Helm, London, U.K.

Hoese, R. 1971. Dolphin feeding out of water in a salt marsh. Journal of Mammalogy 52: 222–223.

Hogger, J. B. 1988. Ecology, population biology, and behaviour; pp. 114–144 in D. M. Holdich and R. S. Lowery (eds.), Freshwater Crayfish: Biology, Management and Exploitation. Croom Helm, London, U.K.

Holland, A. F., R. G. Zingmark, and J. M. Dean. 1974. Quantitative evidence concerning the stabilization of sediments by marine benthic diatoms. Marine Biology 27: 192–196.

Hölldobler, B., and E. O. Wilson. 1990. The Ants (3rd Edition). Harvard University Press, Cambridge, Massachusetts, 732 pp.

Horch, K. 1975. The acoustic behavior of the ghost crab *Ocypode cordimana* Latreille, 1818 (Decapoda, Brachyura). Crustaceana 29:193–205.

Horwitz, P. H. J., and A. M. M. Richardson. 1986. An ecological classification of the burrows of Australian freshwater crayfish. Australian Journal of Marine and Freshwater Research 37:237–242.

Howard, J. D. 1968. X-ray radiography for examination of burrowing in sediments by marine invertebrate organisms. Sedimentology 11:249–258.

———. 1969. Radiographic examination of variations in barrier island facies; Sapelo Island, Georgia. Transactions of the Gulf Coast Association Geological Society 19: 217–232.

Howard, J. D., and J. Dörjes. 1972. Animal-sediment relationships in two beach-related tidal flats: Sapelo Island, Georgia. Journal of Sedimentary Research 42:608–623.

Howard, J. D., and C. A. Elders. 1970. Burrowing patterns of haustoriid amphipods from Sapelo Island, Georgia; pp. 243–262 in T. P. Crimes and J. C. Harper (eds.), Trace Fossils. Seel House Press, Liverpool, U.K.

Howard, J. D., and R. W. Frey. 1973. Characteristic physical and biogenic sedimentary structures in Georgia estuaries. A.A.P.G. Bulletin 57:1169–1184.

———. 1975. Estuaries of the Georgia coast, U.S.A.: sedimentology and biology. II. Regional animal-sediment characteristics of Georgia estuaries. Senckenbergiana Maritima 7:33–103.

———. 1980. Holocene depositional environments of the Georgia coast and continental shelf; pp. 66–134 in J. D. Howard, C. B. DePratter, and R. W. Frey (eds.), Excursions in Southeastern Geology: The Archaeology–Geology of the Georgia Coast. Georgia Geological Society Guidebook 20.

———. 1984. Characteristic trace fossils in nearshore to offshore sequences, Upper Cretaceous of east-central Utah. Canadian Journal of Earth Sciences 21:200–219.

———. 1985. Physical and biogenic aspects of backbarrier sedimentary sequences, Georgia coast. Marine Geology 63:77–127.

Howard, J. D., and H.-E. Reineck. 1972a. Georgia coastal region, Sapelo Island, U.S.A: sedimentology and biology. IV. Physical and biogenic sedimentary structures of the nearshore shelf. Senckenbergiana Maritima 4:81–123.

———. 1972b. Georgia coastal region, Sapelo Island, U.S.A.: sedimentology and biology. VIII. Conclusions. Senckenbergianna Maritima 4:217–222.

Howard, J. D., and R. M. Scott. 1983. Comparison of Pleistocene and Holocene barrier island beach-to-offshore sequences, Georgia and northeast Florida coasts, U.S.A. Sedimentary Geology 34:167–183.

Howard, J. D., R. W. Frey, and H.-E. Reineck. 1972. Georgia coastal region, Sapelo Island U.S.A. Sedimentology and biology I: introduction. Senckenbergiana Maritima 4: 3–14.

———. 1973. Holocene sediments of the Georgia coastal area; pp. 1–58 in R. W. Frey (ed.), The Neogene of the Georgia Coast, Georgia Geological Society Field Guidebook 8.

Howard, J. D., T. V. Mayou, and R. W. Heard. 1977. Biogenic sedimentary structures formed by rays. Journal of Sedimentary Petrology 47:339–346.

Howell, J. 2006. Hey, Bug Doctor! The Scoop on Insects in Georgia's Homes and Gardens. University of Georgia Press, Athens, Georgia, 220 pp.

Hoyt, J. H., and J. R. Hails. 1967. Pleistocene shoreline sediments in coastal Georgia: deposition and modification. Science 155:1541–1543.

Hoyt, J. J., and V. J. Henry Jr. 1967. Influence of island migration on barrier-island sedimentation. Geological Society of America Bulletin 78:77–86.

Hoyt, J. H., R. J. Weimer, and V. J. Henry Jr. 1964. Late Pleistocene and recent sedimentation on the central Georgia coast, U.S.A; pp. 170–176. in L. M. J. U. van Straaten (ed.), Deltaic and Shallow Marine Deposits, Developments in Sedimentology I. Elsevier, Amsterdam.

Hubbard, H. G. 1894. The Insect Guests of the Florida Land Tortoise. Harvard University, Cambridge, Massachusetts, 14 pp.

Huggett, R. J. 1995. Geoecology: An Evolutionary Approach. Routledge, New York, 320 pp.

Hulbert, R. C., and A. E. Pratt. 1998. New Pleistocene (Rancholeabrean) vertebrate faunas from coastal Georgia. Journal of Vertebrate Paleontology 18:412–429.

Hunter, J. M. 1973. Geophagy in Africa and the United States: a culture-nutrition hypothesis. Geographical Review 63:170–195.

Hyland, J., C. Cooksey, W. L. Balthisa, M. Fulton, D. Bearden, G. McFall, and M. Kendall. 2006. The soft-bottom macrobenthos of Gray's Reef National Marine Sanctuary and nearby shelf waters off the coast of Georgia, U.S.A. Journal of Experimental Biology and Ecology 330:307–326.

Ipser, R. M., M. A. Brinkman, W. A. Gardner, and H. B. Peeler. 2004. A survey of ground-dwelling ants (Hymenoptera: Formicidae) in Georgia. The Florida Entomologist 87:253–260.

Jacobsen, A. R. 1998. Feeding behaviour of carnivorous dinosaurs as determined by tooth marks on dinosaur bones. Historical Biology 13:17–26.

Jacobsen, A.R., and R. G. Bromley. 2009. New ichnotaxa based on tooth impressions on dinosaur and whale bones. Geological Journal 53:373–382.

Jackson, D. R., and E. R. Milstrey. 1989. The fauna of gopher tortoise burrows; pp. 86–98 in J. E. Diemer (ed.), Proceedings of the Gopher Tortoise Relocation Symposium, State of Florida, Game and Freshwater Fish Commission, Tallahassee, Florida.

Jansen, K. P., A. P. Summers, and P. R. Delis. 2001. Spadefoot toads (*Scaphiopus holbrookii*) in an urban landscape: effects of nonnatural substrates on burrowing in adults and juveniles. Journal of Herpetology 35:141–145.

Jensen, J. B., C. D. Camp, W. Gibbons, and M. J. Elliott (eds.). 2008. Amphibians and Reptiles of Georgia. University of Georgia Press, Athens, Georgia, 575 pp.

Johnson, C. N. 2009a. Ecological consequences of Late Quaternary extinctions of megafauna. Proceedings of the Royal Society of London B 276:2509–2519.

Johnson, M. N. 2009b. Sapelo Island's Hog Hammock. Arcadia Publishing, Mt. Pleasant, South Carolina, 128 pp.

Jones, C. A., and R. Franz. 1990. Use of gopher tortoise burrows by Florida mice (*Podomys floridanus*) in Putnam County, Florida. Florida Field Naturalist 18:45–68.

Jones, C. C. 2000. Gullah Folktales from the Georgia Coast. University of Georgia Press, Athens, Georgia, 192 pp.

Kaplan, S. L. 1999. A Field Guide to Southeastern and Caribbean Seashores: Cape Hatteras to the Gulf Coast, Florida, and the Caribbean. Peterson Field Guides, Houghton Mifflin Harcourt, New York, 480 pp.

Karban, R., and A. A. Agrawal. 2002. Herbivore offense. Annual Review of Ecology and Systematics 33:641–664.

Kaufmann, J. H. 1986. Stomping for earthworms by wood turtles, *Clemmys insculpta:* a newly discovered foraging technique. Copeia 4:1001–1004.

Kellert, S. R., and E. O. Wilson. 1993. The Biophilia Hypothesis. Island Press, Washington, D.C., 494 pp.

Kelley, P. H., M. Kowalewski, and T. A. Hansen. 2003. Predator–Prey Interactions in the Fossil Record. Springer, Berlin, 464 pp.

Kennedy, W. J. 1975. Trace fossils in carbonate rocks; pp. 377–398 in R. W. Frey (ed.), The Study of Trace Fossils, Springer-Verlag, Berlin.

Kennedy, W. J., M. E. Jakobson, and R. T. Johnson. 1969. A *Favreina–Thalassinoides* association from the Great Oolite of Oxfordshire. Palaeontology 12:549–554.

Kent, B. W. 1981. Prey dropped by herring gulls (*Larus argentatus*) on soft sediments. The Auk 98:350–354.

Kent, D. M., M. A. Langston, D. W. Hanf, and P. M. Wallace. 1997. Utility of a camera system for investigating gopher tortoise burrows. Florida Scientist 60:193–196.

Kern, J. P. 1978. Paleoenvironment of new trace fossils from the Mission Valley Formation, California. Journal of Paleontology 52:186–194.

Keyes, R. 2007. The Quote Verifier: Who Said What, Where, and When. Macmillan, New York, 387 pp.

King, A. 2004. Over a century of explorations at Etowah. Journal of Archaeological Research 11:279–306.

Kneib, R. T. 1997. The role of tidal marshes in the ecology of estuarine nekton; pp. 163–220. in A. Ansell and M. Barnes (eds.), Oceanography and Marine Biology: An Annual Review 35.

Knepton, J. C., Jr. 1954. A note on the burrowing habits of the salamander *Amphiuma means means*. Copeia 1954:68.

Knisley, C. B., and T. D. Schultz. 1997. Tiger Beetles and a Guide to the Species of the South Atlantic States. Virginia Museum of Natural History, Martinsville, Virginia, 209 pp.

Kobluk, D. R., S. G. Pemberton, M. Karolyi, and M. J. Risk. 1977. The Silurian–Devonian disconformity in southern Ontario. Bulletin of Canadian Petroleum Geology 25: 1157–1186.

Kogel, J. E., and E. Shelobolina. 2007. The Georgia Kaolins: Geology and Utilization. Society of Mining, Metallurgy, and Exploration, Littleton, Colorado, 84 pp.

Koretsky, C. M., C. Meile, and P. Van Cappelen. 2002. Quantifying bioirrigation using ecological parameters: a stochastic approach. Geochemical Transactions 3:17–30.

Kowalewski, M., T. M. Demko, S. T. Hasiotis, and D. Newell. 1998. Quantitative ichnology of Triassic crayfish burrows (*Camborygma eumekenomos*): ichnofossils as linkages to population paleoecology. Ichnos 6:5–20.

Kraeuter, J. N. 1976. Biodeposition by salt marsh invertebrates. Marine Biology 35: 215–223.

Krause, J. M., T. M. Bown, E. M. Bellosi, and J. F. Genise. 2008. Trace fossils of cicadas in the Cenozoic of central Patagonia, Argentina. Palaeontology 51:405–418.

Krech, S., III, 2009. Spirits of the Air: Birds and American Indians in the South. University of Georgia Press, Athens, Georgia, 245 pp.

Kurtén, B., and E. Anderson. 1980. Pleistocene Mammals of North America. Columbia University Press, New York, New York, 442 pp.

Kwasna, H. 2002. Changes in microfungal communities in roots of *Quercus robur* stumps and their possible effects on colonization by *Armillaria*. Journal of Phytopathology 150:403–411.

Labandeira, C. C., B. A. LePage, and A. H. Johnson. 2001. A *Dendroctonus* bark engraving (Coleoptera: Scolytidae) from a middle Eocene *Larix* (Coniferales: Pinaceae): early or delayed colonization? American Journal of Botany 88:2026–2039.

Labandeira, C. C., P. Wilf, K. R. Johnson, and F. Marsh. 2007. Guide to Insect (and Other) Damage Types on Compressed Plant Fossils (Version 3.0). Smithsonian Institution, Washington, D.C., 25 pp.

Laerm, J., T. C. Carter, M. A. Menzel, T. S. McCay, J. L. Boone, W. M. Ford, L. T. Lepardo, D. M. Krishon, G. Balkcom, N. L. Van Der Maath, and M. J. Harris. 1999. Amphibians, reptiles, and mammals of Sapelo Island, Georgia. Journal of the Elisha Mitchell Scientific Society 115:104–126.

Landes, D. A., M. S. Obin, A. B. Cady, and J. H. Hunt. 1987. Seasonal and latitudinal variation in spider prey of the mud dauber *Chalybion californicum* (Hymenoptera, Sphecidae). Journal of Arachnology 15:249–256.

Larsen, C. S. 1990. The Archaeology of Mission Santa Catalina de Guale. Anthropological Papers of the American Museum of Natural History 68, 150 pp.

Lawfield, A. M. W., and R. K. Pickerell. 2006. A novel contemporary fluvial ichnocoenose: unionid bivalves and the *Scoyenia-Mermia* transition. Palaios 21:391–396.

Laza, J. H. 2006. Dung-beetle fossil brood balls: the ichnogenera *Coprinisphaera* Sauer and *Quirogaichnus* (Coprinisphaeridae). Ichnos 13:217–235.

Lazell, J. D., and P. J. Auger. 1981. Predation on diamondback terrapin (*Malaclemys terrapin*) eggs by dunegrass (*Ammophila breviligulata*). Copeia 1981:723–724.

Lee, K. E. 1985. Earthworms: Their Ecology and Relationships with Soils and Land Use. Academic Press, Sydney, 411 pp.

Lee, S.-H., R.-L. Yang, and N.-Y. Su. 2008. Tunneling response of termites to a preformed tunnel. Behavioural Processes 79:192–194.

Lenz, R. J. 2002. Highroad Guide to Georgia and Okefenokee. John F. Blair Publishing, Winston-Salem, North Carolina, 320 pp.

Leong, T., and L. Waller. 2008. Evaluating the impact of fishing pier construction on total emergences for two species of sea turtles nesting in Palm Beach County, Florida, 1997–2000. NOAA Technical Memorandum 569:95.

LePage, B. A., R. S. Currah, R. A. Stockey, and G. W. Rothwell. 1997. Fossil ectomycorrhizae in Eocene *Pinus* roots. American Journal of Botany 84:401–412.

Letzsch, W. S., and R. W. Frey. 1980. Deposition and erosion in a Holocene salt marsh, Sapelo Island, Georgia. Journal of Sedimentary Research 50:529–542.

Levey, D. J., R. S. Duncan, and C. F. Levins. 2004. Animal behaviour: use of dung as a tool by burrowing owls. Nature 431:39.

Levin, P. S., J. Ellis, R. Petrik, and M. E. Hay. 2002. Indirect effects of feral horses on estuarine communities. Conservation Biology 16:1364–1371.

Lewis, D. W., and A. A. Ekdale. 1992. Composite ichnofabric of a mid-Tertiary unconformity on a pelagic limestone. Palaios 7:222–235.

Li, C., X.-C. Wu, O. Rieppel, L.-T. Wang, and L.-J. Zhao. 2008. An ancestral turtle from the Late Triassic of southwestern China. Nature 456:497–501.

Liebenberg, L. 2001. The Art of Tracking: The Origin of Science (2nd Edition). David Philip, Cape Town, South Africa, 176 pp.

Lin, N. 1978. Contributions to the ecology of the cicada killer, *Sphecius speciosus* (Hymenoptera: Sphecidae). Journal of the Washington Academy of Science 68: 75–82.

Linsenmair, E. 1967. Konstruktion und Signalfunktion der Sandpyramide der Reiterkrabbe *Ocypode saraten* Forsk (Decapoda, Brachyura, Ocypodidae). Zeitschrift für Tierpsychologie 24:403–456.

Linsley, D. L., G. A. Bishop, and H. B. Rollins. 2008. Stratigraphy and geologic evolution of St. Catherines Island, Georgia; pp. 26–41 in D. H. Thomas (ed.), Native American Landscapes of St. Catherines Island. Anthropological Papers of the American Museum of Natural History 88.

Lips, K. R. 1991. Vertebrates associated with tortoise (*Gopherus polyphemus*) burrows in four habitats in south central Florida. Journal of Herpetology 25:477–481.

Lockley, M. G. 1991. Tracking Dinosaurs: A New Look at an Ancient World. Cambridge University Press, Cambridge, U.K., 238 pp.

Lockley, M. G., and A. P. Hunt. 1995. Dinosaur Tracks and Other Fossil Footprints of the Western United States. Columbia University Press, New York, 338 pp.

Lockley, M. G., and C. A. Meyer. 1999. Dinosaur Tracks and Other Fossil Footprints of Europe. Columbia University Press, New York, 323 pp.

Lockley, M. G., C. A. Meyer, and V. F. dos Santos. 1998. *Megalosauripus* and the problematic concept of megalosaur footprints. Gaia 15:313–337.

López-Martínez, N., J. J. Moratalla, and J. L. Sanz. 2000. Dinosaurs nesting on tidal flats. Palaeogeography, Palaeoclimatology, Palaeoecology 160:153–163.

Louv, R. 2005. Last Child in the Woods: Saving Our Children from Nature-Deficit Disorder. Algonquin Books, Chapel Hill, North Carolina, 336 pp.

Lucas, J. R. 1982. The biophysics of pit construction by antlion larvae (Myrmeleon, Neuroptera). Animal Behaviour 30:651–664.

Luken, J. O., and P. J. Kalisz. 1989. Soil disturbance by the emergence of periodical cicadas. Soil Science Society of America Journal 53:310–313.

Lundgren, S. A. B. 1891. Studier öfver fossilförande lösa block. Geologiska Föreningens i Stockholm Förhandlingar 13:111–121.

Lutz, P. L., J. A. Musick, and J. Wyneken. 1997. The Biology of Sea Turtles. Vol. 1. CRC Press, Boca Raton, Florida, 432 pp.

MacDonald, J. F., and R. W. Matthews. 1981. Nesting biology of the eastern yellowjacket, *Vespula maculifrons* (Hymenoptera: Vespidae). Journal of the Kansas Entomological Society 54:433–457.

———. 1984. Nesting biology of the southern yellow jacket, *Vespula squamosa* (Hymenoptera: Vespidae): social parasitism and independent founding. Journal of the Kansas Entomological Society 57:134–151.

MacEachern, J. A., S. G. Pemberton, M. K. Gingras, and K. L. Bann. 2007a. The ichnofacies paradigm: a fifty-year perspective; pp. 52–77 in W. M. Miller III (ed.), Trace Fossils: Concepts, Problems, Prospects. Elsevier, Amsterdam, Netherlands.

MacEachern, J. A., S. G. Pemberton, M. K. Gingras, K. L. Bann, and L. T. Dafoe. 2007b. Uses of trace fossils in genetic stratigraphy; pp.110–134 in W. M. Miller III (ed.), Trace Fossils: Concepts, Problems, Prospects. Elsevier, Amsterdam, Netherlands.

MacGown, J. A., J. G. Hill, L. C. Majure, and J. L. Seltzer. 2008. Rediscovery of *Pogonomyrmex badius* (Latreille) (Hymenoptera: Formicidae) in mainland Mississippi, with an analysis of associated seeds and vegetation. Midsouth Entomologist 1: 17–28.

Magallón, S. A., and M. J. Sanderson. 2005. Angiosperm divergence times: the effect of genes, codon positions, and time constraints. Evolution 59:1653–1670.

Mángano, M. G., L. A. Buatois, R. R. West, and C. G. Maples. 1999. The origin and paleoecologic significance of the trace fossil *Asteriacites* in the Pennsylvanian of Kansas and Missouri. Lethaia 32:17–30.

Mann, R., and S. M. Gallagher. 1984. Physiology of the wood-boring mollusk *Martesia cuneiformis* Say. Biology Bulletin 166:167–177.

Manning, R. B., and D. L. Felder. 1989. The *Pinnixa cristata* complex in the Western Atlantic, with a description of two new species (Crustacea: Decapoda: Pinnotheridae). Smithsonian Contributions to Zoology 473:1–36.

Manucy, A. C. 1952. Tapia or tabby. The Journal of the Society of Architectural Historians 11:32–33.

Manzi, J. J. 1970. Combined effects of salinity and temperature on the feeding, reproductive, and survival rates of *Eupleura cadata* (Say) and *Urosaplinx cinerea* (Say) (Prosobranchia: Muricidae). Biology Bulletin 138:35–46.

Marcy, B. C., D. E. Fletcher, F. D. Martin, M. H. Paller, and M. J. M Reichert. 2005. Fishes of the Middle Savannah River Basin: With Emphasis on the Savannah River Site. University of Georgia Press, Athens, Georgia, 462 pp.

Marscher, B., and F. Marscher. 2004. The Great Sea Island Storm of 1893. Mercer University Press, Macon, Georgia, 134 pp.

Marshall, S. D., W. R. Hoeh, and M. A. Deyrup. 2000. Biogeography and conservation biology of Florida's *Geolycosa* wolf spiders: threatened spiders in endangered ecosystems. Journal of Insect Conservation 4:11–21.

Martin, A. J. 1992. Semiquantitative and statistical analysis of bioturbate textures in the Sequatchie Formation (Upper Ordovician), Georgia and Tennessee, U.S.A. Ichnos 2:117–136.

———. 1999. Fossil upogebiid burrows and their geologic significance: Grotto Beach Formation (Pleistocene), San Salvador, Bahamas; pp. 81–92 in H. A. Curran and J. E. Mylroie (eds.), Proceedings of the 9th Symposium on the Geology of the Bahamas and Other Carbonate Regions, Gerace Research Centre, San Salvador (Bahamas).

———. 2006a. Resting traces of *Ocypode quadrata* associated with hydration and respiration: Sapelo Island, Georgia, U.S.A. Ichnos 13:57–67.

———. 2006b. A composite trace fossil of decapod and hymenopteran origin from the Rice Bay Formation (Holocene), San Salvador, Bahamas; pp. 99–112 in D. Gamble and R. L. Davis. (eds.), 12th Symposium of the Geology of the Bahamas and Other Carbonate Regions, Gerace Research Centre, San Salvador, Bahamas.

———. 2006c. Trace Fossils of San Salvador. Gerace Research Centre, San Salvador, Bahamas, 80 pp.

———. 2006d. Introduction to the Study of Dinosaurs. Wiley-Blackwell Publishing, Oxford, U.K., 560 pp.

———. 2009a. Neoichnology of an Arctic fluvial point bar, Colville River, Alaska (U.S.A.). Geological Quarterly 53:383–396.

———. 2009b. Dinosaur burrows in the Otway Group (Albian) of Victoria, Australia, and their relation to Cretaceous polar environments. Cretaceous Research 30:1223–1237.

———. 2009c. Applications of trace fossils to interpreting paleoenvironments and sequence stratigraphy; pp. 35–42 in M. S. Duncan and R. L. Kath (eds.), Fall Line Geology of East Georgia: With a Special Emphasis on the Upper Eocene. Georgia Geological Society Guidebook 29.

Martin, A. J., and N. Pyenson. 2005. Behavioral significance of vertebrate trace fossils from the Union Chapel Mine site. Pennsylvanian Footprints in the Black Warrior Basin of Alabama. Alabama Paleontological Society Monograph No. 1:59–73.

Martin, A. J., and A. K. Rindsberg. 2007. Arthropod tracemakers of *Nereites*? Neoichnological observations of juvenile limulids and their paleoichnological applications; pp. 478–491 in W. M. Miller III (ed.), Trace Fossils: Concepts, Problems, Prospects. Elsevier, Amsterdam, Netherlands.

———. 2011. Ichnological diagnosis of ancient storm washover fans, Yellow Banks Bluff, St. Catherines Island, Georgia (U.S.A.); pp. 113–127 in G. A. Bishop, H. B. Rollins, and D. H. Hurst (eds.), Geoarchaeology of St. Catherines Island, Georgia. Anthropological Papers of the American Museum of Natural History 94.

Martin, A. J., and D. J. Varricchio. 2008. Composite trace fossils in terrestrial environments and their resultant ichnofabrics. Ichnia 2008, Abstracts and Program, Krakow, Poland 2:76.

———. 2011. Paleoecological utility of insect trace fossils in dinosaur nesting sites of the Two Medicine Formation (Campanian), Choteau, Montana. Historical Biology 23: 15–25.

Martin, A. J., T. H. Rich, G. C. B. Poore, M. B. Schultz, C. M. Austin, L. Kool, and P. Vickers-Rich. 2008. Fossil evidence from Australia for oldest known freshwater crayfish in Gondwana. Gondwana Research 14:287–296.

Martin, A. J., G. M. Vazquez-Prokopec, and M. Page. 2010. First known feeding trace of the Eocene bottom-dwelling fish *Notogoneus osculus* and its paleontological significance. PLoS One 5: http://dx.plos.org/10.1371/journal.pone.0010420.

Martin, D. J. 1973. Selected aspects of burrowing owl ecology and behavior. Condor 75: 446–456.

Martin, L. D., and D. K. Bennett. 1977. The burrows of the Miocene beaver *Palaeocastor*, western Nebraska, U.S.A. Palaeogeography, Palaeoclimatology, Palaeoecology 22: 173–193.

Martin, L. D., and D. L. West. 1995. The recognition and use of dermestid (Insecta, Coleoptera) pupation chambers in paleoecology. Palaeogeography, Palaeoclimatology, Palaeoecology 113:303–310.

Martin, P. S. 2005. Twilight of the Mammoths: Ice Age Extinctions and the Rewilding of America. University of California Press, Berkeley, California, 269 pp.

Martinell, J., J. M. de Gilbert, R. Domenech, A. A. Ekdale, and P. P. Steen. 2001. Cretaceous ray traces?: an alternative interpretation for the alleged dinosaur tracks of La Posa, Isona, NE Spain. Palaios 16:409–416.

Martino, R. L., and S. F. Greb. 2007. Walking trails of the giant terrestrial arthropod *Arthropleura* from the Upper Carboniferous of Kentucky. Journal of Paleontology 83: 140–146.

Mason, T. R., and V. V. Bruu. 1978. Mudcracks initiated by wader birds. South African Journal of Science 74: 224–225.

Mauseth, J. D. 2008. Botany: An Introduction to Plant Biology (4th Edition). Jones & Bartlett, Subdbury, Massachusetts, 624 pp.

Mayor, J. J., Jr., and I. L. Brisbin. 2008. Wild Pigs in the United States: Their History, Comparative Morphology, and Current Status. University of Georgia Press, Athens, Georgia, 336 pp.

Mazin, J.-M., J.-P. Billon-Bruyat, and K. Padian. 2009. First record of a pterosaur landing trackway. Proceedings of the Royal Society of London B 276:3881–3886.

McBee, K., and R. J. Baker. 1982. *Dasypus novemcinctus*. Mammalian Species 162: 1–9.

McCrone, J. D. 1963. Taxonomic status and evolutionary history of the *Geolycosa pikei* complex in the southeastern United States (Araneae, Lycosidae). American Midland Naturalist 70:47–73.

McDermott, J. J. 2005. Biology of the brachyuran crab *Pinnixa chaetopterana* Stimpson (Decapoda: Pinnotheridae) symbiotic with tubicolous polychaetes along the Atlantic coast of the United States, with additional notes on other polychaete associations. Proceedings of the Biological Society of Washington 118:742–764.

McGavin, G. C. 2001. Essential Entomology: An Order-by-Order Introduction. Oxford University Press, Oxford, U.K., 318 pp.

McIlroy, D. 2008. Ichnological analysis: the common ground between ichnofacies workers and ichnofabric analysts. Palaeogeography, Palaeoclimatology, Palaeoecology 270:332–338.

Means, D. B. 1982. Responses to winter burrow flooding of the gopher tortoise (*Gopherus polyphemus* Daudin). Herpetologica 38:521–525.

———. 2009. Effects of rattlesnake roundups on the Eastern diamondback rattlesnake (*Crotalus adamanteus*). Herpetological Conservation and Biology 4:132–141.

Meeker, J. R., W. N. Dixon, and J. L. Foltz. 1995. The Southern pine beetle *Dendroctonus frontalis* Zimmerman (Coleoptera: Scolytidae). Florida Department of Agricultural and Consumer Services, Entomology Circular, 369, 4 pp.

Melchor, R. N., J. F. Genise, and S. E. Miquel. 2002. Ichnology, sedimentology and paleontology of Eocene calcareous paleosols from a palustrine sequence, Argentina. Palaios 17:16–35.

Melchor, R. N., E. Bedatou, S. de Valais, and J. F. Genise. 2006. Lithofacies distribution of invertebrate and vertebrate trace-fossil assemblages in an Early Mesozoic ephemeral fluvio-lacustrine system from Argentina: implications for the *Scoyenia* ichnofacies. Palaeogeography, Palaeoclimatology, Palaeoecology 239:253–285.

Meng, Q., J. Liu, D. J. Varricchio, T. Huang, and C. Gao. 2004. Parental care in an ornithischian dinosaur. Nature 431:145–146

Menzies, R. J., and D. Frankenberg. 1966. Handbook on the Common Marine Isopod Crustacea of Georgia. University of Georgia Press, Athens, Georgia, 93 pp.

Merritt, R. W., and K. W. Cummins. 1996. An Introduction to the Aquatic Insects of North America. Kendall Hunt, Dubuque, Iowa, 862 pp.

Messina, M. G., and W. H. Conner. 1996. Southern Forested Wetlands: Ecology and Management. CRC Press, Boca Raton, Florida, 616 pp.

Metz, R. 1990. Tunnels formed by mole crickets (Orthoptera: Gryllotalpidae): paleoecological implications. Ichnos 1:139–141.

Meyers, A. C. 1970. Some palaeoichnological observations on the tube of *Diopatra cuprea* (Bosc.): Polychaeta, Onuphidae; pp. 331–334 in T. P. Crimes and J. C. Harper (eds.), Trace Fossils. Seel House Press, Liverpool, U.K.

Mihail, J. D., J. N. Bruhn, and T. D. Leininger. 2002. The effects of moisture and oxygen availability on rhizomorph generation by *Armillaria tabescens* in comparison to *A. gallica* and *A. mellea*. Mycological Research 106:697–704.

Mikuláš, R. 1990. The ophiuroid *Taeniaster* as a tracemaker of *Asteriacites,* Ordovician of Czechoslovakia. Ichnos 1:133–137.

———. 1998. Two different meanings of the term "bioglyph" in the geological literature: history of the problem, present-day state, and possible resolution. Ichnos 6: 211–213.

———. 2001. Modern and fossil traces in terrestrial lithic substrates. Ichnos 8:177–184.
Mikuláš, R., and B. Zasadil. 2004. A probable fossil bird nest, ?*Eocavum* isp., from the Miocene wood of the Czech Republic. Fourth International Bioerosion Workshop, Prague, Czech Republic, Abstracts Book 49–51.
Michener, C. D. 2000. Bees of the World. Johns Hopkins University Press, Baltimore, Maryland, 913 pp.
Michener, W. K., E. R. Blood, K. L. Bildstein, M. M. Brinson, and L R. Gardner. 1997. Climate change, hurricanes and tropical storms, and rising sea level in coastal wetlands. Ecological Applications 7:770–801.
Mihail, J. D., J. N. Bruhn, and T. D. Leininger. 2002. The effects of moisture and oxygen availability on rhizomorph generation by *Armillaria tabescens* in comparison to *A. gallica* and *A. mellea*. Mycological Research 106:697–704.
Milanich, J. T. 1999. The Timucua. Blackwell Publishing, Oxford, U.K., 235 pp.
Minter, N. J., and S. J. Braddy. 2006. The fish and amphibian swimming traces *Undichna* and *Lunichnium*, with examples from the Lower Permian of New Mexico, U.S.A. Palaeontology 49:1123–1142.
Mitra, O., M. A. Callaham Jr., and J. E. Yack. 2009. Grunting for worms: seismic vibrations cause *Diplocardia* earthworms to emerge from the soil. Biology Letters 2009:16–19.
Montalvo, C. I. 2002. Root traces in fossil bones from the Huayquerian (Late Miocene) faunal assemblage of Telén, La Pampa, Argentina. Acta Geológica Hispánica 37: 37–42.
Moore, J. 2002. Parasites and the Behavior of Animals. Oxford University Press, Oxford, U.K., 315 pp.
Moore, J., and R. Overhill. 2001. An Introduction to the Invertebrates. Cambridge University Press, Cambridge, U.K., 355 pp.
Moore, J. M., and M. D. Picker. 1991. Heuweltjies (earth mounds) in the Clanwilliam district, Cape Province, South Africa: 4000-year-old termite nests. Oecologia 86: 424–432.
Moreria, A. A., L. C. Forti, A. P. A. Andrade, M. A. C. Boaretto, and J. F. S. Lopes. 2004. Nest architecture of *Atta laevigata* (F. Smith, 1858) (Hymenoptera: Formicidae). Studies on Neotropical Fauna and Environment 39:109–116.
Morgan, D. L., and G. W. Frankie. 1982. Biology and control of the mealy-oak gall. Journal of Arboriculture 8:230–233.
Morgan, P. (ed.). 2010. African-American Life in the Georgia Low Country. University of Georgia Press, Athens, Georgia, 320 pp.
Morris, R. W., and H. B. Rollins. 1977. Observations on intertidal organism associations on St. Catherines Island, Georgia. I. General description and paleoecological implications. Bulletin of the American Museum of Natural History 159:87–128.
Morrisey, L. B., and S. J. Braddy. 2004. Terrestrial trace fossils from the Lower Old Red Sandstone, southwest Wales. Geological Journal 39:315–336.
Morse, B. W., M. L. McElroy, and K. V. Miller. 2009. Seasonal diets of an introduced population of fallow deer on Little St. Simons Island, Georgia. Southeastern Naturalist 8:571–586.
Müller, A. H. 1956. Über problematische Lebensspuren aus dem Rotliegenden von Thüringen. Geologische Gesellschaft DDR, Berichte (Berichte der Geologischen Gesellschaft in der Deutschen Demokratischen Republik für das Gesamtgebiet der Geologischen Wissenschaften) 1:147–155.

Nara, M. 2006. Reappraisal of *Schaubcylindrichnus:* a probable dwelling/feeding structure of a solitary funnel feeder. Palaeogeography, Palaeoclimatology, Palaeoecology 240:439–452.
Nebelsick, J. H. 1999. Taphonomic comparison between Recent and fossil sand dollars. Palaeogeography, Palaeoclimatology, Palaeoecology 149:349–358.
Neff, B. D., L. M. Cargnelli, and I. M. Cote. 2004. Solitary nesting as an alternative breeding tactic in colonial nesting bluegill sunfish (*Lepomis macrochirus*). Behavioral Ecology and Sociobiology 56:381–387.
Nesbitt, E. A., and K. A. Campbell. 2006. The paleoenvironmental significance of *Psilonichnus*. Palaios 21:187–196.
Netto, R. G. 2004. *Skolithos*-dominated piperock in nonmarine environments; an example from the Triassic Caturrita Formation, southern Brazil; pp. 109–121 in R. G. Bromley, L. A. Buatois, G. M. Mángano, J. F. Genise, and R. N. Melchor (eds.), Sediment-Organism Interactions: A Multifaceted Ichnology. Society for Sedimentary Geology, Special Publication 88.
Netto, R. G., L. A. Buatois, M. G. Mángano, and P. Balistieri. 2007. *Gyrolithes* as a multipurpose burrow: an ethologic approach. Revista Brasileira de Paleontologica 10: 157–168.
Nickerson, J. C., D. E. Synder, and C. C. Oliver. 1979. Acoustical burrows constructed by mole crickets. Annals of the Entomological Society of America 72:438–440.
Niedźwiedzki, G., P. Szrek, K. Narkiewicz, M. Narkiewicz, and P. E. Ahlberg. 2010. Tetrapod trackways from the early Middle Devonian Period of Poland. Nature 463: 43–48.
Nobel, P. S. 2009. Physiochemical and Environmental Plant Physiology (Fourth Edition). University of Chicago Press, Chicago, Illinois, 582 pp.
Odum, E. P. 1968. Energy flow in ecosystems; a historical review. American Zoologist 8: 11–18.
———. 2000. Tidal marshes as outwelling/pulsing systems; pp. 3–7 in M. P. Weinstein and D. A. Kreeger (eds.), Concepts and Controversies in Tidal Marsh Ecology. Springer, Amsterdam, Netherlands.
Odum, E. P., and A. E. Smalley. 1959. Comparison of population energy flow of a herbivorous and a deposit-feeding invertebrate in a salt marsh ecosystem. Proceedings of the National Academy of Sciences 45:617–622.
O'Geen, A. T., P. A. McDaniel, and A. J. Busacca. 2002. Cicada burrows as indicators of paleosols in the inland Pacific northwest. Soil Science of America Journal 66:1584–1586.
O'Mahoney, P. M., and R. J. Full. 1984. Respiration of crabs in air and water. Comparative Biochemistry and Physiology 79:275–282.
O'Neill, K. M. 2001. Solitary Wasps: Behavior and Natural History. Cornell University Press, Ithaca, New York, 406 pp.
Osgood, R. G. 1975. The history of invertebrate ichnology; pp. 3–12 in R. W. Frey (ed.), The Study of Trace Fossils. Springer-Verlag, New York.
Paik, I. S. 2000. Bone chip–filled burrows associated with bored dinosaur bone in floodplain paleosols of the Cretaceous Hasandong Formation, Korea. Palaeogeography, Palaeoclimatology, Palaeoecology 157:213–225.
Palmer, W. M., and C. L. Cordes. 1988. Habitat suitability index models: diamondback terrapin nesting, Atlantic Coast. U.S. Fish & Wildlife Service Biological Report 82 (10.151): 18 pp.

Parga, J. A., and R. G. Lessnau. 2008. Dispersal among male ring-tailed lemurs (*Lemur catta*) on St. Catherines Island. American Journal of Primatology 70:650–660.
Parr, C. S., T. Jones, and N. B. Songer. 2004. Evaluation of a handheld data collection interface for science learning. Journal of Science Education and Technology 13:233–242.
Parrish, J. W., Jr., G. Beaton, and G. Kennedy. 2006. Birds of Georgia. Lone Pine Publishing International, Auburn, Washington, 384 pp.
Pearson, D. L., and A. P. Vogler. 2001. Tiger Beetles: The Evolution, Ecology, and Diversity of the Cicindelids. Cornell University Press, Ithaca, New York, 333 pp.
Pearson, D.L., C. B. Knisley, and J. Kazilek. 2006. A Field Guide to the Tiger Beetles of the United States and Canada. Oxford University Press, Oxford, U.K., 227 pp.
Pearson, P. N. 1992. Walking traces of the giant myriapod *Arthropleura* from the Strathclyde Group (Lower Carboniferous) of Fife. Scottish Journal of Geology 28: 127–133.
Pemberton, S. G. 1992. Applications of Ichnology to Petroleum Exploration: A Core Workshop. Society of Economic Paleontologists and Mineralogists, Tulsa, Oklahoma, Core Workshop 17, 429 pp.
Pemberton, S. G., and R. W. Frey. 1985. The *Glossifungites* ichnofacies: modern examples from the Georgia coast, U.S.A.; pp. 237–259 in H. A. Curran (ed.), Biogenic Structures: Their Use in Interpreting Depositional Environments. Society of Economic Paleontologists and Mineralogists Special Publication 35.
———. 1990. Darwin on worms: the advent of experimental neoichnology. Ichnos 1: 65–71.
Pemberton, S. G., and M. K. Gingras. 2005. Classification and characterizations of biogenically enhanced permeability. A.A.P.G. Bulletin 89:1493–1517.
Pemberton, S. G., D. R. Kobluk, R. K. Yeo, and M. R. Risk. 1980. The boring *Trypanites* at the Silurian-Devonian disconformity in southern Ontario. Journal of Paleontology 54:1258–1266.
Pemberton, S. G., M. K. Gingras, and J. A. MacEachern. 2007a. Edward Hitchcock and Roland Bird: two early titans of vertebrate ichnology in North America; pp. 32–51 in W. M. Miller III (ed.), Trace Fossils: Concepts, Problems, Prospects. Elsevier, Amsterdam, Netherlands.
Pemberton, S. G., J. A. MacEachern, and M. K. Gingras. 2007b. The antecedents of invertebrate ichnology in North America: the Canadian and Cincinnati schools; pp. 14–31 in W. M. Miller III (ed.), Trace Fossils: Concepts, Problems, Prospects. Elsevier, Amsterdam, Netherlands.
Pemberton, S. G., R. McCrea, and M. K. Gingras. 2008a. History of ichnology: the correspondence between the Reverend Henry Duncan and the Reverend William Buckland and the discovery of the first vertebrate footprints. Ichnos 15:5–17.
Pemberton, S. G., J. A. MacEachern, M. K. Gingras, and T. D. A. Saunders. 2008b. Biogenic chaos: cryptobioturbation and the work of sedimentologically friendly organisms. Palaeogeography, Palaeoclimatology, Palaeoecology 270:273–279.
Peñalver, E., D. A. Grimaldi, and X. Delclòs. 2006. Early Cretaceous spider web with its prey. Science 312:1761.
Perrichot, V., D. Néraudeau, D. Azar, J.-J. Menier, and A. Nel. 2002. A new genus and species of fossil mole cricket in the Lower Cretaceous amber of Charente-Maritime, SW France (Insecta: Orthoptera: Gryllotalpidae). Cretaceous Research 23:307–314.

Petersen, K. E., and T. L. Yates. 1980. *Condylura cristata*. Mammalian Species 129:1–4.

Petti, F. M., M. Avanzini, M. Belvedere, M. De Gasperi, P. Ferretti, S. Girardi, F. Remondino, and R. Tomasoni. 2008. Digital 3D modelling of dinosaur footprints by photogrammetry and laser scanning techniques: integrated approach at the Coste dell'Anglone tracksite (Lower Jurassic, Southern Alps, northern Italy); pp. 303–315 in M. Avanzini and F. M. Petti (eds.), Italian Ichnology: Studi Trentini di Scienze Naturali. Acta Geologica 83.

Pfennig, D. W., and H. K. Reeve. 1993. Nepotism in a solitary wasp as revealed by DNA fingerprinting. Evolution 47:700–704.

Pickford, M. 2000. Toiles d'araignées fossiles du désert du Namib et l'antiquité du genre Seothyra (Araneae, Eresidae) [Fossil spider's webs from the Namib Desert and the antiquity of *Seothyra* (Araneae, Eresidae)]. Annales de Paléontologie 68:147–155.

Pienkowski, G., A. J. Martin, and C. A. Meyer. (eds.). 2009. Preface. Special Issue, Second International Congress on Ichnology (Ichnia 2008). Geological Quarterly 53:369–371.

Pilkey, O., and M. E. Fraser. 2003. A Celebration of the World's Barrier Islands. Columbia University Press, New York, 309 pp.

Pinder, A. W., K. B. Storey, and G. R. Ultsch. 1992. Estivation and hibernation; pp. 250–274 in M. E. Feder and W. W. Burggren (eds.), Environmental Physiology of the Amphibians. University of Chicago Press, Chicago, Illinois.

Pirkle, W. A., and F. L. Pirkle. 2007. Introduction to heavy-mineral sand deposits of the Florida and Georgia Atlantic coastal plain; pp. 129–135 in F. J. Rich (ed.), Guide to Fieldtrips—56th Annual Meeting, Southeastern Section of the Geological Society of America. Georgia Southern University, Department of Geology and Geography Contribution, Series 1.

Pisani, D., L. D. Poling, M. Lyons-Weiler, and S. B. Hedges. 2004. The colonization of land by animals: molecular phylogeny and divergence times among arthropods. BioMed Central 2: doi:10.1186/1741-7007-2-1.

Pitts-Singer, T. L., and B. T. Forschler. 2000. Influence of guidelines and passageways on tunneling behavior of *Reticulitermes flavipes* (Kollar) and *R. viginicus* (Banks) (Isoptera: Rhinotermitidae). Journal of Insect Behavior 13:273–290.

Platt, B. F., S. T. Hasiotis, and D. R. Hirmas. 2010. Use of low-cost multistripe laser triangulation (MLT) scanning technology for three-dimensional, quantitative paleoichnological and neoichnological studies. Journal of Sedimentary Research 80:590–610.

Plaziat, J.-C., and M. Mahmoudi. 1988. Trace fossils attributed to burrowing echinoids: a revision including new ichnogenus and ichnospecies. Geobios 21:209–233.

Pollitzer, W. S. 1999. The Gullah People and Their African Heritage. University of Georgia Press, Athens, Georgia, 298 pp.

Prestwich, K. N. 2006. Anaerobic metabolism and maximum running speed in the scorpion *Centruroides hentzi* (Banks) (Scorpiones, Buthidae). Journal of Arachnology 34:351–356.

Prezant, R. S., R. B. Toll, H. B. Rollins, and E. J. Chapman. 2002. Marine macroinvertebrate diversity of St. Catherines Island, Georgia. American Museum Novitates 3367:1–31.

Proffitt, C. E., S. E. Travis, and K. R. Edwards. 2003. Genotype and elevation influence *Spartina alterniflora* colonization and growth in a created salt marsh. Ecological Applications 13:180–192.

Pryor, W. A. 1975. Biogenic sedimentation and alteration of argillaceous sediments in shallow marine environments. Geological Society of America Bulletin 86: 1244–1254.

Ratcliffe, B. C., and J. A. Fagerstrom. 1980. Invertebrate lebensspuren of Holocene floodplains: their morphology, origin and paleontological significance. Journal of Paleontology 54:614–630.

Reading, H. G. 1996. Sedimentary Environments: Processes, Facies, and Stratigraphy. Wiley-Blackwell, Oxford, U.K., 688 pp.

Reineck, H.-E. 1963. Sedimentgefüge im Bereich der südlichen Nordsee. Abhandlungen Senckenbergische Naturforschende Gesellschaft 505:1–138.

———. 1981. Lebensspur of a bird starting to fly in snow on the German–Austrian border. Journal of Sedimentary Petrology 51:699.

Reineck, H.-E., and J. D. Howard. 1978. Alligatorfährten. Natur und Museum 108:10–15.

Reineck, H.-E., and I. B. Singh. 1975. Depositional Sedimentary Environments. Springer-Verlag, Berlin, 439 pp.

Reitz, E. J., D. Linsley, G. A. Bishop, H. B. Rollins, D. H. Thomas, and R. H. Hayes. 2008. A brief natural history of St. Catherines Island; pp. 48–61 in D. H. Thomas (ed.), Native American Landscapes of St. Catherines Island. Anthropological Papers of the American Museum of Natural History 88.

Retallack, G. J. 2001. *Scoyenia* burrows from Ordovician palaeosols of the Juniata Formation in Pennsylvania. Palaeontology 44:209–235.

———. 2004. Late Oligocene bunch grassland and early Miocene sod grassland paleosols from central Oregon, U.S.A. Palaeogeography, Palaeoclimatology, Palaeoecology 207:203–237.

Retallack, G. J., and C. R. Feakes. 1987. Trace fossil evidence for Late Ordovician animals on land. Science 235:61–63.

Rezendes, P. 1999. Tracking and the Art of Seeing: How to Read Animal Tracks and Signs (2nd Edition). Harper Collins, New York, 336 pp.

Rhoads, D. C., and J. D. Germano. 1982. Characterization of organism-sediment relations using sediment profile imaging: an efficient method of remote ecological monitoring of the seafloor (Remote™ System). Marine Ecology Progress Series 8: 115–128.

Rich, F. J. 2007. Biological and sedimentological dynamics of the Okefenokee Swamp; pp. 163–176 in F. J. Rich (ed.), Guide to Fieldtrips, 56th Annual Meeting, Southeastern Section of the Geological Society of America. Georgia Southern University, Department of Geology and Geography Contribution Series, 1.

Rich, F. J., A. Semratedu, J. Elzea, and L. Newsom. 1999. Palynology and paleoecology of a wood-bearing clay deposit from Deepstep, Georgia. Geological Society of America Abstracts with Programs 31(3):61.

Rindsberg, A. K., and A. J. Martin. 2003. *Arthrophycus* and the problem of compound trace fossils. Palaeogeography, Palaeoclimatology, Palaeoecology 192:187–219.

———. 2009. Take continental ichnology to the masses. North American Paleontological Convention (NAPC) Abstracts 9:21.

Riskin, D. K., S. Parsons, W. A. Schutt Jr, G. G. Carter, and J. W. Hermanson. 2006. Terrestrial locomotion of the New Zealand short-tailed bat *Mystacina tuberculata* and the common vampire bat *Desmodus rotundus*. Journal of Experimental Biology 209:1725–1736.

Roberts, E. M., R. R. Rogers, and B. Z. Foreman. 2007. Continental insect borings in dinosaur bone: examples from the Late Cretaceous of Madagascar and Utah. Journal of Paleontology 81:201–208.

Robertson, J. R., and S. Y. Newell. 1982. A study of particle ingestion by three fiddler crab species foraging on sandy sediments. Journal of Experimental Marine Biology and Ecology 65:11–17.

Robertson, J. R., and W. J. Pfeiffer. 1982. Deposit feeding by the ghost crab *Ocypode quadrata*. Journal of Experimental Marine Biology and Ecology 56:165–177.

Robinson, W. H. 2005. Handbook of Urban Insects and Arachnids. Cambridge University Press, Cambridge, U.K., 472 pp.

Rodgers, J. C., III. 2002. Effects of human disturbance on the dune vegetation of the Georgia Sea Islands. Physical Geography 23:79–94.

Rogers, R. R. 1992. Non-marine borings in dinosaur bones from the Upper Cretaceous Two Medicine Formation, northwestern Montana. Journal of Vertebrate Paleontology 12:528–531.

Rona, P. A., A. Seilacher, C. de Vargas, A. J. Gooday, J. M. Bernhard, S. Bowser, C. Vetriani, C. O. Wirsen, L. Mullineaux, R. Sherrell, J. F. Grassle, S. Low, and R. A. Lutz. 2009. *Paleodictyon nodosum:* a living fossil on the deep-sea floor. Deep Sea Research Part II: Topical Studies in Oceanography 56:1700–1712.

Ruckdeschel, C., and C. R. Shoop. 2006. Sea Turtles of the Atlantic and Gulf Coasts of the United States. University of Georgia Press, Athens, Georgia, 126 pp.

Rue, L. L., III, and L. L. Rue. 2004. The Deer of North America. Globe Pequot, Guilford, Connecticut, 560 pp.

Runge, M. C., C. A. Sanders-Reed, C. A. Langtimm, and C. J. Fonnesbeck. 2007. A quantitative threats analysis for the Florida manatee (*Trichechus manatus latirostris*). U.S. Geological Survey Open-File Report 2007–1086, 34 pp.

Ruppert, E. E., and R. S. Fox. 1988. Seashore Animals of the Southeast. University of South Carolina Press, Columbia, South Carolina, 429 pp.

Russo, M. 2004. Measuring shell rings for social inequality; pp. 26–70 in J. L. Gibson and P. J. Carr. (eds.), Signs of Power: The Rise of Cultural Complexity in the Southeast. University of Alabama Press, Tuscaloosa, Alabama.

———. 2006. Archaic Shell Rings of the Southeast U.S.: National Historic Landmarks Historic Context. Southeast Archeological Center, U.S. National Park Service, Tallahassee, Florida, 173 pp.

Rützler, K. 1975. The role of burrowing sponges in bioerosion. Oecologia 19:203–216.

Sadler, C. J. 1993. Arthropod trace fossils from the Permian De Chelly Sandstone, northeastern Arizona. Journal of Paleontology 67:240–249.

Sarzetti, L. C., C. C. Labandeira, and J. F. Genise. 2008. A leafcutter bee trace fossil from the Middle Eocene of Patagonia, Argentina, and a review of megachilid (Hymenopteran) ichnology. Palaeontology 51:933–941.

Sassaman, K. E. 1993. Early Pottery in the Southeast: Tradition and Innovation in Cooking Technology. University of Alabama Press, Tuscaloosa, Alabama, 312 pp.

Savazzi, E. 1999. Functional Morphology of the Invertebrate Skeleton. John Wiley & Sons, New York, 712 pp.

Savrda, C. E., and L. L. Nanson. 2003. Ichnology of fair-weather and storm deposits in an Upper Cretaceous estuary (Eutaw Formation, western Georgia, U.S.A.). Palaeogeography, Palaeoclimatology, Palaeoecology 202:67–83.

Savrda, C. E., K. Ozalas, T. H. Demko, R. A. Huchison, and T. D. Scheiwe. 1993. Loggrounds and the ichnofossil *Teredolites* in transgressive deposits of the Clayton Formation (lower Paleocene), western Alabama. Palaios 8:311–324.

Schäefer, W. 1972. Ecology and Paleoecology of Marine Environments. University of Chicago Press, Chicago, Illinois, 584 pp. [Reprinted and translated from German. Originally published as: Aktuo-Paläontologie nack, Studien in der Nordsee, Waldemar Kramer, Frankfurt am Main, 666 pp.]

Scheffrahn, R. H., N.-Y. Su, J. A. Chase, and B. T. Forschler. 2001. New termite (Isoptera: Kalotermitidae, Rhinotermitidae) records from Georgia. Journal of Entomological Science 36:109–113.

Schmidt, A. R., H. Dörfelt, and V. Perrichot. 2007. Carnivorous fungi from Cretaceous. Science 318:1743.

Schnell, D. E. 2002. Carnivorous Plants of the United States and Canada. Timber Press, Portland, Oregon, 468 pp.

Schult, M. F., and J. O. Farlow. 1992. Vertebrate trace fossils; pp. 34–63 in C. G. Maples and R. R. West. (eds.), Trace Fossils. Short Courses in Paleontology, Paleontological Society 5.

Schwimmer, D. R. 2002. King of the Crocodylians: The Paleobiology of *Deinosuchus*. Indiana University Press, Bloomington, Indiana, 220 pp.

———. 2004. Late Cretaceous dinosaurs of the eastern Gulf Coast, and their relationships with Atlantic Coast taxa. Geological Society of America Abstracts with Programs 36(2):117.

Schwimmer, D. R., G. D. Williams, J. L. Dobie, and W. G. Siesser. 1993. Upper Cretaceous dinosaurs from the Blufftown Formation, western Georgia and eastern Alabama. Journal of Paleontology 67:288–296.

Schwimmer, D. R., J. D. Stewart, and G. D. Williams. 1997. Scavenging by sharks of the genus *Squalicorax* in the Late Cretaceous of North America. Palaios 12:71–83.

Scott, A. C. 1992. Trace-fossils of plant–arthropod interactions; pp. 197–223 in C. G. Maples, and R. R. West (eds.), Paleontological Society, Short Courses in Paleontology 5.

———. 2008. Evidence for plant–arthropod interactions in the fossil record. Geology Today 7:58–61.

Scott, C. 2004. Endangered and Threatened Animals of Florida and Their Habitats. University of Texas Press, Austin, Texas, 315 pp.

Scott, J. J., R. W. Renaut, L. A. Buatois, and R. B. Owen. 2009. Biogenic structures in exhumed surfaces around saline lakes: an example from Lake Bogoria, Kenya Rift Valley. Palaeogeography, Palaeoclimatology, Palaeoecology 272:176–198.

Sealey, N. E. 2007. Bahamian Landscapes: An Introduction to the Geography of the Bahamas (3rd Edition). Macmillan, New York, 184 pp.

Seilacher, A. 1953: Studien zur paläontologie: 1. Über die methoden der palichnologie. Neues Jahrbuch fur Geologie und Paläontologie, Abhandlungen 96:421–452.

———. 1967. Bathymetry of trace fossils. Marine Geology 5:413–428.

———. 1973. Fabricational noise in adaptive morphology. Systematic Zoology 22: 451–465.

———. 2007. Trace Fossil Analysis. Springer, Berlin, 226 pp.

Seiple, W., and Salmon, M. 1982. Comparative social behavior of two grapsid crabs, *Sesarma reticulatum* (Say) and *S. cinereum* (Bosc). Journal of Experimental Marine Biology and Ecology 62:1–24.

Sfakiotakis, M., M. D. Lane, and B. C. Davies. 1999. Review of fish swimming modes for aquatic locomotion. IEEE Journal of Oceanic Engineering 24:237–358.

Shaheen, N., K. Patel, P. Priyank, M. Moore, and M. A. Harrington. 2005. A predatory snail distinguishes between conspecific and heterospecific snails and trails based on chemical cues in slime. Animal Behaviour 70:1067–1077.

Sharp, H. F., Jr. 1967. Food ecology of the rice rat, *Oryzomys palustris* (Harlan), in a Georgia salt marsh. Journal of Mammalogy 48:557–563.

Shelley, R. M., and W. D. Sissom. 1995. Distributions of the scorpions *Centruroides vittatus* (Say) and *Centruroides hentzi* (Banks) in the United States and Mexico (Scorpiones, Buthidae). The Journal of Arachnology 23:100–110.

Shinn, E. A. 1968. Burrowing in recent lime sediments of Florida and the Bahamas. Journal of Paleontology 42:879–894.

Shipman, P. 1981. Life History of a Fossil: An Introduction to Taphonomy and Paleoecology. Harvard University Press, Cambridge, Massachusetts, 222 pp.

Shook, R. S. 1978. Ecology of the wolf spider, *Lycosa carolinensis* Walckenaer (Araneae). Journal of Arachnology 6:53–64.

Shoop, C. R., and C. A. Ruckdeschel. 1990. Alligators as predators on terrestrial mammals. American Midland Naturalist 124:407–412.

Shuster, C. N., Jr., P. B. Barlow, and H. J. Brockmann (eds.). 2003. The American Horseshoe Crab. Harvard University Press, Cambridge, Massachusetts, 427 pp.

Sibley, D. A. 2001. The Sibley Guide to Bird Life and Behavior. Alfred A. Knopf, New York, 608 pp.

Sickles-Taves, L. B. 1997. Understanding historic tabby structures: their history, preservation, and repair. APT Journal 28:22–29.

Sickles-Taves, L. B., and M. S. Sheehan. 2002. Specifying historic materials: the use of lime; pp. 3–22 in S. Throop and R. E. Klingner (eds.), Masonry: Opportunities for the 21st Century. ASTM 1(1432).

Silliman, B. R., and M. D. Bertness. 2002. A trophic cascade regulates salt marsh primary production. Proceedings of the National Academy of Sciences 99:10500–10505.

Silliman, B. R., C. A. Layman, and A. H. Alteiri. 2003. Symbiosis between an alpheid shrimp and a xanthoid crab in salt marshes of mid-Atlantic states, U.S.A. Journal of Crustacean Biology 23:876–879.

Silliman, B. R., C. A. Layman, K. Geyer, and J. C. Zieman. 2004. Predation by the black-clawed mud crab, *Panopeus herbstii*, in mid-Atlantic salt marshes: further evidence for top-down control of marsh grass production. Estuaries 27:188–196.

Simpkins, D. L. 1975. A preliminary report on test excavations at the Sapelo Island shell ring. Early Georgia 3:15–37.

Simpkins, D. L., and D. J. Allard. 1986. Isolation and identification of Spanish moss fiber from a sample of Stallings and Orange Series ceramics. American Antiquity 51:102–117.

Sinclair, A. R. E., and J. E. Parkes. 2008. On being the right size: food-limited feedback on optimal body size. Animal Behaviour 77:635–373.

Slake, A., and J. Gate. 2000. Carnivorous Plants. MIT Press, Cambridge, Massachusetts, 240 pp.

Sleper, D. A., and J. M. Poehlman. 2006. Breeding Field Crops. Wiley-Blackwell, Oxford, U.K., 424 pp.

Sloss, L. L. 1963. Sequences in the cratonic interior of North America. Geological Society of America Bulletin 74:93–113.

Smith, J. J., and S. T. Hasiotis. 2008. Traces and burrowing behaviors of the cicada nymph *Cicadetta calliope:* neoichnology and paleoecological significance of extant soil-dwelling insects. Palaios 23:503–513.

Smith, J. J., S. T. Hasiotis, M. J. Kraus, and D. T. Woody. 2008. *Naktodemasis bowni:* new ichnogenus and ichnospecies for adhesive meniscate burrows (AMB), and paleoenvironmental implications, Paleogene Willwood Formation, Bighorn Basin, Wyoming. Journal of Paleontology 82:267–278.

Smith, J. J., B. F. Platt, G. A. Ludvigson, and J. R. Thomasson. 2009. Exceptionally well-preserved ant nest fossils in calcic paleosols of the Ogallala Formation (Miocene), Scott County, Kansas, U.S.A. Geological Society of America, Abstracts with Program 41(7):161–162.

Smith, J. M., and R. W. Frey. 1985. Biodeposition by the ribbed mussel *Geukensia demissa* in a salt marsh, Sapelo Island, Georgia. Journal of Sedimentary Research 55:817–825.

Smith, S. E., and D. J. Read. 1997. Mycorrhizal Symbiosis (2nd Edition). Academic Press, London, U.K., 605 pp.

Snyder, E. Y., and J. F. Loring. 2006. Beyond fraud: stem-cell research continues. The New England Journal of Medicine 354:321–324.

Solomon, J. D. 1995. Guide to Insect Borers in North American Broadleaf Trees and Shrubs. Agriculture Handbook, United States Department of Agriculture AH-706, 735 pp.

Spotila, J. R. 2004. Sea Turtles: A Complete Guide to Their Biology, Behavior, and Conservation. Johns Hopkins University Press, Baltimore, Maryland, 227 pp.

Stashower, D. 2001. Teller of Tales: The Life of Arthur Conan Doyle. Macmillan, New York, 472 pp.

Steen, D. A., S. C. Sterrett, S. A. Miller, and L. L. Smith. 2007. Terrestrial movements and microhabitat selection of overwintering subadult eastern mud turtles (*Kinosternon subrubrum*) in southwest Georgia. Journal of Herpetology 41:532–535.

Steinbock, R. T. 1989. Ichnology of the Connecticut River Valley; a vignette of American science in the early 19th century; pp. 27–32 in D. D. Gillette and M. G. Lockley (eds.), Dinosaur Tracks and Traces. Cambridge University Press, Cambridge, U.K.

Stenzler, D., and J. Atema. 1977. Alarm response of the marine mud snail, *Nassarius obsoletus:* specificity and behavioral priority. Journal of Chemical Ecology 3:159–171.

Stevens, R. T., T. L. Ashwood, and J. M. Sleeman. 1997. Fall–early winter home ranges, movements, and den use of male mink, *Mustela vison* in eastern Tennessee. Canadian Field-Naturalist 111:312–314.

Stewart, M. A. 2002. What Nature Suffers to Grow: Life, Labor, and Landscape on the Georgia Coast, 1680–1920. University of Georgia Press, Athens, Georgia, 370 pp.

Stockey, R. A., G. W. Rothwell, H. D. Addy, and R. S. Currah. 2001. Mycorrhizal association of the extinct conifer *Metasequoia milleri*. Mycological Research 105:202–205.

Stone, E. C., and P. J. Kalisz. 1991. On the maximum extent of tree roots. Forest Ecology and Management 46:59–102.

Suiter, D. R. 2009. Biology and management of carpenter ants. University of Georgia Cooperative Extension Bulletin 1225:1–8.

Sullivan, B. 2000. Sapelo Island. Arcadia Publishing, Mt. Pleasant, South Carolina, 128 pp.

———. 2008. Ecology as history in the Sapelo Island National Estuarine Research Reserve. Occasional Papers of the Sapelo Island NERR 1:1–28.

Summers, A. P. 2000. Stiffening the stingray skeleton: an investigation of durophagy in Myliobatid stingrays (Chondricthyes, Batoidea, Myliobatidae). Journal of Morphology 243:113–126.

Summers, A. P., and P. R. Delis. 1994. Burrowing performance and kinematics in the eastern spadefoot toad, *Scaphiopus holbrooki*. American Zoologist 34:14A.

Sumner, F. B., and J. J. Karol. 1929. Notes on the burrowing habits of *Peromyscus polionotus*. Journal of Mammalogy 10:213–215.

Swanton, J. R. 1922. Early History of the Creek Indians and their Neighbors. Bureau of American Ethnology Bulletin 73, Smithsonian Institution, Washington, D.C., 492 pp.

Taplin, L. E. 1988. Osmoregulation in crocodilians. Biological Reviews 63:333–463.

Taplin, L. E., and G. C. Grigg. 1989. Historical zoogeography of the eusuchian crocodilians: a physiological perspective. American Zoologist 29:885–901.

Taulman, J. F. 1994. Observations of nest construction and bathing behaviors in the nine-banded armadillo *Dasypus novemcinctus*. The Southwestern Naturalist 39: 378–380.

Taylor, A. M., and R. Goldring. 1993. Description and analysis of bioturbation and ichnofabric. Journal of the Geological Society, London 150:141–148.

Taylor, A. M., R. Goldring, and S. Gowland. 2003. Analysis and application of ichnofabric. Earth Science Reviews 60:227–259.

Taylor, T. N., and M. Krings. 2008. Paleobotany: The Biology and Evolution of Fossil Plants. (2nd Edition). Academic Press, New York, 1230 pp.

Teal, J. M. 1958. Distribution of fiddler crabs in Georgia salt marshes. Ecology 39:186–193.

———. 1962. Energy flow in the salt marsh ecosystem of Georgia. Ecology 43:614–624.

Teal, M., and J. M. Teal. 1964. Portrait of an Island. Atheneum, New York, 167 pp. [Reprinted by University of Georgia Press in 1997, 184 pp.]

Teper, D. 2006. Food plants of *Bombus terrestris* L. as determined by pollen analysis of faeces. Journal of Apicultural Science 50:101–108.

Terry, R. C. 2004. Owl pellet taphonomy: a preliminary study of the postregurgitation taphonomic history of pellets in a temperate forest. Palaios 19:497–506.

Thomas, D. H. 1987. The archaeology of mission Santa Catalina de Guale: 1. Search and discovery. Anthropological Papers of the American Museum of Natural History 63: 47–161:

———. 1996. The Archaeology of Mission Santa Catalina de Guale; pp. 82–109 in C. E. Orster Jr. (ed.), Images of the Recent Past: Readings in Historical Archaeology. Altamira Press, Walnut Creek, California.

———. 2008a. History of archaeological research on St. Catherines Island; pp. 9–21 in D. H. Thomas (ed.), Native American Landscapes of St. Catherines Island. Anthropological Papers of the American Museum of Natural History 88.

———. 2008b. Terrestrial foraging on St. Catherines Island; pp. 136–197 in D. H. Thomas (ed.), Native American Landscapes of St. Catherines Island. Anthropological Papers of the American Museum of Natural History 88.

Thomas, K. R. 1974. Burrow systems of the eastern chipmunk (*Tamias striatus pipilans* Lowery) in Louisiana. Journal of Mammalogy 55:454–459.

Thompson, E. P. 1965. The Making of the English Working Class. Victor Gollancz, London, U.K., 848 pp.

Thompson, V. D. 2007. Articulating activity areas and formation processes at the Sapelo Island shell ring complex. Southeastern Archaeology 26:91–107.

Tobler, W. R. 1970. A computer movie simulating urban growth in the Detroit region. Economic Geography 46:234–240.

Tomkins, I. R. 1935. The marsh rabbit: an incomplete life history. Journal of Mammalogy 16:201–205.

———. 1944. Wilson's plover in its summer home. Auk 61:259–269.

Trueman, E. R., A. R. Brand, and P. Davis. 1966. The dynamics of burrowing of some common littoral bivalves. Journal of Experimental Biology 44:469–492.

Tschinkel, W. R. 2003. Subterranean ant nests: trace fossils past and future? Palaeogeography, Palaeoclimatology, Palaeoecology 192:321–333.

———. 2004. The nest architecture of the Florida harvester ant, *Pogonomyrmex badius*. Journal of Insect Science 4:1–19.

———. 2005. The nest architecture of the ant, *Camponotus socius*. Journal of Insect Science 5:1–9.

———. 2006. The Fire Ants. Harvard University Press, Cambridge, Massachusetts, 723 pp.

Tucker, A. D. 2009. Nest site fidelity and clutch frequency of loggerhead turtles are better elucidated by satellite telemetry than by nocturnal tagging efforts: implications for stock estimation. Journal of Experimental Marine Biology and Ecology 383:48–55.

Tucker, A. D., N. N. Fitzsimmons, and J. W. Gibbons. 1995. Resource partitioning by the estuarine turtle *Malaclemys terrapin*: trophic, spatial, and temporal foraging constraints. Herpetologica 51:167–181.

Turner, M. G. 1988. Simulation and management implications of feral horse grazing on Cumberland Island, Georgia. Journal of Range Management 41:441–447.

Turner, M. G., and S. P. Bratton. 1987. Fire, grazing, and the landscape heterogeneity of a Georgia barrier island; pp. 87–101 in M. G. Turner (ed.), Landscape Heterogeneity and Disturbance. Ecological Studies 64, Springer-Verlag, New York.

Turner, R. L., and C. E Meyer. 1980. Salinity tolerance of the brackish water echinoderm *Ophiophragmus filograneus* (Ophiuruoidea). Marine Ecology Progress Series 2: 249–256.

Uchman, A. 2007. Deep-sea ichnology: development of major concepts; pp. 248–267 in W. M. Miller III (ed.), Trace Fossils: Concepts, Problems, Prospects. Elsevier, Amsterdam, Netherlands.

Uchman, A., and A. K. Rindsberg. 2010. Advances in marine ichnology: foreword. Acta Geologica Polonica 60:1–2.

Uchman, A., P. B. Milena, and P. A. Hochuli. 2004. Oligocene trace fossils from temporary fluvial plain ponds: an example from the freshwater molasse of Switzerland. Eclogae Geologicae Helvetiae 97:133–148.

Uchman, A., V. Kazakaukas, and A. Galgalas. 2009. Trace fossils from Late Pleistocene varved lacustrine sediments in eastern Lithuania. Palaeogeography, Palaeoclimatology, Palaeoecology 272:199–211.

Ulagaraj, S. M. 1976. Sound production in mole crickets (Orthoptera: Gryllotalpidae: *Scapteriscus*). Annals of the Entomological Society of America 69:299–306.

Urquhart, B. S. 2002. Hoofprints in the Sand: Wild Horses of the Atlantic Coast. Eclipse Press, Lexington, Kentucky, 224 pp.

Vail, P. R., R. M. Mitchum Jr., R. G. Todd, J. M. Widmier, S. Thompson III, J. B. Sangree, J. N. Bubb, and W. G. Hatleilid. 1977. Seismic stratigraphy and global changes of sea level; pp. 49–212 in C. E. Payton (ed.), Seismic Stratigraphy: Applications to Hydrocarbon Exploration. American Association of Petroleum Geologists Memoir, 26.

Van Noy, R. 2008. A Natural Sense of Wonder: Connecting Kids with Nature through the Seasons. University of Georgia Press, Athens, Georgia, 164 pp.

Vance, K. R., and F. L. Pirkle. 2007. An overview of titanium concentration in heavy mineral sand ore deposits; pp. 177–185 in F. J. Rich (ed.), Guide to Fieldtrips, 56th Annual Meeting, Southeastern Section of the Geological Society of America. Georgia Southern University, Department of Geology and Geography Contribution Series 1.

Varricchio, D. J., F. Jackson, J. J. Borkowski, and J. R. Horner. 1997. Nest and egg clutches of the dinosaur *Troodon formosus* and the evolution of avian reproductive traits. Nature 385:247–250.

Varricchio, D. J., A. J. Martin, and Y. Katsura. 2007. First trace and body fossil evidence of a burrowing, denning dinosaur. Proceedings of the Royal Society of London B 274:1361–1368.

Verde, M., M. Ubilla, J. J. Jiménez, and J. F. Genise. 2007. A new earthworm trace fossil from paleosols: aestivation chambers from the Late Pleistocene Sopas Formation of Uruguay. Palaeogeography, Palaeoclimatology, Palaeoecology 243:339–347.

Vollrath, F., and P. Selden. 2007. The role of behavior in the evolution of spiders, silks, and webs. Annual Review of Ecology, Systematics, and Evolution 38:819–846.

Wada, K., and I. Murata. 2000. Chimney building in the fiddler crab *Uca arcuata*. Journal of Crustacean Biology 20:505–509.

Wagner, M. R. 2002. Mechanisms and Deployment of Resistance in Trees to Insects. Springer, Berlin, 332 pp.

Wahl, A., and A. J. Martin. 1998. Coprolites happen: post-defecation history of vertebrate coprolites as a study in ichnology and taphonomy. Geological Society of America Abstracts with Programs 30(4):64.

Walker, S. E. 1989. Hermit crabs as taphonomic agents. Palaios 4:439–452.

———. 1990. Biological taphonomy and gastropod temporal dynamics; pp. 391–421 in W. M. Miller III (ed.), Paleocommunity Temporal Dynamics: The Long-Term Development of Multispecies Assemblages. Paleontological Society Special Publication 5.

———. 1992. Criteria for recognizing marine hermit crabs in the fossil record using gastropod shells. Journal of Paleontology 66:535–558.

———. 2007. Traces of gastropod predation on molluscan prey in tropical reef environments; pp. 324–344 in W. M. Miller III (ed.), Trace Fossils: Concepts, Problems, Prospects. Elsevier, Amsterdam, Netherlands.

Walker, S. E., and C. E. Brett. 2002. Post-Paleozoic patterns in marine predation: was there a Mesozoic and Cenozoic marine predatory revolution?; pp. 119–193 in M. Kowalewski and P. H. Kelley (eds.), The Fossil Record of Predation. Paleontological Society Special Publication 8.

Walker, S. E., S. M. Holland, and L. Gardiner. 2003. *Coenobichnus* new ichnogenus, a walking trace from a land hermit crab, early Holocene, San Salvador Island, Bahamas. Journal of Paleontology 77:576–582.

Walker, T. J., and D. A. Nickle. 1981. Introduction and spread of pest mole crickets: *Scapteriscus vicinus* and *S. acletus* reexamined. Annals of the Entomological Society of America 74:158–163.

Wallace, H. K. 1942. A revision of the burrowing spiders of the genus *Geolycosa* (Araneae, Lycosidae). American Midland Naturalist 27:1–62.

Wallace, J. E., and M. Wallace. 1992. Mudcracks initiated on bird footprints. Journal of Sedimentary Petrology 62:751.

Waller, L., and T. Leong. 2008. Evaluating local spatial nesting impacts within and adjacent to a beach nourishment project, Juno Beach, Florida, 1999–2002. NOAA Technical Memorandum 569:113.

Walls, E. A., J. Berkson, and S. A. Smith. 2002. The horseshoe crab, *Limulus polyphemus*: 200 million years of existence, 100 years of study. Reviews of Fisheries Science 10: 39–73.

Walters, D. R., and D. J. Keil. 1996. Vascular Plant Taxonomy (4th Edition). Kendall Hunt, Dubuque, Iowa, 608 pp.

Wappler, T., and M. S. Engel. 2003. The Middle Eocene bee faunas of Eckfield and Messel, Germany (Hymenoptera: Apoidea). Journal of Paleontology 77:908–921.

Waring, A. J., Jr., and L. H. Larson Jr. 1968. The shell ring on Sapelo Island; pp. 263–278 in S. Williams (ed.), The Waring Papers: The Collected Works of Antonio Waring Jr., Papers of the Peabody Museum of Archaeology and Ethnology. Harvard University, Cambridge, Massachusetts 58.

Weber, J. N., and H. E. Hoekstra. 2009. The evolution of burrowing behaviour in deer mice (genus *Peromyscus*). Animal Behaviour 77:603–609.

Weimer, R. J., and J. H. Hoyt. 1964. Burrows of *Callianassa major* Say, geologic indicators of littoral and shallow neritic environments. Journal of Paleontology 38:761–767.

Wells, H. W., M. J. Wells, and I. E. Gray. 1961. Food of the sea star *Astropecten articulatus*. Biology Bulletin 120:265–271.

West, D. L., and S. T. Hasiotis. 2009. Trace fossils in an archaeological context: examples from bison skeletons, Texas, U.S.A; pp. 545–561 in W. M. Miller III (ed.), Trace Fossils: Concepts, Problems, Prospects. Elsevier, Amsterdam, Netherlands.

Wetzel, A. 1983. Biogenic structures in modern abyssal slope to deep-sea sediments in the Sulu Sea Basin (Philippines). Palaeogeography, Palaeoclimatology, Palaeoecology 42:285–304.

———. 2002. Modern *Nereites* in the South China Sea: ecological association with redox conditions in the sediment. Palaios 17:507–515.

Whitaker, J. O., Jr., and W. J. Hamilton. 1998. Mammals of the Eastern United States. University of Chicago Press, Chicago, Illinois, 583 pp.

Whitaker, J. O., Jr., and C. Ruckdeschel. 2009. Diet of *Amphiuma means* from Cumberland Island, Georgia, U.S.A. Herpetological Review 40:154–156.

White, B., and H. A. Curran. 1997. Are the plant-related features in Bahamian Quaternary limestones trace fossils? Discussion, answers, and a new classification system; pp. 47–54 in H. A. Curran (ed.), Guide to Bahamian Ichnology: Pleistocene, Holocene and Modern Environments. Gerace Research Centre, San Salvador, Bahamas.

Whitfield, J. B. 1998. Phylogeny and evolution of host–parasitoid interactions in Hymenoptera. Annual Review of Entomology 43:129–151.

Whiting, N. H., and G. A. Moshiri. 1974. Certain organism-substrate relationships affecting the distribution of *Uca minax* (Crustacea: Decapoda). Hydrobiologia 44:481–493.

Whyte, M. A. 2005. A gigantic fossil arthropod trackway. Nature 438:576.

Wilkinson, D. M. 1997. Plant colonization: are wind-dispersed seeds really dispersed by birds at larger spatial and temporal scales? Journal of Biogeography 24:61–65.

Williams, M., and V. D. Thompson. 1999. A guide to Georgia Indian pottery types. Early Georgia 27:1–167.

Williams, R. B. 2007. Unusual mobility of tube-dwelling pennate diatoms. Journal of Phycology 4:145–146.

Williams, S. C. 1987. Scorpion bionomics. Annual Review of Entomology 32:275–295.

Wilson, D. S., H. R. Mushinsky, and E. D. McCoy. 1999. Nesting behavior of the striped mud turtle, *Kinosternon baurii* (Testudines: Kinosternidae). Copeia 1999:958–968.

Wilson, E. O. 1951. Variation and adaptation in the imported red fire ant. Evolution 5:68–79.

———. 1984. Biophilia: The Human Bond with Other Species. Harvard University Press, Cambridge, Massachusetts, 157 pp.

Witz, B. W., D. S. Wilson, and M. D. Palmer. 1991. Distribution of *Gopherus polyphemus* and its vertebrate symbionts in three burrow categories. American Midland Naturalist 126:152–158.

Wolcott, T. G. 1978. Ecological role of ghost crabs, *Ocypode quadrata* (Fabricius) on an ocean beach: scavengers or predators? Journal of Experimental Marine Biology and Ecology 31:67–82.

———. 1984. Uptake of interstitial water from soil: mechanisms and ecological significance in the ghost crab *Ocypode quadrata* and two gecarcinid land crabs. Physiological Zoology 57:161–184.

———. 1988. Ecology; pp. 55–96 in W. W. Burggren and B. R. McMahon (eds.), Biology of the Land Crabs. Cambridge University Press, New York.

Wolfe, J. L. 1982. *Oryzomys palustris*. Mammalian Species 176:1–5.

Young, J., and T. Morgan. 2007. Animal Tracking Basics. Stackpole Books, Mechnicsburg, Pennsylvania, 298 pp.

Zach, R. 1978. Selection and dropping of whelks by northwestern crows. Behaviour 67:134–147.

Zimmerman, T. L., and D. L. Felder. 1991. Reproductive ecology of an intertidal brachyuran crab, *Sesarma* sp. (nr. *reticulatum*), from the Gulf of Mexico. Biological Bulletin, 181:387–401.

Zomlefer, W. B., D. E. Giannasi, K. A. Bettinger, S. L. Echols, and L. M. Kruse. 2008. Vascular plant survey of Cumberland Island National Seashore, Camden County, Georgia. Castanea 73:251–282.

Index

Page numbers in *italics* represent illustrations.

Acanthohaustorius millsi (amphipod), 288–291
Acer rubrum (red maple), 173
acorn worms: burrows, *110*; castings, *113*; golden acorn worm (*Balanoglossus aurantiactus*), 326–328, *327*; helical acorn worm (*Saccoglossus kowalevskii*), *327*, 328
acrobat ant (*Crematogaster clara*), 191
Actinaria, 257
Actitis macularius (spotted sandpiper), 461
actualism, 17, 511. *See also* uniformitarianism
Adventures of Tom Sawyer, The (Twain), 197–198
aestivation, 235
African presence, 64–65
Agapostemon virescens (ground-nesting bee), 31, 198
Agkistrodon piscivorus (cottonmouth snake), 101
Albunea paretii (beach mole crab), 312
algae, 95–96, 112, 558; as a source of oxygen, 539; as a substrate, 10
alligator (*Alligator mississippiensis*), 359–360, *361*, 454–455; burrow formation, 99, *100*, 125; Cabretta Island trackways and trace find, 331–333, *334*; dens, *334*, 336–338; high walk, *335*; keystone species, 28–29, 98–99; prey, 70, 336; resting trace, *334*; salt water/freshwater flexibility, 23–24, *334*, 454–455; tail drag trace, *361*; trace through water and algae, 10, *11*; track identification, 359–360; trails, *334*, 413
Alligator mississippiensis. *See* alligator (*Alligator mississippiensis*)
Altamaha River, 39
Ambystoma talpoideum (mole salamander), 348
Ameghinichnus, 534
Amelia Island, 43
American Association of Petroleum Geologists Bulletin (journal), 527
American Chemical Society, 527
American Civil War, 64
American holly (*Ilex opaca*), 105
American Museum of Natural History, 60, 75
American oystercatcher (*Haematopus palliatus*), 111, 462, 475, 477
American ribbed fluke snail (*Pseudosuccinea columella*), 237. *See also* gastropods (snails)
American robin (*Turdus migratorius*), 236, 388
amphibians: about, 342–343; salamanders, 347–349; toads and frogs, 349–353; tracking concepts, 343–347
Amphipholis gracillima (burrowing brittle star), 319, 321–322
amphipod burrows, *110*, 112
amphipods: about, 33, 288–290

Amphiuma means (congo eel), 348–349
anatomy of a tracemaker, 10–12, *13*, 20–22
anemones, sea, 257–259
angiosperms, 143
anhinga (*Anhinga anhinga*), 468
Anhinga anhinga, 468
anisodactyl tracemakers, 467–468
anisodactyl tracks, 458
anisodactyls, 381, 382
annelids. *See* oligochaetes (earthworms)
Anolis carolinensis (green anole), 363
Anomura, 310
ant lions, 195–198; burrows, *110*; traces, *197*
ant nest, *110*
anthozoans, 255–259
ants, 189–195, 573
Anurans. *See* toads and frogs
Apalone ferox (Florida softshell turtle), 360
Apatosaurus, 518
Aphaenogaster miamiana (funnel ant), 191
Aphaenogaster rudis (woodland ant), 230
Aphodius. *See* dweller beetles
Appalachian Mountains, 40–42, 379, 557
aquatic plants, 101
arachnids, 183–189
aragonite, 557
Archaeopteryx lithographica, 380–381
Archelon ischyros, 454
Archodontes melanopus (live oak root or stump borer), 157, 221
Arctodus simus (short-faced bear), 59
Ardea alba (great egret), 35, 88, 387, 401
Ardea herodias (great blue heron), 35, 88, 385, 458
Arenaria interpres (ruddy turnstone), 23, 462
Arenicolites, 326, 329, 525
Aristida stricta (wiregrass), 125, 356
armadillo (*Dasypus novemcinctus*), 90, 106, 372, 416, 595; breath holding capability, 421; burrow size, 12, 108; burrows, 420–421; track, *406*, *412*; tunneling habit, 85

Armases cinereum (marsh or wharf crab), 91, 182
arrowhead (*Sagittaria latifolia*), 100
Asteriacites, 325, 329
Astropecten articulatus (margined sea star), 319
Athene cunicularia (burrowing owl), 393
Atlantic cockle (*Dinocardium robustum*), 21, 276; anatomical traits, 21; mysterious circumstances of demise, 1–5, *2*
Atlantic jackknife (*Ensis directus*), 257
Atlantic mud piddock (*Barnea truncata*), 281
Atlantic oyster drill (*Urosalpinx cinerea*), 269
Atlantic sharpnose shark (*Rhizoprionodon terraenovae*), 450
Atlantic stingray (*Dasyatis sabina*), 446
Attaichnus sp., 195
auditory traces, 10, 396
Australia, 6, 334, 358, 453, 578; crayfish, 170, 574; extinctions, 217

baby's ear (*Sinum perspectivum*), 272
back-dune meadows, *109*, *110*; ecosystem plants, 153–154
Bahamas, 557, 558; San Salvador Island, 152–153, 559
Bailey, Cornelia, 379
bald eagle (*Haliaeetus leucocephalus*), 29, 458; feeding trace, 481
bald-faced hornet (*Dolichovespula maculata*), 210
ballast islands, 66–68, *67*
bank swallow (*Riparia riparia*), 394
bark sloughing, 400
barn swallow (*Hirundo rustica*), 398–399
Barnea truncata (Atlantic mud piddock), 281
barrier islands: map, *39*, *45*; rarity of, 44. *See also* Georgia barrier islands
Batis maritima (saltwort), 146
bats, 430–431
beach elder (*Iva imbricata*), 111, 153
beach mole crab (*Albunea paretii*), 312
"beach renourishment," 172–173

beaches, 111–116; "beach renourishment," 172–173; erosion and manmade structures, 172–173; rack line, 112; sedimentary structures, *114*, 115–116; subsurface, 115–116
Beaconites, 242
beak mark traces, 477
Beals, Jennifer, 481
beaver (*Castoroides ohioensis*), 36, 59, 101, 416; importance to ecosystem, 102–103
bee nests, *110*
bee, *110*, 198–204, 515, 536, 594; traces, *200*
bees (*Agapostemon virescens*), 31, 198
beetles, 213–221, 358; traces, *214*
behavior, 12–13; importance of, 12; role of genetics, 23–24
belted kingfisher (*Ceryle alcyon*), 394
benthic organisms, 48
Biffarius biformis (Georgia ghost shrimp), 112, 307
bioerosion, 119
BioKids, University of Michigan, 573
bioturbation, 79, 80, *100*, 233–234, 528, 588–589; about, 276–283; cryptobioturbation, 112, 288, 290
bird beak marks, 399–403
bird feeding traces, 4–5, 399–403
birds, 380–381; beak marks, 399–403; dust baths, 403–405; feather piles, 399; feces, 400, 402; feeding traces, 399–403; gait patterns, 386–391, *387*; identification tools, 467; nests, 111, 391–399, *392*; species, list of common, *383*; trackways, 381–391, *383*. *See also* shorebirds
Bishop, Gale, 454, 508
bison (*Bison bison*), 38, 40, 59
Bison bison (bison), 38, 40, 59
bitter panic grass (*panicum amarum*), 153
black bear (*Ursus americanus*), 69, 415
black needle rush (*Juncus roemerianus*), 86, 87
black skimmer (*Rynchops niger*), 465
black vulture (*Coragyps atratus*), 372, 458; and a dead opossum, 371–379
black-and-yellow mud dauber (*Sceliphron caementarium*), 208, 211–213

black-backed gull (*Larus marinus*), 467, 468
Blackbeard Island, 45
Blackbeard the Pirate, 62
black-bellied plover (*Pluvialis squatarola*), 461
blacknose shark (*Carcharhinus acronotus*), 450
blacktip shark (*Carcharhinus limbatus*), 450
Blarina carolinensis (short-tailed shrew), 416
Blind Men and the Elephant, The (Saxe), 16
blood brittle star (*Hemipholis elongata*), 319, 322
blue crab (*Callinectes sapidus*), 54, 316
blue dauber (*Chalybion californicum*), 211–213
bluegill fish (*Lepomis gulosus*), 101, 338, 341
bluntnose stingray (*Dasyatis sayi*), 446
boat-tailed grackle (*Quiscalus major*), 152, 387, 390, 458
bobcat (*Felis rufus*), 70, 416
bobwhite (*Colinus virginianus*), 382
Bombus spp. (bumblebees), 201–202. *See also* bees
boring bird nest, 392
Borrichia frutescens (sea-oxeye daisy), 87
Bos taurus (feral cattle), 24, 69, 375, 405, 590–592, *591*; courting story, 432; scat, 428
bottlenose dolphin (*Tursiops truncatus*), 483, 495–497
box core study observations, 120–121
box cores, 79, 117, 120–121, 588–589
Brachidontes exustus (scorched muscle), 119
Brachidontes recurvus (hooked mussel), 282
Brachyura, 310
Branta canadensis (Canada goose), 382, 468
broad-headed skink (*Eumeces laticeps*), 363
broken shells, 476
bromeliads, 106
Bromley, Richard, 132, 513, 586
Bronx Zoo, 75
brown bat (*Myotis lucifugus*), 430–431

brown pelican (*Pelecanus occidentalis*), 384–385, 468, 469
brownouts, 486
Bubo virginianus (great horned owl), 382, 402
Buckland, William, 18
buffalo (*Bison* spp.), 38, 40, 59
Bufo terrestris (southern toad), 349
bulrushes (*Scirpus* spp.), 100
burial cemetery, 60–61
burial mounds, 56–57
burrow bird nest, 392, 395
burrowing beaver (*Palaeocastor* spp.), 249
burrowing brittle star (*Amphipholis gracillima*), 319, 321–322
burrowing mud shrimp (*Upogebia*), 48
burrowing owl (*Athene cunicularia*), 393
burrowing water beetles (*Hydrocanthus* spp.), 219
Busycon carica (knobbed whelk), 266, 270, 317
Busycon spp. (whelks), 33, 46, 54; clionaid-sponge borings, 253
"*Buthus vittatus*" (striped scorpion), 183
Butorides virescens (green heron), 458
Buzzard Blast (festival), 379

cabbage palm (*Sabal palmetto*), 29, 61, 105; nests in, 397; roots, 161–163, 162; wasps and, 210
Cabretta Beach, Sapelo Island, 123, 127–130
caddis flies, 227
calcite, 557
Calidris alba (sanderling), 461, 463, 464, 472; trackway, 473
Calidris alpina (dunlin), 461, 462
Calidris canutus (red knot), 23
Calidris minutilla (least sandpiper), 461, 463, 464
callianassid shrimp, 180, 181; burrow compared to *Ophiomorpha*, 519 (*see also* callianassid shrimp [*Glypturus acanthochirus*])
callianassid shrimp (*Glypturus acanthochirus*), 558
Callichirus major (Carolina ghost shrimp), 112, 307, 519

Callinectes sapidus (blue crab), 54, 316
Campbell, Kathleen, 161
Camponotus floridanus (Florida carpenter ant), 191, 193–194
Canada goose (*Branta canadensis*), 382, 468
Canadian Journal of Petroleum Geology (journal), 527
Canis dirus (dire wolf), 59
Canis latrans (coyote), 70, 416, 595; track, 406, 412
Canis lupus (gray wolf), 69
Canthon. *See* roller beetles
capitellid thread worm (*Heteromastus filiformis*), 260, 262
capitellid worm, 119, 260, 262
Carcharhinus acronotus (blacknose shark), 450
Carcharhinus limbatus (blacktip shark), 450
Carcharhinus milberti (sandbar shark), 450
Carcharhinus obscurus (dusky shark), 450
Cardisoma spp., 179
Caretta caretta (loggerhead sea turtle), 31, 253; nests, 111; trackway, 441
Carmelopodus, 515
Carolina ghost shrimp (*Callichirus major*), 112, 307, 519
Carolina sand wasp (*Stictia carolina*), 206, 208
carpenter bee (*Xylocopa micans*), 203
carrion beetles, 221
Castoroides ohioensis (beaver), 36, 59, 101, 416; importance to ecosystem, 102–103
Castrichnus, 236
Cathartes aura (turkey vulture), 372, 458
cattle. *See* feral cattle (*Bos taurus*)
cavity bird nest, 392, 395, 397
Cellicalichnus, 201
Celliforma (bee), 201, 515, 536
Cemophora coccinea (northern scarlet snake), 368
Centropomus pectinatus (tarpon snook), 338
Centruroides hentzi (striped scorpion), 183–184, 185

Centruroides vittatus (striped scorpion), 183–184, *185*
ceramic tempering, 57
Ceratophaga vicinella (moth), 358
Cerebratulus lacteus (silky ribbon worm), 264
Ceriantharia, 257
Ceriantheopsis (tube anemone), 257–259, *258*
Ceryle alcyon (belted kingfisher), 394
Chaetopterus variopedatus (parchment worm), 260, 264
Chalybion californicum (blue dauber), 211–213
Charadrius melodus (piping plover), 461
Charadrius semipalmatus (semipalmated plover), 461, 462
Charadrius vociferus (killdeer), 461, 475
Charadrius wilsonia (Wilson's plover), 111, 461, 472, 474, 475
Chelonia mydas (green sea turtle), 440
Chelydra serpentina (snapping turtle), 102; egg laying, 102
Cherokee vulture lore, 379
China, 381
chipmunk (*Tamias striatus*), 416
Chondrites, 263, 555
cicada (*Magicicada septendecim*). *See* cicadas
cicada (*Tibicen auletes*). *See* cicadas
cicada killer wasp (*Sphecius speciosus*), 206–209, *208*
cicadas, 30, 221–224; nymphs, 107, 110, *222*
Cicindela spp., *208*, 219
Cistothorus palustris (marsh wren), 391
Cladium jamaicense (sawgrass), 100
clapper rail (*Rallus longirostris*), 450
clay minerals: composition, 42–43; particle size, 42
Clibanarius vittatus (striped hermit), 311
climate change, global, 71, 596–597; and dune formation, 154–155; methane, 229
Cliona celata (sponge), 119, 253. *See also Cliona* spp. (sponges)
Cliona spp. (sponges), 252–255, *253*
cnidarians, 255–256

coachwhip (*Masticophis flagellum*), 367, 369
Cochlichnus, 534
Coffin, Howard, 497
coleopterans, 188
Colinus virginianus (bobwhite), 382
Coluber constrictor (southern black racer), 365–366
Columbina passerina (ground dove), 385, 387, *395*, 458
commensal species, 125, 357, 358
common Atlantic auger (*Terebra dislocata*), 266, 272
common clam worm (*Nereis succinea*), 262
common mole crab (*Emerita talpoida*), 263, 312
common moon snail (*Neverita duplicata*), 266, 524
common oyster (*Crassostrea virginica*), 35, 49, 54, 91, 281
complex traces, 512–513
composite trace fossils, *131*, *133*, 523; overprinted substrates, 192, 357, 365
composite traces, 124, 125, 133, 512–513
compound traces, 512–513
Condylura cristata (star-nosed mole), 30, 483, 494
congo eel (*Amphiuma means*), 348–349
Conichnus, 258
Coprinisphaera, 536–537
Coprinisphaera ichnofacies, 529, *534*; about, 536–537; as defined by Frey et al., *531*; table with representative ichnogenera, *530*
Copris gopheri (scarab beetle), 358
coquina clam (*Donax variabilis*), 259, 263, 276, 279–280
Coragyps atratus (black vulture), 372, 458; and a dead opossum, 371–379
Corvus ossiraus (fish crow), 5, 390, 458
cottonmouth snake (*Agkistrodon piscivorus*), 101
cotyledons, 143
cough pellets, 402–403, 479
coyote (*Canis latrans*), 70, 416, 595; track, 406, 412
coyote lore, 70

Crassostrea virginica (common oyster), 35, 49, 54, 91, 281
crayfish (*Procambarus lunzi*), 175, 177–178. See also crayfish (*Procambarus* spp.)
crayfish (*Procambarus* spp.), 31, 169–183, *174*, *175*, *595*; Australian fossil, *574*; burrow identification, 179–183, *180*, *494*; burrows, 101, *174*; burrows and burrowing, 170–171; flood time behavior, 171; partially eaten with beak mark trace, *174*; as prey, 428, 487; rudimentary gills, 171; towers, *174*, *180*; trace fossils, *494*, *566*; trackways, 179, *182*, *344*
crayfish (*Procambarus talpoides*), 175, 177–178. See also crayfish (*Procambarus* spp.)
Crayfishes of Georgia, The (Hobbs), 175
Crematogaster clara (acrobat ant), 191
crevice bird nest, 392
crocodile (*Crocodylus acutus*), 28–29, 334, 359
crocodiles (*Crocodylus* spp.), 453
crossing facies, 183
Crotalus adamanteus (eastern diamondback rattlesnake), 366, 368
Cruziana, 546–547, *550*
Cruziana ichnofacies, 525
cryptobioturbation, 112, 288, 290
Cryptotermes brevis, 228
Cumberland Island, 39, 43, *45*; bobcat presence, 70; crayfish, 175; Lake Whitney, 98; Timucuan tribe, 60
cup bird nest, 392, 398
cusk eel (*Rissola marginata*), 450
cutgrass (*Zizaniopsis miliacea*), 100
CyberTracker, 590, 593
cynipid wasps, 166
Cyprinodon variegatus (sheepshead minnow), 338

Dama dama (fallow deer), 425, 497
damselflies, 101
Darwin, Charles, 232–233, 236, 589
Dasyatis americana (southern stingray), 32, 92; feeding pits, 115, 120; feeding traces, *447*
Dasyatis sabina (Atlantic stingray), 446

Dasyatis sayi (bluntnose stingray), 446
Dasypus novemcinctus (nine-banded armadillo), 90, 106, 372, 416; breath holding capability, 421; burrow size, 12, 108; burrows, 420–421; track, *406*, *412*; tunneling habit, 85
death spiral, 246, *247*
decapods, 178–183, 291–329
Dendroctonus frontalis (southern pine beetle), *214*, 220
density of fresh water, 42
density of saltwater, 48
dermestid beetles, 107, 214, 221, 427
Dermochelys coriacea (leatherback sea turtle), 31, 440, 454; trackway, *441*
Desmodus rotundus (vampire bat), 431
Devil's Corkscrew, 249
diamondback terrapin (*Malaclemys terrapin*), 456–457
dicots, 143
Didelphis virginiana (opossum), 371; carcass of and black vultures, 371–379; pes tracks of, *373*; playing dead, *374*; track, *406*, *412*
Dinocardium robustum. See Atlantic cockle (*Dinocardium robustum*)
dinosaur burrows, 250, 358
dinosaur coprolites, 216, 221, 534
dinosaur trace fossils, 18, 533, 537
dinosaur tracks, 18
dinosaurs: bird-like features, 380–381
Diopatra cuprea (tube worm), 33–34, 113, 259–261, 266, 280, 322
Diplichnites, 242
Diplocraterion, 326
Diplopoda, 239
dipterans, 188
dire wolf (*Canis dirus*), 59
Disholcaspis cinerosa (gall wasp), 213
disk dosinia (*Dosinia discus*), 276–277, 479
Dolichovespula maculata (bald-faced hornet), 210
dolphins. See bottlenose dolphin (*Tursiops truncatus*)
Donax variabilis (coquina clam), 259, 263, 266, 276, 279–280

doodlebugs. *See* ant lions
Dorymyrmex bureni (pyramid ant), 192, 193
Dosinia discus (disk dosinia), 276–277, 479
double-crested cormorant (*Phalacrocorax auritus*), 383, 385, 468
downy woodpecker (*Picoides pubescens*), 386
dragonflies, 101
Drymarchon couperi (eastern indigo snake), 357
Dryocopus pileatus (pileated woodpecker), 30, 386, 395, 396–397
Dufour's gland, 200
Dugong dugon (dugong), 495
dugong (*Dugong dugon*), 495
dunes, 108–111, *109*, *110*; rising/falling sea levels, 154–156; sea oat stabilizers, *149*; *Spartina* role, *150*, 150–151
dung beetles, 216–217; baited by owls, 394; behavioral categories, 214
dunlin (*Calidris alpina*), 461, 462
Duplin River National Estuarine Sanctuary, 73
durophagy, 315–316
dusk-calling cicada (*Tibicen auletes*), 221
dusky shark (*Carcharhinus obscurus*), 450
dust baths, 403–405
dwarf surf clam (*Mulinia lateralis*), 259, 260, 276, 278
dweller beetles, *214*, 216

earthworms. *See* oligochaetes (earthworms)
eastern carpenter bee (*Xylocopa virginica*), 200, 203
eastern chipmunk (*Tamias striatus*), 107, 416, 418–419
eastern diamondback rattlesnake (*Crotalus adamanteus*), 366, 368
eastern gray squirrel (*Sciurus carolinensis*), 415, 569–570; feeding signs, 425
eastern indigo snake (*Drymarchon couperi*), 357
eastern massa. *See* mud snail (*Ilyanassa obsoleta*)
eastern mole (*Scalopus aquaticus*), 11–12, 30, 180, 236; burrows, 106, 416–418, *417*

eastern mosquito fish (*Gambusia holbrooki*), 338
eastern mud snake (*Farancia abacura*), 366
eastern mud turtle (*Kinosternon subrubrum*), 361, 363
eastern newt (*Notophthalmus viridescens*), 345, 347
eastern oyster (*Urosalpinx cinema*), 119
eastern screech owl (*Otus asio*), 402
eastern spadefoot toad (*Scaphiopus holbrookii*), 350–352
eastern subterranean termite (*Reticulitermes flavipes*), 229–230
eastern yellow jacket (*Vespula maculifrons*), 210–211
Echinocardium cordatum (heart urchin), 320
echinoderms, 318–328
echinoids, 322–323
ecology, development, 72–73
ecotones, 85–86, *86*, 92
ectomycorrhizae, 138; Paleozoic fossils, 139
Edward John Noble Jr. Foundation, 75
Egretta thula (snowy egret), 458
Emerita talpoida (common mole crab), 263, 312
Emory University, 80, 132, 647
Enhydra lutris (sea otter), 484
Ensis directus (Atlantic jackknife), 257
entisols, 163
Entobia, 255, 329, 552
equilibrichnia, 249
Equisetum spp. (scouring rushes), 100
Equus cabullus (feral horse), 40, 69, 405, 497, 499
Eretmochelys imbricata (hawksbill sea turtle), 440
erosion and man-made structures, 172–173
estivation, 349; amphibian, 348–349; earthworm, 235
estuaries, 120–121; continental and marine ichnofacies, 533, 535; formation, 48; mud deposition, 49–51; nutrient availability, 49–50; riverine, 48–49; salt marsh, 49–50
estuary formation, 48
Etowah mounds, 56

Eubrontes, 515
Eucinostomus lefroyi (mottled mojarra), 338
Eudocimus albus (white ibis), 382
Eumeces egregious (mole skink), 364–365, 365
Eumeces fasiatus (five-lined skink), 363
Eumeces laticeps (broad-headed skink), 363
Eumeces spp., 363
Eupleura caudata (thick-lipped oyster), 119, 269
European arrival, 60–68, 61
euryhaline animals, 48
eurypterid, 183
Eutrichota gopheri (fly), 358
extinctions, 247; Australia, 217; beaver, 102; coincide with human appearance, 59; megafauna, 59; others, 69–70

facies, 92–93, 183
fall line, 41–42
fallow deer (*Dama dama*), 425, 497
false angelwing (*Petricola pholadiformis*), 119, 278, 281–282
Farancia abacura (eastern mud snake), 366
feather piles, 399
feces, 476; birds, 400, 402; ghost shrimp spp., 114; mammals, 427–430; ribbed muscles, 94; shorebirds, 480–481
Felis domestica (feral cats), 591
Felis rufus (bobcats), 70, 416
Feoichnus, 224
feral cat (*Felis domestica*), 591
feral cattle (*Bos taurus*), 24, 69, 375, 405, 590–592, 591; courting story, 432; scat, 428
feral hog (*Sus scrofa*), 69, 375, 405, 590; dig marks, 421; egg predation, 442; track, 406, 412; trampling, 103; wallows of, 422
feral horse (*Equus cabullus*), 40, 69, 405, 497, 499
feral mammals, 497–500; traces, 498
fiddle-leaf morning glory (*Ipomoea stolonifera*), 111, 151
fiddler crab (*Uca major*), 558
fiddler crabs (*Uca* spp.), 31, 35, 122, 558. See also fiddler crab (*Uca major*); mud fiddler crab (*Uca pugnax*); sand fiddler crab (*Uca pugilator*)
field techniques: electronic files, 588–589; laser scanning, 585–586; technological aids, 592–593; uncomplicated, 577–579; underground data, 586–587
field trips, 83–93
firestorms, 28
firmground ichnofacies, 526
firmgrounds, 124
fish crow (*Corvus ossiraus*), 5, 390, 458
five-lined skink (*Eumeces fasiatus*), 363
Flashdance (1983), 481
flat-clawed hermit crab (*Pagarus pollicaris*), 311
flies (*Eutrichota gopheri*), 358
flies (*Machinmus polyphemus*), 358
Flint River, 171–172
Florida carpenter ant (*Camponotus floridanus*), 191, 193–194
Florida harvester ant (*Pogonomyrmex badius*), 191, 192, 193
Florida manatee (*Trichechus manatus latirostris*), 495
Florida mouse (*Peromyscus floridanus*), 357
Florida softshell turtle (*Apalone ferox*), 360
flower petals, 140
flying traces, 468, 470–471
foredune berms, 110
Forelius pruinosus (lawn ant), 192, 193
Formation of Vegetable Mould through the Action of Worms with Observation on their Habits, The (Darwin), 232–233, 589
Forschungsstation Senchenberg am Meer (SAM), 19
Forster's tern (*Sterna forsteri*), 466
Foster, Stephen, 40
freshwater aquifers, 71; present-day, 98
freshwater environments, 70–71; formation, 28–29; Okefenokee swamp, 40; pond formation, 11, 125; role of beaver, 102–103; seasonal, 335–336
freshwater fish, 338–342, 340
freshwater turtles, 360–362
Frey, Robert, W., 77, 78, 80, 116, 120–121, 267, 329, 527, 578, 595

fugichnia (escape traces), 249
Fundulus majalis (striped killifish), 338
fungi, 106, 107, 138, 220
fungi/plant root interactions, 138–140
funnel ant (*Aphaenogaster miamiana*), 191

gait patterns, 345, 346, 441; bird, 386–391, 387; mammal, 407–413; tetrapod, 345, 346–347
gall wasp (*Disholcaspis cinerosa*), 213
galls (insect trace), 105, 165, 221
Gambusia affinis (western mosquitofish), 338
Gambusia holbrooki (eastern mosquito fish), 338
Garbisch, Jon, 176, 331
Gastrochaenolites, 552
gastropods (snails), 237–239, 238
Gecarcinus spp., 179
Geechee, 64, 379
genetic predetermination, 23–24
Geological Society of America, 503, 510
Geolycosa spp. burrows, 185
Georgia barrier islands: early life, 27–37; ecological research, 72–76; geological history, 37–53; geological research, 76–81; ichnological appropriateness, 19–25; map, 39, 45; modern human presence, 68–72; Native American presence, 53–60; settlement and agriculture, 60–68; trace sampling, 7
Georgia Bight, 51, 53, 286
Georgia ghost shrimp (*Biffarius biformis*), 112, 307. See also ghost shrimp spp.
Georgia Southern University, 77
Georgia State University, 77
Geukensia demissa (ribbed mussel), 35, 49, 54, 94, 281; byssae, 128; in relict marsh, 123
ghost crab (*Ocypode quadrata*), 91, 111, 152, 182, 298–307, 300; behaviors, 15; burrows, 15, 17, 110, 299–303; range, 15; resting traces, 304–306; signaling, 396; trackways, 13, 17, 303–304
ghost shrimp spp., 11, 32, 307–310; burrows, 110, 112–113, 118, 308; fecal pellets, 114

giant ground sloth (*Megalonyx*), 59
giant wolf spider (*Hogna carolinensis*), 187
Ginglymostoma cirratum (nurse shark), 450
glass lizard (*Ophisaurus* spp.), 364
glasswort (*Salicornia virginica*), 86, 87
Glaucomys volans (southern flying squirrel), 107; trails, 415
glenoacetabular distance of a tetrapod track, 345, 346
Glossifungites, 552
Glossifungites ichnofacies, 526, 552
Glyptodon (glyptodont), 28, 59
glyptodont (*Glyptodon*), 28, 59
God, Dr. Buzzard, and the Bolito Man, (Bailey), 379
golden acorn worm (*Balanoglossus aurantiactus*), 326–328, 327
gopher frog (*Rana capito*), 357
gopher tortoise (*Gopherus polyphemus*), 125, 354–358, 355, 365, 368, 595; burrow size, 12, 108
Gopherus polyphemus (gopher tortoise), 125, 354–358, 355, 365, 368, 595; burrow size, 12, 108
Graduate, The (1967), 594
Grallator, 515
Gray's Reef, 38, 40; National Marine Sanctuary, 75; sponges, 254
great blue heron (*Ardea herodias*), 35, 88, 385, 458
great egret (*Ardea alba*), 35, 88, 387, 401
great horned owl (*Bubo virginianus*), 382, 402
greater yellowlegs (*Tringa melanoleuca*), 462
green anole (*Anolis carolinensis*), 363
green heron (*Butorides virescens*), 458
green jackknife (*Solen viridis*), 277
green sea turtle (*Chelonia mydas*), 440
green-arrow arum (*Peltandra virginica*), 100
Gregory, Murray, 80, 132, 161
ground dove (*Columbina passerina*), 385, 387, 395, 458
ground-nesting bee (*Agapostemon virescens*), 31

ground skink (*Scincella lateralis*), 363, 365
Grovner, Yvonne, 65
Grus canadensis (sandhill crane), 35, 458, 515; trackway, 460
Guale language, 60
Guale tribe, 60
Gullah. *See* Geechee
gulls (*Larus* spp.), 466, 467–468, 477; stomping behavior, 481–482. *See also under individual species names*

habitat, 13–16, 497–500
Haematopus palliatus (American oystercatcher), 111, 462, 475, 477
Haliaeetus leucocephalus (bald eagle), 29, 458; feeding trace, 481
hallux, 382, 384–385
halophytes, 35
Hänstzschel, Walter, 78
hardground, 123–124
hardground ichnofacies, 526
hawksbill sea turtle (*Eretmochelys imbricata*), 440
Hayes, Royce, 57, 176
heart urchin (*Echinocardium cordatum*), 320
heavy minerals, 42, 45
helical acorn worm (*Saccoglossus kowalevskii*), 327, 328
hemichordates, 326–328
hemimetabolous insects, 188–189
Hemipholis elongata (blood brittle star), 319, 322
Henderson, Steve, 83–84, 132
Henry, V. J. "Jim," Jr., 77
hermit crab (*Pagarus* spp.), 113, 310–312, 311
herring gull (*Larus argentatus*), 466, 466, 467, 479
Hesperotettix floridensis (marsh grasshopper), 35
Heteromastus filiformis (capitellid thread worm), 259, 260, 262, 539
Hexapodichnus, 534
hibernation, 349
high intertidal, 110
high marsh, 86, 87–91, 96
Hilburn, Jen, 393, 465

Hill, Robert, 515–516
Hillichnus lobosensis, 515, 516
Hilton Head Island, 43
hirudineans, 232
Hirundo rustica (barn swallow), 398–399
histosols, 163
HMS *Beagle*, 233
Hobbs, Horton, Jr., 175, 177
Hoffman, Dustin, 594
Hog Hammock, 65, 73, 372
Hog Hammock dump site, 372
Hogna carolinensis (giant wolf spider), 187
Hogna sp. burrows, 185
hogs, wild. *See* feral hog (*Sus scrofa*)
Holocene Epoch: relict dunes, 92
holometabolous insects, 188
holothurians, 324–326
holotype, 518
Homo sapiens, 28
hooked mussel (*Brachidontes recurvus*), 282
horseflies, 210
horses, wild. *See* feral horse (*Equus cabullus*)
horseshoe crab (*Limulus polyphemus*), 283–291; behaviors, 22–23, 33; egg fertilization, 22–23, 31; egg laying, 22–23, 247; juvenile and adult traces, 284; mating traces, 287; not a death spiral tale, 245–252, 246; resting trace, 446; size, 53; trackways, 13, 246, 247, 284
Houstorius (amphipod), 288–291
Howard, James D., 75, 77, 78, 79, 80, 116, 120–121, 329, 527, 595
Hoyt, John H., 76, 77
Human Bond with Other Species, The (Wilson), 572
human trace fossils, 53
hurricanes, 53, 154–155; post storm colonization, 66; Sea Island of 1893, 66
Huxley, Thomas, 251, 507
Hydrocanthus spp. (burrowing water beetles), 101, 219
hydrocarbon exploration, 526–527
Hydrocotyle bonariensis (pennywort), 111, 153, 499

Hyla spp. (tree frogs), 350
hymenopterans, 188, 255

ichnofabric analysis, 527–528
ichnofacies: about, 524–525, *525*, 540–541, 550–551, 553, 565–567; *Coprinisphaera*, 536–537; *Cruziana*, 546–547, 550; defined, 524; *Glossifungites*, 552; *Mermia*, 539–540; *Nereites*, 553, 555–556; nonmarine, 526; *Psilonichnus*, 541–545, *542*, *544*, *548*; *Scoyenia*, 537–539; *Skolithos*, 545–546, *548*–549; *Teredolites*, 552; *Trypanites*, 549, 552; *Zoophycos*, 553, 555
ichnofacies concept, 524–529, *525*
ichnofossils, 6
ichnology, 8, 9–13, 517; anatomy, 10–12; behavior, 12–13; definition, 6; habitat, 13–16; life history, 12, 15–16
Ichnos (journal), 79
ichnotaxonomy, 514–523
ichnotourism, 445
Idia gopheri (moth), 358
Ilyanassa obsoleta (mud snail), 113, 266, 274–275
Incisitermes minor (termite), 228
Incisitermes snyderi (termite), 228
incongruous ark (*Anadara brasiliana*), 277, 278
incumbent feet, 382, 384
Indiana University, 77
insect-plant interactions, 105–106, 163–167; passive insect traps, 400
insects, 101; about, 188–189; ant lions, 195–198; ants, 189–195; bees, 198–204; caddis flies, 227; cicadas, 221–224; maritime forest, 157; mole crickets, 224–227; of primary dunes, 111; research opportunities, 231; termites, 227–231; as tracemakers, 594; wasps, 205–213
interference ripples, 114
International Ichnological Congress (*Ichnia*), 514, 528
International Taxonomy Workshop, 522
intertidal runnel, *110*
intertidal sediment deposition, 113
invasive species, 103, 155, 590–593, *591*
invertebrate taxa, 178

Ipomoea pes-caprae (railroad vine), 111, 151, 152–153, *152*
Ipomoea stolonifera (fiddle-leaf morning glory), 111, 151
irrigation, 71
isopods (*Limnoria* spp.), 30, 290–291
Isoptera (termites), 193, 227–231
Iva imbricata (beach elder), 111, 153
ivory-billed woodpecker (*Campehilus principalis*), 30, 397

Jekyll Island, 39, 43, 45; crayfish, 171–175; sediment redistribution, 51, 53
Juncus roemerianus (black needle rush), 86, 87
Juniperus virginiana (red cedar), 50, 87, 401

Kalotermes approximatus, 228
kaolinite, 43
Kaopectate, 43
Kelly, Walt, 40
Kemp's ridley sea turtle (*Lepidochelys kempii*), 440
keyhole sand dollar (*Mellita isometra*), 319, 322–323
killdeer (*Charadrius vociferus*), 461, 475
Kinosternon baurii (striped mud turtle), 360–361
Kinosternon spp. (mud turtles), 102
Kinosternon subrubrum (eastern mud turtle), 361, 363
knobbed whelk (*Busycon carica*), 266, 270, 317
Krakow, Poland, 514

lagoons, 558–559
Lake Whitney, 98, 338
Larus altricilla (laughing gull), 21, 466, 467; anatomical traits, 22; cough pellets, 479; mystery of the broken bivalve, 1–5; tracks, 3, *3*
Larus argentatus (herring gull), 466, *466*, 467, 479
Larus delawarensis (ring-billed gull), 467, 479
Larus marinus (black-backed gull), 467, 468

Larus spp. (gulls), 466, 467–468, 477; stomping behavior, 481–482. *See also under individual species names*
Last Child in the Woods (Louv), 572–573
laughing gull (*Larus altricilla*), 21, 466, 467; anatomical traits, 22; cough pellets, 479; mystery of the broken bivalve, 1–5; tracks, 3, 3
laurel oak (*Quercus laurifolia*), 105
lawn ant (*Forelius pruinosus*), 192, 193
Lawrence, T. E., 8
leaf impressions, 7–8
leaf mines, 105, 165, 165–167
leaf-cutter bees (*Osmia*), 204, 594. *See also* bees
leaf-cutting ant (*Trachymyrmex septentrionalis*), 192, 193
least sandpiper (*Calidris minutilla*), 461, 463, 464
least tern (*Sterna antillarum*), 466, 466
leatherback sea turtle (*Dermochelys coriacea*), 31, 440, 454; trackway, 441
Lemur catta (ring-tailed lemur), 75
Lepidactylus dytiscus, 288–291
Lepidochelys kempii (Kemp's ridley sea turtle), 440
Lepidopa websteri (square-eyed mole crab), 312–313, 313
lepidopterans, 188
Lepomis gulosus (bluegill fish), 101, 338, 341
Lepomis macrochirus (warmouth fish), 101, 338, 341
lettered olive (*Oliva sayana*), 266, 270–272
life history, 15–16
Life Savers (candy), 75
life traces, 8
lime cement, 62–64, 63
Limnodromus griseus (short-billed dowitcher), 462
Limosa fedoa (marbled godwit), 462
limulids (*Limulus polyphemus*). *See* *Limulus polyphemus* (horseshoe crab)
Limulus polyphemus (horseshoe crab), 283–291; behaviors, 22–23, 33; egg fertilization, 22–23, 31; egg laying, 22–23, 247; juvenile and adult traces, 284; mating traces, 287; not a death spiral tale, 245–252, 246; resting trace, 446; size, 53; trackways, 13, 246, 247, 284
Lindberg, Charles, 72
lined sea star (*Luidia clathrata*), 319, 320
Little Cumberland Island, 45
Little St. Simons Islands, 45
Little Tybee Island, 45
Littoraria irrorata (marsh periwinkle), 35, 87, 127, 273–274, 456; contribution to organic detritus, 88
live oak (*Quercus virginiana*), 10, 29, 86, 104; boring nest in, 395; root architecture, 158–159, 158; wasp trace, 213, 214
live oak root or stump borer (*Archodontes melanopus*), 157, 221
lizard burrow, 110
lizards, 362–365
loblolly pine (*Pinus taeda*), 22, 91, 155; low hammock, 105; root architecture, 131
Lockeia, 534
loggerhead sea turtle (*Caretta caretta*), 31, 253; nests, 111; trackway, 441
logging, 69, 156
long leaf pine (*Pinus palustris*), 155, 356
long-billed curlew (*Numenius americanus*), 462, 465
longshore drift, 46, 51–53
long-wristed hermit crab (*Pagarus longicarpus*), 311
"Lord God Bird," 397. *See also* ivory-billed woodpecker (*Campehilus principalis*)
Louv, Richard, 572
low hammock forests, 103
low-marsh, 96
Luidia clathrata (lined sea star), 319, 320
Lundgren, S.A.B., 518
Lutra canadensis (river otter), 36, 101, 483, 484–487, 485

Machinmus polyphemus (flies), 358
Magicicada septendecim (periodical cicada), 221
Magnolia grandiflora (southern magnolia), 105; roots, 157
Magnolia virginiana (sweet bay), 105

Malaclemys terrapin (diamondback terrapins), 456–457
mammal tracks, classification, 405–413, *406, 408, 410, 412*
mammals: burrows, 416–421, *417*; classification (table), *406*; feces, 427–430; feces as olfactory markers, 429–430; feeding signs, 424–427, *425*; feral, 69, 375, 497–500, *498, 591*; lays, 421–424; marginal marine or marine, 483–501; tooth marks, 424–427, *425*; trails, 413–415, *414*
mammoth (*Mammuthus*), 40, 59
Mammut (mastodon), 59
Mammuthus (mammoth), 40, 59
manatees. *See* West Indian manatee (*Trichechus manatus*); Florida manatee (*Trichechus manatus latirostris*)
mantis shrimp (*Squilla empusa*), 118, 313–314, *314*
manus, 343, 345
marbled godwit (*Limosa fedoa*), 462
margined sea star (*Astropecten articulatus*), 319
maritime forests, 105–108, *110*, 157; composition, 87; trees and shrubs, 155–163; types, 103, 105
marsh crab (*Armases cinereum*), 91
marsh crab (*Sesarma reticulatum*), 182
marsh grasshopper (*Hesperotettix floridensis*), 35
marsh periwinkle (*Littoraria irrorata*), 35, 87, 127, 273–274, 456; contribution to organic detritus, *88*
marsh rabbit (*Sylvilagus palustris*), 483, 489–491, *490*
marsh rice rat (*Oryzomys palustris*), 483, 491
marsh wren (*Cistothorus palustris*), 391
Martesia cuneiformis (wedge piddock), 29, 115–116, 278, 281, 282–283
Martin, Anthony J., 80, 510, *571*; about, 647
Martin, Wayne, Dr., 529
Masticophis flagellum (coachwhip), 367, 369
mastodon (*Mammut*), 59
mayflies, 101

McIntosh County, Georgia, 177
meadow vole (*Microtus pennsylvanicus*), 419–420
Megalonyx (giant ground sloth), 59
Megalosouripus, 516
Melanerpes carolinus (red-bellied woodpecker), 29, *395*, 397
Meleagris gallopavo (wild turkey), 382, 458
Mellita isometra (keyhole sand dollar), 32, 92, 319, 322–323, *323*
Memento (2000), 249
Menippe mercenaria (stone crab), 316
Mercenaria mercenaria (southern quahog), 54, 276
Mermia ichnofacies, 529, 534; about, 539–540; as defined by Frey et. al., *531*; representative ichnogenera, *530*
Mesozoic Era, 198; dinosaur death spiral, 250; gastropod predation, 275; termites, 231
Metasequoia milleri, 139
methane, 229
Microphiopholis gracillima, 322
Microtus pennsylvanicus (meadow vole), 419–420
millipedes, 239–242, *240*
minerals, heavy, 42
mink (*Mustela vison*), 101, 483, 487
Miocene Epoch: beaver death spiral, 249–250; spider burrows, 187
Mischocyttarus mexicanus (paper wasps), 210
missionaries, Spanish, 61
mistletoe (*Phoradendron* spp.), 106
Moira atropos (mud heart urchin), 319, *320*, 323–324
mole crickets (*Scapteriscus* spp.), 224–227, *225*
mole salamander (*Ambystoma talpoideum*), 348
mole skink (*Eumeces egregious*), 364–365, *365*
molehill, *417*
monocots, 143
Moody, Pliny, 18
moon snail (*Neverita duplicata*), 119, 267–268, 277

morning glories, 151
moth (*Ceratophaga vicinella*), 358
moth (*Idia gopheri*), 358
mottled mojarra (*Eucinostomus lefroyi*), 338
mud anchorage: algal films, 95–96
mud deposition, 93–98; anchorage, 95–96; by mussels, 128; salt water estuaries, 49–51; species responsible for, 93, 94
mud fiddler crab (*Uca pugnax*), 34–35, 509; about, 94–95; burrow identification, 180–181, *180*; pseudofeces, 95
mud heart urchin (*Moira atropos*), 319, 320, 323–324
mud shrimp (*Upogebia affinis*), 118, *314*, 315
mud snail (*Ilyanassa obsoleta*), 113, 266, 274–275
mud turtles (*Kinosternon* spp.), 102
Mugil cephalus (striped mullet), 338
Muhlenbergia filipes (sweetgrass), weaving of, 65
Mulinia lateralis (dwarf surf clam), 259, 260, 266, 276, 278
Mustela vison (mink), 101, 483, 487
Mustelidae, 484
Mustelis canis (smooth dogfish), 450
mutualism, 574
mycorrhizal fungi, 133; Paleogene fossils, 139
Myotis lucifugus (brown bat), 430–431
myriapods, 239–242
Myrica cerifera (wax myrtle), 91, 153
Myrmeleon crudelis. *See* ant lions

Naktodemasis, 224
narrow-leafed cattail (*Typha angustifolia*), 100
National Marine Sanctuary: Gray's Reef, 75
National Oceanographic and Atmospheric Administration (NOAA), 75
Native Americans: burial mounds, 56–60; coyote lore, 70; diet, 58–59, 143; European arrival, 60–68, *61*; pottery patterns, *58*; pottery shards, 57–58, *58*; shell rings, 53–56, *55*; vulture lore, 379
neap tides, 49
Neofiber alleni (round-tailed muskrat), 416

Neohaustorius schmitzi, 288–291
neoichnology, 6, 16; and uniformitarianism, 19
Nephtys bucera (shimmy worm), 263
Nereis succinea (common clam worm), 259, 262
Nereites, 553, 555–556
Nereites ichnofacies, 525
Nerodia fasciata (water snake), 102
nests, bird, 111, 391–399
net making tradition, 64
Neverita duplicata (common moon snail), 119, 267–268, *266*, 277, 524
New Zealand, 161, 507; tree-root trace fossils, 132
Newell, Steve, 173
Newton, Isaac, 589
Nifong, James, 176
nikau palm tree (*Rhopalostylis sapida*), 161
nine-banded armadillo (*Dasypus novemcinctus*), 90, 106, 372, 416; breath holding capability, 421; burrow size, 12, 108; burrows, 420–421; track, *406*, *412*; tunneling habit, 85
Noble, Edward John, Jr., 75
northern fence lizard (*Sceloporus undulatus*), 363
northern scarlet snake (*Cemophora coccinea*), 368
Notophthalmus viridescens (eastern newt), 345, 347
Numenius americanus (long-billed curlew), 462, 465
nurse shark (*Ginglymostoma cirratum*), 450
nutrient enrichment, 62
Nymphaea stellata (water lily), 100

Ocypode quadrata (ghost crab), 91, 111, 152, 182, 298–307, *300*; acoustic signaling, 396; behaviors, 15; burrows, 15, *17*, *110*, 299–303; range, 15; resting traces, 304–306; trackways, *13*, *17*, 303–304
Odocoileus virginiana (white-tailed deer), 70, 500; lays, 421, 423; scat, 107, 428; tooth marks, 425; track, *406*, *412*; trails, 414, *414*
Odum, Eugene "Gene," 72, 73, 93

Odum, Howard T., 72
Ogeechee River, 39
Oglethorpe, James, 64
Oichnus, 275, 329, 523, 524
Okefenokee swamp, 40
"Old Folks at Home," (Foster), 40
olfactory markers, 10; mammal droppings, 429–430
oligochaetes (earthworms), 232–237; estivation chambers, 110, 234; impact, 112; traces, 234; trails of, 235–236
Oliva sayana (lettered olive), 92, 266, 270–272
"One Ichnology," 513
Onthophagus. See tunneler beetles
Onuphis microcephala (parchment worm), 113, 260
Ophichthus spp. (snake eels), 120, 449–450
Ophiomorpha annulata, 519
Ophiomorpha nodosa, 519
Ophiomorpha spp., 310, 329, 519
Ophiophragmus filograneus (brittle star), 322
Ophisaurus spp. (glass lizard), 364
organ-pipe mud dauber (*Trypoxylon politum*), 208, 211–213
Orthoporus ornatus (millipede), 240, 241
Oryctodromeus cubicularis (dinosaur), 250
Oryzomys palustris (marsh rice rat), 483, 491
Osmia (leaf-cutting bee), 204, 594. *See also* bees
osprey (*Pandion haliaetus*), 29, 395, 458
owls, 382, 394, 402, 462
Oxford College of Emory, 83

Pagarus annulipes (hairy hermit crab), 311
Pagarus longicarpus (long-wristed hermit crab), 311
Pagarus pollicaris (flat-clawed hermit crab), 311
Pagarus spp. (hermit crabs), 113, 310–312, 311
Palaeocastor spp. (burrowing beavers), 249
Palaeohelcura, 184
paleoichnology, 6

Paleoscolytus, 220
Paleozoic Era, 183
palmate tracemakers, 465–467
palmate tracks, 384
Palmiraichnus, 534
palp worm (*Emerita talpoida*), 263
Pamlico Island, 37–40, 39
Pandion haliaetus (osprey), 29, 395, 458
Panicum amarum (bitter panic grass), 153
paper wasp (*Mischocyttarus mexicanus*), 210
paracoprids. *See* tunneler beetles
Parahaustorius longimerus (amphipod), 288–291
Paranthus rapiformis (sea onion anemone), 256–257
parasitic flatworms, 274
parasitoids, 205
parchment worm (*Chaetopterus variopedatus*), 260, 264
parchment worm (*Diopatra cuprea*), 266
parchment worm (*Onuphis microcephala*), 113, 260
Parowanichnus sp., 195
Pectinaria gouldii (trumpet worm), 260
peepers (*Pseudacris* spp.), 350
Pelecanus erythrorhynchos (white pelican), 468
Pelecanus occidentalis (brown pelican), 384–385, 468, 469
Peltandra virginica (green-arrow arum), 100
Pemberton, George, 79, 329, 595
Penholoway Island, 37–40, 39
peninsular ribbon snake (*Thamnophis sauritus*), 366
pennywort (*Hydrocotyle bonariensis*), 111, 153, 499
periodical cicada (*Magicicada septendecim*), 221
Peromyscus floridanus (Florida mouse), 357
Peromyscus polionotus (southeastern beach mouse), 492, 493
Peromyscus spp., burrowing, 494
Persea borbonia (red bay), 105, 220–221
pes, 343, 345
Petricola pholadiformis (false angelwing), 119, 278, 281–282

petroleum exploration, 526–527
Petroleum Research Fund (American Chemical Society), 527
Phalacrocorax auritus (double-crested cormorant), 383, 385, 468
Phanaeus. *See* tunneler beetles
Phoebichnus, 161
Phoradendron spp. (mistletoe), 106
phytoturbation, 154
pickerel weed (*Pontedaria*), 100
Picoides pubescens (downy woodpecker), 386
pileated woodpecker (*Dryocopus pileatus*), 30, 386, 395, 396–397
Pilkey, Orrin, Jr., 46, 48, 77
pine pollen, 155, 157
pine tree roots, 110, 130
pines (*Pinus* spp.), 29
Pinnixa chaetopterana (crab), 310
Pinnixa cristata (crab), 309
Pinus echinata (short leaf pine), 155
Pinus elliottii (slash pine), 155
Pinus palustris (long leaf pine), 155, 356
Pinus serotina (pond pine), 100, 155
Pinus spp. (pines), 29, 155–156; root system description, 156–157
Pinus taeda (loblolly pine), 22, 91, 155; low hammock, 105; root architecture, 131
piping plover (*Charadrius melodus*), 461
pitcher plant (*Sarracenia* spp.), 100
palmate, 381, 382
plant identification, 143
plant responses: basic responses, 164; to fungal attacks, 166; galls, 165; to insects, 105–106, 163–167; various, 165
plantations, 62–64
plants: fecal seed dispersion, 427; trace of, 7–8
plants, salt tolerant, 35
platform bird nest, 392, 395, 398
Pleistocene Epoch, 54, 170
Pleistocene islands, 46
Pluvialis squatarola (black-bellied plover), 461
Poecilia latipinna (sailfin molly), 338
Pogo (Kelly), 40

Pogonomyrmex badius (Florida harvester ant), 191, 192, 193
Point Lobos State Park, California, 516
polychaete worms, 232, 259–265, 260, 539
Polyergus lucidus (slave raider ant), 191
polyp, 255–256
Polypodium polypodioides (resurrection fern), 106, 157
pond pine (*Pinus serotina*), 100, 155
ponds, freshwater. *See* freshwater environments
Pontedaria (pickerel weed), 100
poriferans, 252
Portland cement, 64
pottery patterns, 58
pottery shards, 57–58, 58
predators/prey: avian, 399–403; diamondback terrapin eggs, 456–457; interactions, 268; present-day, 98–99; sea turtle eggs, 441–443; sharks, 450; spider evidence in amber, 186
Princess Anne Island, 37–40, 39
Procambarus spp. *See* crayfish (*Procambarus* spp.)
Procyon lotor (raccoon), 35–36, 87–88, 356, 372, 483; egg predation, 442, 456; scat, 7, 107; trace and tracks, 487–489, 488
protodunes, 150–151, 150
Pseudacris spp. (peepers), 350
pseudofeces, 95
Pseudosuccinea columella. *See* American ribbed fluke snail (*Pseudosuccinea columella*)
Psilonichnus, 329, 508, 541–545, 542, 544, 548
Psilonichnus ichnofacies, 526
pterosaurs, 431
puma (*Puma concolor*), 69
Puma concolor (puma), 69
pyramid ant (*Dorymyrmex bureni*), 192, 193

quartz, 42
Quercus laurifolia (laurel oak), 105
Quercus virginiana (live oak), 10, 29, 86, 104; boring nest in, 395; root architecture, 158–159, 158; wasp trace, 213, 214

Quiscalus major (boat-tailed grackle), *152, 387, 390, 458*

raccoon (*Procyon lotor*), 35–36, 87–88, 356, 372, 483; egg predation, 442, 456; scat, 7, 107; trace and tracks, 487–489, *488*
Raccoon Bluff, Sapelo Island, 130–135, 507, *519*
radiography, 57, 117
Raffaelea lauricola (fungus), 220
railroad vine (*Ipomoea pes-caprae*), 111, 151, 152–153, *152*
Rallus longirostris (clapper rail), 450
Rana capito (gopher frog), 357
Rana sphenocephala (southern leopard frog), 349, *351*
razor clam (*Tagelus plebeius*), 277
red bay (*Persea borbonia*), 105, 220–221
red bay ambrosia beetle (*Xyleborus glabratus*), 220
red cedar (*Juniperus virginiana*), 50, 87, 401
red cockaded woodpecker, 397
red fire ant (*Solenopsis invicta*), 68, 191, 194–195; introduction, 69; and sea turtle eggs, 443
red knot (*Calidris canutus*), 23
red maple (*Acer rubrum*), 173
red-bellied woodpecker (*Melanerpes carolinus*), 29, 395, 397
Reed-Bingham State Park, 379
regurgitants, 402–403, 476; shorebird, 480
Reineck, Hans, 78, 588–589
relict dunes, 91, 92, *110*
relict marshes, 121–124, 127, 129–130, *509*
Renilla reniformis (sea pansy), 254, 255–259
reptiles, 353–354; alligators, 359–360 (*see also* alligator [*Alligator mississippiensis*]); freshwater turtles, 360–362; gopher tortoises, 354–358; lizards, 362–365; marginal-marine, 451–452; snakes, 365–369. *See also* diamondback terrapin (*Malaclemys terrapin*); sea turtles
Republic of Korea, 80
research opportunities, 231, 593–597; freshwater environments, 523; freshwater fish, 342; insects, 231; root traces, 167
research techniques: digital documentation, 580–583; electronic files, 588–589; laser scanning, 585–586; measurements, 584; technological aids, 592–593; uncomplicated, 577–579; underground data, 586–587
resurrection fern (*Polypodium polypodioides*), 106, 157
Reticulitermes flavipes (eastern subterranean termite), 229–230
Reticulitermes spp., 228–229
retroflexed hallux, 382
Reynolds, R. J., Jr., 72, 73
Rhizocorallium, 552
rhizomes, 142
rhizomorph, 144
Rhizoprionodon terraenovae (Atlantic sharpnose shark), 450
Rhopalostylis sapida (nikau palm tree), 161
ribbed mussel (*Geukensia demissa*), 35, 49, 54, 94, 281; byssae, 128; in relict marsh, *123*
ribbon snake (*Thamnophis sauritus*), 102
rice production, 62
Rindsberg, Andrew, 80, 245, 506, 510
ring-billed gull (*Larus delawarensis*), 467, 479
ring-tailed lemur (*Lemur catta*), 75
Riparia riparia (bank swallow), 394
Rissola marginata (cusk eel), 450
river otter (*Lutra canadensis*), 36, 101, 483, 484–487, *485*
Road-Kill Run, 379
rock art, 6
Roebling, Dorothy, 75
Roebling, Robert, 75
roller beetles, 214, 216
roly-polies, 290
root trace fossils, 129, *131*, 144, *504*, 553, 595; describing, 140–145; New Zealand, 132; physical description, 130; in sandy substrates, 163; suggested research, 167; Yellow Banks Bluff, *504*
roots, 107, *110*; organisms that interact with, 138

rough-winged swallow (*Stelgidopteryx serripennis*), 394, 395, 553
round-tailed muskrat (*Neofiber alleni*), 416
royal tern (*Sterna maxima*), 466
Ruckdeschel, Carol, 102, 175
ruddy turnstone (*Arenaria interpres*), 23, 462
rutile, 42, 68
Rynchops niger (black skimmer), 465

Sabal palmetto (cabbage palm), 29, 61, 105, 161–163, *162*; nests in, 397; wasps and, 210
saber-toothed cat (*Smilodon*), 59
Saccoglossus kowalevskii (helical acorn worm), 327, 328
sailfin molly (*Poecilia latipinna*), 338
salamanders, 347–349
Salicornia virginica (glasswort), 86, 87
salt barrens, 89–91
salt cedar (*Tamarix gallica*), 68
salt-meadow cordgrass (*Spartina patens*), 87
saltwater, density of, 48
saltwater marshes, 93–96, 559–560; flora and fauna of, 87–88, 451–452 (*see also* alligator [*Alligator mississippiensis*]; diamondback terrapin [*Malaclemys terrapin*]; sea turtles); high productivity, 73, 75; organic flux of, 73; rice production, 62; sediments used as fertilizer, 62
saltwort (*Batis maritima*), 146, 457
San Jose de Zapala, 61
San Salvador Island, 152–153, 559
sand: Appalachian origins, 40–41; Bahamian, 153; composition, 42; grain size, 42; stabilizers, *149*; and wave energy, 51
sand dollar (*Mellita isometra*), 32, 92, 323
sand fiddler crab (*Uca pugilator*), 34–35, 87; distinguished from *Uca pugnax*, 91; traces, 90
sandbar shark (*Carcharhinus milberti*), 450
sanderling (*Calidris alba*), 461, 463, 464, 472; trackway, 473
sandhill crane (*Grus canadensis*), 35, 458, 515; trackway, 460

Santa Catalina de Guale (mission), 60–61, 61
Sapelo Island, 39, 43, 45, 46; Cabretta Beach, 123; formation of UGAMI, 73; oldest shell ring on, 56; prominent Americans of, 73; Raccoon Bluff, 130–135
Sapelo Island Research Foundation, 73
Sapelo Nature Trail: field trips, 83–93
saprophytes, 138
sapwells, 400
Sarracenia spp. (pitcher plants), 100
Satilla River, 39, 40
saucer bird nests, 398
Savannah River, 39
saw palmetto (*Serenoa repens*), *104*; about, 159–161, *160*; rootlets, *160*; stem traces, *160*
sawgrass (*Cladium jamaicense*), 100
Saxe, John Godfrey, 16
scalloped hammerhead shark (*Sphyrna lewini*), 450
Scalopus aquaticus (eastern mole), 11–12, 30, 180, 236; burrows, 106, 416–418, *417*
Scaphiopus holbrookii (eastern spadefoot toad), 350–352
Scapteriscus abbreviatus (short-winged mole cricket). See mole crickets *Scapteriscus* spp.)
Scapteriscus borellii (southern mole cricket). See *Scapteriscus* spp. (mole crickets)
Scapteriscus spp. (mole crickets), 224–227, *225*
scarab beetle (*Copris gopheri*), 358
Scarabaeidae (dung beetles), 216
Sceliphron caementarium (black-and-yellow mud dauber), *208*, 211–213
Sceloporus undulatus (northern fence lizard), 363
scent markings, 10
Schaubcylindrichnus, 328
Schowalter, Ruth, 133, 173, 245, 435, 464, 570, 571
Scincella lateralis (ground skink), 363, 365
Scirpus spp. (bulrushes), 100
Sciurus carolinensis (eastern gray squirrel), 415, 569–570; feeding signs, 425

Sciurus niger (southern fox squirrel), 107
scorched mussel (*Brachidontes exustus*), 119
scorpions, 183–186
Scotland, 18
scouring rushes (*Equisetum* spp.), 100
Scoyenia, 537–539
Scoyenia ichnofacies, 525, 526, 529, 534; about, 537–539; as defined by Frey et. al., 531; representative ichnogenera, 530
sea cucumber (*Synapta inhaerens*), 319
sea cucumber (*Thyone briareus*), 319, 320
sea cucumber burrows, 110
Sea Islands, 43, 62
Sea Islands hurricane (1893), 66
sea levels: dune formation, 154–155; effects of changing on trace fossil deposition, 560, 561, 562–565; lowering, 36–37; Pleistocene Epoch, 54; rising, 28–36, 71–72, 121; sediment migration and deposition, 44, 46
sea lice, 290–291
sea oats (*Uniola paniculata*), 91, 110–111, 149, 457
sea onion anemone (*Paranthus rapiformis*), 256–257
sea otter (*Enhydra lutris*), 484
sea pansy (*Renilla reniformis*), 254, 255–259
sea turtles, 91, 435–436, 452–454; egg predation, 441–443; hatching, 444–445; nesting, 31, 441–445; nesting factors, 443–444; nonnesting crawlway, 435–441; resident species, 440; sex determination, 444; trackways, 60. *See also* leatherback sea turtle (*Dermochelys coriacea*); loggerhead sea turtle (*Caretta caretta*)
sea-oxeye daisy (*Borrichia frutescens*), 87
sediment: deposition, 79–80 (*see also* mud deposition); moisture, 13
sedimentary structures, 115–117, 121, 141, 289, 393, 505, 512, 565; and biogenic material deposition, 114, 115
sedimentary substrates, 9–10, 57, 116–117
Seilacher, Dolf, 78, 524, 526, 532, 578
semipalmated plover (*Charadrius semipalmatus*), 461, 462

Senckenberg am Meer (SAM), 76
Senckenbergiana Maritima, 78
Serenoa repens (saw palmetto), 104; about, 159–161, 160; rootlets, 160; stem traces, 160
Sesarma reticulatum (marsh crab), 182
sheepshead minnow (*Cyprinodon variegatus*), 338
shell rings, 53–56, 55; age of, 56; exploration methods, 57; molluscan contributors, 54; raided for shells, 63–64; significance, 55; size, 54
shimmy worm (*Nephtys bucera*), 263
ship ballast, 66–68, 67
shorebirds: common species and tracks, 459t; feeding traces, 476–483; flying traces, 469–471; mating traces, 472–476; nesting traces, 472–476
short leaf pine (*Pinus echinata*), 155
short-billed dowitcher (*Limnodromus griseus*), 462
short-faced bear (*Arctodus simus*), 59
short-tailed shrew (*Blarina carolinensis*), 416
short-winged mole cricket (*Scapteriscus abbreviatus*). *See* mole crickets (*Scapteriscus* spp.)
silky ribbon worm (*Cerebratulus lacteus*), 264
Silver Bluff islands, 37–40, 39, 45; sea level high, 44
Sinum perspectivum (baby's ear), 272
six-lined racer (*Cnemidophorus sexlineatus*), 363
Skidaway Institute of Oceanography (SKIO), 75–76
Skidaway Island, 46, 75–76
SKIO (Skidaway Institute of Oceanography), 75–76
Skolithos, 262, 263, 545–546, 548–549
Skolithos ichnofacies, 525
slash pine (*Pinus elliottii*), 155
"slave oaks," 69
slave raider ant (*Polyergus lucidus*), 191
slavery, 64–65, 69
sloppy gut anemone, 257
slugs, 237–238

Smilodon (saber-toothed cat), 59
smooth cordgrass (*Spartina alterniflora*), 24, 35, 62; about, 145–151; manatee feeding, 496; mud anchor, 95; mud deposition, 49; natural selection example, 127; in rack line, 112, *150*; relationships, 128; remnants, *123*, *148*; root detail, *147*
smooth dogfish (*Mustelis canis*), 450
snake eels (*Ophichthus* spp.), 120, 449–450
snake in the path account, 83, 87
snakes, 365–369, *367*; land activities, 102
snapping turtle (*Chelydra serpentina*), 102; egg laying, 102
snowy egret (*Egretta thula*), 458
Solen viridis (green jackknife), 277
Solenopsis invicta (red fire ant), 68, 191, 194–195; introduction, 69; and sea turtle eggs, 443
South Africa, 6
southeastern beach mouse (*Peromyscus polionotus*), 492, 493
southern black racer (*Coluber constrictor*), 365–366
southern devil scorpion (*Vejovis carolinianus*), 183
southern flying squirrel (*Glaucomys volans*), 107; trails, 415
southern fox squirrel (*Sciurus niger*), 107
southern leopard frog (*Rana sphenocephala*), 349, *351*
southern magnolia (*Magnolia grandiflora*), 105; roots, 157
southern mole cricket (*Scapteriscus borellii*). *See* mole crickets (*Scapteriscus* spp.)
southern pine beetle (*Dendroctonus frontalis*), 214, 220
southern quahog (*Mercenaria mercenaria*), 54, 276
southern stingray (*Dasyatis americana*): feeding traces, 447
southern toad (*Bufo terrestris*), 349
southern yellow jacket (*Vespula squamosa*), 211
Spalding, Randolph, 64
Spalding, Thomas, 64

Spanish bayonet (*Yucca alifolia*), 111; importance as root tracemaker, 153–154
Spanish colonization, 60–62
Spanish moss (*Tillandsia usneoides*), 57, 103, *104*, 157; about, 106; nesting material, 398
sparrow (*Spizella* sp.), 387
Spartina alterniflora (smooth cordgrass), 24, 35, 62; about, 145–151; manatee feeding, 496; mud anchor, 95; mud deposition, 49; natural selection example, 127; in rack line, 112, *150*; relationships, 128; remnants, *123*, *148*; root detail, *147*
Spartina patens (salt-meadow cordgrass), 87
Sphecius speciosus (cicada killer wasp), 31, 206–209, *208*
spherical bird nest, 392
Sphyrna lewini (scalloped hammerhead), 450
spiders, 185–188, *185*
Spizella sp. (sparrow), 387
spodosols, 163
sponge (*Cliona celata*), 119, 253. *See also* sponges (*Cliona* spp.)
Spongeliomorpha, 520, 521
sponges (*Cliona* spp.), 252–255, *253*
spotted sandpiper (*Actitis macularius*), 461
spreite, 257
square-eyed mole crab (*Lepidopa websteri*), 312–313, *313*
Squilla empusa (mantis shrimp), 118, 313–314, *314*
squirrel nests (dreys), 415, 423–424
St. Catherines Island, *39*, 43, 45, 47, 75; burial mounds, 56–60; crayfish, 176–177; shell rings, 53–56, *55*, *56*; Spanish presence on, 60–63; Yellow Banks Bluff, 503–511, *504*, *509*
St. Marys River, 40
St. Simons Island, *39*, 43; predicted traces, 596; sediment redistribution, 51, 53
star-nosed mole (*Condylura cristata*), 30, 483, 494
Stelgidopteryx serripennis (rough-winged swallow), 394, *395*, 553

Stellar's sea cow (*Hydrodamalis gigas*), 495
stenohaline organisms, 319
Sterna antillarum (least tern), 466, *466*
Sterna forsteri (Forster's tern), 466
Sterna maxima (royal tern), 466
Stictia carolina (Carolina sand wasp), 31, 206, *208*
stillstand, 505, 506, 563
stingrays, 445–449
stingray (*Dasyatis americana*), 32, 92; feeding pits, 115, 120
stomping behavior, 236, 481–482
stone crab (*Menippe mercenaria*), 316
storytelling tradition, 65
straddle of a tetrapod trackway, 345, *346*
strand forests, 103–104
striped hermit (*Clibanarius vittatus*), 311
striped killifish (*Fundulus majalis*), 338
striped mud turtle (*Kinosternon baurii*), 360–361
striped mullet (*Mugil cephalus*), 338
striped scorpion ("*Buthus vittatus*"), 183
striped scorpion (*Centruroides hentzi*), 183–185, *185*
striped scorpion (*Centruroides vittatus*), 183–185, *185*
Study of Trace Fossils, The (Frey), 77
substrate dependent ichnofacies, 550–553. *See also* ichnofacies
substrates: ichnofabric analysis, 527–528; re-purposed, 20, 46; subsurface, 107, 110; vertical zones on, 103
substrates of the Georgia coast, 20
subtidal sediments, 116–121; box-core observations, 120–121
Sus scrofa (feral hog), 69, 375, 405, 590; dig marks, 421; egg predation, 442; track, *406*, *412*; trampling, 103; wallows of, 422
Suwanee River, 40
sweet bay (*Magnolia virginiana*), 105
sweetgrass (*Muhlenbergia filipes*), weaving of, 65
Sylvilagus floridanus (eastern cottontail rabbit), 483, 489
Sylvilagus palustris (marsh rabbit), 414, 483, 489–491, *490*

Synapta inhaerens (sea cucumber), 319, 324

tabanid flies, 210
tabby construction, 62–64, *63*, 68
Taenidium, 224, 236, 504
Tagelus plebeius (razor clam), 277
Talbot Island, 37–40, *39*
Tamarix gallica (salt cedar), 68
Tamias striatus (eastern chipmunk), 107, 416, 418–419
taphichnia, 19, 91, 249, 251, 281, 367, 390
taproots, *140*, 143
tarpon snook (*Centropomus pectinatus*), 338
tawny mole cricket (*Scapteriscus vicinus*). *See* mole crickets (*Scapteriscus* spp.)
Taxodium spp. (Virginia chain ferns), 173
tempering, ceramic, 57
Terebra dislocata (common Atlantic auger), 266, 272
Teredo navilis (clam), 283
Teredolites, 283, 329, 552
Teredolites ichnofacies, 526
termites, 193, 227–231
terns, 465–467, *466*
Terrindusia, 227
tetrapod digit numbering system, 343–344
Thalassinoides, 310, 329, 418, 504, 508, 520, 521, 534, 552
Thamnophis sauritus (ribbon snake), 102, 366
theropod tracks, 515
thick-lipped oyster (*Eupleura caudata*), 119, 269
Three Pillars of Ichnologic Wisdom, 8, 9–13, 517; anatomy, 10–12; behavior, 12–13; habitat, 13–16; life history, 12, 15–16; sedimentary substrates, 9–10
Thyone briareus (sea cucumber), 319, 320, 324
Tibicen auletes (dusk-calling cicada), 221
tidal creeks and channels, 47, 50; ebb-tidal deltas, 51; observations, 120–121
tidal effects: point impacts and feeding example, 4
tidal range, 49

tides: influence on coastal features, 50; lowering, 51; rising, 51
tiger beetle (*Cicindela* sp.), 214, 217–219
tiger salamander (*Ambystoma tigrinum*), 347
tiger shark (*Galeocerdo cuvier*), 450
Tillandsia usneoides (Spanish moss), 57, 103, *104*, 157; about, 106; nesting material, 398
Timucuan tribe, 56; Cumberland Island, 60
toads and frogs, 349–353
Tobler's Law, 511, 518
tooth marks: mammal distinguished from bird, 426; shark, 450. *See also* bird beak marks
totipalmate trackmakers, 468
totipalmate tracks, 381, *382*, 384, 458
trace fossils, 6; *vs.* animal fossils distinguished, 517–518; bees, 536, 594; bird nests, 397–398; birds, 380–381; bivalves, 283; in carbonate sediments and limestone, 556–560; Devonian Period amphibian, 343; dinosaur, 18, 533; ichnogenera characteristic of continental environments, *534*; ichnotaxonomy and, 514–523; marginal-marine fish, 450–451; ray feeding traces, 448–449; reptilian tracemakers, 358; sea level changes and, 560, 562–565; sea turtle, 454; *Thalassinoides*, 419; tortoise tracks, 18. *See also* ichnofacies; ichnofossils
traces: about, 6–9; distinguished from other signs, 6–9; examples, 7; not without oxygen, 539; Three Pillars of Ichnologic Wisdom, 8, 9–13; types, 512–513; water as a medium, 339
Trachymyrmex septentrionalis (leaf-cutting ant), 192, *193*
track floor of a tetrapod, 344, *345*
tracking concepts, basic, 343–347
Trail Ridge, 40
tree boneyards, 155, *156*
tree frogs (*Hyla* spp.), 350
Trelew, Argentina, 514
trematodes, 274

Trichechus manatus (West Indian manatee), 483, 495–497
Trichechus manatus latirostris (Florida manatee), 495
Trichoptera, 227
trilobites, 284–285, 556
Tringa melanoleuca (greater yellowlegs), 462
trumpet worm (*Pectinaria gouldii*), 260
Trypanites, 526, 549, 552
Trypoxylon politum (organ-pipe mud dauber), 208, 211–213
tube anemone (*Ceriantheopsis*), 257–259, *258*
tube worm (*Diopatra cuprea*), 113, 259–261, 280
tubers, 143–144
tunneler beetles, 214, 216–217
Turdus migratorius. *See* American robin (*Turdus migratorius*)
turkey vulture (*Cathartes aura*), 372, 458
Tursiops truncatus (bottlenose dolphin), 483, 495–497
Twain, Mark, 197
Tybee Island, 45
Typha angustifolia (narrow-leafed cattail), 100

Uca major (fiddler crab), 558. *See also* fiddler crabs (*Uca* spp.)
Uca pugilator (sand fiddler crab), 34–35, 87; distinguished from *Uca pugnax*, 91; traces, 90
Uca pugnax (mud fiddler crab), 34–35, 509; about, 94–95; burrow identification, 180, 180–181; pseudofeces, 95
Uca punax (mud fiddler crab): distinguished from *Uca pugnax*, 91
Uchman, Alfred, 132
ultisols, 163
Undichna, 534
Undichna britannica, 520
Undichna simplicitas, 520
Undichna unisulca, 520
uniformitarianism, 17–19, 76–77, 511–512
Uniola paniculata (sea oats), 91, 110–111, 149, 457

University of Auckland, 80
University of Florida, 176
University of Georgia, 77, 79
University of Georgia (Athens) Marine Institute (UGAMI), 73, 74, 80, 331; Sapelo Nature Trail, 83–84
University of Michigan, 573
University of West Alabama, 80
upland hammock forests, 103–104
Upogebia (burrowing mud shrimp), 48
Upogebia affinis (mud shrimp), 118, *314*, 315
Upogebia vasquezi (upogebiid shrimp), 558
upogebiid shrimp (*Upogebia vasquezi*), 558
Upper Floridan Aquifer, 71
uric acid, 400
Urosalpinx cinema (eastern oyster), 119
Urosalpinx cinerea (Atlantic oyster drill), 119, 269
Ursus americanus (black bear), 69, 415
U.S.S. Constitution (Old Ironsides), 69

vampire bat (*Desmodus rotundus*), 431
Vejovis carolinianus (southern devil scorpion), 183
Vella americana. *See* ant lions
Vespula maculifrons (eastern yellow jacket), 210–211
Vespula squamosa (southern yellow jacket), 211
Virginia chain fern (*Taxodium* spp.), 173
volichnia, 468, 470
vulture lore, 379
vultures, olfactory senses, 22

Wadden Sea, 19, 482
Walker, Sally, 80
wallows, 423
warmouth fish (*Lepomis macrochirus*), 101, 338, 341
Warvichnium, 219–220
washover fans, *509*
wasps, 31, 205–213
Wassaw Island, *45*, 46
water, density of, 42
water bugs, 101
water density, 42
water lilly (*Nymphaea stellata*), 100

water snake (*Nerodia fasciata*), 102
wax beetles, 101
wax myrtle (*Myrica cerifera*), 91, 153
weaving, 64, *65*
wedge piddock (*Martesia cuneiformis*), 29, 115–116, *278*, 281, *282–283*
West Indian manatee (*Trichechus manatus*), 483, 495–497
western mosquitofish (*Gambusia affinis*), 338
wharf crabs (*Armases cinereum*), 182
whelks (*Busycon* spp.), 33, 46, 54; clionaid-sponge borings, 253
white ibis (*Eudocimus albus*), 382
white pelican (*Pelecanus erythrorhynchos*), 468
white-tailed deer (*Odocoileus virginiana*), 70, 500; lays, 421, 423; scat, 107, *428*; tooth marks, *425*; track, *406*, *412*; trails, *414*, *414*
Wicomico Island, 37–40, *38*, *39*
wild rice (*Zizania aquatica*), 100
wild turkey (*Meleagris gallopavo*), 382, 458
Wildlife Conservation Society, 75
Wilhelmshaven, Germany, 19
willet (*Tringa semipalmata*), 465
Wilson, E. O., 189–190, 572
Wilson's plover (*Charadrius wilsonia*), 111, 461, 472, 474, 475
wiregrass (*Aristida stricta*), 125, 356
wolf, dire (*Canis dirus*), 59
wolf, gray (*Canis lupus*), 69
Wolf Island, 45
wolf spider burrows, *185*
wood-boring beetles, 220
woodground ichnofacies, 526
woodland ant (*Aphaenogaster rudis*), 230
worms, polychaetes, 33–34, 119, 232, 259–265, *260*, 539
Wormsloe Plantation, 62

Xyleborus glabratus (red bay ambrosia beetle), 220
Xylocopa micans (carpenter bee), 203
Xylocopa virginica (eastern carpenter bee), 200, 203

yaupon holly (*Ilex vomitoria*), 91, 153
Yellow Banks Bluff, St. Catherines Island, 503–511, *504*, *509*
yellow-crowned night heron (*Nyctanassa violacea*), 445
yellow fever, 68
yellow rat snake (*Elaphe obsoleta*), 367, 369
yellow rat snake (*Elaphe obsoleta quadrivittata*), 367, 369

yellow-bellied sapsucker (*Sphyrapicus varius*), 400, *401*
Yucca alifolia (Spanish bayonet), 111; importance as root tracemaker, 153–154

Zizania aquatica (wild rice), 100
Zizaniopsis miliacea (cutgrass), 100
Zoophycos ichnofacies, *525*, *553*, *555*
zygodactyls, 381, 382, 385

ANTHONY (TONY) J. MARTIN is an ichnologist, paleontologist, and geologist. He is Professor of Practice in the Department of Environmental Studies at Emory University in Atlanta, Georgia, where he has taught for more than 20 years. He has studied modern traces and trace fossils made by a wide variety of organisms—plants, invertebrates, and vertebrates—in rocks from many parts of the world ranging from 500 million to 5,000 years old, and has presented many public lectures on ichnology, paleontology, and geology.